# DFG

Essential
MAK Value
Documentations

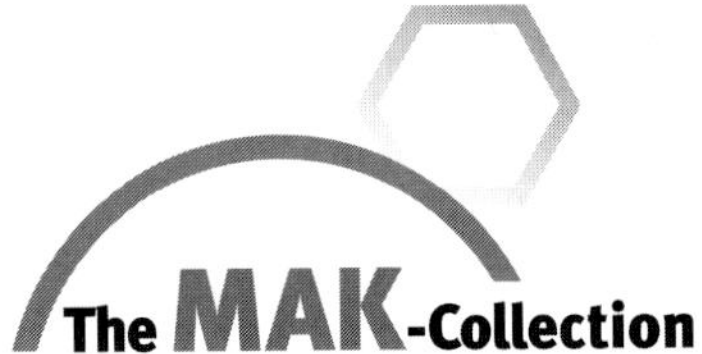

Current Volumes

Part I: MAK Value Documentations, Volume 21
2005. 3-527-31134-3

Part II: BAT Value Documentations, Volume 4
Drexler, H. (ed.)
2005. ISBN 3-527-27049-3

Part III: Air Monitoring Methods, Volume 9
Parlar, H. (ed.)
2005. ISBN 3-527-31138-6

Part IV: Biomonitoring Methods, Volume 10
Angerer, J. (ed.)
2006. ISBN 3-527-31137-8

The MAK-Collection online
www.mak-collection.com


A special selection to start with

Essential BAT Value Documentations
Drexler, H. / Greim, H. (eds.)
2006. ISBN 3-527-31477-6

Essential Air Monitoring Methods
Parlar, H. / Greim, H. (eds.)
2006. ISBN 3-527-31476-8

Essential Biomonitoring Methods
Angerer, J. / Greim, H. (eds.)
2006. ISBN 3-527-31478-4

**DFG** Deutsche Forschungsgemeinschaft

# Essential MAK Value Documentations

## from the MAK-Collection
## for Occupational Health and Safety

Edited by Helmut Greim

Commission for the Investigation of Health
Hazards of Chemical Compounds in the Work Area

**WILEY-
VCH**

WILEY-VCH Verlag GmbH & Co. KGaA

*Prof. Dr. Helmut Greim*
Commission
for the Investigation of Health Hazards
of Chemical Compounds in the Work Area
Hohenbachernstr. 15–17
85354 Freising
Germany

**Important Notice**
This volume presents a selection of
documentations or methods taken from
published volumes of the MAK-Collection.
These include occupational exposure
values (MAK and BAT values, EKA, BLW)
or classifications which may have changed
since the publication of the respective
volume. Readers are advised to consult the
latest edition of the "List of MAK and BAT
Values", which is published annually both in
print and online (www.mak-collection.com),
for current occupational exposure values
and classifications.



**Library of Congress Card No.:** applied for

**British Library Cataloguing-in-
Publication Data:** A catalogue record
for this book is available from the British
Library

**Bibliographic information published by
Die Deutsche Bibliothek**
Die Deutsche Bibliothek lists this publication in the Deutsche Nationalbibliografie;
detailed bibliographic data is available in
the Internet at <http://dnb.ddb.de>

© 2006 WILEY-VCH Verlag GmbH & Co.
KGaA, Weinheim, Germany



**Typesetting**  prosatz Unger, Weinheim
**Printing**  betz-druck GmbH, Darmstadt
**Binding**  Litges & Dopf Buchbinderei
GmbH, Heppenheim

Printed in the Federal Republic of Germany
Printed on acid-free paper

Printed on acid-free paper

**ISBN-13:** 978-3-527-31394-5
**ISBN-10:** 3-527-31394-X

# Preface

The year 2005 has marked both the 50th anniversary of the *Commission for the Investigation of Health Hazards of Chemical Compounds in the Work Area* of the Deutsche Forschungsgemeinschaft (DFG, German Research Foundation) as well as the launch of the new international reference series, the *MAK-Collection for Occupational Health and Safety*. Together with the annual *List of MAK and BAT Values*, this series comprises the Commission's English language publications, which are available both in print and online. The series has four parts, each part continuing an established book series:

- MAK Value Documentations (prior title: Occupational Toxicants)
- BAT Value Documentations (prior title: Biological Exposure Values for Occupational Toxicants)
- Air Monitoring Methods (prior title: Analyses of Hazardous Substances in Air)
- Biomonitoring Methods (prior title: Analyses of Hazardous Substances in Biological Materials)

For those unfamiliar with these publications and the wealth and quality of information offered through the *MAK Collection*, we have an additional new design feature on offer: special editions, named *Essentials*, with 20 highlights from each of the four parts of the Collection. These *Essentials* are particularly relevant for professionals and researchers in the fields of occupational health and safety. This *Essentials* features a selection of *MAK Value Documentations* taken from previous volumes of the *MAK Collection*. The selected documentations are exemplary for the standards maintained by the Commission, which works strictly according to scientific criteria. Also included are general chapters which, for example, describe the significance, use and derivation of MAK-Values. I hope that this selection will prove useful for fostering occupational health and safety as well as toxicological research around the globe.

I would like to thank the members and guests of the Commission, who have contributed with all their expertise to the respective documentations. Thanks also go to the Commission's scientific secretariat who have thoroughly reviewed the documentations and for the assistance of the translators, all of whom have made them ready for publication with outstanding accuracy. I also wish to thank the publisher Wiley-VCH for editorial guidance, swift production and world-wide distribution. Last, but not least, I am thankful for the continuing support of the Deutsche Forschungsgemeinschaft (DFG).

December 2005

H. Greim
Chairman of the Commission for the
Investigation of Health Hazards of
Chemical Compounds in the Work Area

*Essential MAK Value Documentations.* DFG, Deutsche Forschungsgemeinschaft
Copyright © 2006 WILEY-VCH Verlag GmbH & Co. KGaA, Weinheim
ISBN: 3-527-31394-X

# Contents and Substance Index

## General Aspects


**Significance, use and derivation of MAK values** . . . . . . . . . . . . . . . . . . . . . . .  3
**Changes in the Classification of Carcinogenic Chemicals in the Work Area** . .  11
**Metal-working fluids** . . . . . . . . . . . . . . . . . . . . . . . . . . . . . . . . . . . . . . . . .  21
**Sensitizing substances** . . . . . . . . . . . . . . . . . . . . . . . . . . . . . . . . . . . . . . . .  25


## Substances


**α-Amylase** . . . . . . . . . . . . . . . . . . . . . . . . . . . . . . . . . . . . . . . . . . . . . . .  39
**Arsenic and its anorganic compounds** . . . . . . . . . . . . . . . . . . . . . . . . . . . .  45
**2-Butoxyethanol** . . . . . . . . . . . . . . . . . . . . . . . . . . . . . . . . . . . . . . . . . . . .  103
**Carbon disulfide** . . . . . . . . . . . . . . . . . . . . . . . . . . . . . . . . . . . . . . . . . . .  109
**Cereal flour dusts** . . . . . . . . . . . . . . . . . . . . . . . . . . . . . . . . . . . . . . . . . .  125
**Chloroform** . . . . . . . . . . . . . . . . . . . . . . . . . . . . . . . . . . . . . . . . . . . . . . .  127
**1,4-Dioxane** . . . . . . . . . . . . . . . . . . . . . . . . . . . . . . . . . . . . . . . . . . . . . . .  167
**General Threshold Limit Value for Dust** . . . . . . . . . . . . . . . . . . . . . . . . . .  197
**Ethanol** . . . . . . . . . . . . . . . . . . . . . . . . . . . . . . . . . . . . . . . . . . . . . . . . . .  229
**Ethylene oxide** . . . . . . . . . . . . . . . . . . . . . . . . . . . . . . . . . . . . . . . . . . . . .  267
**Formaldehyde** . . . . . . . . . . . . . . . . . . . . . . . . . . . . . . . . . . . . . . . . . . . . . .  271
**Germ Cell Mutagens** . . . . . . . . . . . . . . . . . . . . . . . . . . . . . . . . . . . . . . . . .  311
**Hexachlorobenzene** . . . . . . . . . . . . . . . . . . . . . . . . . . . . . . . . . . . . . . . . . .  319
**Ozone** . . . . . . . . . . . . . . . . . . . . . . . . . . . . . . . . . . . . . . . . . . . . . . . . . . . .  343
**Passive Smoking** . . . . . . . . . . . . . . . . . . . . . . . . . . . . . . . . . . . . . . . . . . . .  373
**Styrene** . . . . . . . . . . . . . . . . . . . . . . . . . . . . . . . . . . . . . . . . . . . . . . . . . . .  411
**Trichloroethylene** . . . . . . . . . . . . . . . . . . . . . . . . . . . . . . . . . . . . . . . . . . .  417

**Contents of Volumes 1–21** . . . . . . . . . . . . . . . . . . . . . . . . . . . . . . . . . . . . . .  461



*Essential MAK Value Documentations.* DFG, Deutsche Forschungsgemeinschaft
Copyright © 2006 WILEY-VCH Verlag GmbH & Co. KGaA, Weinheim
ISBN: 3-527-31394-X

# General Aspects

*Essential MAK Value Documentations.* DFG, Deutsche Forschungsgemeinschaft
Copyright © 2006 WILEY-VCH Verlag GmbH & Co. KGaA, Weinheim
ISBN: 3-527-31394-X

# Significance, use and derivation of MAK values

## Definition

The MAK value ("maximale Arbeitsplatz-Konzentration": maximum workplace concentration) is defined as the maximum concentration of a chemical substance (as gas, vapour or particulate matter) in the workplace air which generally does not have known adverse effects on the health of the employee nor cause unreasonable annoyance (e. g. by a nauseous odour) even when the person is repeatedly exposed during long periods, usually for 8 hours daily but assuming on average a 40-hour working week. As a rule, the MAK value is given as an average concentration for a period of up to one working day or shift. MAK values are established on the basis of the effects of chemical substances; when possible, practical aspects of the industrial processes and the resulting exposure patterns are also taken into account. Scientific criteria for the prevention of adverse effects on health are decisive, not technical and economic feasibility.

For the establishment of a MAK value,

**the carcinogenicity** (see Section III)
**the sensitizing effects** (see Section IV)
**the contribution to systemic toxicity after percutaneous absorption** (see Section VII)
**the risks during pregnancy** (see Section VIII)
**the germ cell mutagenicity** (see Section IX)

of a substance are evaluated and the substance classified or designated accordingly. Descriptions of the procedures used by the Commission in the evaluation of these end points may be found in the appropriate sections of the *List of MAK and BAT Values*, in the "Toxikologisch-arbeitsmedizinischen Begründungen von MAK-Werten" (available in English translation in the series *Occupational Toxicants*)[1] and in scientific journals.[2,3,4,5,6]

---

[1]  obtainable from the publisher: WILEY-VCH, D-69451 Weinheim

[2]  Adler ID, Andrae U, Kreis P, Neumann HG, Thier R, Wild D (1999) Vorschläge zur Einstufung von Keimzellmutagenen. Arbeitsmed Sozialmed Umweltmed 34: 400–403.

[3]  Drexler H (1998) Assignment of skin notation for MAK values and its legal consequences in Germany. Int Arch Occup Environ Health 71: 503–505.

[4]  Hofmann A (1995) Fundamentals and possibilities of classification of occupational substances as developmental toxicants. Int Arch Occup Environ Health 67: 139–145.

[5]  Neumann HG, Thielmann HW, Filser JG, Gelbke HP, Greim H, Kappus H, Norpoth KH, Reuter U, Vamvakas S, Wardenbach P, Wichmann HE (1998) Changes in the classification of carcinogenic chemicals in the work area. (Section III of the German List of MAK and BAT Values). J Cancer Res Clin Oncol 124: 661–669.

[6]  Neumann HG, Vamvakas S, Thielmann HW, Gelbke HP, Filser JG, Reuter U, Greim H, Kappus H, Norpoth KH, Wardenbach P, Wichmann HE (1998) Changes in the classification of carcinogenic chemicals in the work area. Section III of the German List of MAK and BAT Values. Int Arch Occup Environ Health 71: 566–574.

*Essential MAK Value Documentations.* DFG, Deutsche Forschungsgemeinschaft
Copyright © 2006 WILEY-VCH Verlag GmbH & Co. KGaA, Weinheim
ISBN: 3-527-31394-X

In line with the so-called "preferred value approach" also used e.g. in the European Union, MAK values are to be established preferentially as the numerical values 1, 2 or 5 ml/m$^3$ or, for non-volatile substances, 1, 2 or 5 mg/m$^3$, multiplied by powers of ten.

In the use of MAK values, the analytical procedures used for sampling and analysis and the sampling strategy are of great importance.

## Purpose

MAK values promote the protection of health at the workplace. They provide a basis for judgement of the toxic potential or safety of the concentrations of substances in the workplace air. However, they do not provide constants from which the presence or absence of a health hazard after longer or shorter periods of exposure can be determined; nor can proven or suspected damage to health be deduced, in an isolated case, from MAK values or from the classification of a substance as carcinogenic. Such deductions can be made only on the basis of medical findings, taking into consideration all the circumstances of the particular case. Therefore, on principle, statements in the *List of MAK and BAT Values* are not to be seen as *a priori* judgements for individual cases. On principle, observation of MAK values does not eliminate the necessity for regular medical examination of the exposed individuals.

MAK values are not suitable for providing constant conversion factors for deduction of health risks associated with long-term exposure to contaminants in the non-occupational atmosphere, e.g., in the vicinity of industrial plants.

## Prerequisites

In principle, the substances are dealt with according to their importance for practical occupational hygiene and the expertise of the members of the Commission. The prerequisite for the establishment of a MAK value is the availability of sufficient data for the substance from the fields of toxicology, occupational medicine or industrial hygiene. Adequate documentation is not always available. The List is revised annually and suggestions for substances to be added and new information on listed substances are welcome.[7]

## Derivation of MAK values

MAK values are derived by the "DFG Commission for the Investigation of Health Hazards of Chemical Compounds in the Work Area" exclusively on the basis of scientific arguments and are published in the *List of MAK and BAT Values* which is issued annually. For the derivation of MAK values, certain rules of procedure have been developed by the Commission on the basis of established toxicological and occupational medical concepts; answers to at least the more common questions are repeatedly sought in the same way. Therefore the usual procedures and the general principles for the derivation of MAK values are described below. Essentially, these principles correspond with those

---

[7] Please contact the Geschäftsstelle der Deutschen Forschungsgemeinschaft, Kennedyallee 40, D-53175 Bonn; or the Sekretariat der Kommission: Technische Universität München, Institut für Toxikologie, Hohenbachernstraße 15–17, D-85350 Freising-Weihenstephan.

published by the European "Scientific Committee on Occupational Exposure Limits, SCOEL".[8]

First the most sensitive parameters described in the available data are to be identified, i. e., those effects which appear first during exposure to increasing concentrations of the substance. To be taken into account in this process are both local effects, that is, the results of effects on surfaces of the organism which are in contact with the environment (e. g. mucous membranes of the respiratory tract and the eyes, skin) and also systemic effects, that is, the results of uptake of the substance into the organism. Generally the concentration-effect relationships for these two kinds of effects are different. The derivation of the MAK value is based on the "no observed adverse effect level" (NOAEL) for the most sensitive effect with relevance for health. It must be decided whether or not such effects may be considered to be adverse effects. At present there is no generally accepted definition for an "adverse" effect, at least in part because of the lack of clarity about the still changing definition for the state of being "healthy";[9, 10] therefore this decision must be made anew in every case.

Fundamentally, known effects of a substance in man are given highest priority in the derivation of the MAK value.

In the evaluation of a substance, known effects of structural analogues may also be taken into account.

If no NOAEL may be derived from the available data, a scientifically founded MAK value cannot be established and the substance is listed in Section IIb of the *List of MAK and BAT Values*.

## a.  Selection of substances and collection of data

For the substances being studied, the epidemiological data published in scientific journals, occupational medical reports, toxicological properties and any other potentially useful information is first assembled by carrying out researches in appropriate databanks. The references found in the literature search are checked for their relevance for the assessment of the substance in question and the original publications of the selected literature are examined. When necessary, unpublished internal company data in the form of complete study reports are also included. These are then identified as such in the reference list at the end of the documentation. The validity of the available information and studies is checked. Whether or not a study is relevant for the current assessment is decided on a case to case basis. Whenever possible, evaluation of the studies is based on the guidelines of the OECD or similar bodies.

The unabridged reports are made available to the Commission and are filed at the Commission's scientific central office. Information required by a third party about the company reports cited in the Commission's documentation is supplied in writing by the chairman of the Commission at his discretion. Access to company reports is not made available to third parties. Copies, even of parts of reports, are not provided.

---


[8]  European Commission (Ed.) (1999) Methodology for the derivation of occupational exposure limits: Key documentation Cat. No. CE-NA-19253-EN-C, ISBN 92-828-8106-7, EUR 7, Office for Official Publications of the European Communities, L-2985 Luxembourg.

[9]  DFG (Deutsche Forschungsgemeinschaft) (Ed.) (1997) Verhaltenstoxikologie und MAK-Grenzwertfestlegungen. Wissenschaftliche Arbeitspapiere. Wiley-VCH, Weinheim.

[10]  Henschler D (1992) Evaluation of adverse effects in the standard-setting process. Toxicology Letters 64/65: 53–57.

### b. Values based on effects in man

For many substances encountered at the workplace, irritation or central nervous depression is the critical effect. Valuable information – at least for these acute effects of single exposures – may be obtained from studies of volunteers exposed under controlled conditions which yield data for concentration-effect relationships and also for concentrations without effects (NOAEC). A detailed review of the methods required of such studies and of the usefulness of various parameters for the establishment of threshold concentrations has been published.[9] Such studies often demonstrate differences in sensitivity between persons who have never been exposed to the test substance and those who have been repeatedly exposed, e. g., at work.

Occupational medical and epidemiological studies provide further information from which the health risks associated with handling particular substances may be evaluated. However, not only the parameters determined in the exposed persons, but also any differences in study design, in the analytical methods and sampling strategies must be considered in evaluating such studies. Various confounders, exposure to mixtures, previous disorders or inadequate exposure records can alter or falsify any detected concentration-effect relationships.

Cross-sectional studies with only single determinations of exposure levels and only single examinations of the exposed persons do not generally permit the association of any observed symptoms with the current exposure situation. This requires information as to past exposure levels.

Therefore longitudinal studies with repeated determination of the workplace and systemic exposure levels and repeated examination of the exposed persons play a decisive role in the establishment of thresholds. Valid epidemiological studies of persons exposed for long periods to concentrations which do not produce adverse effects provide a reliable basis for the establishment of threshold levels for the workplace, especially when the study design permits statements as to both local and systemic effects.

The diverse sensitivities of individual employees (as determined by age, constitution, nutrition, climate, etc.) are taken into consideration in the establishment of MAK values. It is currently not possible to take sex-specific differences in toxicokinetics and toxicodynamics into account when establishing MAK and BAT values because of the lack of appropriate scientific data.

When the NOAEL has been determined from effects of the substance in man, the MAK value is generally established at the level of this NOAEL.

### c. Values based on effects on animals

Because the effects in man are not known for many substances, MAK values are often derived from results obtained with experimental animals. This is carried out in the clear understanding of the problems associated with extrapolation between species and of the much smaller group sizes than is usual in epidemiological studies. On the other hand, animal studies carried out according to modern principles also offer advantages including precise characterization of exposure levels, the wide range of parameters that can be studied, and the possibility of determining dose-response relationships and NOAELs. The minimum database for the derivation of a MAK value is generally considered to be a NOAEL from a valid 90-day inhalation study with experimental animals. Of the re-

---

[9] DFG (Deutsche Forschungsgemeinschaft) (Ed.) (1997) Verhaltenstoxikologie und MAK-Grenzwertfestlegungen. Wissenschaftliche Arbeitspapiere. Wiley-VCH, Weinheim.

sults of studies in which substances were administered to experimental animals by the oral or dermal route, mostly only the systemic effects may be considered to be relevant for persons exposed at the workplace. Therefore, in the documentation of a MAK value such results must be accompanied by information about the local effects of the substance, especially the effects on the respiratory tract.

When the NOAEL has been determined from effects of the substance in animals, the MAK value is generally established at the level of half of this NOAEL. However, in some cases species differences in sensitivity to the substance must be taken into account and here the toxicokinetic data are particularly important.

### d. Exceptional workplaces

During exposure to gaseous substances which are metabolized rapidly and for which the blood/air distribution coefficient is larger than 10, it must be taken into account that the concentrations of the substances in blood and tissues are positively correlated with the level of physical activity.

Likewise, the concentrations of inhaled gaseous substances in blood and tissues of persons working under hyperbaric pressure have been shown to correlate positively with the pressure.

This dependence of the body burden on the workplace conditions must be taken into account in the establishment of MAK and BAT values.

### e. Odour, irritation and annoyance

Exposure of persons to substances at the workplace can cause smells (*nervi olfactorii*) or sensory irritation (*nervus trigeminus*). Such effects must be differentiated according to their relevance for health. This differentiation can cause difficulties because the parameters of interest can still not be determined with sufficient objectivity. Smells are mostly detected at lower concentrations than is sensory irritation. In general, if the smell and irritation are unpleasant and powerful enough, both can cause annoyance. When assessing these effects on the well-being of exposed persons, the physiological process of habituation (adaptation) must also be taken into account. In particular the sense of smell is affected markedly by habituation processes so that during constant exposure even to high concentrations the smell of a substance may no longer be noticed after a while. Excessive annoyance of workers by sensory irritation or persistent intensive or nauseous smells is taken into account when establishing thresholds.

### f. Habituation

Even at constant exposure concentrations, persons can become accustomed to sensory irritation, effects on the sense of well-being or smells so that these no longer function adequately as warning signals. At present, however, not enough is known about the mechanisms and dose-response relationships involved. With many substances, on the other hand, habituation is the result of toxic effects such as inactivation of enzymes or inhibition of receptor molecules. Especially in such cases, and when the warning signals produced by sensory irritation are reduced or cancelled out by disorders of well-being or the sense of smell, it is necessary to take habituation into account when establishing thresholds.

## Documentation

A detailed scientific documentation of each decision is published in the series *Toxikologisch-arbeitsmedizinische Begründung von MAK-Werten*, also available in English translation in the series *Occupational Toxicants.*[1] Annual supplements are planned. These documents present, clearly and in detail, the scientific data and the reasons for the establishment of a MAK value. Because of this system, it is sufficient to establish only general principles for the derivation of MAK values. The assessment of individual substances on the basis of all the available toxicological and occupational medical data yields a more differentiated and specific evaluation than would the observance of stringently formulated rules.

The published data for the toxicity and effects of a substance in man and animals and all other relevant information are organized according to the kind of effect and presented in the form of a review. This review of the toxicological and epidemiological data for a substance serves initially as a basis for the discussion within the Commission for the derivation of a MAK value and for detailed evaluation of the physicochemical properties, percutaneous absorption, sensitizing effects, carcinogenic effects, prenatal toxicity and germ cell mutagenicity of the substance. When new data become available, the MAK value, classification and designation of the substance are reassessed and, when necessary, altered.

## Publication

Prospective changes and new entries are announced each year in the *List of MAK and BAT Values*[11] and in the periodical *Zentralblatt für Arbeitsmedizin* and in the *Bundesarbeitsblatt*. The periodical *Arbeitsmedizin, Sozialmedizin, Umweltmedizin* also publishes a detailed discussion of the list with notes of changes and new entries. Following ratification of the annual List, the organizations listed below are officially informed of the planned changes: "Länderausschuß für Arbeitsschutz und Sicherheitstechnik (LASI)" (Federal Committee for Occupational Safety and Technical Security), the "Bundesverband der Deutschen Industrie" (Federation of German Industries), the "Hauptverband der gewerblichen Berufsgenossenschaften" (Central Association of Industrial Injuries Insurance Institutes) and the "Deutsche Gewerkschaftsbund" (the German Trade Union Federation). The purpose of this measure is to give these organizations enough time to send to the Commission any available scientific documentation relevant to the planned changes and additions to the *List of MAK and BAT Values*.

## Mixtures of substances

In general, the MAK value is only valid for exposure to a single, pure substance. It cannot be applied unconditionally to one component of a mixture in the workplace air or to a technical product which might contain more toxic impurities. Simultaneous or successive exposure to several substances may be much more or, in isolated cases, even less dangerous than the exposure to one of the substances on its own. A MAK value for a mixture of substances cannot be satisfactorily determined by simple calcula-

---

[1]  obtainable from the publisher: WILEY-VCH, D-69451 Weinheim
[11]  see yellow pages

tion because the components of the mixture generally have very different kinds of effect; MAK values can presently be established for such mixtures only after specific toxicological examination or studies of the particular mixture of substances. Given the inadequacy of the currently available data, the Commission decidedly refrains from calculating MAK values for mixtures, particularly for liquid solvent mixtures. However, it is willing, on the basis of its own investigations, to provide values for defined vapour mixtures of practical relevance.

## Analytical controls

The observance of the MAK, BAT and EKA values (that is, keeping the exposure levels below these values) is intended to protect the health of persons exposed to dangerous substances at work. This objective is only to be attained by regular monitoring of the concentration of the dangerous substances in the workplace air or of the concentration of the substances, their metabolites or other parameters of intermediary metabolism in the body fluids of exposed persons. For this purpose it is necessary to use analytical methods which have been tested for reliability and practicability. The Commission's analytical chemistry group has developed such methods and published them in the series *Luftanalysen* and *Analysen in biologischem Material.*[12] New issues of these publications appear regularly in both German and English. The methods are conceived as so-called standard operating procedures (SOP) which are intended to ensure comparability of the analytical results from laboratory to laboratory and with the above-mentioned thresholds. Thus they contribute to the quality control of the results. They also provide a good basis for the health protection which is the objective of the threshold values.

In the development and selection of these analytical methods, the accuracy and reliability of the results they yield is the most important factor. The methods reflect the current state of technology. They are repeatedly modified in the light of recent developments.

The methods for analyses in biological material are, whenever possible, designed so that their analytical range includes the concentration range relevant in environmental studies. This makes it also possible to differentiate the occupational from the environmental concentration range and to evaluate any differences.

---

[12] *Analytische Methoden zur Prüfung gesundheitsschädlicher Arbeitsstoffe*, available from the publisher: WILEY-VCH, D-69451 Weinheim
Volume 1: *Luftanalysen* (*Analyses of Hazardous Substances in Air*)
Volume 2: *Analysen in biologischem Material* (*Analyses of Hazardous Substances in Biological Materials*)
Supplementary reports at yearly intervals are anticipated. The Commission welcomes suggestions for inclusion of new chemical compounds as well as analytical methods. Contact the group "Analytische Chemie" at the Deutsche Forschungsgemeinschaft, D-53170 Bonn. Analytical methods for carcinogenic workplace substances are published in cooperation with the analytical chemistry group from the association of German industrial chemists, 'Arbeitsgruppe Analytik des Fachausschusses Chemie der BG Chemie' *Von den Berufsgenossenschaften anerkannte Analysenverfahren zur Feststellung der Konzentration krebserzeugender Arbeitsstoffe in der Luft in Arbeitsbereichen*, Carl Heymanns Verl. KG, D-50939 Köln.

# Changes in the Classification of Carcinogenic Chemicals in the Work Area

## 1 Introduction

The system for classification of carcinogenic substances by the DFG Commission for the Investigation of Health Hazards of Chemical Compounds in the Work Area (subsequently called the Commission) was developed in the 1970s. In the *List of MAK and BAT Values*, the following sections were established: IIIA1 – carcinogenic for man, IIIA2 – carcinogenic for animals, IIIB – suspected carcinogenic potential. By the year 1997 the Commission had assigned 21 substances to section IIIA1, 107 substances to section IIIA2 and 94 to section IIIB (DFG 1997). This classification, like that established by other international bodies, was based on relatively inflexible criteria. In the meantime, however, cancer research has made great progress and carcinogenic substances can now be better differentiated on the basis of their mode of action and potency. One of the Commission's working groups suggested in 1988 (Bolt *et al.* 1988) that section IIIB should be subdivided on the basis of whether or not further studies of the substances are necessary. This suggestion has been largely adopted by the European Union (EU).

The present document presents the classification scheme used until May 1998 by the Commission together with the new classification scheme, which is based on new findings.

## 2 Categories used previously

### IIIA1

*Substances shown to induce malignant tumours in humans*

A substance was assigned to section IIIA1 (carcinogenic to humans) if it had been shown in epidemiological studies to cause an increased incidence of tumours in man. However, sufficient epidemiological data are available for only a few substances. Frequently the cohorts studied have been too small to demonstrate effects unequivocally or there have

*Essential MAK Value Documentations.* DFG, Deutsche Forschungsgemeinschaft
Copyright © 2006 WILEY-VCH Verlag GmbH & Co. KGaA, Weinheim
ISBN: 3-527-31394-X

been problems in associating the effects with an individual substance. Therefore, substances for which carcinogenic effects have been demonstrated in man are substances with marked carcinogenic potency which have induced a few cases of rare tumours or to which a large number of persons have been exposed.

Epidemiological data that, taken by themselves, would not sufficiently justify assignment to section IIIA1 can, however, be used for an assignment to the new Category 1 if they are supported by data that show that the mode of action of the substance is relevant for humans.

## IIIA2

*Substances shown to be clearly carcinogenic only in animal studies but under conditions indicative of carcinogenic potential at the workplace*

Substances which were assigned to this section are considered on principle to be carcinogenic for man and, when handled at the workplace, necessitate the same protective measures as do the substances of section IIIA1. This approach is supported by a number of studies that show that the results obtained in animal studies largely concur both qualitatively and quantitatively with those obtained for man.

However, there are also substances that have caused tumours in animals but the tumours are considered not to be relevant for man. The definition required that such substances be assigned to section IIIA2. This implied a hazard that may not really exist. In the new classification scheme, such substances are no longer classified.

The results of epidemiological studies were only taken into account in the past for the assignment of a substance to section IIIA1. However, in the light of the recent improvement in epidemiological methods, it now seems appropriate to use limited evidence from epidemiological studies also for assignment to the new Category 2, if this evidence substantiates a suspicion arising from animal studies.

The previous limitation of section IIIA2 to "conditions indicative of carcinogenic potential at the workplace" has repeatedly caused problems in the classification of substances. It became evident that administration routes or exposure concentrations in animal experiments that are not comparable with those found at the workplace can indeed demonstrate the carcinogenic potential of a substance for man. This was the case with some man-made mineral fibres that, when administered intraperitoneally or intrapleurally, elicited tumours in animals but did not give rise to tumours when the administration was by inhalation, the exposure route most directly comparable with that at the workplace. In order to avoid inconsistency with the above definition of section IIIA2, these man-made mineral fibres were designated "as if IIIA2". Such man-made mineral fibres are classfied as Category 2 substances in the new classification system.

## IIIB

*Substances suspected of having carcinogenic potential*

Section IIIB contained a variety of substances (Bolt *et al.* 1988), including some that have been studied in detail and for which it is currently not expected that further studies would be able to dispel or confirm the suspicion of carcinogenicity, and others for which the carcinogenic potential must still be clarified. Therefore, a working group of the Commission (Bolt *et al.* 1988) suggested that section IIIB be divided into two groups. This suggestion has already been adopted by the EU so that the EU Category 3 comprises two subcategories defined by the following criteria:

- substances which are well-investigated but for which the evidence of a tumour-inducing effect is insufficient for classification in Category 2. Additional experiments would not be expected to yield further relevant information with respect to classification;
- substances which are insufficiently investigated. The available data are inadequate but they raise concern for man. This classification is provisional; further experiments are necessary before a final decision can be made.

A main objective for the development of the new classification scheme was to be able to remove well-studied substances from category IIIB and to evaluate them conclusively.

## Further differentiation of carcinogens

The classification of carcinogenic substances into the three categories used until 1998 (equivalent to sections IIIA1, IIIA2, and IIIB in the *List of MAK and BAT values*) was based on the certainty with which a carcinogenic potential could be established. The mode of action and potency of the substance were either not taken into account or, at best, were used in supporting arguments. This purely qualitative approach is also common to other international systems of classification. The progress that has been made in research on the multistage process of tumour development now permits a more differentiated approach.

## Mode of action

The historical development of risk assessment for carcinogens began with genotoxic substances. It is, however, increasingly necessary to assess chemicals shown to be carcinogenic in long-term animal studies but to be only weakly genotoxic or not genotoxic at all. The substances defined by exclusion with the term "non-genotoxic substances" include carcinogens that are known to induce cancer or to promote cancer by a large variety of mechanisms. They have been shown, for example, to be capable of amplifying the effects of genotoxic carcinogens in the sense of promotion. Effects on receptor-dependent regulatory processes or cytotoxic effects can play an important role. In some cases non-linear dose-response relationships are observed; in others it is not possible to distinguish the effects of low doses from variation in the physiological range. In contrast to the irreversible, additive damage caused by genotoxic substances, these effects, especially in the low dose range, are primarily reversible processes with little potential to cause damage. Even for substances that cause oxidative DNA damage via secondary reactions involving activation of oxygen, at low levels of exposure it is difficult to demonstrate damage in excess of the background resulting from endogenous processes. Similarly for cytotoxic substances it can be shown that slight changes in biochemical equilibria can be compensated by counter-regulating processes. The cell is not damaged until its adaptation mechanisms have been overloaded.

In all the above-mentioned cases, the biochemical parameters may increase linearly with the dose—as, for example, with 2,3,7,8-tetrachlorodibenzo-$p$-dioxin. Therefore the term "threshold", below which no (biochemical) effects occur, should be avoided and, wherever possible, the dose-dependence of the changes in the relevant biochemical parameters should be determined and compared with changes associated with physiological activity or background exposures. Common to such non-genotoxic substances is that, after low level exposures, they produce primarily reversible changes in the range of the normal physiological variability. Therefore the DFG Commission for the Investigation of Health Hazards of Chemical Compounds in the Work Area considers it justified to classify them in a single group and to define the exposure levels at which no significant contribution to cancer risk is to be expected.


## Potency of carcinogens

In the classification of carcinogens by the Commission or by other relevant international bodies, potency has so far not played a role or has only been of subsidiary importance. This is due to controversy over how to arrive at cancer risk estimates for man and to the difficulty that those estimates that have been obtained with disputed methods are prone to considerable uncertainty. This is true both for risk estimates derived from epidemiological data and, in particular, for those deduced from the results of animal studies. In this unsatisfactory situation it is necessary to improve the methods of risk estimation used at present in order to reduce the uncertainty attached to the results and to clarify whether relative risks could be used as the basis for establishing regulations.

The main difficulties in determining the level of cancer risk for man lie in the estimation of values for a heterogeneous human population on the basis of data obtained from animal studies, in the practical difficulty of determining the uncertainty associated with such a value and in the fundamental impossibility of testing such a hypothesis once it has been conceived. In practice, the data available are also generally inadequate. The incidence of tumours in animal studies is frequently found to be increased at only one dose, namely at the highest dose that is within the range of the maximum tolerated dose, or occasionally at two doses. Such data allow only a rough estimate of the carcinogenic potency in an animal study; they do not allow reliable risk assessment for man. For risk characterization, data for the mode of action, dose-response relationships and exposure are required. For exposure assessment, data for the exposure of humans to exogenous and endogenous carcinogens, which together make up the so-called background exposure, are increasingly available both for genotoxic and non-genotoxic substances. Data obtained by biochemical effect monitoring, e.g. in the form of levels of DNA and protein adducts, serve as individual biochemical effect parameters for genotoxic substances. Therefore, for specific situations, e.g. at the workplace, it is possible to determine whether external exposures lead to an increase in the background levels of biomarkers of exposure or biomarkers of response (e.g. biochemical effect markers). The decisive advantage of this approach lies in the fact that assessments can be made on the basis of information obtained under real exposure conditions from the affected persons themselves. Thus a quantitative element is introduced into the assessment. It must then be decided how much increase in the biomarkers of exposure or, preferentially, the biomarkers of response is to be considered significant.

## Extra categories for the classification of carcinogens

Because of the considerations mentioned above, additional categories for the classification of carcinogens have been introduced. To achieve this it was necessary to combine scientifically different classification criteria. Sections IIIA1, IIIA2 and IIIB of the *List of MAK and BAT Values* were retained as Categories 1, 2 and 3 to correspond with the EU Categories. The certainty with which a carcinogenic potential for man can be established continues to be decisive for the classification of a substance. In the documentation, however, information about the mode of action and the potency should be presented and evaluated. This information can then be taken up in the discussion that precedes the establishment of TRK values (technical exposure limits: limits for concentrations of carcinogenic substances at workplaces in Germany).

In addition, two new categories have been established for substances with carcinogenic potential for which the carcinogenic potency can be assessed and a MAK value established by considering all available evidence. Classification of a substance in Category 4 requires the demonstration that the mode of action is a result of non-genotoxic properties and that the effects at low doses are reversible. The main criteria for classification in Category 5 are the low carcinogenic potency of genotoxic substances and the possibility of monitoring internal exposure and response, e.g. by determining biochemical effect markers.

It is now possible to re-evaluate substances that have been proven to be carcinogenic for man (section IIIA1 and the new Category 1) or carcinogenic for animals (section IIIA2 and the new Category 2) for their potency at low exposure concentrations at the workplace and to reclassify them in the new categories as carcinogens with MAK values provided the data justify this procedure. In addition, substances that were classified in section IIIB (new Category 3) because their effects were weak or not detectable with the usual methods can be better assessed according to the new criteria. Category 3 will then become a genuine list of suspected carcinogens.

The new Categories 4 and 5, together with their categorization criteria have been established to include both the mode of action and the carcinogenic potency in the evaluation of carcinogens.

# 3 The New Classification Categories

### Category 1

*Substances that cause cancer in man* and can be assumed to make a significant contrib
tion to cancer risk. Epidemiological studies provide adequate evidence of a positive cc
relation between the exposure of humans and the occurence of cancer. Limited epiden
ological data can be substantiated by evidence that the substance causes cancer by
mode of action that is relevant to man.

### Category 2

*Substances that are considered to be carcinogenic for man* because sufficient data fro
long-term animal studies or limited evidence from animal studies substantiated by ev
dence from epidemiological studies indicate that they can make a significant contributi
to cancer risk. Limited data from animal studies can be supported by evidence that t
substance causes cancer by a mode of action that is relevant to man and by results of
*vitro* tests and short-term animal studies.

### Category 3

*Substances that cause concern that they could be carcinogenic for man but cannot
assessed conclusively because of lack of data. In vitro* tests or animal studies ha
yielded evidence of carcinogenicity that is not sufficient for classification of the su
stance in one of the other categories. The classification in Category 3 is provision:
Further studies are required before a final decision can be made. A MAK value can
established provided no genotoxic effects have been detected.

### Category 4

*Substances with carcinogenic potential for which genotoxicity plays no or at most
minor role. No significant contribution to human cancer risk is expected provided t
MAK value is observed.* The classification is supported especially by evidence tl
increases in cellular proliferation or changes in cellular differentiation are important
the mode of action. To characterize the cancer risk, the manifold mechanisms contribu
ing to carcinogenesis and their characteristic dose-time-response relationships are tak
into consideration.

### Category 5

*Substances with carcinogenic and genotoxic potential, the potency of which is considered to be so low that, provided the MAK value is observed, no significant contribution to human cancer risk is to be expected.* The classification is supported by information on the mode of action, dose-dependence and toxicokinetic data pertinent to species comparison.

Table 1 presents these new criteria for the classification of carcinogenic chemicals in comparison with those from international bodies.

# 4 Possible Criteria for Classification

For the new Categories 4 and 5, criteria are proposed on which classification can be based. They are intended as guidelines and are neither exhaustive nor to be used as a check-list. Not all the criteria for a particular category must be met; however, fulfilment of only a single criterion is not sufficient for classification. In a case-to-case approach, the classification decision should be based on a well-founded and comprehensible combination of criteria.

## Category 4

*Substances with carcinogenic potential for which genotoxicity plays no or at most a minor role. No significant contribution to human cancer risk is expected provided the MAK value is observed.*

### Criteria indicating that genotoxic effects do not play a decisive role

- Lack of genotoxicity in *in vivo* and *in vitro* tests; the latter yielding negative results even though appropriate activation systems were used; genotoxicity was seen only at cytotoxic concentrations
- No reactive metabolites detectable in metabolic studies or presence of reactive metabolites only when metabolic detoxification systems are saturated
- *In vivo* DNA binding cannot be detected when investigated with sensitive methods, e.g., postlabelling with $^{32}$P.

**Table 1.** The new criteria for classification of carcinogens compared with those from international bodies

| EU | DFG/MAK | ACGIH/TLV | IARC |
| --- | --- | --- | --- |
| **1** | **1** | **A1** | **1** |
| Substances known to be carcinogenic to man | Substances that cause cancer in man | Confirmed human carcinogen | The agent is carcinogenic to humans |
| **2** | **2** | **A2** | **2A** |
| Substances which should be regarded as if they are carcinogenic to man | Substances that are considered to be carcinogenic for man | Suspected human carcinogen | The agent is probably carcinogenic to humans |
| **3** | **3** | **A3** | **2B** |
| Substances which cause concern for man owing to possible carcinogenic effects:<br>**3a**<br>substances which are well-investigated<br>**3b**<br>substances which are insufficiently investigated | Substances that cause concern that they could be carcinogenic for man but cannot be assessed conclusively because of lack of data | Animal carcinogen | The agent is possibly carcinogenic to humans |
|  | **4** | **A4** | **3** |
|  | Substances with carcinogenic potential for which genotoxicity plays no or at most a minor role. No significant contribution to human cancer risk is expected provided the MAK value is observed | Not classifiable as a human carcinogen | The agent is not classifiable as to its carcinogenicity to humans |
|  | **5** |  |  |
|  | Substances with carcinogenic and genotoxic potential, the potency of which is considered to be so low that, provided the MAK value is observed, no significant contribution to human cancer risk is to be expected |  |  |
|  |  | **A5** | **4** |
|  |  | Not suspected as a human carcinogen | The agent is probably not carcinogenic to humans |

## Criteria indicative of tumour-promoting properties

- Positive results in two-stage carcinogenesis systems
- Perturbation of gene regulation, e.g., alterations in the expression of genes relevant for carcinogenesis such as oncogenes or tumour-suppressor genes
- Hormonal or hormone-like effects, e.g., disruption or deregulation of endocrine homeostasis
- Receptor-mediated effects
- Enhanced cell proliferation in the target tissue, e.g., stimulation of growth of initiated cells
- Interference with apoptosis
- Inhibition of intercellular communication
- Positive results with cell transformation tests
- Evidence that regenerative cell proliferation, as a consequence of toxicity, plays an essential role
- Enhanced tumour formation in organs exhibiting a high incidence of spontaneous tumours.

## Indications of non-linear dose-effect relationships

Development of tumours, e.g., after administration of doses for which first order toxico-kinetics do not apply (e.g., saturation of metabolism).

## Criteria indicating low carcinogenic potency

- Tumour incidence is increased only slightly above the control values
- The number of tumours per organ is small
- The tumours are species-specific, strain-specific or sex-specific
- The proposed EU definition of low potency may be adopted (T25 > 100 mg per kg body weight and day).

# Category 5

*Substances with carcinogenic and genotoxic potential, the potency of which is considered to be so low that, provided the MAK value is observed, no significant contribution to human cancer risk is to be expected.*

## Classification criteria

- The substance or one of its metabolites is mutagenic.
- Low carcinogenic potency; the proposed EU definition of low potency may be adopted (T25 > 100 mg per kg body weight and day).
- Physiological-toxicokinetic modelling on the basis of experimental data indicates a low cancer risk for man.


## Criteria characterizing a non-significant contribution to human cancer risk

- Various biochemical or biological endpoints may be used to characterize the contribution to cancer risk. Typically the contribution is considered not significant if an external exposure at the workplace leads to internal exposures in the range of background levels in a reference population not specifically exposed to the substance.
- Biochemical effect markers (e.g., DNA or protein adducts) are not increased significantly above background levels under workplace conditions.


# 5 References

Bolt HM, Gelbke HP, Greim H, Kimmerle G, Laib RJ, Neumann HG, Norpoth KH, Pott F, Steinhoff D, Wardenbach P (1988) Stoffe mit begründetem Verdacht auf krebserzeugendes Potential (Abschnitt III, Gruppe B der MAK-Werte-Liste): Probleme und Lösungsmöglichkeiten. *Arbeitsmed Sozialmed Praeventivmed 23*: 139–144

completed 01.04.1998

## Metal-working fluids

Metal-working fluids are used to cool and/or lubricate metallic workpieces during cutting and shaping. They are subdivided into water-immiscible (ca. 60%) and water-miscible (ca. 40%) metal-working fluids. The water-immiscible fluids are used undiluted, whereas water-miscible fluids are used in concentrations of 1–20% in water. Individual components may be present in both water-miscible and water-immiscible metal-working fluids.

Of importance in this context is the formation of nitrosamines in alkaline metal-working fluids which contain secondary amines and also formaldehyde (see Section III "Amines which form carcinogenic nitrosamines on nitrosation"). Such conditions are found, for example, in some metal-working fluids and so in these fluids the formation of significant levels of nitrosamines is possible. However, appropriate formaldehyde-releasers which do not contain or release nitrosatable compounds can, as a result of the bacteriostatic effects of the trace amounts of formaldehyde that they release, prevent the bacterial reduction of nitrate to nitrite and so inhibit nitrosamine formation. Because nitrogen oxides can also act as nitrosating agents and the nitrite concentration does not always correlate with the nitrosamine concentration, it is more reliable to determine the nitrosamines than the nitrites in metal-working fluids containing secondary amines (these do not comply with TRGS 611 unless Paragraph 3.4 applies). In particular, the absence of nitrite in a particular sample at a particular time does not exclude the possibility that nitrosamines are present. *N*-Nitroso-oxazolidines may be formed in small amounts from biocides which contain the necessary precursors (see documentation of MAK values[74] "Kühlschmierstoffe" (Metal-working fluids)).

In addition, a differentiation is made between petroleum-bearing and nonpetroleum-bearing metal-working fluids, and also between doped (additive type) and straight (no additive type) metal-working fluids.

The toxicological assessment of metal-working fluids is dependent on the composition and the type of components, which vary greatly in number and proportion according to the intended use. There are no fixed formulations. New components and compositions are to be expected with advancing technology.

When metal-working fluids are in use, it is possible for vapours and aerosols to pass into the air of the work area. To date, hardly any animal-experimental or epidemiological data are available for long-term effects following uptake in the lungs. It can, however, be assumed that the toxic effects on the whole organism after inhalation of mists and vapours, as well as the effects of liquids on the skin – in the sense of a toxic irritation or allergic sensitization – are mainly effects of the additives.

The mineral oil component alone is therefore not representative of the potential effect. The MAK value of 5 mg/m$^3$ previously established for pure mineral oil is, as a result, not applicable to present-day metal-working fluids since, as a rule, one is dealing with mixtures whose composition varies considerably according to the intended use. For this reason, it is not possible to establish a single MAK value for all types of metal-working fluids.[86]

---

[74] *Toxikologisch-arbeitsmedizinische Begründung von MAK-Werten* and in English Translation in *Occupational Toxicants* Volumes 2, 5, 9, 10, available from the publisher: WILEY-VCH, D-69451 Weinheim

[86] The AGS (Committee for Hazardous Substances) has established a threshold value of 10 mg/m$^3$ I (sum of vapour and aerosol) for water-miscible and water-immiscible metal-working fluids with a flash point above 100°C (TRGS 900).

*Essential MAK Value Documentations.* DFG, Deutsche Forschungsgemeinschaft
Copyright © 2006 WILEY-VCH Verlag GmbH & Co. KGaA, Weinheim
ISBN: 3-527-31394-X

Since there is no regulation requiring declaration of components of metal-working fluids it is difficult to attain systematic registration. A list compiled according to the best information available was published by the Commission in 1990 and brought up to date in 1994 and 2000.[87] This list includes substances having proved or suspected acute and/or chronic toxic effects or carcinogenic potential. There is also the possibility that carcinogenic or otherwise toxic substances be formed secondarily during technical use of metal-working fluids.

The list, which is subject to revision, should assist in the case to case assessment of the effects of metal-working fluids and in taking any necessary action for health protection.

The Commission prepares toxicological/occupational-medical reviews of individual components with the aim of making practicable recommendations, possibly in the form of MAK values.

Documentation has been published for the following substances:[87]

Abietic acid [514-10-3]
Alkyl ether carboxylic acids
2-(2-Aminoethoxy)ethanol [929-06-6]
2-Amino-2-ethyl-1,3-propanediol [115-70-8]
2-Amino-2-methyl-1-propanol [124-68-5]
1-Amino-2-propanol [78-96-6]
1,2-Benzisothiazol-3(2*H*)-one [2634-33-5]
1*H*-Benzotriazole [95-14-7]
Benzyl alcohol mono(poly)hemiformal [18] [14548-60-8]
1,3-Bis(hydroxymethyl)urea [18] [140-95-4]
Bithionol [97-18-7]
Boric acid [10043-35-3]
2-Bromo-2-nitro-1,3-propanediol [19] [52-51-7]
4-*tert*-Butylbenzoic acid [98-73-7]
Chloroacetamide-*N*-methylol [18] [2832-19-1]
*p*-Chloro-*m*-cresol [59-50-7]
5-Chloro-2-methyl-2,3-dihydroisothiazol-3-one [26172-55-4] and
  2-Methyl-2,3-dihydroisothiazol-3-one [2682-20-4] mixture in ratio 3:1
Citric acid [77-92-9] and its alkali metal salts
Decyl oleate [3687-46-5]
Dibenzyl disulfide [150-60-7]
1,2-Dibromo-2,4-dicyanobutane [35691-65-7]
2,6-Di-*tert*-butylphenol [128-39-2]
4-(Diiodomethylsulfonyl)-toluene [20018-09-1]
1,3-Dimethylol-5,5-dimethyl hydantoin [6440-58-0]
4,4'-Dioctyldiphenylamine [101-67-7]
2,2'-Dithiobis(*N*-methylbenzamide) [2527-58-4]
1-Dodecanol [112-53-8]
5-Ethyl-3,7-dioxa-1-azabicyclo[3.3.0]octane (EDAO) [18] [7747-35-5]
2-Ethyl-1,3-hexanediol [94-96-2]

---

[87]  published in *Toxikologisch-arbeitsmedizinische Begründung von MAK-Werten* and in English in *Occupational Toxicants*, available from the publisher: WILEY-VCH, D-69451 Weinheim
[18]  releases formaldehyde
[19]  Use in metal-working fluids is not permitted: see TRGS 611.

2-Ethylhexanol [104-76-7]
1-Hexadecanol [36653-82-4]
Hexamethylenetetramine [18] [100-97-0]
1-Hexanol [111-27-3]
2-Hexyl-1-decanol [2425-77-6]
Hexylene glycol [107-41-5]
1-Hydroxyethyl-2-heptadecenyl-imidazoline [21652-27-7]
2-Hydroxymethyl-2-nitro-1,3-propanediol [19] [126-11-4]
3-Iodo-2-propynyl butylcarbamate [55406-53-6]
Isodecyl oleate [59231-34-4]
Isotridecanol [27458-92-0]
Lauric acid [143-07-7]
2-Mercaptobenzothiazole [149-30-4]
Methyl-1*H*-benzotriazole [29385-43-1]
Methyldiethanolamine [105-59-9]
*N,N'*-Methylene-bis(5-methyloxazolidine) [66204-44-2]
4,4'-Methylenedimorpholine [18] [5625-90-1]
Myristic acid [544-63-8]
4-(2-Nitrobutyl)morpholine [2224-44-4] (70% w/w) and 4,4'-(2-Ethyl-2-
  nitro-1,3-propanediyl)bismorpholine [1854-23-5] (20% w/w)
  (mixture) [52, 19]
1-Octadecanol [112-92-5]
1-Octanol [111-87-5]
2-Octyl-1-dodecanol [5333-42-6]
2-Octyl-4-isothiazolin-3-one [26530-20-1]
Oleic acid [112-80-1]
Oleoyl sarcosine [110-25-8]
Oleyl alcohol [143-28-2]
Palmitic acid [57-10-3]
2-Phenoxyethanol [122-99-6]
1-Phenoxy-2-propanol [770-35-4]
*N*-Phenyl-1-naphthylamine [90-30-2]
*o*-Phenylphenol [90-43-7] and Sodium *o*-phenylphenol [132-27-4]
Piperazine [55, 56] [110-85-0]
Polyethylene glycol (PEG)
Polypropylene glycol (PPG) [25322-69-4]
Sodium diethyldithiocarbamate [27] [148-18-5]
Sodium pyrithione [3811-73-2; 15922-78-8]
Stearic acid [57-11-4]
1-Tetradecanol [112-72-1]

---

[19]  Use in metal-working fluids is not permitted: see TRGS 611.
[18]  releases formaldehyde
[52]  In this mixture formaldehyde can be released and nitrosamines formed.
[55]  Reaction with nitrosating agents can result in the formation of carcinogenic *N,N'*-dinitrosopiperazine, see
    Section III "Amines which form carcinogenic nitrosamines on nitrosation", p.139.
[56]  Use in metal-working fluids is restricted: see TRGS 611.
[27]  Reaction with nitrosating agents can result in the formation of carcinogenic *N*-nitrosodiethylamine, see
    Section III "Amines which form carcinogenic nitrosamines on nitrosation", p.139.

Tetrahydrobenzotriazole [6789-99-7]
Triazinetriyltriiminotrishexanoic acid [80584-91-4]
Triethanolamine [102-71-6]
Triethylene glycol *n*-butyl ether [143-22-6]
1,3,5-Triethylhexahydro-1,3,5-triazine [18] [7779-27-3]
Triphenylphosphate [115-86-6]
*N,N',N''*-Tris(β-hydroxyethyl)-hexahydro-1,3,5-triazine [18] [4719-04-4]
Zinc, *O,O'*-di-2-ethylhexyl dithiophosphate [4259-15-8]


## Hydraulic fluids and lubricants

Hydraulic fluids and lubricants are groups of compounds of heterogeneous and often complex composition, whose use often results in skin contact. Skin contact with components of hydraulic fluids is mainly expected due to their entry into metal-working fluid solutions during the metalworking process.

Lubricants are differentiated as fats, oils and emulsions, hydraulic fluids only as oils and emulsions. Lubricating greases are oils with thickeners. In both groups the oils can be mineral oils, natural oils or synthetic liquids. Primarily metal soaps are used as thickeners, but minerals, polymers and other substances can be used as well.

Significant technical properties such as antiwear, corrosion protection, load-carrying ability, increased viscosity index and decreased pour point are only achieved by means of additives. Among other things, antioxidants prevent the decomposition of lubricating greases, while metal deactivators inhibit the catalytic activity and corrosion of non-ferrous metals.

In hydraulic oils, additives for improving load-carrying ability and fluidity, increasing the viscosity index, and decreasing corrosion, ageing and wear play an important role, as do additives with detergent, demulsifying and dispersant effects.

The Commission put forward a list of current components in 2003 (see "Hydraulikflüssigkeiten und Schmierstoffe"[88]), which is subject to revision in collaboration with the association of German lubricant producers ("Verband Schmierstoff-Industrie e.V."). There is some overlap with the list of metal-working fluid components.

Because of the intensive skin contact, effects on health are expected most frequently after absorption by the skin or as a result of a sensitizing, irritative reaction on the skin or also on the respiratory tract following exposure to metal-working fluid solutions after aerosolization.

Documentation has been published for the following substances:

Zinc, *O,O'*-di-2-ethylhexyl dithiophosphate [4259-15-8]

---

[18]  releases formaldehyde

[88]  published in *Toxikologisch-arbeitsmedizinische Begründung von MAK-Werten*, available from the publisher: WILEY-VCH, D-69451 Weinheim (not yet available in English)

# Sensitizing substances

The allergies caused by substances at the workplace affect mostly the skin (contact eczema, contact urticaria), the respiratory passages (rhinitis, asthma, alveolitis) and the conjunctiva (blepharoconjunctivitis). The kind of allergy is determined by the route of uptake, the chemical properties of the substance and its aggregation state.

Contact allergies are generally manifested as contact eczema, the pathogenesis of which involves a T lymphocyte-mediated immune reaction of delayed type. Contact eczema is almost always caused by reactive substances of low molecular weight. Immunologically, these low molecular weight substances must be seen as haptens or prohaptens. They become complete allergens in the organism either by binding to peptides or proteins as such (haptens) or after metabolism (prohaptens).

The development of a contact allergy of delayed type is determined by several factors, by the sensitization potential resulting from the chemical properties of the substance or the metabolites produced from it in the organism, by the exposure concentration and the duration and manner of exposure, by the genetic disposition of the person and, not least, by the state of the tissue with which the substance makes contact. Under workplace conditions particular cofactors are also relevant and must be taken into account; attention is drawn to these in the documentation. The s e n s i t i z a t i o n   p o t e n t i a l of a substance is not the same as the i n c i d e n c e   o f   s e n s i t i z a t i o n s which it causes because the clinical significance of a contact allergen is not only determined by its sensitization potential but also by the distribution of the substance and the possibilities of exposure to it. At present, the sensitization potential of a substance can best be estimated in animal studies.

Other allergic skin reactions, e.g., urticaria, involve immune reactions mediated by specific antibodies. Similar symptoms can also be produced, however, by mechanisms not involving immune reactions (see below).

Most respiratory allergens are macromolecules, mainly peptides or proteins. But low molecular weight substances can also produce specific immunological reactions in the airways (see List of allergens). Some of the low molecular weight respiratory allergens are also contact allergens.

The allergic reactions of the airways and conjunctiva which take the form of bronchial asthma or rhinoconjunctivitis mostly involve reaction of the allergen with specific IgE antibodies and belong to the manifestations of immediate type. However, in the deeper airways they can also first appear after a delay of several hours. Exogenous allergic alveolitis is induced generally by allergen-specific immune complexes of IgG type and by cell-mediated reactions. Allergic reactions of immediate type can also cause systemic reactions and even anaphylactic shock.

The development of allergies of the respiratory passages, like that of contact allergies, is dependent on a number of factors. In addition to the substance-specific potential for causing sensitization, the amount of the allergen, the exposure period and the genetically determined disposition of the exposed person play a decisive role. Factors which increase the sensitivity of the mucosa may play a role in predisposing a person to allergy; they may be genetically determined or acquired, for example, during infections or exposure to irritant substances. Particular attention should be drawn to atopic diathesis which is characterized by an increased susceptibility to atopic eczema (neurodermatitis) or to allergic rhinitis and allergic bronchial asthma and is often associated with increased IgE synthesis.

*Essential MAK Value Documentations.* DFG, Deutsche Forschungsgemeinschaft
Copyright © 2006 WILEY-VCH Verlag GmbH & Co. KGaA, Weinheim
ISBN: 3-527-31394-X

In addition, there are also a number of relatively rare disorders of quite different kinds which are immunologically induced and so belong with the allergic phenomena such as, for example, manifestations involving granuloma formation (e.g. berylliosis) and certain exanthematous skin disorders.

A number of substances only induce the formation of antigens and then contact sensitization when they have previously been put into an energetically excited state by the absorption of light (photocontact sensitization, "photoallergization"). Likewise, many other substances can cause a skin reaction after exposure to light but without proof of an immunological mechanism (phototoxicity). Differentiation between photo-toxicity and immunological photocontact sensitization can be difficult, as the classical factors that distinguish between (photo)allergic and (photo)toxic effects are not always found. In Anglo-American usage the expression "photosensitization" is used for both mechanisms. Although photocontact sensitization and phototoxicity involve primarily the physical activation (photosensitization) of a chromophore, both types of reactions are in principle clinically and diagnostically distinguishable.

It is still not possible to determine (and document scientifically) generally applica-ble threshold concentrations either for the induction of an allergy (sensitization) or for triggering the allergic reaction in an already sensitized person. The likelihood of induction increases with the concentration of the allergen to which persons are ex-posed. The concentrations required for the triggering of acute symptoms are generally lower than those required for sensitization. Even when MAK values are observed, the induction or triggering of an allergic reaction cannot be excluded with any certainty.

Sensitizing substances are indicated in the List of MAK and BAT Values in the column "H;S" by "Sa" or "Sh". This designation refers only to the organ or organ system in which the allergic reaction is manifested. The pathological mechanism pro-ducing the symptoms is not taken into account. "Sh" designates substances which can cause allergic reactions of the skin and the mucosa close to the skin (skin-sensitizing substances). The designation "Sa" (substances causing airway sensitization) indicates that a sensitization can involve symptoms of the airways and also of the conjunctiva, and that other effects associated with reactions of immediate type are also possible. These include systemic effects (anaphylaxis) and local effects on the skin (urticaria). The latter reactions only result in the additional designation with "Sh" when the skin symptoms are relevant under workplace conditions. Substances which increase photo-sensitivity in exposed persons by mechanisms not involving immune reactions (e.g. fu-rocoumarins) are not separately designated. On the other hand, substances which cause photocontact sensitization are designated with "SP" (e.g. bithionol). Separate criteria for their evaluation are not necessary because, in essence, the criteria for the evaluation of substances causing contact sensitization may be applied.

Some substances can cause local or systemic reactions the symptoms of which are entirely or largely identical to those of allergic reactions but which do not involve spe-cific immunological mechanisms; they involve, e.g., the release of various mediators by mechanisms which are not immunological. These reactions are not a result of anti-gen-antibody interactions and can therefore also appear on the very first contact with the substance. Such reactions are induced, e.g., by sulfites, by benzoic acid and acetyl-salicylic acid and their derivatives and by a variety of dyes such as tartrazine. Such substances are not designated with an "S". However, attention is drawn to their po-tential for producing such non-immunological reactions in the documentation and in some cases in the List of MAK and BAT Values as well.

The criteria which are used in the evaluation of substances causing contact and airway sensitization are described below.

## a. Criteria for the assessment of contact allergens

The allergological evaluation is based on a variety of information which must be seen as providing different qualities of evidence.

**1) Sufficient evidence** of an allergenic effect is provided by valid results from **either** i) **or** ii):

   i)  effects in man
- studies in which numerous clinically relevant cases of sensitization (i.e. association of symptoms and exposure) were observed in tests with large collectives of patients in at least two independent centres, or
- epidemiological studies which reveal a relationship between sensitization and exposure, or
- case reports of clinically relevant sensitization (association of symptoms and exposure) for more than one patient from at least two independent centres

   **or**

   ii)  results of animal studies
- at least one positive result in an animal study without adjuvant carried out according to accepted guidelines, or
- positive results from at least two less well-documented animal studies carried out according to accepted guidelines, one of which did not use adjuvant.

**2)** An allergenic effect can be considered **probable** on the basis described in i) **and** ii) below:

   i)  effects in man
- studies in which numerous clinically relevant cases of sensitization (association of symptoms and exposure) were observed in tests in just one centre, or
- studies in which numerous cases of sensitization without details of clinical relevance were observed in tests with large collectives of patients in at least two independent centres

   **and**

   ii)  results of animal studies
- a positive result in an animal study with adjuvant carried out according to accepted guidelines, or
- positive results from in vitro studies, or
- evidence from structural considerations based on sufficiently valid results for structurally closely related compounds.

**3)** An allergenic effect is **not sufficiently documented**, but also not excluded, when only the data listed below are available:

- insufficiently documented case reports, or
- only one positive result in an animal study with adjuvant carried out according to accepted guidelines, or
- positive results in animal studies which were not carried out according to accepted guidelines, or
- evidence from studies of structure-effect-relationships or from *in vitro* studies.

**Commentary:**
Effects in man
The data obtained in numerous clinics and allergy centres from serial patch tests give us a useful picture of the frequency of skin sensitization and of the practical importance of the individual contact allergens. In contrast, useful results of reliable epidemiological studies are available for only few allergens.

The allergens which are observed most frequently, e.g., nickel, are not always the most potent sensitizers. On the other hand, very strongly sensitizing substances such as 2,4-dinitrochlorobenzene play quantitatively only a small role because only a small number of persons have sufficiently intensive contact with them. A number of highly potent contact allergens have been identified from clinical observations obtained in only a few patients, often after the first and only exposure (sometimes even after the first patch test). Examples of such substances are chloromethylimidazoline, diphenylcyclopropenone, quadratic acid diethyl ester, *p*-nitrobenzoyl bromide. For such exceptional substances, given proper scientific data, the evidence could be considered as 'probable' (category a2) even when the data come from just one centre.

The results of use tests in man with substances found at the workplace – often internal studies of the producing company – are of considerable value when they are carried out properly. Nowadays, the use of experimental sensitization tests must be rejected for ethical reasons, but historical results can be of importance in the evaluation of a substance.

Results of animal studies
Animal studies to determine the potential of a substance to cause allergic sensitization are carried out mainly with guinea pigs. These tests can be carried out with or without the use of Freund's complete adjuvant (FCA). The most frequently used methods are the maximization test of Magnusson and Kligman (with FCA), the Buehler test (without FCA) and the open epicutaneous test (without FCA). The methods with FCA are generally more sensitive and their use can therefore occasionally result in overrating of the risk of sensitization. For this reason, a positive result in a test without adjuvant is considered to be better evidence than a positive result in one with adjuvant.

The tests with experimental animals generally yield useful data; that is, for most substances they have yielded results in agreement with the data obtained in man. One advantage of the experimental animal methods is that dose-response relationships can be studied. Sensitization tests can also be carried out with mice. These methods, especially the local lymph node assay (LLNA), are being used increasingly and are included in the evaluation of sensitizing effects in the category of tests without adjuvant.

Substances to which people have not yet been exposed or known to be exposed (e. g. because they have been newly synthesized or are newly available products) and for which, therefore, clinical data cannot be available (neither negative nor positive results of clinical observations can be applied as a criterion) can also be classified as probably sensitizing (category a2) simply on the basis of positive results from animal studies with adjuvant carried out according to accepted guidelines. Even positive results from animal studies not carried out according to accepted guidelines can be accepted, provided the methods used were plausible and theoretical chemical structural considerations indicate that the substance is so closely related to known allergens that it may be expected to have analogous properties.

Theoretical considerations require practical confirmation; therefore they are generally of less importance in the final evaluation and cannot be the only criterion for a potential sensitizing effect in the absence of other clinical or experimental data.

## b. Criteria for the evaluation of respiratory allergens

The kinds of data which may be used in the evaluation of respiratory allergens are listed below; here too the different kinds of data provide different qualities of evidence.

1) Sufficient evidence of allergenic effects of a substance in the airways or the lungs is provided by valid data from:

   - studies or case reports of a specific hyperreactivity of the airways or the lungs which are indicative of an immunological mechanism from more than one patient and at least two independent testing centres. In addition, the (clinical) symptoms or adverse effects on the function of the upper or lower airways or the lungs must be shown to be associated with the exposure to the substance.

2) An allergenic effect can be considered probable on the basis of the results listed below:

   - one single case report of a specific hyperreactivity of the airways or the lungs

   **and**

   - other indications of sensitizing effects, e.g., a close structure-effect relationship with known airway allergens.

3) An allergenic effect is not sufficiently documented, but also not excluded, when only the data listed below are available:

   - epidemiological studies which demonstrate an increased incidence of symptoms or impaired function in exposed persons, or
   - studies or case reports of a specific hyperreactivity of the airways or the lungs in only one patient, or
   - studies or case reports of sensitization (e.g. detection of IgE) without accompanying symptoms or impairment of function causally associated with the exposure, or
   - positive results of animal studies, or
   - structure-effect relationships with known respiratory allergens.

**Commentary:**
Generally the classification is based on the results of epidemiological studies. Case reports do not always withstand critical examination, not least because of the difficulty or impossibility of carrying out adequate control studies. This is particularly true of inhalation-provocation tests. In addition, it is not always possible to produce adequate exposure data.

Symptoms are usually not a sufficient criterion for the designation of a substance as a respiratory allergen; generally it is necessary to demonstrate sensitization and record objective changes such as exposure-related impairment of lung function or bronchial hyperreactivity to specific stimuli. An immunological mechanism can generally be recognized on the basis of *in vivo* (e.g. prick test) or in *vitro* test results, ideally by detection of a specific antibody after proved exposure.

For many substances, an immunological mechanism has not yet been demonstrated directly. Therefore indirect evidence of immunological mechanisms can also be taken into account in classification. These include:

- the existence of a latency period between the start of the exposure and the appearance of the first symptoms (sensitization period),
- the triggering of symptoms with low concentrations of the substance which do not cause symptoms in appropriate controls,
- occasional delayed reactions or sequential immediate and delayed reactions (dual reactions) in the inhalation-provocation test,
- associated cutaneous symptoms such as urticaria or Quincke's oedema.

An allergenic effect is not sufficiently documented, but also not excluded, when evidence of airway sensitization is available but the conditions described in the criteria are not fulfilled. In particular, epidemiological studies which demonstrate an increased incidence of symptoms or of impaired function in exposed persons (even with demonstrated dose-response relationships) do not provide sufficient evidence of sensitizing properties if no indications of a specific immunological mechanism are available. Likewise, studies or case reports which merely document workplace-related variations in lung function or bronchial hyperreactivity are not sufficient.

To date there is no thoroughly validated method to induce and detect respiratory allergies in an animal model.

In guinea pig models, sensitizing substances cause reactions like those seen in man. Antibody-mediated sensitization can be induced by inhalation or by intradermal, subcutaneous injection but, unlike in man, IgG antibodies play the main role. In these tests the respiratory hyperreactivity (respiration rate, tidal volume, respiratory minute volume, inhalation and exhalation times, exhalation rate) and the antibody levels are measured. In the mouse IgE test, the potential of a substance to cause sensitization in BALB/c mice is determined as a function of the increase in the level of total IgE but not, to date, as a function of substance-specific IgE.

With these models no observed effect levels (NOEL) can be established but it is questionable whether they apply for man. Systematic comparative tests have not yet been carried out.

To date, with few exceptions, validated sensitive and specific standard *in vitro* methods for detecting low molecular weight allergens of immediate type are not available.

## c. Designation of a substance as an allergen

Whether or not it is necessary to designate a substance as an allergen in the List of MAK and BAT Values is determined on the basis of the available evidence of allergenic effects and, when possible, also on the basis of the expected levels of exposure.

- The substances characterized according to the criteria in Section IVa) or IVb) as belonging in Categories 1) or 2) are generally designated as allergens with "Sa", "Sh", "Sah" or "SP".
- Substances for which these criteria are fulfilled are also designated with an "S" when the observed sensitization is associated mainly with cofactors which are (only) relevant under workplace conditions (e.g. (previous) damage to the skin caused by chemical or physical agents).
- On the other hand, substances are not designated with an "S" when
- in spite of extensive handling of the substance, very few (well-documented) cases of sensitization are observed, or
- the observed cases of sensitization are mainly associated with cofactors which are not relevant under workplace conditions (e.g. the presence of eczema on the lower leg), or
- the criteria of Section IVa) or IVb) resulted in classification of the substance in category 3).

The criteria are to be seen as guidelines for an intelligible evaluation of the data but in certain special cases their strict application may not be obligatory.

## d. List of allergens

The list below shows the substances in Section IIa which are designated with Sa, Sh, Sah or SP. It does not claim to be a complete list of sensitizing substances and is subject to continual revision and extension.

Abietic acid[14] [514-10-3] (Sh)
Acrylic acid *n*-butyl ester [141-32-2] (Sh)
Acrylic acid ethyl ester [140-88-5] (Sh)
Acrylic acid 2-ethylhexyl ester [103-11-7] (Sh)
Acrylic acid 2-hydroxyethyl ester [818-61-1] (Sh)
Acrylic acid hydroxypropyl ester (all isomers) [25584-83-2] (Sh)
Acrylic acid methyl ester [96-33-3] (Sh)
Acrylonitrile [107-13-1] (Sh)
Alkali persulfates (Sah)
Allyl glycidyl ether [106-92-3] (Sh)
*o*-Aminoazotoluene [97-56-3] (Sh)
2-Aminoethanol [141-43-5] (Sh)

---

[14] An immunological genesis of the asthma often seen in persons working with materials containing abietic acid has not been proved.

3-Aminomethyl-3,5,5-trimethyl-cyclohexylamine
  (Isophorone diamine) [2855-13-2] (Sh)
*p*-Aminophenol [123-30-8] (Sh)
Ammonium persulfate [7727-54-0] (Sah)
α-Amylase (Sa)
α-Amylcinnamaldehyde [122-40-7] (Sh)
Animal hair, epithelia and other materials derived from animals (Sah)
Antibiotics
Benomyl [17804-35-2] (Sh)
1,2-Benzisothiazol-3(2*H*)-one [2634-33-5] (Sh)
Benzyl alcohol mono(poly)hemiformal [18] [14548-60-8] (Sh)
Beryllium [7440-41-7] and its compounds (Sah)
Bisphenol A [80-05-7] (SP)
Bisphenol A diglycidyl ether [1675-54-3] (Sh)
Bisphenol A diglycidyl methacrylate [1565-94-2] (Sh)
Bithionol [97-18-7] (SP)
Bromelain [9001-00-7] (Sa)
2-Bromo-2-nitro-1,3-propanediol [19] [52-51-7] (Sh)
1,4-Butanediol diacrylate [1070-70-8] (Sh)
1,4-Butanediol dimethacrylate [2082-81-7] (Sh)
Butanone oxime [96-29-7] (Sh)
*tert*-Butyl acrylate [1663-39-4] (Sh)
*p-tert*-Butylcatechol [98-29-3; 27213-78-1] (Sh)
*n*-Butyl glycidyl ether (BGE) [2426-08-6] (Sh)
*tert*-Butyl glycidyl ether [7665-72-7] (Sh)
*n*-Butyl methacrylate [97-88-1] (Sh)
*p-tert*-Butyl phenol [98-54-4] (Sh)
Butynediol [110-65-6] (Sh)
*N*-Carboxyanthranilic anhydride [118-48-9] (Sh)
Cellulases (Sa)
Cereal flour dusts (Sa)
  Rye
  Wheat
Chloroacetamide-*N*-methylol [2832-19-1] (Sh)
Chloroacetic acid methyl ester [96-34-4] (Sh)
*m*-Chloroaniline [108-42-9] (Sh)
*p*-Chloro-*m*-cresol [59-50-7] (Sh)
1-Chloro-2,4-dinitrobenzene [97-00-7] (Sh)
5-Chloro-2-methyl-2,3-dihydroisothiazol-3-one [26172-55-4] and
  2-Methyl-2,3-dihydroisothiazol-3-one [2682-20-4] mixture in ratio 3:1 (Sh)
Chlorothalonil [1897-45-6] (Sh)
Chlorpromazine (2-Chloro-10-(3-dimethylaminopropyl)phenothiazine)
  [50-53-3] (SP)
Chromium(VI) compounds (Sh)
Cinnamaldehyde [104-55-2] (Sh)

---

[18]  releases formaldehyde
[19]  Use in metal-working fluids is not permitted: see TRGS 611.

Cinnamyl alcohol [104-54-1] (Sh)
Cobalt and cobalt compounds (as inhalable dusts/aerosols): (Sah)
  Metallic cobalt [7440-48-4]
  Cobalt(II) carbonate [513-79-1]
  Cobalt(II) oxide [1307-96-6]
  Cobalt(II,III) oxide [1308-06-1]
  Cobalt(II) sulfate [10026-24-1] and similar salts,
  Cobalt(II) sulfide [1317-42-6]
Cyanamide [420-04-2] (Sh)
4,4'-Diaminodiphenylmethane [101-77-9] (Sh)
1,5-Diaminonaphthalene [2243-62-1] (Sh)
1,2-Dibromo-2,4-dicyanobutane [35691-65-7] (Sh)
3,4-Dichloroaniline [95-76-1] (Sh)
1,3-Dichloropropene (*cis* and *trans*) [542-75-6] (Sh)
Dicyclohexylcarbodiimide [538-75-0] (Sh)
Diethanolamine [26] [111-42-2] (Sh)
Diethylene glycol diacrylate [4074-88-8] (Sh)
Diethylene glycol dimethacrylate [2358-84-1] (Sh)
Diethylenetriamine [111-40-0] (Sh)
Diglycidyl resorcinol ether [101-90-6] (Sh)
Dimethylcarbamoyl chloride [79-44-7] (Sh)
1,1-Dimethylhydrazine [31] [57-14-7] (Sh)
1,2-Dimethylhydrazine [31] [540-73-8] (Sh)
Dimethylol dihydroxyethyleneurea [1854-26-8] (Sh)
1,3-Dimethylol-5,5-dimethyl hydantoin [6440-58-0] (Sh)
Disulfiram [27] [97-77-8] (Sh)
2,2'-dithiobis(*N*-methylbenzamide) [2527-58-4] (Sh)
Epichlorohydrin [106-89-8] (Sh)
5-Ethyl-3,7-dioxa-1-azabicyclo[3.3.0]octane (EDAO) [18] [7747-35-5] (Sh)
Ethylenediamine [107-15-3] (Sah)
Ethylene glycol dimethacrylate [97-90-5] (Sh)
Eugenol [97-53-0] (Sh)
Farnesol [4602-84-0] (Sh)
Formaldehyde [50-00-0] (Sh)
Formaldehyde condensation products with *p-tert*-butylphenol (low-molecular) (Sh)
Fragrance components
Geraniol [106-24-1] (Sh)
Glutaraldehyde [111-30-8] (Sah)
Glyceryl monothioglycolate [30618-84-9] (Sh)
Glycidyl trimethylammonium chloride [3033-77-0] (Sh)
Glyoxal [107-22-2] (Sh)

---

[26]  Reaction with nitrosating agents can result in the formation of carcinogenic *N*-nitrosodiethanolamine, see Section III "Amines which form carcinogenic nitrosamines on nitrosation", p. 139.

[31]  no longer produced in Germany

[27]  Reaction with nitrosating agents can result in the formation of carcinogenic *N*-nitrosodiethylamine, see Section III "Amines which form carcinogenic nitrosamines on nitrosation", p. 139.

[18]  releases formaldehyde

Gold [7440-57-5] and its inorganic compounds [39] (Sh)
Hard metal containing tungsten carbide and cobalt (inhalable fraction) (Sah)
Hexahydrophthalic anhydride [85-42-7] (Sa)
1,6-Hexamethylene diisocyanate [822-06-0] (Sah)
Hexamethylenetetramine [100-97-0] (Sh)
1,6-Hexanediol diacrylate [13048-33-4] (Sh)
Hydrazine [302-01-2] (Sh)
Hydrazine hydrate [7803-57-8] and Hydrazine salts (Sh)
Hydroxycitronellal [107-75-5] (Sh)
2-Hydroxyethyl methacrylate [868-77-9] (Sh)
Hydroxylamine [7803-49-8] and its salts (Sh)
4-(4-Hydroxy-4-methyl pentyl)-3-cyclohexene-1-carboxaldehyde [31906-04-4] (Sh)
3-Iodo-2-propynyl butylcarbamate [55406-53-6] (Sh)
Isoeugenol [97-54-1] (Sh)
  *cis*-Isoeugenol [5912-86-7]
  *trans*-Isoeugenol [5932-68-3]
Isophorone diisocyanate [4098-71-9] (Sah)
4-Isopropylphenyl isocyanate [31027-31-3] (Sh)
D,L-Limonene [138-86-3] and similar mixtures (Sh)
D-Limonene [5989-27-5] (Sh)
L-Limonene [5989-54-8] (Sh)
Maleic anhydride [108-31-6] (Sah)
Manganous ethylenebis(dithiocarbamate) (Maneb) [12427-38-2] (Sh)
Merbromin [129-16-8] (Sh)
2-Mercaptobenzothiazole [149-30-4] (Sh)
Mercury (metallic mercury [7439-97-6] and inorganic mercury
  compounds) (Sh)
Mercury compounds, organic (Sh)
Methacrylic acid ethyl ester [97-63-2] (Sh)
Methacrylic acid 2-hydroxypropyl ester [923-26-2] (Sh)
Methacrylic acid methyl ester [80-62-6] (Sh)
*N*-Methyl-bis(2-chloroethyl)amine [51-75-2] (Sh)
*N,N'*-Methylene-bis(5-methyloxazolidine) [66204-44-2] (Sh)
4,4'-Methylene diphenyl isocyanate (MDI) [101-68-8] (Sah)
Methyl isocyanate [624-83-9] (Sh)
Methyltetrahydrophthalic anhydride [11070-44-3] (Sa)
*N*-Methyl-2,4,6-*N*-tetranitroaniline (Tetryl) [479-45-8] (Sh)
Methyl vinyl ketone [78-94-4] (Sh)
Monomethylhydrazine [60-34-4] (Sh)
1,8-Naphthalic anhydride [81-84-5] (Sh)
1,5-Naphthylene diisocyanate [3173-72-6] (Sa)
Natural rubber latex [9006-04-6] (Sah)
Nickel and nickel compounds:[51] (Sah)
  Metallic nickel [7440-02-0]

---

[39]  only soluble gold compounds
[51]  Regarding compounds which have been found to be unequivocally carcinogenic in man, see *Toxikologisch-arbeitsmedizinische Begründung von MAK-Werten*, available from the publisher: WILEY-VCH, D-69451 Weinheim.

Nickel acetate [373-02-4] and similar soluble salts,
Nickel carbonate [3333-67-3]
Nickel chloride [7718-54-9]
Nickel dioxide [12035-36-8]
Nickel hydroxide [12054-48-7]
Nickel monoxide [1313-99-1]
Nickel sesquioxide [1314-06-3]
Nickel subsulfide [12035-72-2]
Nickel sulfate [7786-81-4]
Nickel sulfide [16812-54-7]
Nickel alloys (Sah)
4-Nitro-4'-aminodiphenylamine-2-sulfonic acid [91-29-2] (Sh)
4-(2-Nitrobutyl)morpholine [2224-44-4] (70% w/w) and
   4,4'-(2-Ethyl-2-nitro-1,3-propanediyl)bismorpholine [1854-23-5]
   (20% w/w) (mixture)[52, 19] (Sh)
*p*-Nitrocumene [1817-47-6] (Sh)
2-Nitro-*p*-phenylenediamine [5307-14-2] (Sh)
Oakmoss extracts (Sh)
2-Octyl-4-isothiazolin-3-one [26530-20-1] (Sh)
Olaquindox (*N*-(2-Hydroxyethyl)-3-methyl-2-quinoxalinecarboxamide
   1,4-dioxide) [23696-28-8] (SP)
4,4'-Oxydianiline [101-80-4] (Sh)
Palladium chloride [7647-10-1] and other bioavailable palladium(II) compounds
(Sh)
Papain [9001-73-4] (Sa)
Pentaerythritol triacrylate [3524-68-3] (Sh)
*o*-Phenylenediamine [95-54-5] (Sh)
*p*-Phenylenediamine [54] [106-50-3] (Sh)
Phenyl glycidyl ether (PGE) [122-60-1] (Sh)
Phenylhydrazine [100-63-0] (Sh)
Phenyl isocyanate [103-71-9] (Sah)
*N*-Phenyl-1-naphthylamine [90-30-2] (Sh)
Phthalic anhydride [85-44-9] (Sa)
Picric acid [88-89-1] (Sh)
Picryl chloride [88-88-0] (Sh)
Piperazine [55,56] [110-85-0] (Sah)

---

[52]  In this mixture formaldehyde can be released and nitrosamines formed.

[19]  Use in metal-working fluids is not permitted: see TRGS 611.

[54]  The "Ursol-Asthma" which used to be observed frequently, especially in persons dyeing furs with *p*-phenylenediamine, has not been demonstrated unequivocally to involve respiratory allergy to *p*-phenylenediamine; see *Toxikologisch-arbeitsmedizinische Begründung von MAK-Werten* (18th issue, 1992) and in English translation in *Occupational Toxicants* Volume 6, VCH-Verlagsgesellschaft mbH, Weinheim 1994

[55]  Reaction with nitrosating agents can result in the formation of carcinogenic *N,N'*-dinitrosopiperazine, see Section III "Amines which form carcinogenic nitrosamines on nitrosation", p.139.

[56]  Use in metal-working fluids is restricted: see TRGS 611.

Platinum compounds (Chloroplatinates) (Sah)
"Polymeric MDI" [9016-87-9] (Sah)
Pyrethrum [60] [8003-34-7] (Sh)
Quinone [106-51-4] (Sh)
Resorcinol [108-46-3] (Sh)
Ricinus protein (Sa)
Rosin (colophony) [62] [8050-09-7] (Sh)
Rubber [9006-04-6] components
  Dithiocarbamates (Sh)
  *p*-Phenylenediamine compounds (Sh)
  Thiazoles (Sh)
  Thiurams (Sh)
Sesquiterpene lactone (Sh)
Sodium diethyldithiocarbamate [27] [148-18-5] (Sh)
Soya bean constituents (Sa)
Subtilisins (Sa)
Tetraethylene glycol diacrylate [17831-71-9] (Sh)
Tetraethylene glycol dimethacrylate [109-17-1] (Sh)
Tetrahydrofurfuryl methacrylate [2455-24-5] (Sh)
Thimerosal [54-64-8] (Sh)
Thiourea [62-56-6] (Sh SP)
Thiram [29] [137-26-8] (Sh)
Toluene-2,4-diamine [95-80-7] (Sh)
Toluene-2,5-diamine [95-70-5] (Sh)
Toluenediisocyanate: (Sa)
  Toluene-2,4-diisocyanate [584-84-9]
  Toluene-2,6-diisocyanate [91-08-7]
p-Toluidine [106-49-0] (Sh)
Triethanolamine [102-71-6] (Sh)
Triethylene glycol diacrylate [1680-21-3] (Sh)
Triethylene glycol dimethacrylate [109-16-0] (Sh)
Triethylenetetramine [112-24-3] (Sh)
Triisobutyl phosphate [126-71-6] (Sh)
Trimellitic anhydride [552-30-7] (Sa)
Trimethylolpropane triacrylate [15625-89-5] (Sh)
Tripropylene glycol diacrylate [42978-66-5] (Sh)
*N,N',N''*-Tris($\beta$-hydroxyethyl)-hexahydro-1,3,5-triazine [4719-04-4] (Sh)
Turpentine [8006-64-2] (Sh)
Vinylcarbazole [1484-13-5] (Sh)

---

[60]  does not apply for the constituents of insecticides (pyrethrins and cinerins) or for synthetic derivatives (pyrethroids) but only for the constituents of the plant drug and its crude extracts, including $\alpha$-methylene sesquiterpene lactones (e.g. pyrethrosin)

[62]  An immunological genesis of the asthma often seen in persons working with materials containing rosin has not been proved.

[27]  Reaction with nitrosating agents can result in the formation of carcinogenic *N*-nitrosodiethylamine, see Section III "Amines which form carcinogenic nitrosamines on nitrosation", p. 139.

[29]  Reaction with nitrosating agents can result in the formation of carcinogenic *N*-nitrosodimethylamine, see Section III "Amines which form carcinogenic nitrosamines on nitrosation", p. 139.

Woods
   *Acacia melanoxylon* R.Br., Australian blackwood (Sh)
   *Brya ebenus* DC., cocus wood (Sh)
   *Chlorophora excelsa* (Welw.) Benth. & Hook, iroko, kambala (Sh)
   *Dalbergia latifolia* Roxb., East Indian rosewood, Bombay blackwood (Sh)
   *Dalbergia melanoxylon* Guill. et Perr., African blackwood (Sh)
   *Dalbergia nigra* Allem., Brazilian rosewood (Sh)
   *Dalbergia retusa* Hemsl., cocobolo, rosewood (Sh)
   *Dalbergia stevensonii* Standley, Honduras rosewood (Sh)
   *Distemonanthus benthamianus* Baill., ayan (Sh)
   *Grevillea robusta* A. Cunn., Australian silky oak (Sh)
   *Khaya anthotheca* C.DC., African mahogany (Sh)
   *Machaerium scleroxylon* Tul., pao ferro, Santos rosewood (Sh)
   *Mansonia altissima* A.Chev., mansonia, pruno, bété (Sh)
   *Paratecoma peroba* (Record) Kuhlm., ipe peroba (Sh)
   *Tectona grandis* L.f., teak (Sh)
   *Terminalia superba* Engl. & Diels, fraké, limba, afara, white afara (Sa)
   *Thuja plicata* (D.Don.) Donn., western red cedar, giant arborvitae,
   shinglewood (Sah)
   *Triplochiton scleroxylon* K.Schum., obeche, wawa, African whitewood (Sah)
Xylanases [37278-89-0] (Sa)
*m*-Xylylenediamine [1477-55-0] (Sh)
Zinc chromate [13530-65-9] (Sh)
Zirconium [7440-67-7] and its insoluble compounds (Sah)
Zirconium, soluble compounds (Sah)

## e. Evaluation of members of specific groups of substances

For numerous substances, a reliable evaluation of the sensitizing effects according to the criteria described above is not possible. Often the substance of interest is one of the many members of a specific group of substances. Valid data for man are generally available only for individual members of such groups of substances, those considered to be typical of the group and commercially available as test substances because they are used for this purpose. With other less frequently used substances or substances for which no reliable data as to the extent of their use is available, patch tests are carried out relatively rarely, with some substances – because of the danger of sensitization – only in special cases. The effects in man are made even more difficult to assess because these substances are often used in mixtures with other members of the same group of substances and so can be involved in concomitant sensitizations and in cross-reactions. Mixtures containing members of other groups of allergenic substances are also often used and then too it is not readily possible to determine the causality of the observed disorder. In addition, it is not always possible to establish all the components of the mixture involved and so an allergologically relevant component can be missed. Therefore, substances which are not listed in the List of MAK and BAT Values and which belong to groups of substances known to be able to cause sensitization should be handled with appropriate care.

It is emphasized that in general there is no danger of sensitization from fully polymerized plastics. A danger of sensitization, but only little, can result from release of residual monomer, e. g. during mechanical processing.

The groups of substances of which numerous members have sensitizing effects on the skin or airways include:

- acrylates and methacrylates
- dicarboxylic acid anhydrides
- diisocyanates
- glycidyl compounds (epoxides)
- dusts containing enzymes
- certain plant or animal proteins

The **antibiotics** are a group of substances which are very heterogeneous in chemical structure and in sensitizing effects. Persons may be exposed to these substances at work during isolation or production of the active principles, during preparation and packing of the medicines, and during their medical use in man and animals. Sensitization of the skin can result in the development of haematogenic contact dermatitis after later parenteral use. Sensitization of the airways and allergic contact dermatitis have often been reported in persons exposed occupationally to β-lactam antibiotics (especially p e n i c i l l i n s  and  c e p h a l o s p o r i n s ). Allergic reactions after medicinal use of these antibiotics (enteral or parenteral), on the other hand, generally take the form of IgE-mediated reactions of immediate type. However, other immunological reactions such as medicinal skin eruptions and, in serious cases, also *erythema exsudativum multiforme*, Stevens-Johnson syndrome or Lyell's syndrome can also develop. Some of the a m i n o g l y c o s i d e  a n t i b i o t i c s  are also conspicuous for the relatively high rates of sensitization which they produce, especially as a result of application of medicaments to (chronically) eroded skin. Sensitization of the skin resulting from occupational contact with aminoglycosides has been reported more rarely. Individual m a c r o l i d e  a n t i b i o t i c s, especially those used in veterinary medicine, can cause immunological reactions in the airways and also (inhalation-mediated) contact dermatitis. Only rare individual cases of contact allergy or allergic airway reactions to most other macrolide antibiotics and to p o l y e n e  or p e p t i d e  a n t i b i o t i c s  and t e t r a - c y c l i n e s  have been reported.

**Components of fragrance mixtures**, another group of substances which differ widely in their structures, allergenic potencies and clinical significance, must also be evaluated individually. This becomes clear even on consideration of the components of the standard fragrance test mixture. For many other fragrance components the clinical findings are inadequate because the substances are never or only very rarely used in patch tests. Non-occupational exposure to these practically ubiquitous fragrance mixtures can rarely be excluded and this makes the demonstration that sensitization was occupational more difficult.

# Substances

*Essential MAK Value Documentations.* DFG, Deutsche Forschungsgemeinschaft
Copyright © 2006 WILEY-VCH Verlag GmbH & Co. KGaA, Weinheim
ISBN: 3-527-31394-X

# $\alpha$-Amylase S

**Classification/MAK value:** see Section IV
of the List of MAK and BAT Values

**Classification dates from:** 1995

# 1 Allergenic Effects

## 1.1 Effects in man

$\alpha$-Amylase obtained from *Aspergillus oryzae* is nowadays frequently added to flour used by bakers to improve the quality of the products. The induction of IgE-mediated reactions to this enzyme in bakers is now well documented (Baur *et al.* 1986, 1987, Bermejo *et al.* 1990, Birnbaum *et al.* 1988, Blanco Carmona *et al.* 1991, Bolm-Audorff *et al.* 1992, Brisman and Belin 1991, Losada *et al.* 1992, Quirce *et al.* 1992, Sander *et al.* 1993, Tarvainen *et al.* 1991, Wüthrich 1994). In one study, contact dermatitis induced by $\alpha$-amylase in bakers has also been described (Morren *et al.* 1993).

Positive results were obtained with $\alpha$-amylase for 14/75 bakers with workplace-related respiratory symptoms in an enzyme-allergen-sorbent test (EAST; controls 2/42) and for 16/71 in a prick test (controls 1/42) (Degens *et al.* 1994). 2 % of persons exposed to flour dust and 21 % of bakers with workplace-related rhinitis or asthmatic symptoms are sensitized to this enzyme (Baur *et al.* 1987). Mono-sensitization to $\alpha$-amylase was also described in bakers with workplace-related symptoms (Baur *et al.* 1987, Baur and Weiß 1988). The clinical relevance of sensitization to $\alpha$-amylase was confirmed in provocation tests; the very low concentrations, 0.01 to 1 mg/ml, required to produce symptoms in these tests indicate that the enzyme is an aggressive respiratory allergen (Baur *et al.* 1987, Birnbaum *et al.* 1988, Quirce *et al.* 1992).

In a study of 259 millers, eye, nose or respiratory tract symptoms were detected in 73 persons (28 %). Skin tests with $\alpha$-amylase and the test for specific IgE antibodies yielded positive results in 12/73 workers. Positive results in the skin test were obtained in 4 of the millers who were free of symptoms and specific IgE antibodies were found in 2. However, in both workers without symptoms and those with symptoms, reactions to other allergens were also detected. Specific IgG antibodies were found in 45 of the workers (Moneo *et al.* 1994).

The question as to whether this baking additive also poses a risk for the consumer with respect to the development of food allergies is currently a matter of controversy (Baur *et al.* 1994b, Kanny and Moneret-Vautrin 1995, Schata and Jorde 1992).

*Essential MAK Value Documentations.* DFG, Deutsche Forschungsgemeinschaft
Copyright © 2006 WILEY-VCH Verlag GmbH & Co. KGaA, Weinheim
ISBN: 3-527-31394-X

In immunoblot studies, a major allergen with a relative molecular weight of 53 kD has been isolated from *Aspergillus oryzae* and has been shown to have $\alpha$-amylase activity (Baur *et al.* 1986, 1994a). The traces of other components and contaminants which are present in the $\alpha$-amylase as a result of the production method are probably of little immunological significance.

$\alpha$-Amylase and $\beta$-amylase from cereal grains have also been shown to be potent allergens and seem to play a role in the allergy to flour dust. There is little cross-reactivity between the specific antibodies against $\alpha$-amylase from fungi and those against $\alpha$-amylase and $\beta$-amylase from cereals (Sandiford *et al.* 1994).

## 1.2 Effects on animals

In a group of 10 BALB/c mice, two 250 µl doses of a 1 % amylase solution injected intraperitoneally 7 days apart caused an increase in the level of specific IgG and IgE antibodies. However, no increase in the serum IgE concentration was found (Hilton *et al.* 1994).

## 2 Manifesto

$\alpha$-Amylase can produce respiratory sensitization in man and is therefore designated with an "S".

## 3 References


Baur X, Fruhmann G, Haug B, Rasche B, Reiher W, Weiss W (1986) Role of Aspergillus amylase in baker's asthma. *Lancet 1*: 43

Baur X, Fruhmann G, Kimm KW, Rasche B, Weiss W (1987) Aspergillus-Amylase als Ursache des Bäcker-Asthmas. *Prax Klin Pneumol 41*: 638

Baur X, Weiß W (1988) Neue Entwicklungen in der Diagnostik des Berufsasthmas. *Prax Klin Pneumol 42*: 6–16

Baur X, Chen Z, Sander J (1994a) Isolation and denomination of an important allergen in baking additives: $\alpha$-amylase from Aspergillus oryzae (Asp o II). *Clin Exp Allergy 24*: 465–470

Baur X, Sander I, Jansen A, Czuppon AB (1994b) Sind Amylasen von Backmitteln und Backmehl relevante Nahrungsmittelallergene? *Schweiz Med Wochenschr. 124*: 846–851

Bermejo N, Maria Y, Gueant JL, Moneret-Vautrin DA (1990) Allergie professionnelle du boulanger à l'alpha-amylase fongique. *Rev Fr Allergol 31*: 56–58

Birnbaum J, Latil F, Vervloet D, Senft M, Charpin J (1988) Role de l'alpha-amylase dans l'asthme du boulanger. *Rev Mal Respir 5*: 519–521

Blanco Carmona JG, Juste Picón S, Garcés Sotillos M (1991) Occupational asthma in bakeries caused by sensitivity to $\alpha$-amylase. *Allergy 46*: 274–276

Bolm-Audorff U, Baur X, Bienfait HG, Fruhmann G (1992) Sensibilisierung gegenüber Aspergillusamylase bei Bäckern. In: Schäcke G, Ruppe K, Vogel-Süring C (Eds) *31. Verhandlungen der Deutschen Gesellschaft für Arbeitmedizin e. V.*, Gentner Verlag, Stuttgart

Brisman J, Belin L (1991) Clinical and immunological responses to occupational exposure to $\alpha$-amylase in the baking industry. *Br J Ind Med 48*: 604–608

Degens PO, Priebeler K, Spiekermann W, Czuppon AB, Baur X (1994) Sensiblisierung von Bäckern und Kontrollpersonen im Hauttest (Prick) und der IgE-Bestimmung mittels EAST. *Allergologie 17*: 160–170

Hilton J, Dearman RJ, Basketter DA, Kimber I (1994) Serological responses induced in mice by immunogenic proteins and by protein respiratory allergens. *Toxicol Lett 73*: 43–53

Kanny G, Moneret-Vautrin DA (1995) $\alpha$-Amylase contained in bread can induce food allergy. *J Allergy Clin Immunol 95*: 132–133

Losada E, Hinojosa M, Quirce S, Sánches-Cano M, Moneo I (1992) Occupational asthma caused by $\alpha$-amylase inhalation: Clinical and immunologic findings and bronchial response patterns. *J Allergy Clin Immunol 89*: 118–125

Moneo I, Alday E, Gonzalez-Munoz M, Maqueda J, Curiel G, Lucena R (1994) $\alpha$-Amylase hypersensitivity in non-exposed millers. *Occup Med 44*: 91–94

Morren MA, Janssens V, Dooms-Goossens A, van Hoeyveld E, Cornelis A, de Wolf-Peeters C, Heremans A (1993) $\alpha$-Amylase, a flour additive: an important cause of protein contact dermatitis in bakers. *J Am Acad Dermatol 29*: 723–728

Quirce S, Cuevas M, Diez-Gomez M, Fernandez-Rivas M, Hinojosa M, Gonzalez R, Losada E (1992) Respiratory allergy to Aspergillus-derived enzymes in bakers' asthma. *J Allergy Clin Immunol 90*: 970–978

Sander I, Baur X, Inringhausen-Bley S, Rozynek P (1993) Aspergillus-Amylase (Asp ol) als Bäckerallergen. *Allergologie 3*: 87–90

Sandiford CP, Tee RD, Taylor AJN (1994) The role of cereal and fungal amylases in cereal flour hypersensitivity. *Clin Exp Allergy 24*: 549–557

Schata M, Jorde W (1992) Allergische Reaktionen durch $\alpha$-Amylase in Backmitteln. *Allergologie 15*: 57

Tarvainen K, Kanerva L, Grenzquist-Norden B, Estlander T (1991) Berfusallergien durch Cellulase, Xylanase and Alpha-Amylase. *Z Hautkr 66*: 964–967

Wüthrich B (1994) Enzyme: potente inhalative und ingestive Allergene—fehlende Deklarationspflicht von Backmitteln und Backmehl. *Schweiz Med Wochenschr 124*: 1361–1363

completed 23.03.1995

# Arsenic and its inorganic compounds

## (with the exception of arsine)

| | |
|---|---|
| **MAK value** | – |
| **Peak limitation** | – |
| **Absorption through the skin** | – |
| **Sensitization** | – |
| **Carcinogenicity (1971)** | **Category 1** |
| **Prenatal toxicity** | – |
| **Germ cell mutagenicity (2002)** | **3A** |

| Substance | CAS No. | Formula | Molecular weight |
|---|---|---|---|
| arsenic | 7440-38-2 | As | 74.92 |
| **trivalent arsenic compounds:** | | | |
| arsenic trioxide, arsenic(III) oxide | 1327-53-3 | $As_2O_3$ | 197.82 |
| arsenous acid, arsenic(III) acid | 13464-58-9 36465-76-6 | $H_3AsO_3$ (ortho form) $HAsO_2$ (meta form) | 125.94 |
| salts of arsenic(III) acid, e.g. sodium arsenite | 13464-37-4 | $Na_3AsO_3$ | 191.89 |
| **pentavalent arsenic compounds:** | | | |
| arsenic pentoxide, arsenic(V) oxide | 1303-28-2 | $As_2O_5$ | 229.82 |
| arsenic acid, arsenic(V) acid | 1327-52-2 7778-39-4 | $H_3AsO_4$ | 141.94 |
| salts of arsenic(V) acid, e.g. sodium arsenate | 13464-38-5 | $Na_3AsO_4$ | 207.89 |

This documentation is based mainly on a few reviews, such as WHO 2001, ATSDR 2000 and IARC 1980. More recent literature and original studies are cited only as far as it is necessary for the evaluation.

*Essential MAK Value Documentations.* DFG, Deutsche Forschungsgemeinschaft
Copyright © 2006 WILEY-VCH Verlag GmbH & Co. KGaA, Weinheim
ISBN: 3-527-31394-X

In this documentation, with the exception of monomethylarsonate (MMA) and dimethylarsinate (DMA), only arsenic and inorganic arsenic compounds are discussed. As, however, MMA and DMA always occur during the metabolism of arsenic, they are taken into consideration for relevant end points, e.g. genotoxicity and carcinogenicity.

# 1 Toxic Effects and Mode of Action

Arsenic and inorganic arsenic compounds can be absorbed via inhalation and ingestion. They are distributed rapidly in all organs. Accumulation is observed particularly in the liver, kidneys and lungs. The metabolism of the inorganic arsenic compounds is independent of the route of absorption. Reduction and oxidation reactions lead to mutual transformation, with the preferred reduction of arsenate to arsenite. Only arsenite can be methylated to form first MMA and then DMA. DMA, the main metabolite of the inorganic arsenic compounds, is excreted with the urine.

In case studies of the acute toxicity of the arsenic compounds, above all after ingestion, the first symptoms of intoxication were reported after 30 to 60 minutes. Depending on the severity of the clinical symptoms, intoxication with arsenic is described by the terms acute paralytical syndrome and acute gastrointestinal syndrome. The acute paralytical syndrome is characterized by cardiovascular collapse, central nervous weakness and death within hours. Characteristic of the gastrointestinal syndrome is a metallic or garlic-like taste, a dry mouth, burning lips, difficulties swallowing, headaches, dizziness and vomiting. This can then lead to multi-organ failure.

Long-term exposure to arsenic in man causes fever, sleep disorders, weight loss, swelling of the liver, dark discoloration of the skin, sensory and motor neuropathy, and encephalopathy with symptoms such as headaches, poor concentration, deficits in the learning of new information, difficulties in remembering, mental confusion, anxiety and depression. Long-term exposure to arsenic can also cause peripheral vascular effects such as acrocyanosis, Raynaud's disease and tissue necrosis on the extremities (blackfoot disease). Also cardiovascular diseases and the occurrence of *diabetes mellitus* are described.

Long-term exposure to dust containing arsenic causes irritation of the conjunctiva and mucous membranes of the nose and throat. The pustular or follicular skin reactions observed after contact with inorganic arsenic compounds are usually attributed to irritative effects and not to sensitizing effects of the arsenic compounds.

In animal experiments, repeated ingestion of MMA induced fertility disorders in mice.

In man there is evidence that long-term exposure to inorganic arsenic compounds significantly increases the incidence of spontaneous abortions, single or multiple malformations and stillbirths. The results of animal experiments confirm these toxic effects on development.

In persons exposed long-term to arsenic, an increase in the incidence of micronuclei was found in epithelial cells of the cheeks and bladder, urothelial cells and lymphocytes, and additionally in lymphocytes an increase in the incidence of chromosomal aberrations and sister chromatid exchange.

Arsenic and inorganic arsenic compounds are carcinogenic in man. The target organs of the carcinogenic effects after inhalation are the lungs and after ingestion the bladder, kidneys, skin and lungs.

The prerequisite for the formation of reactive, systemically effective, genotoxic and carcinogenic arsenic species is the reduction of the arsenic compound to the arsenite and possibly its subsequent methylation. The relevant enzymes for these reactions are subject to genetic polymorphism.


# 2 Mechanism of Action


As inorganic arsenic compounds, above all arsenic(III) compounds, react readily with SH groups, the reaction with proteins is regarded as a mechanism for the toxicity of arsenic and inorganic arsenic compounds. It has only become clear over recent years that the mechanism of arsenic toxicity is much more complex than the unspecific reaction with proteins. Arsenic(III) compounds inactivate, possibly as a result of the reaction with SH groups, the phosphatases responsible for the inactivation of the *jun-N*-terminal kinases (JNKs). This results in an increase in the activity of the JNKs and of proto-oncogenes of the *c-jun* family. This activation of JNKs was also discussed in the context of the carcinogenicity of arsenic(III) compounds (Cavigelli *et al.* 1996).

The metabolism of inorganic arsenic compounds, above all their methylation, was thought for a long time to be a route of detoxification. In recent years it has become clear, however, that in particular the trivalent methylated metabolites are of greater toxicity in various systems. For DMA there is, in addition, clear evidence of co-carcinogenic and possibly also carcinogenic effects in animal experiments (Kenyon and Hughes 2001, Kitchin 2001, Thomas *et al.* 2001; see Section 5.7.2). This question has not, however, been systematically investigated. At least for the genotoxic effects of inorganic arsenic compounds there is an increasing body of evidence from experiments that DNA-damaging metabolites are also formed as a result of the methylation. While in earlier investigations methylated arsenic metabolites were less mutagenic and clastogenic (Moore 1997b), in the comet assay (*in vitro*) practically only the methylated arsenic(III) compounds were found to be DNA-damaging (Mass *et al.* 2001). The different composition of oxidoreductases in the cell systems used *in vitro* for the reduction of arsenic(V) to arsenic(III) and of methyltransferases for the subsequent methylation may explain the many contradictory results. As these enzymes are also subject to genetic polymorphism (Vahter 2000), this could explain the species and interindividual differences in the carcinogenesis of the arsenic compounds (Kenyon and Hughes 2001). The animal experiments with inorganic arsenic compounds to date, however, were not carried out under optimum conditions (Huff *et al.* 2000; see Section 5.7.2).

A suggested mechanism for the genotoxicity of the substance is the generation of reactive oxygen species (Liu *et al.* 2001a, Nordenson and Beckman 1991, Wang and Huang 1994). It was concluded from various studies that DNA damage caused by arsenic(III) compounds is the result of the calcium-mediated formation of peroxynitrite, hypochlorous acid and hydroxyl radicals (Wang *et al.* 2001). DMA forms a radical, possibly intracellularly, and is transformed in the presence of molecular oxygen into a peroxy radical (Kitchin 2001), which could be responsible for the observed oxidative DNA damage. Other studies show inorganic arsenic compounds to inhibit DNA repair processes (Bau *et al.* 2001, Hartwig *et al.* 1997, Li and Rossman 1989, Yager and Wiencke 1997). To date, however, no isolated DNA repair enzyme could be identified which is inhibited by arsenic(III) compounds in low concentrations. Most likely the regulation of repair is affected (Hu *et al.* 1998).

Also the interference of inorganic arsenic compounds with DNA methylation is described (Mass and Wang 1997, Goering *et al.* 1999, Zhao *et al.* 1997). Whether this occurs as a result of the depletion of the methyl pool in the cell is unclear (Kitchin 2001). The result is changes in gene expression, as determined for the tumor supressor gene *p53* and the oncogene *c-myc* (Hsu *et al.* 1999, Mass and Wang 1997, Menéndez *et al.* 2001, Salnikow and Cohen 2002, Vogt and Rossman 2001).

In addition it was shown that inorganic arsenic compounds make certain cells more sensitive to mitogenic stimulus and that cell proliferation is increased, but cell differentiation inhibited (Germolec *et al.* 1998, Trouba *et al.* 2000, Wauson *et al.* 2002). This imbalance in cellular control is thought to be the reason for the carcinogenicity of inorganic arsenic compounds (Salnikow and Cohen 2002).

Also the changes in cell cycle regulation are of importance. In various cells exposure to arsenic causes the induction of stress proteins, the stimulation of growth factors and hormone receptors, changes in cytokine production and signal transduction, and the expression of various oncogenes (Bernstam and Nriagu 2000, Chen *et al.* 2001, Liu and Huang 1997, Liu *et al.* 2001b). It was only recently, however, that a correlation between the toxicity of the substance and its carcinogenicity was investigated (Chen *et al.* 2001). On the other hand, in recent years arsenic trioxide has been used successfully in chemotherapy for promyelocytic leukaemia (Chen *et al.* 1996, Huang *et al.* 1998, Shen *et al.* 1997, Soignet *et al.* 1998). The anti-carcinogenic effects are thought to be the result e.g. of the induction of apoptosis and the inhibition of the formation of micro-tubules. The induction of apoptosis was detected in leukaemia cells (Chen *et al.* 1996, Dai *et al.* 1999, Ishitsuka *et al.* 1998) and in lymphoid cell lines (Bazarbachi *et al.* 1999, Zhang *et al.* 1998), and the inhibition of microtubule formation was demonstrated for arsenic(III) compounds (Huang and Lee 1998, Li and Brome 1999).

The false regulation of certain interleukins is also held responsible for impairment of the immune system in patients exposed to arsenic (Salnikow and Cohen 2002).

# 3 Toxicokinetics and Metabolism

## 3.1 Absorption, distribution, elimination

### Absorption

**Inhalation:** In air, inorganic arsenic compounds are mainly found particle-bound. These are usually arsenites [As(III)] and arsenates [As(V)]. The levels in the outdoor ambient air are about $0.001$ µg/m$^3$, and near the source of emission even $> 0.01$ µg/m$^3$ (EC 2000). Data for the levels of arsenic in workplace air vary considerably (WHO 2001). In studies from the 1950s, values of $>> 50$ µg/m$^3$ were found. Arsenic concentrations in air of $> 10$ µg/m$^3$ are still found today. Cigarette smoking is thought to cause additional inhalation exposure to arsenic (ATSDR 2000). Smoking, however, was not found to have an influence on the levels of arsenic excreted with the urine (Seiwert *et al.* 1999).

The absorption of arsenic after inhalation takes place in two steps: deposition of the particles and their subsequent absorption. The extent of absorption varies considerably depending on the solubility of the arsenic compounds and the size of the particles. It is estimated to be in the range of 30 % to 90 % (ATSDR 2000). A more exact quantification of absorption is not possible (WHO 2001).

**Ingestion:** The main sources of exposure of the general population not occupationally exposed to arsenic are food and drinks. Around 25 % is taken up in the form of inorganic arsenic compounds. In regions with high levels of arsenic in the drinking water ($> 100$ µg/l), drinking water is the main source of inorganic arsenic compounds. Gastrointestinal absorption is rapid and the amounts absorbed are in the range of 45 % to 75 %. Of environmental-medical relevance is also the ingestion of arsenic in young children as a result of contaminated soils (WHO 2001).

**Dermal absorption:** Dermal absorption of individual arsenic compounds at the workplace is possible. Arsenic acid penetrates the skin *in vivo* and *in vitro* in amounts of less than 10 % and sodium arsenate from aqueous solution or as the solid *in vitro* in amounts of about 30 % to 60 % (WHO 2001). Several studies have shown that the internal exposure to arsenic correlates well with the external exposure (ACGIH 1999, Greim and Lehnert 1994).

### Distribution

The half-life for the elimination of inorganic arsenic compounds from the blood is short. The arsenic concentration in blood is therefore not a suitable parameter for biological monitoring (WHO 2001).

Arsenic compounds can pass the blood-brain barrier and the placental barrier. Studies with animals showed that the substance is distributed in all organs, with accumulation particularly in the liver, kidneys, bladder and tissues rich in keratin and sulfhydryl groups, such as hair, nails and skin (WHO 2001). Inorganic arsenic compounds have a tendency to accumulate in the epididymis, thyroid gland and lenses of the eyes (Lindgren *et al.* 1982). With increasing age an increase in the accumulation of arsenic in the various tissues is observed (WHO 2001).

## Elimination

Inorganic arsenic compounds are mainly eliminated as DMA via the kidneys (WHO 2001). In 101 male test persons the amounts of the following four arsenic species were determined in urine by means of anion exchange chromatography with atomic absorption spectroscopy: DMA 88.1 %, As(III) 11.9 %; the levels of MMA and As(V) were below the detection limit (Heinrich-Ramm *et al.* 2001).

Small amounts of absorbed arsenic are eliminated via the skin, hair, nails and breast milk. Even with low-level internal exposure the substance enters the placenta (WHO 2001).

## 3.2 Metabolism

Metabolism takes place via reduction and oxidation reactions with the mutual transformation of arsenite and arsenate—preferred is the reduction of As(V) to As(III). Via methylation arsenite is transformed into MMA(V), DMA(V) and DMA(III) (see Figure 1). These steps take place independent of the route of absorption. As the capacity of this enzymatic reaction is limited, the methylation capacity must be exceeded after a certain arsenic dose. Exact data are, however, not available. Habituation resulting from long-term exposure to arsenic can lead to more efficient metabolism (WHO 2001).

Arsenate can be reduced to arsenite by glutathione. Methylation takes place only with As(III). As(V) is first of all reduced to As(III) and then methylated. Methylation takes place mainly in the liver with the involvement of enzymes which use *S*-adenosylmethionine (SAM) as a methyl donor and form *S*-adenosylhomocysteine (SAH), and use GSH as an essential cofactor and form oxidated glutathione (GSSG) (WHO 2001).

$$As^{V}O_4^{3-} \xrightarrow[-GSSG]{+2GSH} As^{III}O_3^{3-} \xrightarrow[-SAH]{+SAM} H_3C-As^{V}O_3^{2-} \xrightarrow[-GSSG]{+2GSH} H_3C-As^{III}O_2^{2-}$$

Arsenate          Arsenite          Methylarsenate          Methylarsenite

$$H_3C-As^{III}O_2^{2-} \xrightarrow[-SAH]{+SAM} (H_3C)_2As^{V}O_2^{-} \xrightarrow[-GSSG]{+2GSH} (H_3C)_2As^{III}O^{-}$$

Methylarsenite          Dimethylarsenate          Dimethylarsenite

**Figure 1.** Metabolism of inorganic arsenic compounds

The methylation capacity is reduced by a reduction in the methyl donors provided in the diet. The quantity and quality of the methylation of inorganic arsenic compounds differs considerably in different species (see Section 2).

# 4 Effects in Man

## 4.1 Single exposures

### 4.1.1 Inhalation

There are various reports that short-term inhalation of arsenic led to nausea, vomiting and diarrhoea. As these symptoms, however, are not typical of intoxication, it was assumed that the arsenic particles were transported to the larynx by mucociliary clearance and then reached the gastrointestinal tract as a result of swallowing. Arsenic was not found to have effects on the blood count, skeletal muscles, liver or kidneys (ATSDR 2000). In other case studies, with no data for the levels of exposure, peripheral neuropathy (loss of muscle reflexes, muscle weakness, tremor) and encephalopathy (hallucinations, increased excitability, emotional lability, memory loss, difficulties in learning new information) were reported after inhalation exposure to arsenic (ATSDR 2000, Beckett *et al.* 1986, Bolla-Wilson and Bleecker 1987). This peripheral neuropathy and encephalopathy were attributed, however, to ingestion of arsenic (ATSDR 2000).

### 4.1.2 Ingestion

Ingestion of arsenic compounds led to the first symptoms usually within 30 to 60 minutes. Depending on the severity of the clinical symptoms, intoxication with arsenic is described by the terms acute paralytical syndrome and acute gastrointestinal syndrome. The acute paralytical syndrome is characterized by cardiovascular collapse, central nervous weakness and death within hours. The most frequent form of acute arsenic intoxication was, however, the gastrointestinal syndrome, which began with a metallic or garlic-like taste, a dry mouth, burning lips, difficulties in swallowing, headaches, dizziness, vomiting and diarrhoea. In some cases multi-organ failure occurred (ATSDR 2000, IARC 1980).

In 4 case studies in which the person died arsenic doses of between 22 and 121 mg/kg body weight were reported (ATSDR 2000).

Severe damage to the respiratory tract, such as shortness of breath, haemorrhagic bronchitis and pulmonary oedema, were described after acute intoxication with arsenic doses of 8 mg/kg body weight and above. These symptoms are probably the result of damage to pulmonary vessels (ATSDR 2000).

Characteristic arsenic-induced effects on the cardiovascular system were changes in the ECG (delayed Q-T interval, unspecific S-T segment) and cardiac irregularity. After a lethal dose of arsenic (93 mg/kg body weight), hypertrophy of the ventricle walls was reported (ATSDR 2000).

There are numerous case reports available that show that a single, high oral dose of arsenic (2 mg/kg body weight and above) causes damage to the nervous system. Usually encephalopathy occurs first with symptoms such as headaches, confusion, lethargy, hallucinations, convulsions and coma, and later peripheral neuropathy. Peripheral neuropathy, however, has also been observed without previous encephalopathy (ATSDR 2000).

In a man who swallowed 8 to 9 g arsenic with suicidal intent, vomiting and diarrhoea were observed after a few hours. Haemodialysis was carried out when he was taken to hospital. 10 days later the patient developed symmetrical polyneuropathy. After 7 weeks Wallerian degeneration of the myelinated nerve fibres was diagnosed and chelate therapy was commenced. Nerve biopsy revealed that arsenic was detectable in the fibular nerve, although it could not be clarified whether the arsenic was situated in the axon or the myelin sheath. In an examination 3 years later, arsenic could no longer be detected in the fibular nerve. The condition of the patient had improved, apart from slight polyneuropathy. The biopsy showed that regeneration and proliferation of the axons had taken place and there was no longer evidence of Wallerian degeneration (Goebel *et al.* 1990).

In another study the electrophysiological profiles of 13 persons with arsenic-induced neuropathy were drawn up; 12 of the persons were exposed to arsenic only once. The conductivity of sensory nerves was impaired as a result of a lack of or low action potentials, the conductivity of motor nerves, however, was only slightly impaired. Biopsy of the fibular nerve revealed axon degeneration (Oh 1991).

Various studies reported that single oral doses of arsenic cause anaemia and leukopenia; this was attributed to direct haemolytic effects or the inhibition of erythropoiesis. There are also, however, other case studies in which no arsenic-induced effects on the blood system were found (ATSDR 2000).

## 4.2 Repeated exposures

### 4.2.1 Inhalation

Various studies indicate that long-term exposure to dust containing arsenic can cause laryngitis, bronchitis and rhinitis (ATSDR 2000). Copper smelters in Anaconda, Montana and Tacoma, Washington, who were exposed to arsenic, died of respiratory diseases such as pulmonary emphysema significantly more frequently than persons not exposed. A clear dependence on the level of arsenic exposure could not, however, be deduced and smoking habits were not taken into account (ATSDR 2000).

In various epidemiological studies, peripheral vascular effects such as acrocyanosis, Raynaud's disease (episodes of ischaemia resulting from spasms in vessels, usually in the arteries of the fingers) and tissue necrosis on the extremities (blackfoot disease) were described after long-term inhalation exposure to arsenic (ATSDR 2000, Lagerkvist *et al.* 1988).

The risk of dying from cardiovascular diseases, such as ischaemia of the heart, or cerebrovascular diseases, was increased in many cohort studies. A clear dose–response relationship was not observed, however. In addition there was exposure to a mixture of substances, including copper, lead and radon (ATSDR 2000).

In a copper smelting plant, peripheral neuropathy was investigated in 70 employees exposed to arsenic trioxide and 41 control persons who were not exposed. The results show that the level of arsenic, which was determined by analysing urine, hair and finger nails, was associated with a higher number of cases of sensory and motor neuropathy and electrophysiological changes (Feldman *et al.* 1979).

In another study, the occurrence of chronic encephalopathy resulting from occupational exposure to arsenic was reported. After repeated exposure to arsenic during wood impregnation, one person developed encephalopathy with such symptoms as concentration difficulties, deficits in learning new information and in short-term memory, mental confusion, anxiety and depression. Another man suffered from irritation, headaches and memory difficulties. The condition of both persons improved after the end of the arsenic exposure (Morton and Caron 1989).

Evidence was found in a Swedish study that long-term inhalation exposure to arsenic can play a role in the development of *diabetes mellitus* (Rahman and Axelson 1995).


## 4.2.2 Ingestion

Case reports showed daily uptake of arsenic with the drinking water of 0.05 to 0.1 mg/kg body weight to be lethal in young children and of 0.014 to 3 mg/kg body weight to be lethal in adults (ATSDR 2000).

In one of two men exposed short-term to arsenic (no other details), indisposition, fever, coughing and a headache developed, and in the other man vomiting and diarrhoea. In both patients a scaly, papular skin rash developed. Blood analyses yielded increased haemoglobin values (1120 g/l; normal value: 140–180 g/l) in one of the exposed men and a reduced haematocrit value (32 %; normal value: 40–50 %) in the other. Both persons developed severe sensory deficits (tremor, proprioception) and motor deficits within 4 weeks. In the first examination in both patients action potentials of various sensory nerves were no longer present. Motor performance decreased in the first three months and then improved. Two years later, however, sensory and motor neuropathy was still detected, which indicates destruction of the sensory nerves (Murphy *et al.* 1981).

As a result of contaminated milk powder, around 12000 Japanese children ingested up to 3.5 mg arsenic daily for 33 days. The most frequent symptoms were fever, sleep disorders, weight loss, swelling of the liver and dark coloured skin. 130 deaths occurred (IARC 1980).

In a family which were exposed to arsenic long-term (no other details), only a 16-year-old girl was found to have extremely delayed thought processes, severe hypotonic paraparesis, generalized areflexia and white finger and toe nails. The girl's symptoms of intoxication were explained at the time, among other things, by a genetic defect in methylene tetrahydrofolate reductase, with the result that methylation of the arsenic compounds, and thus detoxification, could not take place (Brouwer *et al.* 1992).

Long-term uptake of arsenic doses of about 0.03–0.1 mg/kg body weight and day resulted in symmetrical peripheral neuropathy, which began with a loss of feeling in the hands and feet and developed into painful paresthesia. Sensory and motor nerves were affected. In some cases muscle weakness developed, which led to paralysis of the wrist and ankle, and changes in reflexes were observed. Histological examinations revealed the death of axons in conjunction with demyelination. After the end of the exposure to arsenic the damage regressed only very slowly and also not completely (ATSDR 2000).

Other studies reported that neurotoxic effects could not yet be observed with daily doses of arsenic of 0.006 mg/kg body weight. Another study from China reported, however, that exposed persons described fatigue, headaches, dizziness, sleep disorders, nightmares or a loss of feeling in the extremities after daily arsenic doses of 0.005 mg/kg body weight and more (ATSDR 2000). Various studies reported a relationship between the repeated oral uptake of arsenic with the drinking water and the increased occurrence of cerebrovascular diseases and circulatory disturbances of the heart (Taiwan), high blood pressure (Bangladesh), Raynaud's disease and cyanosis of the fingers and feet (with daily arsenic doses of 0.02 to 0.06 mg/kg body weight; Chile) (ATSDR 2000). Long-term oral uptake of arsenic caused damage to the peripheral vascular system. The characteristic example of this is blackfoot disease, which is endemic above all in Taiwan in those areas in which the level of arsenic in the drinking water varies between 0.17 and 0.8 mg/l, which corresponds to a daily dose of about 0.014 to 0.065 mg/kg body weight. The disease is characterized by a progressive loss of blood circulation in the hands and feet, and finally leads to necrosis and gangrene (ATSDR 2000). In a retrospective cohort study of 789 persons affected by blackfoot disease, mortality from peripheral vascular or cardiovascular diseases was significantly increased after an observation period of 15 years (Chen *et al.* 1988).

The uptake of arsenic (about 0.02 mg/kg body weight) with the drinking water over many years led also to a reduction in body weight (ATSDR 2000). Other studies showed a relationship between repeated oral uptake of arsenic with the drinking water and the increased occurrence of *diabetes mellitus* (Lai *et al.* 1994, Rahman *et al.* 1998).

## 4.3 Local effects on skin and mucous membranes

### Skin

In an employee of a glass factory, who had contact with arsenic trioxide while mixing the raw materials, papulo-vesicular skin reactions developed on the hands, the bends of the arms, the hollows of the knees, and the neck (Paschoud 1964).

In a very extensive investigation of employees of a Swedish company for refining arsenic, toxic and irritative effects on the skin, above all caused by arsenic trioxide, were found in 71 employees. The skin changes were observed particularly on the face, while the hands, which were protected by gloves, were affected less often. Although the course of the skin diseases was acute in most cases, in some cases it was chronic. A latency period for the development of skin diseases of 1 week to 10 years was given. In view of the variable clinical picture with erythema, oedema, the formation of papula or vesicles, or even purely follicular reactions, it was often not possible to differentiate between an irritative and an allergic genesis (Holmqvist 1951; see Section 4.4).

Other pustular or follicular skin reactions which occurred after contact with inorganic arsenic compounds were attributed to the irritative effects of the arsenic compounds (Birmingham *et al.* 1965, Eberhartinger *et al.* 1969, Goncalo *et al.* 1989, Mohamed 1998).

## Eyes and mucous membranes

Long-term exposure to dust containing arsenic caused irritation of the mucous membranes of the nose and throat and the conjunctiva (ATSDR 2000).

# 4.4 Allergenic effects

In several earlier reports sensitization resulting from skin contact with inorganic (but above all organic) arsenic compounds was described.

One case was an employee of a glass factory who had contact with arsenic trioxide while mixing the raw materials and as a result developed papulo-vesicular skin reactions on the hands, the bends of the arms, the hollows of the knees, and the neck. Patch tests were carried out with arsenic trioxide powder and with 1 % potassium arsenite in water and yielded positive results. No details were given, however, about the test conditions and the reading procedure (Paschoud 1964).

After long-term occupational contact with sodium arsenite, 3 persons developed suspected allergic contact dermatitis. In the patch test 2 of the 3 patients produced a weak reaction and 1 patient a marked reaction to 1 % sodium arsenite in water, yet no reaction to 5 % sodium arsenate in water (Schulz 1962).

Patch tests with arsenic trioxide produced papulous reactions in 71 of 163 exposed employees of a Swedish company that refined arsenic. Only 12 of 145 new employees and 17 of 105 employees with low-level exposure to arsenic produced such reactions, however. In the three groups skin reactions to arsenic trioxide were observed in 131 of 163 employees (80.4 %), 44 of 145 employees (30.3 %) and 34 of 105 employees (32.4 %), respectively. A 1 % sodium arsenite preparation caused a marked reaction in one of the affected employees, which persisted for at least 14 days and was regarded as allergic. Of 23 persons tested who previously did not react to arsenic trioxide, none reacted to 1 % sodium arsenite, but all of them to 10 % sodium arsenite. In patch tests with 1 % to 40 % sodium arsenate 4 of 10 persons not exposed also produced (weak) irritative reactions, in 2 cases even with 1 % sodium arsenate (Holmqvist 1951). Nevertheless, the severity and course of the reactions described for several employees indicate an allergic genesis rather than an irritative one. Further evidence for an immunological mechanism is the flaring up of the original skin reactions during the patch test or oral provocation. In some cases reactions that occurred only after several tests indicate the possibility of sensitization as a result of the testing. All in all, however, despite extensive documentation of the test results, no decisive statement can be made at present about the sensitizing effects of inorganic arsenic compounds on the basis of these investigations, as no standardized preparations were used for testing and no details are available about the purity of the test substances. In addition, it is not stated whether the tests were carried out during a period in which the skin had been free of symptoms for a longer period. These limitations are very probably true also for earlier findings listed by the author in a historical review and are therefore not taken into consideration here.

Later studies seem to indicate that even the lowest test concentrations can yield false positive results, so that 0.1 % sodium arsenate and 0.05 % sodium arsenite were used in preliminary comparative tests in exposed employees and for the consecutive testing of patients of a dermatological clinic. Two of 379 clinic patients produced marked reactions to both preparations. Previous exposure to arsenic was not evident from the persons' working history (Wahlberg and Boman 1986). Test results from the investigation of the employees were evidently not published.

In another earlier study the increased occurrence of eczematous or follicular skin reactions was reported in employees of a smelting plant and their families in connection with an inadequately functioning filter system. Only in 14 persons (12 children, 2 employees) with eczematous or follicular skin reactions were patch tests carried out 6 months later with 5 % arsenic trioxide in starch. This produced oedematous erythema in one case and oedematous-vesicular reactions in 2 cases. Only one of the persons who produced a reaction previously had eczematous skin changes but tolerated further contact with small amounts of arsenic without recurrence of the symptoms. The persons who reacted to arsenic trioxide also reacted to 1 % sodium arsenite and 10 % sodium arsenate in water, but some control persons (no other details) produced weak irritative reactions with these preparations (Birmingham *et al.* 1965).

Also in an arsenic-processing works erythematous, eczematous and bullous skin changes were observed, as was a specific manifestation of pustulous reactions ("arsenic scabies"). Allergological investigations, however, were not carried out (Rossberg 1967). Such pustulous or follicular reactions occur also after contact or testing with other metal salts, are not always reproducible and their genesis is usually difficult to evaluate. Between 1966 and 1968, 1638 patients were tested with standard substances, which also included an arsenic compound (it is unclear whether this was 1 % sodium arsenite or 1 % sodium arsenate). Pustulous or follicular reactions were observed in 109 patients, in 88 of these persons to the arsenic compound, in 33 to 10 % nickel sulfate and in 17 to 2 % formalin solution. In none of the 33 patients who reacted in such a way to nickel sulfate was evidence of a nickel allergy found in subsequent investigations (Eberhartinger *et al.* 1969).

In an employee who had previously never suffered from skin complaints, skin changes were observed on the hands, arms and legs after employment for 3 months in a crystal factory. In a patch test with various substances with which the employee had contact at work, 1 % sodium arsenate in water was the only substance which yielded a positive result, with reactions after 48 and 96 hours. 22 control persons did not react to the test preparation (Barbaud *et al.* 1995).

In a tin smelting works, in which 11 employees suffered from dry, itching, hyperpigmented skin with folliculitis, the concentration of arsenic trioxide in the air was between 5.2 and 14.4 µg/m$^3$. The authors attributed the complaints to probably irritative contact dermatitis resulting from dust containing arsenic. Patch tests were not carried out (Mohamed 1998).

Three workers in the glass industry suffered among other things from pustular and follicular skin changes on exposed areas of skin. In patch tests with the arsenic trioxide powder to which the workers were exposed, a clear pustular-follicular reaction was observed, while the workers produced only weak reactions to a 5 % arsenic trioxide preparation. The authors attributed the skin changes to irritative effects of the arsenic trioxide (Goncalo *et al.* 1989).

In the more recent literature there is only a report about a 26-year-old female worker who was occupationally exposed to DMA and developed dermatitis on the face. In patch tests with 0.1 % and 1 % DMA in water, reactions were observed after 3 days at both concentrations (Bourrain *et al.* 1998).

One study investigated the immunological effects after inhalation of inorganic arsenic compounds. In workers exposed to arsenic in a coal-firing power plant, the serum levels of the immunoglobulins were not unusual (Bencko *et al.* 1988).

## 4.5 Reproductive and developmental toxicity

### 4.5.1 Fertility

There are no studies available of the influence of arsenic and inorganic arsenic compounds on the fertility of man after inhalation or oral exposure (ATSDR 2000).

### 4.5.2 Developmental toxicity

#### Single exposures

A 17-year-old, who swallowed arsenic trioxide by drinking rat and mouse poison (about 0.39 mg/kg body weight) in week 30 of pregnancy, was admitted to hospital 24 hours later with acute kidney failure. On day 3 her condition worsened and she became progressively disoriented. She was delivered of a 1.1 kg baby girl and underwent haemodialysis (no other details). After the first minute of life the baby had an APGAR score of 4 (a measure of heart rate, respiratory effort, muscle tone, reflex irritability and colour; optimum score: 9–10), had increasing difficulties breathing, a reduced heart rate, hypoxia, hypercabia and acidosis and died after 11 hours. Autopsy revealed in addition to the symptoms of premature birth generalized skin haemorrhage and severe intra-alveolar bleeding (Lugo *et al.* 1969).

## Repeated exposure

Numerous studies investigated the influence of arsenic exposure on intrauterine development, in particular the frequency of spontaneous abortions, congenital anomalies and stillbirths.

To test the hypothesis that the developmental toxicity of arsenic is due to its great affinity to thiol groups and glutathione and the resulting increase in lipid peroxidation in maternal and foetal compartments, the glutathione levels in blood and the lipid peroxidases in the placenta or blood were investigated in 49 mother–child pairs in hospitals near a copper mine and in a control group. The exposure was characterized by high levels of arsenic in the air (0.119 mg/m$^3$) and low levels in the water (about 12 µg/l). The incidence of complications during pregnancy, e.g. maternal toxaemia and foetal mortality as a result of congenital malformations, was higher in the region of the copper mines than in the national average. Taking into account maternal parity, smoking habits and the sex of the children, the average body weights and body lengths at birth were reduced. The average arsenic levels in the placentas of the exposed mothers were three times higher than those in the mothers not exposed (0.023 compared to 0.007 mg/kg, p < 0.01). In the group with the highest exposure the reduced glutathione was significantly decreased and the lipid peroxidases increased (Tabacova and Hunter 1995).

In a region in southeast Hungary with increased levels of arsenic in the water (> 100 µg/l, 25648 persons), a 1.4-fold increase in spontaneous abortions and a 2.7-fold increase in stillbirths (p < 0.05) were determined between 1980 and 1987 compared to in a region with low arsenic levels in the water (no data for the levels of arsenic, 20836 persons). There was no significant difference between the groups for the frequency of premature births and stillbirths (Borzsonyi *et al.* 1992).

To determine the relationship between the uptake of chemicals with the drinking water of the mother and congenital heart diseases of the offspring, all live births (270 cases, 665 controls) between 01.04.1980 and 31.03.1983 were investigated. The arsenic concentrations in the drinking water were between 0.8 and 22 µg/l. No relationship was found between the arsenic level in drinking water and all congenital heart diseases, but a three-fold increase in the frequency of coarctation of the aorta (OR = 3.4; 95 % CI = 1.3–8.9) was determined. There was no relationship with the frequency of patent *ductus arteriosus*, cono-truncal defects and ventricular septal defects. Non-differential errors in the analyses and the classification of the exposure, and the low exposure could be responsible in the opinion of the authors for the lack of positive findings (Zierler *et al.* 1988).

In a hospital-based study of the relationship between the quality of communal drinking water and the risk for spotaneous abortions, an increase in the frequency of spontaneous abortions (taking confounders into consideration) which was not significant was found for arsenic concentrations in drinking water of between 0.8 and 1.9 µg/l. As a result of the small amount of data, the authors state that further investigations are necessary to confirm these findings (Aschengrau *et al.* 1989).

In a hospital-based case–control study with an atmospheric distribution model to estimate the arsenic concentration in the air of a region in which agricultural products on an arsenic basis were produced for over 60 years, an increased risk was found for stillbirths with exposure to increased levels of arsenic ($> 100$ ng/m$^3$) after adjustment for demographic factors (POR = 4.0; 95 % CI = 1.2–13.7). In a separate analysis of the interaction between race/ethnic provenance and exposure, a significant relationship was found for Spanish inhabitants, who were exposed to the highest concentration ($> 10$ ng/m$^3$) (POR = 8.4; CI = 1.4–50.1). The risk for stillbirths was higher among African Americans in the adjusted model, but not significantly increased in the interaction model. Further studies are necessary to clarify the role of inhalation exposure via the environment in addition to the role of nutrition and other factors (Ihrig *et al.* 1998).

The subject of numerous studies was the possible influence of the smelting works in Ronnskar, Sweden, on the health of the employees and the local population. In addition to copper and lead, the smelting works produced arsenic, arsenic trioxide and a few other metals and their compounds. The arsenic emissions were given as 50 tons per year (Nordström *et al.* 1978, 1979a, 1979b). Significantly reduced birth weights and increased incidences of spontaneous abortions were determined in the employees and the local population. If women worked in the immediate area of the smelting processes, there was a significant increase in the frequency of abortions, which increased further if their husbands were employed in the smelting works (Nordström *et al.* 1979a). An increase in the incidence of spontaneous abortions and stillbirths together was found in women whose husbands worked in a smelter, relative to in women whose husbands were not exposed (Beckman and Nordstrom 1982). In children whose mothers worked near the smelting processes during pregnancy, unlike in children of mothers who lived in the vicinity of the smelting works, the frequency of single malformations (e.g. cleft palates) or multiple malformations was significantly increased (Nordström *et al.* 1979b). These studies are, however, unsuitable for evaluating developmental toxicity, as the exposure to arsenic was not sufficiently characterized.

In a retrospective study infant mortality was investigated in two regions of Chile between 1950 and 1996. In Antofagasta contamination of the drinking water with arsenic was documented, while in Valparaiso the levels were comparatively low. Investigation of the temporal development of late foetal mortality, mortality of newborn babies and mortality in early childhood revealed a relationship with the arsenic level in drinking water (Hopenhayn-Rich *et al.* 2000).

In another study, in Bangladesh two groups of 96 women aged between 15 and 49 were compared. One group had consumed $\geq 0.1$ mg arsenic per litre drinking water (43.8 % of the women for 5 to 10 years), and the other group had not. The two groups were matched for age, social status, education and age at marriage. Only in the group of exposed women were the frequencies of spontaneous abortions, stillbirths and premature births significantly higher than in the control group (Ahmad *et al.* 2001).

## 4.6 Genotoxicity

Studies of the genotoxicity of arsenic compounds were carried out above all in persons who had taken up arsenic long-term with the drinking water. It was not differentiated between the different arsenic species in drinking water. As inorganic arsenic compounds can be transformed by micro-organisms into organic arsenic species by methylation, organic arsenic compounds are found in drinking water in addition to the inorganic arsenic compounds. Also in the human organism methylation of the absorbed inorganic arsenic compounds takes places to form MMA and DMA. The internal exposure in man after absorption of arsenic via the drinking water is therefore a mixture of inorganic and organic arsenic compounds. Studies of these exposed persons revealed clastogenic effects of arsenic (Table 1).

**Table 1**. Genotoxicity of arsenic compounds in man

| Study population | Number of persons (age) | Exposure | Effects | References |
|---|---|---|---|---|
| single exposures | | | | |
| test persons | 1 volunteer per dose | ingestion (1×) 0.15 g $KAsO_2$ or 1, 10 or 20 g $As_2O_3$ | in L: SCE significantly increased at 20 g $As_2O_3$ | Hantson *et al.* 1996 |
| long-term ingestion | | | | |
| general population, Mexico | 11 (21–62) | drinking water exposure; exposed persons: 390 µg/l, controls: 26 µg/l; As in urine: 1565 ± 916 µg/l (controls: 121 ± 90 µg/l) | in L: CA increased, no SCE, mutations at the HPRT locus increased in those with the highest exposure | Ostrosky-Wegman *et al.* 1991 |
| general population, Mexico | 35 (39.0) | drinking water exposure; exposed persons: 408 µg/l, controls: 30 µg/l; As in urine: 740 ± 361 µg/l (controls: 34 ± 35 µg/l) | in L: CA increased, in CC und UC: MN increased | Gonsebatt *et al.* 1997 |
| general population, Nevada, USA | 98 (not stated) | drinking water exposure; exposed persons: 109 µg/l, controls: 12 µg/l; ≥ 5 years | in L: no CA, no SCE | Vig *et al.* 1984 |
| general population, Nevada, USA | 18 (37.5) | drinking water exposure; exposed persons: 1313 µg/l, controls: 16 µg/l; ≥ 1 year, As in urine: 750 ± 160 µg/l (controls: 68 ± 23 µg/l) | in BC: MN increased | Warner *et al.* 1994 |

**Table 1**. continued

| Study population | Number of persons (age) | Exposure | Effects | References |
|---|---|---|---|---|
| general population, Chile | 70 (42.0) | drinking water exposure; exposed persons: 600 µg/l, controls: 15 µg/l; As in urine: 616, 84–1893 µg/l (controls: 66, 4–267 µg/l) | in BC: MN increased | Moore *et al.* 1997a |
| general population, Finland | 32 (15–83) | drinking water exposure; exposed persons: 410 µg/l, controls: < 1 µg/l; As in urine: 180, 7–500 µg/l (controls: 7, 4–44 µg/l) | in L: CA significant correlations with level of As in urine | Maki-Paakkanen *et al.* 1998 |
| general population, Argentina | 22 (8–66) | lifelong drinking water exposure; 205 µg/l, As in urine: 260, 120–440 µg/l (controls: 8.4, 5.2–27 µg/l) | in L: MN significantly increased, no CA, no SCE | Dulout *et al.* 1996 |
| general population, Argentina | 282 (27–82) | drinking water exposure (> 20 years) (exposed persons: 130 µg/l, controls: 20 µg/l), As in urine: 160, 60–410 µg/l (controls: 70, 10–600 µg/l) | in L: SCE significantly increased | Lerda 1994 |
| general population, Inner Mongolia | 19 (38 ± 15) | drinking water exposure (exposed persons: 527.5 µg/l, controls: 4.4 µg/l) | in SC, CC and UC: MN significantly increased | Tian *et al.* 2001 |
| patients with arsenic-induced skin tumours | (not stated) | (no other details) | 8-OH-dG adducts in skin tumours significantly increased | Matsui *et al.* 1999 |

long-term inhalation

| Study population | Number of persons (age) | Exposure | Effects | References |
|---|---|---|---|---|
| smelters | 9 | exposure to a mixture of arsenic, lead and selenium | in L: CA significantly increased | Beckman *et al.* 1977 |
| smelters | 39 (21–62) | arsenic exposure 4–22 years, As in urine: 167–290 µg/l | in L: CA significantly increased | Nordenson *et al.* 1978 |

**Table 1**. continued

| Study population | Number of persons (age) | Exposure | Effects | References |
|---|---|---|---|---|
| smelters | 33 (20–65) | exposure probably to a mixture of substances; As in urine: 200 µg/l (1976), 80 µg/l (1979) | in L: CA significantly increased | Nordenson and Beckman 1982 |
| smelters, Chile | 15 (24–66) | copper factory; As in urine of employees with highest exposure > 260 µg/l | in L: HPRT mutations increased | Harrington-Brock *et al.* 1999 |
| long-term dermal absorption | | | | |
| tumour patients | 6 (41–84) | patients treated with Fowler's solution (1 % KAsO$_2$ in H$_2$O) (no other details) | in L: SCE significantly increased, no chromosomal breaks | Burgdorf *et al.* 1977 |

BC = bladder epithelial cells, CA = chromosomal aberration,  CC = epithelial cells of the cheek, HPRT= hypoxanthine phosphoribosyl transferase, L = lymphocytes, MN = micronuclei, SC = cells from sputum,  SCE = sister chromatid exchange,  UC = urothelial cells

In buccal and bladder epithelial cells, urothelial cells and lymphocytes an increase in the incidence of micronuclei was found and additionally in lymphocytes an increase in chromosomal aberrations and sister chromatid exchange (Dulout *et al.* 1996, Gonsebatt *et al.* 1997, Hantson *et al.* 1996, Lerda 1994, Maki-Paakkanen *et al.* 1998, Moore *et al.* 1997a, Ostrosky-Wegman *et al.* 1991, Tian *et al.* 2001, Vig *et al.* 1984, Warner *et al.* 1994). In another study, however, no genotoxic effects could be found in persons exposed to arsenic (Vig *et al.* 1984).

A significant increase in chromosomal aberrations in peripheral lymphocytes was detected in workers of the copper smelting plant in Ronnskar, Sweden, who were exposed to unspecified concentrations of arsenic trioxide, (Beckman *et al.* 1977, Nordenson and Beckman 1982, Nordenson *et al.* 1978). It is questionable whether there was additional exposure to other genotoxic substances.

A significant increase in sister chromatid exchange, but no chromosomal breaks were found in the peripheral lymphocytes of tumour patients who had applied a solution containing arsenic to the skin long-term (Burgdorf *et al.* 1977). A slight increase in HPRT mutations was described in peripheral lymphocytes of workers in a copper mine in Chile, in whom arsenic levels in urine of more than 260 µg/l were determined (Harrington-Brock *et al.* 1999).

# 4.7 Carcinogenicity

On the basis of epidemiological studies, national and international organisations have classified arsenic and arsenic compounds (with the exception of arsine) as carcinogenic in man (ACGIH 1999, EPA 2001, IARC 1980, 1987).

## 4.7.1 Inhalation

In several large cohort studies, an increased risk for lung cancer was observed in gold and tin miners and copper smelters in various countries and under different exposure conditions. The possibility of significant distortion of the increased lung cancer risk as a result of co-exposure to sulfur dioxide, various dusts and above all cigarette smoke could be ruled out in several studies (e.g. Enterline 1987b, Järup and Pershagen 1991). An extensive description of the epidemiological studies can be found in IARC (1980) and WHO (2001).

In view of the verified epidemiological evidence for the carcinogenicity of arsenic in man, the question arises whether a threshold concentration for the carcinogenic effects of arsenic can be deduced from the epidemiological studies. Only the relevant epidemiological studies are described below.

The dose–response relationship between exposure to arsenic and lung cancer risk was investigated in six published studies with quantitative data for the arsenic exposure. The studies investigated smelters in Tacoma, Washington (Enterline 1987a), Anaconda, Montana (Lee-Feldstein 1986), Ronnskar, Sweden (Järup and Pershagen 1991, Järup *et al.* 1989) and six smaller smelting works in the USA (Enterline 1987b), a cohort of workers from insecticide production in Midland, Michigan (Ott *et al.* 1974) and a cohort of miners and a few smelters in China (Taylor *et al.* 1989). In only two of the six studies (Lee-Feldstein 1986, Järup *et al.* 1989) was the relationship between arsenic exposure and lung cancer compatible with a linear dose–response relationship, while the relationship in the other studies was supralinear. The reasons suggested for this possible supralinear dose–response relationship were dose-dependent synergisms with smoking habits, competing dose-dependent causes of death or dose-dependent false classification of the arsenic exposure, but not distortion as a result of age or occupational co-exposures (Hertz-Picciotto and Smith 1993).

In the analysis of an extended follow-up to the Tacoma study (Enterline 1987a) the previously found supralinear dose-response relationship could be confirmed (Enterline *et al.* 1995).

An extended follow-up study of the cohort of 8014 smelters in Anaconda, Montana (see Lee-Feldstein 1986) investigated the relationship between arsenic exposure and lung cancer risk (Lubin *et al.* 2000). It was already seen in earlier analyses that sulfur dioxide and smoking habits probably did not represent relevant confounders for the relationship between arsenic exposure and lung cancer risk for this cohort (Lubin *et al.* 1981, Welch *et al.* 1982). Using internal controls and different weighting factors for the duration of work in areas with high levels of exposure to arsenic, the authors found a linear dose–response relationship between inhalation exposure to arsenic and lung cancer risk. They suggested that the supralinear dose–response relationships in earlier analyses of this cohort and in the analyses of the extended follow-up study of the Tacoma cohort (Enterline *et al.* 1987a) may be the result of a differential false classification of high exposure to arsenic (Lubin *et al.* 2000). Further data for the above-mentioned studies can be found in Table 2.

All in all, the data of the studies of the relationship between occupational inhalation exposure to arsenic and the lung cancer risk indicate there is most probably a linear, possibly also a supralinear, but not a sublinear dose–response relationship with a deducible threshold dose.


### 4.7.2 Ingestion

The evaluation of the EPA suggests that the re-analysis of a Taiwanese study of the relationship between oral exposure to arsenic via contaminated drinking water and skin cancer risk may indicate a non-linear dose–response relationship in the sense of a sublinear curve (EPA 1997). Even if the EPA do not attach much importance to this statement themselves, occupational mortality studies are not suitable for identifying increased skin cancer risks.

The study of Lubin *et al.* (2000) indicates a clear dose–response relationship between inhalation exposure to arsenic and lung cancer risk; an increase in other types of cancer was not observed. The cumulative inhalation exposure to arsenic in the cohort of Lubin *et al.* (2000) was much higher than in the Tacoma cohort and corresponded more or less to the long-term oral uptake of arsenic in endemic regions, e.g. Taiwan. Studies in these regions consistently associated the increased uptake of arsenic with an increased risk for cancer of the bladder, kidneys, liver and lungs. The incidences for lung cancer observed in these studies were increased to a much smaller extent than those for bladder, kidney and skin cancer (Bates *et al.* 1992).


### 4.7.3 Dermal absorption

Many early studies reported that the long-term dermal application of solutions containing arsenic, e.g. Fowler's solution (1 % $KAsO_2$ in $H_2O$), caused skin carcinomas (IARC 1980, WHO 2001).

**Table 2**. Epidemiological studies of exposure to arsenic

| Study population (male) | Number of persons | Type of study | Exposure | Effects | References |
|---|---|---|---|---|---|
| workers in the packing of insecticides, USA | 173 exposed persons, 1809 persons not exposed | proportional mortality study: causes of death of exposed persons compared with the cause of death of persons not exposed | IAEM < 1 to > 96 | cancer of the respiratory passages increased two-fold (from 1–2 IAEM) to seven-fold (over 96 IAEM); dose–response relationship supralinear | Ott *et al.* 1974 |
| smelters, Utah, USA | 2288 | cohort study: workers compared with the white male population of the USA, the state or the region | mean exposure 959.7 $\mu g/m^3 \times$ year; mean duration of employment 21.1 years | SMR for lung cancer 119.7 (not significant), 226.9 (p < 0.01) and 210.9 (p < 0.01); great differences between USA and regional values possibly as a result of the different smoking habits of the workers compared with the rest of the population in Utah | Enterline *et al.* 1987b |
| smelters, different companies, USA | unclear, several thousand (?) | cohort study: only male persons exposed to arsenic, comparison of different concentrations | < 100 to > 1000 $\mu g/m^3 \times$ year | lung cancer RR < 100 $\mu g/m^3$: 0.58; 100–249 $\mu g/m^3$: 0.85; 250–999 $\mu g/m^3$: 1.21; > 1000 $\mu g/m^3$: 1.60; p = 0.059 for trend; dose–response relationship supralinear | Enterline *et al.* 1987b |
| mine workers and smelters, China | 107 exposed persons, 107 persons not exposed | embedded case–control study: workers with lung cancer compared with workers without lung cancer | IAEM median for cases: 45.9 (0–255.6); IAEM median for controls: 7.9 (0–132.9) | lung cancer in smokers with low cigarette consumption: OR 5.0 (IAEM 58.6–255.6); reference category: IAEM 0–10.4; no interaction with smoking | Taylor *et al.* 1989 |

**Table 2.** continued

| Study population (male) | Number of persons | Type of study | Exposure | Effects | References |
|---|---|---|---|---|---|
| smelters, Sweden | 3916 | cohort study: workers compared with the population of the region | $< 250$ to $> 100000$ $\mu g/m^3 \times$ year | SMR of 271 (lowest cumulative exposure) up to 1137 (highest cumulative exposure) | Järup *et al.* 1989 |
| smelters, Sweden | 107 exposed persons, 214 persons not exposed | embedded case–control study: workers with lung cancer compared with dead workers without lung cancer | $< 250$ to $> 100000$ $\mu g/m^3 \times$ year | OR 1.0 for up to 5000 $\mu g/m^3$ per year, then increased up to 8.7 for $> 100000$ $\mu g/m^3 \times$ year (age and smoking checked) | Järup and Pershagen 1991 |
| smelters, Tacoma, Washington, USA | 2802 | cohort study: workers compared with the white male population of the state of Washington | $< 750$ to $> 45000$ $\mu g/m^3 \times$ year | SMR for cancer of the respiratory passages (bronchi, trachea, lungs): 188.1 after $< 20$ years exposure; SMR after $\geq 20$ years 217.1; SMR also significantly increased for intestinal and bone cancer; relationship between the arsenic dose and cancer of the respiratory passages supralinear | Enterline *et al.* 1995 |
| smelters, Anaconda, Montana, USA | 8014 | cohort study: workers compared with the population of the USA | time-weighted exposure 0.29, 0.58 and 11.3 mg/m$^3$ | SMR for cancer of the respiratory passages 183; RR for low dose group, exposure $\geq 35$ years, 1.98; RR for highest dose group, exposure $\geq 10$ years, 3.68 | Lubin *et al.* 2000 |

RR: relative risk, OR: odds ratio, IAEM: time-weighted arsenic exposure multiplied by the duration of exposure (months)

# 5 Animal experiments and *in vitro* Studies

## 5.1 Acute toxicity

### 5.1.1 Inhalation

After mice inhaled arsenic trioxide (0.27, 0.50 or 0.94 mg/m$^3$) for 3 hours, pulmonary defence against bacteria was reduced in a dose-dependent manner, probably as a result of damage to the alveolar macrophages, and the susceptibility to infection was correspondingly increased (Aranyi *et al.* 1985). After mice were given intratracheal doses of sodium arsenite (5.7 mg arsenic per kg body weight), the humoral immune response to antigens was reduced, but not the resistance to bacteria or tumour cells (Sikorski *et al.* 1989).

In rats given intratracheal arsenic trioxide doses of 17 mg/kg body weight, increased lung weights, an increase in the ratio of lung weights to body weights, and an increase in the levels of pulmonary protein, 4-hydroxyproline and DNA were found. This indicates fibrogenic effects in the lungs. Intratracheal doses of gallium arsenide (100 mg/kg body weight) caused an increase in the lipid, protein and DNA levels in the lungs of rats, and the lung weights were increased. Smaller gallium arsenide particles caused greater effects, probably as a result of their better solubility. The effects of 65 mg/kg body weight were minor despite longer presence in the organism. All substances caused inflammatory changes in the lungs. The strength of the observed effects decreased in the order GaAs > As$_2$O$_3$ >> Ga$_2$O$_3$ (WHO 2001).

### 5.1.2 Ingestion

Sodium arsenite was found to be of higher acute toxicity than arsenic trioxide and the arsenic(V) compounds sodium arsenate, calcium arsenate and lead arsenate (see Table 3).

The symptoms of intoxication with arsenic were cramps, nausea and haemorrhage in the intestinal tract. In the case of arsenic trioxide the extent of toxicity depended on the purity of the substance. The pure substance caused less severe side effects, but the lethal effects were more pronounced (WHO 2001).

### 5.1.3 Dermal absorption

The dermal LD$_{50}$ for calcium arsenate and lead arsenate in rats was > 2400 mg/kg body weight (Gaines 1960).

## 5.1.4 Other routes of absorption

$LD_{50}$ values of 21 and 8 mg arsenic per kg body weight were determined in mice for sodium arsenate and sodium arsenite after intramuscular injection of the substances (Bencko *et al.* 1978).

Dose-dependent nephrotoxic effects of sodium arsenate were determined in dogs given intravenous doses of the substance. A dose of 0.73 mg/kg body weight (0.26 mg arsenic per kg body weight) caused slight degeneration and the formation of vacuoles of the renal tubular epithelium, but no functional changes. 7.33 mg/kg body weight (2.64 mg arsenic per kg body weight) led to damage to the glomeruli and renal tubules. Vacuolar changes and tubular necrosis were also observed. After 14.77 mg/kg body weight (5.28 mg arsenic per kg body weight) one of three dogs died. In the kidneys severe damage to the glomerulus, proximal and distal tubules and collecting duct was found. Some of the damage in the surviving dogs was, however, reversible (WHO 2001).

80 % of starved rats did not survive a subcutaneous dose of arsenic trioxide of 10 mg/kg body weight, while the rats which were fed normally suffered no damage (Szinicz and Forth 1988). *In vitro* studies in which arsenic trioxide caused the dose-dependent inhibition of gluconeogenesis in isolated hepatocytes and kidney tubules, and arsenic pentoxide produced the same effects at higher concentrations, confirmed that acute intoxication with arsenic can cause disturbances in carbohydrate metabolism (Szinicz and Forth 1988).

Subcutaneous injection of arsenic trioxide doses of 10 mg/kg body weight caused a reduction in the level of fructose-1,6-diphosphate and glycerolaldehyde-3-phosphate and an increase in the concentration of phosphoenolpyruvate and pyruvate in the liver of guinea pigs (Reichl *et al.* 1988). The survival of mice given glucose or glucose and insulin together with 12 mg arsenic trioxide per kg body weight was significantly higher than that of mice given arsenic trioxide alone. All the mice in this last group died. The hepatic glucose and glycogen values were reduced in this group (Reichl *et al.* 1990).

**Table 3**. Acute toxicity of the arsenic compounds

| Arsenic compounds | Species | $LD_{50}$ [mg/kg body weight] | $LD_{50}$ [mg arsenic/kg body weight] | References |
|---|---|---|---|---|
| **ingestion** | | | | |
| arsenic trioxide | rat | 20–385 | 15–293 | Dieke and Richter 1946, Done and Peart 1971, Harrison *et al.* 1958 |
| | mouse | 34–53 | 26–39 | Harrison *et al.* 1958, Kaise *et al.* 1985 |
| sodium arsenite | rat | 11–42 | 6–24 | Done and Peart 1971, Schröder and Balassa 1966 |
| | mouse | 11 | 6 | Schröder and Balassa 1966 |
| sodium arsenate | rat | 112 | 40 | Schröder and Balassa 1966 |
| | mouse | 112 | 40 | Schröder and Balassa 1966 |
| calcium arsenate | rat | 298 | 53 | Gaines 1960 |
| lead arsenate | rat | 1050 | 231 | Gaines 1960 |
| **dermal absorption** | | | | |
| calcium arsenate | rat | > 2400 | > 400 | Gaines 1960 |
| lead arsenate | rat | > 2400 | > 500 | Gaines 1960 |
| **intramuscular absorption** | | | | |
| sodium arsenate | mouse | 87 | 21 | Bencko *et al.* 1978 |
| sodium arsenite | mouse | 14 | 8 | Bencko *et al.* 1978 |

## 5.2 Subacute, subchronic and chronic toxicity

The results of studies of the toxicity of arsenic compounds after repeated administration are summarized in Table 4.

### 5.2.1 Inhalation

Repeated inhalation of arsenic trioxide caused effects on the immune system in mice after arsenic concentrations of 0.3 mg/m$^3$ and above (Aranyi *et al.* 1985); in rats the first lung damage and reductions in body weights occurred after arsenic concentrations of 7.6 mg/m$^3$ and above. After rats were given arsenic concentrations of 19.7 mg/m$^3$ their general condition was greatly impaired, with a dry, bloody exudate around the eyes, nose and urogenital region, yellow discoloration of the urogenital region and rales, dyspnoea and mortality (Holson *et al.* 1999). In hamsters, 80 % of the animals died after intratracheal administration of arsenic trioxide (0.3 mg arsenic) or arsenic trisulfide (0.5 mg arsenic), and 30 % of the animals died after administration of calcium arsenate (0.5 mg arsenic) (Pershagen *et al.* 1982).

## 5.2.2 Ingestion

Arsenic trioxide (11 mg arsenic per kg body weight) administered to rats for up to 14 days caused diarrhoea, bloody faeces and decreased mobility (Bekemeier and Hirschelmann 1989).

Dogs reacted most sensitively to repeated doses of sodium arsenite. After arsenic doses of about 1.2 mg/kg body weight and above, dose-dependent reductions in body weights were observed. The activity of alanine aminotransferase was increased after arsenic doses of about 1.2 mg/kg body weight and above and that of aspartate amino-transferase after about 4.6 mg/kg body weight. Histological examination, however, did not reveal liver damage (Neiger and Osweiler 1989). In another study sodium arsenite (2.4 mg arsenic per kg body weight) induced, in addition to a marked reduction in body weights, intestinal bleeding, and all 6 treated dogs died (Byron *et al.* 1967). In mice sodium arsenite (about 1.2 mg arsenic per kg body weight) caused increased mortality (Schroeder and Balassa 1967). In rats, after arsenic doses of about 7.2 mg/kg body weight, the bile ducts were found to be enlarged with thickened walls as a result of fibrosis, sometimes with infiltration of inflammatory cells and sometimes with weak to moderate hyperplasia (Byron *et al.* 1967).

Administration of sodium arsenate (up to 4 mg arsenic per kg body weight) for 4 to 6 weeks caused structural and functional changes in the liver of rats and mice (Fowler *et al.* 1977, Hughes and Thompson 1996). Repeated absorption of sodium arsenate by rats caused slight histological changes in the glomerulus and tubulus of the kidney and swollen hepatocytes in the area of the centrilobular vein after arsenic doses of about 2.5 mg/kg body weight (Carmignani *et al.* 1983). After arsenic doses of about 4.5 mg/kg body weight enlarged bile ducts were found and after 21 mg/kg body weight, reduced body weights and mortality (Byron *et al.* 1967, Kroes *et al.* 1974). This damage was induced by lead arsenate only after arsenic doses of about 93 mg/kg body weight and above (Kroes *et al.* 1974).

Of 3 rhesus monkeys given an arsenic(V) complex (3.7 mg arsenic per kg body weight) suspended in milk daily from the age of 3 days, one animal died from bronchopneumonia with severe bleeding, oedema and necrosis in the lungs, and with various aggregations of acute inflammatory cells in the brain and spinal cord after uptake of the substance for 7 days. The other animals survived the exposure to arsenic for one year with no damage visible on gross pathological examination. In 4 other rhesus monkeys, which were given this arsenic(V) complex (3.7 mg arsenic per kg body weight) from week 8 of life, no damage was found on gross pathological examination. One animal died after 273 days (Heywood and Sortwell 1979).

Of 20 cynomolgus monkeys given daily oral doses of sodium arsenate of 0.1 mg arsenic per kg body weight on 5 days a week for 17 years, 11 animals survived. 3 monkeys were found to have endometriosis and 3 others hyalinized islets of Langerhans of the pancreas. In one of these animals there was evidence of diabetes (Thorgeirsson *et al.* 1994).

**Table 4**. Effects of arsenic compounds after repeated administration to laboratory animals

| Arsenic compound | Species | Duration | Dose/Concentration | Findings | References |
|---|---|---|---|---|---|
| inhalation | | | | | |
| arsenic trioxide | rat | 6 h/day, 33 days (whole body inhalation chamber) | 7.6 mg As/m$^3$ 19.7 mg As/m$^3$ | 7.6 mg/m$^3$: rattling sounds in the lungs, body weights decreased 19.7 mg/m$^3$: dry red exudate around the eyes, nose, urogenital region; yellow discoloration of the urogenital region; sporadic wheezing, dyspnoea; mortality: 4/9; in one moribund animal: severe hyperaemia and discharge of plasma into the intestinal lumen | Holson *et al.* 1999 |
| | mouse | 3 h/day, 5 days/week, 1 or 4 weeks | 0.3 mg As/m$^3$ 0.5 mg As/m$^3$ | pulmonary defence against bacteria decreased in a concentration-dependent manner, susceptability to infection increased | Aranyi *et al.* 1985 |
| | hamster | 1 ×/week, 4 weeks | 0.3 mg As, intratracheal | 80 % mortality after 3 to 4 weeks | Pershagen *et al.* 1982 |
| arsenic trisulfide | hamster | 1 ×/week, 4 weeks | 0.5 mg As, intratracheal | 80 % mortality after 3 to 4 weeks | Pershagen *et al.* 1982 |
| calcium arsenate | hamster | 1 ×/week, 4 weeks | 0.5 mg As, intratracheal | 30 % mortality after 4 weeks | Pershagen *et al.* 1982 |
| ingestion | | | | | |
| arsenic trioxide | rat | 5 days/week, 4–14 days | 11 mg As/kg body weight | diarrhoea, bloody faeces, mobility decreased | Bekemeier and Hirschelmann 1989 |

**Table 4**. continued

| Arsenic compound | Species | Duration | Dose/Concentration | Findings | References |
|---|---|---|---|---|---|
| sodium arsenite | rat | 2 years | up to 250 mg As/kg diet up to about 7.2 mg As/kg body weight | at about 7.2 mg As/kg body weight: bile ducts enlarged, with thickened walls as a result of fibrosis, sometimes with infiltration of inflammatory cells, sometimes with weak to moderate hyperplasia | Byron *et al.* 1967 |
| | mouse | 540 days | 5 mg As/l drinking water about 1.2 mg As/kg body weight | survival decreased, mortality increased | Schroeder and Balassa 1967 |
| | dog | 1 ×/day, 58 days then 1 ×/day, 124 days | about 0.6 or 1.2 mg As/kg body weight then about 1.2 or 2.3 mg As/kg body weight | $\geq$ 0.6/1.2 mg As/kg body weight: dose-dependent decrease in food consumption, body weight decreased, ALT activity increased | Neiger and Osweiler 1989 |
| | dog | 1 ×/day, 58 days then 1 ×/day, 61 days | about 2.3 then about 4.6 mg As/kg body weight | activity of AST and ALT increased, no lesions found in the liver after histological examination | Neiger and Osweiler 1989 |
| | dog | 2 years | up to 125 mg As/kg diet up to about 2.4 mg As/kg body weight | body weights decreased by 44 % to 61 %, intestinal bleeding, mortality: 6/6 | Byron *et al.* 1967 |
| sodium arsenate | rat | 6 weeks | up to 85 mg As/l drinking water up to about 4 mg As/kg body weight | $\geq$ about 1 mg As/kg body weight: mitochondrial membrane: activities of mono-oxidase and cytochromoxidase increased $\geq$ about 2 mg As/kg body weight: mitochondria: swelling at about 4 mg As/kg body weight: body weights decreased, liver: deposits in connective tissue, large lipid vacuoles in hepatocytes | Fowler *et al.* 1977 |

**Table 4**. continued

| Arsenic compound | Species | Duration | Dose/Concentration | Findings | References |
|---|---|---|---|---|---|
| | rat | 320 days | 50 mg As/l drinking water about 2.5 mg As/kg body weight | liver and kidneys: slight histological changes | Carmignani *et al.* 1983 |
| | rat | 2 years | 145 mg As/kg diet up to about 7.2 mg As/kg body weight | $\geq$ 4.5 mg As/kg body weight: bile ducts enlarged, with thickened walls as a result of fibrosis, sometimes with infiltration of inflammatory cells, sometimes with weak to moderate hyperplasia | Byron *et al.* 1967 |
| | rat | 29 months | about 21 mg As/kg body weight | body weights decreased, mortality increased | Kroes *et al.* 1974 |
| | mouse | 28 days | about 0.3 mg As/kg body weight about 3.0 mg As/kg body weight | dose-dependent effects: liver: vacuoles in hepatocytes increased, non-protein-sulfhydryl levels decreased; plasma: glucose decreased, creatinine increased, trigylceride values increased | Hughes and Thompson 1996 |
| | dog | 2 years | up to about 2.4 mg As/kg body weight | at 2.4 mg As/kg body weight: body weights decreased; mortality: 1/6 | Byron *et al.* 1967 |
| | cynomolgus monkey | 5 days/week, 17 years | 0.1 mg As/kg body weight | endometriosis, hyalinized islets of Langerhans of the pancreas, mortality 11/20 | Thorgeirsson *et al.* 1994 |
| lead arsenate | rat | 29 months | about 93 mg As/kg body weight | body weights decreased, mortality increased, bile ducts: enlarged, sometimes widened with inflammation and abscesses | Kroes *et al.* 1974 |

**Table 4**. continued

| Arsenic compound | Species | Duration | Dose/Concentration | Findings | References |
|---|---|---|---|---|---|
| arsenic(V) complex | rhesus monkey | 1 ×/day, 13 days | 6 mg As/kg body weight | vomiting, changed faeces, weakness, increased salivation, uncontrolled shaking of the head | Heywood and Sortwell 1979 |
|  | rhesus monkey | 1 year | 3.7 mg As/kg body weight | after ingesting the arsenic complex suspended in milk, a 10-day-old rhesus monkey died after 7 days from bronchopneumonia with marked bleeding, oedema and necrosis of the lungs, with various aggregations of acute inflammatory cells in the brain and spinal cord; mortality: 2/7 | Heywood and Sortwell 1979 |
| parenteral uptake |  |  |  |  |  |
| arsenic trioxide | guinea pig | 2 ×/day, 5 days, subcutaneous | 10 mg As/kg body weight | liver: glucose and glycogen levels decreased; mortality: 10/10 | Reichl *et al.* 1988 |

ALT = alanine aminotransferase, AST = aspartate aminotransferase

## 5.2.3 Other routes of absorption

In guinea pigs given subcutaneous injections of arsenic trioxide of 2.5 mg/kg body weight (1.9 mg arsenic per kg body weight) twice a day for 5 days, the carbohydrate level in the liver was significantly decreased one hour and even 16 hours after the last dose. The level of glycogen was decreased most (Reichl *et al.* 1988). The carbohydrate depletion caused by arsenic was attributed to the inhibition of gluconeogenesis (Szinicz and Forth 1988).

Groups of 25 male and 25 female Swiss mice were given intravenous doses of sodium arsenate of 0.5 mg/kg body weight (0.2 mg arsenic per kg body weight) once a week for 20 weeks. After a follow-up period of a further 76 weeks there were no differences in the survival and body weights of the animals compared with the control animals, but a significant increase in hyperkeratosis was found in the male and female animals and in nephropathy only in the females (Waalkes *et al.* 2000). The preneoplastic and neoplastic effects that occurred are described in Section 5.7.2.3.

There are no data available for dermal absorption.

## 5.3 Local effects on skin and mucous membranes

### 5.3.1 Skin

There are no data available for the effects of inorganic arsenic compounds on the skin. After it was applied to the skin of rabbits, the arsenic metabolite MMA was described as slightly irritative (WHO 2001). Erythematous lesions developed on the skin of rats after exposure for two hours to the metabolite DMA ($3746$ mg/m$^3$) (ATSDR 2000).

### 5.3.2 Eyes

After rats were exposed to DMA concentrations of $2172$ mg/m$^3$ for two hours, scab formation was observed round the eyes (ATSDR 2000).

## 5.4 Allergenic effects

In a maximization test, the sensitizing effects of disodium hydrogen arsenate ($Na_2HAsO_4$) and sodium arsenite were tested in Dunkin-Hartley guinea pigs. For intradermal induction 4 % disodium hydrogen arsenate and 0.2 % sodium arsenite, and for topical induction 20 % disodium hydrogen arsenate (after previous non-occlusive treatment for 24 hours with 10 % sodium dodecyl sulfate in petrolatum) and 0.5 % sodium arsenite were used, each in water. After provocation on day 21, 2 of 19 treated animals and 1 of 20 control animals reacted only after 24 hours to 1 % disodium hydrogen arsenate in physiological saline. At the 48-hour reading and on testing with 0.5 % and 0.1 % of the test substance none of the animals produced a reaction. After 24 hours 4 of 20 treated animals and 4 of 20 controls reacted to 0.1 % sodium arsenite in physiological saline, and after 48 hours only 1 of 20 control animals. None of the animals reacted to 0.05 % and 0.01 % sodium arsenite (Wahlberg and Boman 1986).

## 5.5 Reproductive and developmental toxicity

### 5.5.1 Fertility

Male hamsters and rats were given intratracheal doses of gallium arsenide of 7.7 mg/kg body weight, indium arsenide of 7.7 mg/kg body weight or arsenic trioxide of 1.3 mg/kg body weight twice a week for up to 8 weeks. Gallium arsenide reduced the sperm count in both species, indium arsenide only in rats. Gallium arsenide also caused sperm anomalies. No effects were observed with arsenic trioxide (WHO 2001).

No effects were seen on the frequency of mating and fertility in female rats after daily oral doses of arsenic trioxide of 8 mg arsenic per kg body weight (14 days before mating to day 19 of gestation) (ATSDR 2000).

Oral doses of MMA (sodium salt) of 0, 11.9 and 119 mg/kg body weight were given to male and female Swiss mice (1 male mouse with 5 female mice per cage; the number of treated animals is not given) every second day for 10 weeks. The animals were separated after 19 days. Body weights, the number of litters and some blood parameters, such as the erythrocyte and leukocyte counts, haematocrit value and total serum protein were determined. In the high dose group the body weight gains were significantly reduced. None of the female animals produced a litter. In the low dose group the number of litters was reduced by 50 %. The female animals in the low dose group which did not produce a litter were all mated with the same male animal. As the authors suspected effects on the male germ cells after the first experiment, another experiment was carried out in which only the male animals were treated. During the treatment with 119 mg/kg body weight every second day three times a week for 19 days, 10 male Swiss mice were mated with untreated female mice (1:1). The fertility, measured as the number of litters, was reduced from 90 % in the control group to 50 % in the group of treated male animals. There were no other significant differences in the litter size or the weights of the 3-week-old offspring (Prukop and Savage 1986).

### 5.5.2 Developmental toxicity

Inorganic arsenic compounds were found to be embryotoxic and foetotoxic in mice and hamsters after inhalation, ingestion and parenteral absorption. Arsenite is of greater toxicity than arsenate by a factor of 3 to 10. Hamsters were more sensitive than mice. Observed were reduced foetal weights and lengths, decreased embryonal protein levels, a reduced number of somites, delayed growth and lethal effects. The lethal effects depended on the dose and the day of gestation on which the dose was given. The most frequent malformations caused by inorganic arsenic compounds were dysrhaphic disturbances with exencephaly, encephalocele and incomplete closure of the neural tube. Less frequent were e.g. fused ribs, renal or gonadal agenesis, micromelia, fascial malformations, contortion of the hind limbs, microphthalmia or anophthalmia. Mice reacted most sensitively between days 7 and 9 of gestation. Exposure on these days led to damage to the neural tube and radioactive labelling revealed the highest arsenic concentrations to be in the neuroepithelium (WHO 2001). Very detailed descriptions of these studies can be found in several reviews (ATSDR 2000, WHO 2001). Even the most recent studies of the developmental toxicity of arsenic compounds, which are discussed below, did not yield any important new insights.

## Inhalation

The toxic effects on reproduction were investigated in female rats after inhalation exposure to arsenic trioxide for six hours with arsenic concentrations of 0.1 to 20 mg/m$^3$ (14 days before mating to day 19 of gestation). At arsenic concentrations of 8 mg/m$^3$ toxic effects (breathing noises, dry, red exudate around the nose, reduced body weight gains) were observed in the dams. At concentrations of 20 mg/m$^3$ impairments in foetal development (early resorption of the foetuses) were found in addition to marked maternal toxicity (Holson *et al.* 1999).

## Ingestion

The most sensitive species was found to be the rabbit. An increase in the incidence of resorption and fewer viable foetuses were found after daily oral doses of arsenic(V) acid during pregnancy even with arsenic doses as low as 1.5 mg/kg body weight (ATSDR 2000).

In Crl:CD(SD)BR rats given daily oral arsenic doses of about 1 to 8 mg/kg body weight as arsenic trioxide from day 14 before mating to the end of pregnancy, reduced foetal weights and various skeletal changes were observed after arsenic doses of 8 mg/kg body weight (Holson *et al.* 2000).

After Crl:CD(SD)BR rats were given oral arsenic doses of 4, 8, 15 or 23 mg/kg body weight as arsenic trioxide on day 9 of gestation, an increase in resorptions and a reduction in the number of foetuses per litter were found at the 23 mg/kg body weight dose (Stump *et al.* 1999).

An increase in resorptions, reduced foetal weights and fewer live offspring were found in CD1 mice given oral arsenic doses of 24 mg/kg body weight as arsenic(V) acid from days 6 to 15 of gestation (Nemec *et al.* 1998).

In a 3-generation study with mice given sodium arsenite with the drinking water in daily doses of 1 mg arsenic per kg body weight, a reduction in the frequency of litters was observed in all generations (ATSDR 2000).

In another publication, the evaluation of studies of reproductive toxicity in animals showed that no damage to the offspring is to be expected after inhalation or oral exposure of the dams to arsenic concentrations as can occur at the workplace (DeSesso 2001).

# 5.6 Genotoxicity

### 5.6.1 *In vitro*

There are extensive data available for the *in vitro* genotoxicity of inorganic arsenic compounds (WHO 2001). A selection of more recent studies can be found in Table 5. Inorganic arsenic(III) compounds caused sister chromatid exchange (Gebel *et al.* 1997, Hsu *et al.* 1997, Rasmussen and Menzel 1997), DNA–protein crosslinks (Dong and Luo 1993, Ramirez *et al.* 2000), inhibition of poly(ADP)-ribosylation (Yager and Wiencke 1997) and induction of DNA repair synthesis (Wiencke *et al.* 1997) and the inhibition of DNA repair processes (Bau *et al.* 2001, Hartwig *et al.* 1997). In addition DNA strand breaks (Dong and Luo 1993, Lynn *et al.* 1997, 2000, Mouron *et al.* 2001, Schaumlöffel and Gebel 1998, Sordo *et al.* 2001, Wang *et al.* 2001) and chromosome breaks (Rupa *et al.* 1997) were observed. Inorganic arsenic(III) compounds induced mutations in different genes (Hei *et al.* 1998, Moore *et al.* 1997b, Wiencke *et al.* 1997), structural and numeric chromosomal aberrations (Lee *et al.* 1985, Yih *et al.* 1997) and micronuclei (Liu and Huang 1997, Schaumlöffel and Gebel 1998) and hyperploidy (Ramirez *et al.* 1997, Rupa *et al.* 1997). There are, however, also some studies which do not confirm these effects of arsenic(III) compounds (Costa *et al.* 1997, Gibson *et al.* 1997, Lee *et al.* 1985). Arsenic(V) compounds were usually not genotoxic in the test systems investigated (Gebel *et al.* 1997, Lee *et al.* 1985, Rasmussen and Menzel 1997) or genotoxic only at cytotoxic concentrations (Lee *et al.* 1985, Moore *et al.* 1997b).

Treatment with 0.25 µM sodium arsenite induced in HeLa and CHO cells a significant increase in DNA breaks when a formamidopyridine-DNA-glycosylase was included (cleaves oxidized bases, such as formamidopyridine or 8-oxyguanine, from DNA), or a protein kinase K (can cut DNA out of DNA–protein crosslinks), or both together. It was shown in another study that catalase and inhibitors of calcium, nitrogen oxide synthase, superoxide dismutase or myeloperoxidase can influence the DNA-damaging effects of 2 µM sodium arsenite. The authors concluded that DNA damage caused by arsenic(III) compounds is the result of the calcium-mediated formation of peroxynitrite, hypochlorous acid and hydroxyl radicals (Wang *et al.* 2001). In another study with HeLa cells (short-term incubation 0.5 to 3 hours) oxidative DNA damage occurred with the addition of formamidopyridine-DNA-glycosylase even after as little as 0.01 µM arsenite (no other details) or 10 µM MMA or DMA (Schwerdtle *et al.* 2002).

### 5.6.2 *In vivo*

### Soma cells

Studies of the genotoxic effects of inorganic arsenic compounds in animals are shown in Table 6.

Increased induction of DNA strand breaks (Banu *et al.* 2001) was observed after single exposures and chromosomal aberrations (Das *et al.* 1993, Jha *et al.* 1992,

Nagymajtenyi *et al.* 1985, Poma *et al.* 1981, RoyChoudhury *et al.* 1996) and micronuclei (Deknudt *et al.* 1986, Tice *et al.* 1997, Tinwell *et al.* 1991) after single and repeated doses of arsenic(III) compounds. In the bone marrow of offspring of the mice, which were given sodium arsenite doses of 0.1 mg/kg body weight once a week for 4 weeks, the number of chromosomal aberrations was still significantly increased (RoyChoudhury *et al.* 1996). A linear dose–response relationship was found for the micronuclei in polychromatic erythrocytes induced by sodium arsenite (Deknudt *et al.* 1986, Tinwell *et al.* 1991).

In mice the arsenic metabolite DMA caused aneuploid bone marrow cells only at very high doses (Kashiwada *et al.* 1998) and DNA strand breaks in lung cells, but not in those of the liver or kidneys (Yamanaka *et al.* 1989, Yamanaka and Okada 1994).

## Germ cells

In the spermatogonia of Swiss albino mice, no increase in chromatid or chromosomal aberrations was found 12, 24, 36 and 48 hours after single intraperitoneal doses of arsenic trioxide (0, 4, 8 or 12 mg arsenic per kg body weight) (Poma *et al.* 1981). Nor was an increase in sister chromatid exchange or chromosomal aberrations in the spermatogonia of the Swiss albino mice found after the administration of 250 mg arsenic trioxide per litre drinking water (about 47 mg arsenic per kg body weight) for 2 to 8 weeks (Poma *et al.* 1987).

In BALB/c mice, tests for dominant lethal mutations after single intraperitoneal doses of sodium arsenite of 5 mg/kg body weight (2.9 mg arsenic per kg body weight) yielded negative results (Deknudt *et al.* 1986). No increase in sperm head anomalies was observed in BALB/c mice after sodium arsenite doses of 2.5, 5 or 7.5 mg/kg body weight (1.4, 2.9 or 4.3 mg arsenic per kg body weight) (Deknudt *et al.* 1986).

The dominant lethal test described in Section 5.5.1, in which mice were given oral MMA doses up to 119 mg/kg body weight (Prukop and Savage 1986), is invalid as a result of inadequate documentation and methodological shortcomings, such as the short duration of treatment and the small number of animals, and cannot, therefore, be used in the evaluation of the germ cell mutagenicity of arsenic and arsenic compounds.

**Table 5.** The genotoxicity of inorganic arsenic compounds *in vitro*

| End point | Test system | Oxidation step | Results | Effective concentration [µg/ml] | Cytotoxicity [µg/ml] | References |
|---|---|---|---|---|---|---|
| SOS chromotest | *Escherichia coli* PQ37 | As(III) | – | – | 403 µM | Lantzsch and Gebel 1997 |
| SCE | human lymphocytes | As(III) | + | 10 µM | > 10 µM | Rasmussen and Menzel 1997 |
| | human lymphocytes | As(III) | + | 0.5 µM | 5 µM | Gebel *et al.* 1997 |
| | human lymphocytes | As(III) | + | 65 | not stated | Hsu *et al.* 1997 |
| | human lymphocytes | As(III) | + | 1 µM | not stated | Jha *et al.* 1992 |
| | lymphoid cells | As(V) | – | – | not stated | Rasmussen and Menzel 1997 |
| | human lymphocytes | As(V) | – | – | 10 µM | Gebel *et al.* 1997 |
| DNA–protein crosslinks | human lymphoma cells | As(III) | – | – | 5 µM | Costa *et al.* 1997 |
| | human foetal lung fibroblasts | As(III) | + | 1 µM | not stated | Dong and Luo 1993 |
| | human liver cell line | As(III) | + | 0.001 µM | not stated | Ramirez *et al.* 2000 |
| inhibition of poly(ADP)-ribosylation | Molt-3 cell line | As(III) | + | > 5 µM | > 7.5 µM | Yager and Wiencke 1997 |
| inhibition of DNA repair synthesis | human fibroblasts | As(III) | + | 2.5 µM | > 10 µM | Hartwig *et al.* 1997 |
| DNA repair synthesis | CHO cells | As(III) | + | 4 µM | not stated | Bau *et al.* 2001 |
| | human fibroblasts | As(III) | + | 1 µM | 2.5 µM | Wiencke *et al.* 1997 |
| DNA strand breaks | CHO cells | As(III) | + | 10 µM | 20 µM | Lynn *et al.* 1997 |
| | human smooth muscle cells | As(III) | + | > 1 µM | not stated | Lynn *et al.* 2000 |
| | human fibroblasts | As(III) | + | 2.5 µM | not stated | Mouron *et al.* 2001 |

**Table 5**. continued

| End point | Test system | Oxidation step | Results | Effective concentration [µg/ml] | Cytotoxicity [µg/ml] | References |
|---|---|---|---|---|---|---|
| | human lymphocytes | As(III) | + | >5µM | not stated | Sordo *et al.* 2001 |
| | human foetal lung fibroblasts | As(III) | + | 1 µM | not investigated | Dong and Luo 1993 |
| | human lymphocytes | As(III) | + | 0.01 µM | 5 µM | Schaumlöffel and Gebel 1998 |
| | NB4 cells, HeLa cells | As(III) | + | > 0.25 µM | 2 µM | Wang *et al.* 2001 |
| | CHO-K1 | As(III) | – | – | > 2 µM | Wang *et al.* 2001 |
| chromosome breaks | human lymphocytes | As(III) | + | 9 µM | not stated | Rupa *et al.* 1997 |
| mutations: TK$^{+/-}$ | L5178Y mouse lymphoma cells | As(III) | (+) | 1–2 | about 0.8 | Moore *et al.* 1997b |
| | | As(V) | (+) | 10–14 | about 7.5 | Moore *et al.* 1997b |
| Na$^+$/K$^+$-ATPase | SHE cells | As(III) | – | – | > 3 µM | Lee *et al.* 1985 |
| | | As(V) | – | – | > 10 µM | Lee *et al.* 1985 |
| HPRT | CHO cells with a copy of human chromosome 11 | As(III) | (+) | 1 µM | > 0.5 µM | Hei *et al.* 1998 |
| | SHE cells | As(III) | – | – | > 3 µM | Lee *et al.* 1985 |
| | | As(V) | – | – | > 10 µM | Lee *et al.* 1985 |
| supF reporter gene | human fibroblasts with shuttle vector pz189 | As(III) | (+) | 5 µM | 2.5 µM | Wiencke *et al.* 1997 |
| CA | human lymphocytes | As(III) | + | 1 µM | not stated | Jha *et al.* 1992 |
| | human skin fibroblasts | As(III) | + | 5 µM | > 1.25 µM | Yih *et al.* 1997 |
| | SHE cells | As(III) | (+) | 6.2 µM | > 3 µM | Lee *et al.* 1985 |
| | | As(V) | (+) | 64 µM | > 10 µM | Lee *et al.* 1985 |

**Table 5**. continued

| End point | Test system | Oxidation step | Results | Effective concentration [µg/ml] | Cytotoxicity [µg/ml] | References |
|---|---|---|---|---|---|---|
| micronuclei | CHO cells | As(III) | + | 80 µM | not stated | Liu and Huang 1997 |
| | SHE cells | As(III) | | – | 7.5 | Gibson *et al.* 1997 |
| | human lymphocytes | As(III) | + | 0.5 µM | 5 µM | Schaumlöffel and Gebel 1998 |
| hyperploidy | human lymphocytes | As(III) | + | 0.01 µM | not stated | Ramirez *et al.* 1997 |
| | human lymphocytes | As(III) | + | 9 µM | not stated | Rupa *et al.* 1997 |

(+) positive at cytotoxic concentrations

# 5.7 Carcinogenicity

## 5.7.1 Short-term studies

Sodium arsenate and arsenite induced a dose-dependent transformation in Syrian hamster embryo cells (Lee *et al.* 1985) and BALB/3T3 cells (Bertolero *et al.* 1987). Arsenite was found to be four to ten times stronger than arsenate in the induction of transformations, which is probably the result of the greater cellular uptake of arsenite compared to arsenate (Bertolero *et al.* 1987). BALB/3T3 cells transformed by arsenite were tumour-promoting in mice after subcutaneous injection of the substance (Saffiotti and Bertolero 1989). The rapidly growing fibrosarcomas, however, did not metastasize. The transforming arsenic compounds are probably the arsenites (Bertolero *et al.* 1987, Saffiotti and Bertolero 1989).

Transgenic, v-Ha-ras oncogene-carrying mice (TG.AC) received arsenic doses of 48 mg/kg body weight with the drinking water for four weeks in the form of sodium arsenite. Afterwards, the tumour promotor 12-*O*-tetradecanoylphorbol-13-acetate (TPA) was applied to the shaved dorsal skin twice a day for 2 weeks. Skin papillomas were not found in the control animals nor in the animals treated only with arsenic. Only with the simultaneous application of sodium arsenite and TPA did the incidence of skin papillomas increase markedly. The skin of the animals treated with arsenic developed hyperkeratosis. In the epidermis an increase was found in the mRNA copies for the growth factor TGF-$\alpha$ and the granulocyte macrophage colony stimulating factor GM-CSF (Germolec *et al.* 1997, 1998).

In some studies with mice, after the administration of inorganic arsenic compounds a decrease was described in urethane-induced lung tumours (Blakley 1987), spontaneous breast tumours (Schrauzer and Ishmael 1974, Schrauzer *et al.* 1976) and tumours caused by the injection of mouse sarcoma cells (Kerkvliet *et al.* 1980). The arsenic compounds, however, decreased the lifetime of the tumour-carrying animals (Kerkvliet *et al.* 1980, Schrauzer and Ishmael 1974). If the animals formed tumours or received tumour transplants during the arsenic treatment, the growth of the tumours was significantly increased (Schrauzer and Ishmael 1974, Schrauzer *et al.* 1976). A tumour-promoting effect of sodium arsenite was demonstrated in Wistar rats after partial hepatectomy and pretreatment with diethylnitrosamine (30 mg/kg body weight) (Shirachi *et al.* 1983).

In a multiorgan carcinogenicity bioassay, the carcinogenic effects of the arsenic metabolite DMA was investigated in rats. Male F344/DuCrj rats were given single doses of diethylnitrosamine (DEN) (intraperitoneal, 100 mg/kg body weight) at the beginning of treatment, then four doses of *N*-methyl-*N*-nitrosourea (MNU) (intraperitoneal, 20 mg/kg body weight on days 5, 8, 11 and 14) and later four doses of 1,2-dimethyl-hydrazine (DMH) (subcutaneous, 40 mg/kg body weight on days 18, 22, 26 and 30). The animals were given 0.05 % *N*-butyl-*N*-(4-hyroxybutyl)-nitrosamine (BBN) with the drinking water for the first two weeks and 0.1 % *N*-bis(2-hyroxypropyl)-nitrosamine (DHPN) for the following two weeks. After a two-week pause, the rats were given 50, 100, 200 or 400 mg DMA per litre drinking water. At the end of week 30 the animals were killed. DMA significantly increased the formation of tumours in the bladder, kidneys, liver and thyroid gland in the pretreated animals. Also preneoplastic damage was markedly increased. In the animals exposed only to DMA no preneoplastic changes or tumours were found (Yamamoto *et al.* 1995).

In another study NBR rats were given 0.05 % BBN with the drinking water for 4 weeks and later 100 mg DMA per litre drinking water for 32 weeks. After week 36 a significant increase in simple, papillary or nodular hyperplasia and an increase in carcinomas were observed in the bladder (Li *et al.* 1998).

Male F344/DuCrj rats were given 0.05 % BBN with the drinking water for 1 to 4 weeks. Afterwards they received up to 100 mg DMA per litre drinking water for another 32 weeks. The administration of DMA after pretreatment with BBN caused a dose-dependent increase in the incidence of bladder papillomas and carcinomas, without pretreatment DMA did not increase the formation of tumours (Wanibuchi *et al.* 1996).

In a two-step carcinogenicity test male F344/DuCrj rats were given single intraperitoneal doses of DEN of 100 mg/kg body weight and 2 weeks later 0, 25, 50 or 100 mg DMA per litre drinking water for 6 weeks. In addition, 3 weeks after the DEN dose a partial hepatectomy was carried out. 8 weeks after the beginning of the study an increase in liver cancer was observed after DMA concentrations of 50 mg/l drinking water and above (Yamamoto *et al.* 1997).

**Table 6.** Genotoxicity of arsenic compounds in the mouse

| End point | Species | Organ or cell type investigated | Arsenic compound | Administration, dose, duration | Results | References |
|---|---|---|---|---|---|---|
| single exposures | | | | | | |
| DNA strand breaks | Swiss albino mouse | leukocytes | arsenic trioxide | oral, 0.1–4.9 mg As/kg body weight | + | Banu *et al.* 2001 |
| | CD1 mouse | aneuploid bone marrow cells | DMA | intraperitoneal, 300 mg/kg body weight | + | Kashiwada *et al.* 1998 |
| | CD1 mouse | lung, liver, kidneys | DMA | oral, 1500 mg/kg body weight | + | Yamanaka *et al.* 1989, Yamanaka and Okada 1994 |
| CA | Swiss albino mouse | bone marrow cells | arsenic trioxide | intraperitoneal, 4–12 mg/kg body weight | – | Poma *et al.* 1981 |
| | Swiss albino mouse | bone marrow cells | sodium arsenite | oral, 25 mg/kg body weight | + | Das *et al.* 1993 |
| | Swiss albino mouse | bone marrow cells | sodium arsenite | subcutaneous, 0.1 mg/kg body weight, 1 ×/week, 4 weeks | + | RoyChoudhury *et al.* 1996 |
| MN | C57BL mouse | polychromatic erythrocytes | sodium arsenite | intraperitoneal, 2.5–10 mg/kg body weight | + | Deknudt *et al.* 1986, Tinwell *et al.* 1991 |
| | CBA mouse | polychromatic erythrocytes | sodium arsenite | intraperitoneal, 2.5–10 mg/kg body weight | + | Deknudt *et al.* 1986, Tinwell *et al.* 1991 |
| | Balb/c mouse | polychromatic erythrocytes | sodium arsenite | intraperitoneal, 2.5–10 mg/kg body weight | + | Deknudt *et al.* 1986, Tinwell *et al.* 1991 |

**Table 6**. continued

| End point | Species | Organ or cell type investigated | Arsenic compound | Administration, dose, duration | Results | References |
|---|---|---|---|---|---|---|
| repeated exposure | | | | | | |
| CA | Swiss albino mouse | bone marrow cells | arsenic trioxide | oral, 250 mg/l drinking water, 2–8 weeks | + | Poma *et al.* 1987 |
| | CFLP mouse | foetal liver cells | arsenic trioxide | inhalation, 0.26–28.5 mg/m$^3$, days 9 to 12 *post coitum*, caesarean section day 18 *post coitum* | (+) | Nagymajtenyi *et al.* 1985 |
| MN | B6C3F$_1$ mouse | polychromatic erythrocytes | sodium arsenite | oral, 2.5–10 mg/kg body weight, 4 days | + | Tice *et al.* 1997 |

(+) positive only at foetotoxic concentrations, CA = chromosomal aberrations, MN = micronuclei

## 5.7.2 Long-term studies

Table 7 gives an overview of the carcinogenicity studies with arsenic compounds.

### 5.7.2.1 Inhalation

In hamsters, intratracheal administration of inorganic arsenic compounds for 15 weeks caused an increase in the frequency of lung tumours during a lifelong observation period. These were lung adenomas (Ishinishi *et al.* 1983, Pershagen and Björklund 1985, Yamamoto *et al.* 1987) and papillomas and carcinomas of the respiratory tract (Pershagen *et al.* 1984). Also many early stages of the lung tumours, such as hyperplasia (Yamamoto *et al.* 1987) and adenomatous changes (Pershagen and Björklund 1985, Pershagen *et al.* 1984, Yamamoto *et al.* 1987) were found. Benzo[a]pyrene amplified the carcinogenic effects on the lung induced by arsenic trioxide (Pershagen *et al.* 1984).

Grave disadvantages of the studies described above are the short duration of treatment and the fact that the investigations concentrated almost exclusively on the lungs of the animals. Valid long-term studies with inhalation exposure to inorganic arsenic compounds are not available.

### 5.7.2.2 Ingestion

A study in which mice were given the arsenic metabolite DMA for 50 weeks in concentrations of 0.05, 0.2 or 0.4 mg/l drinking water (about 0.01, 0.05 or 0.1 mg arsenic per kg body weight) showed that the tumour size and the total number of adenomas and carcinomas in the lungs increased in a dose-dependent manner. Administration for only 25 weeks did not yet induce lung tumours (Hayashi *et al.* 1998).

A dose-dependent increase in the incidence of bladder carcinomas was observed in F344 rats after administration of DMA with the drinking water (about 2.5 or 10 mg DMA per kg body weight) for 104 weeks. Other organs were not investigated (Wei *et al.* 1999).

Female C57BL/6J mice and female transgenic metallothionine knock-out mice were given sodium arsenate for 26 months with the drinking water (0.5 mg arsenic per litre). The daily arsenic dose was 2 to 2.5 µg arsenic per day, corresponding to 0.07 to 0.08 mg arsenic per kg body weight. In 41 % of the C57BL/6J mice and 26 % of the knock-out mice one or more tumours were found per animal. The tumours were situated in the gastrointestinal tract, lungs, liver, spleen, skin and gonads (Ng *et al.* 1999).

No malignant tumours were found in 11 of 20 cynomolgus monkeys which survived treatment with sodium arsenate in daily oral doses of 0.1 mg arsenic per kg body weight on 5 days a week for 17 years. 2 animals were found to have adenomas of the renal cortex (see Section 5.2.2) (Thorgeirsson *et al.* 1994).

Valid long-term studies with ingestion of inorganic arsenic compounds are not available.

**Table 7.** Carcinogenicity studies with arsenic compounds

| Species | Type of administration, duration | Exposure* | Tumour incidence | Type of tumour | References |
|---|---|---|---|---|---|
| **arsenic trioxide** | | | | | |
| Syrian hamster ♀ | intratracheal, 1 ×/week, 15 weeks, lifelong observation | controls<br><br>0.25 mg (3 mg/kg body weight) | 1/20<br>1/20<br><br>2/20[1]<br>1/20<br>1/20 | adenocarcinoma (kidneys)<br>papilloma (forestomach)<br><br>adenomas (lungs)<br>fibrous histiocytoma<br>fibrosarcoma in subcutis with metastases in the ovary and lungs | Ishinishi *et al.* 1983 |
| Syrian hamster ♀ | intratracheal, 1 ×/week, 15 weeks, lifelong observation | controls<br><br>0.35 mg (4 mg/kg body weight) | 0/20<br><br>3/10[2]<br>1/10 | <br><br>adenomas (lungs)<br>leiomyoma in the uterus | Ishinishi *et al.* 1983 |
| Syrian hamster ♂ | intratracheal, 1 ×/week, 15 weeks, lifelong observation | controls<br><br>3 mg/kg body weight | 3/53<br><br>21/47<br><br><br>3/47[3] | adenomas, adenomatous changes, and papillomas (respiratory tract)<br>adenomas, adenomatous changes, papillomas (respiratory tract)<br>carcinomas (respiratory tract) | Pershagen *et al.* 1984 |
| Syrian hamster ♀ | intratracheal, 1 ×/week, 15 weeks, lifelong observation | controls<br><br><br>0.25 mg (3 mg/kg body weight) | 1/21<br>1/21<br>1/21<br><br>8/17<br>1/17<br>1/17<br>1/17 | adenocarcinoma (lungs)<br>adenoma (adrenal gland)<br>adenocarinoma (adrenal gland)<br>hyperplasia (lungs)<br>adenocarcinoma (lungs)<br>adenoma (adrenal gland)<br>haemangiosarcoma (liver) | Yamamoto *et al.* 1987 |
| **arsenic trisulfide** | | | | | |
| Syrian hamster ♂ | intratracheal, 1 ×/week, 15 weeks, lifelong observation | controls<br><br>3 mg/kg body weight | 2/26<br><br>12/28<br><br>1/28 | adenomatous changes (lungs)[#]<br>adenomatous changes (lungs)<br>adenoma (lungs) | Pershagen and Björklund 1985 |

**Table 7**. continued

| Species | Type of administration, duration | Exposure* | Tumour incidence | Type of tumour | References |
|---|---|---|---|---|---|
| Syrian hamster ♀ | intratracheal, 1 ×/week, 15 weeks, lifelong observation | controls | 1/21 1/21 1/21 | adenocarcinoma (lungs) adenoma (adrenal gland) adenocarcinoma (adrenal gland) | Yamamoto *et al.* 1987 |
| | | 0.25 mg (3 mg/kg body weight) | 3/22 1/22 1/22 1/22 | hyperplasia adenoma (lungs) adenoma (adrenal gland) nephroblastoma (kidneys) | |

**sodium arsenate**

| Species | Type of administration, duration | Exposure* | Tumour incidence | Type of tumour | References |
|---|---|---|---|---|---|
| Swiss mice ♂ | intravenous, 1 ×/week, 20 weeks, observation for a further 76 weeks | controls | 8/25 | interstitial cell hyperplasia (testes) | Waalkes *et al.* 2000 |
| | | 0.2 mg/kg body weight | 16/25 | interstitial cell hyperplasia (testes)[7] | |
| Swiss mice ♀ | intravenous, 1 ×/week, 20 weeks, observation for a further 76 weeks | controls | 5/25 0/25 1/25 0/25 | cystic hyperplasia (uterus) adenocarcinoma (uterus) preneoplastic foci (liver) adenoma (liver) | Waalkes *et al.* 2000 |
| | | 0.2 mg/kg body weight | 14/25 1/25 5/25 1/25 | cystic hyperplasia (uterus)[7] adenocarcinoma (uterus) preneoplastic foci (liver)[7] adenoma (liver) | |
| C57BL/6J mice ♀ | oral, 1 ×/day, up to 26 months | controls | 0/90 | | Ng *et al.* 1999 |
| | | 0.07–0.08 mg/kg | 14.4 % 17.5 % 7.8 % 3.3 % 2.2 % 3.3 % 3.3 % 1.1 % | tumours in gastrointestinal tract lungs liver spleen bones skin reproductive system eyes (no other details) | |
| transgenic metallothionine knock-out mice ♀ | oral, 1 ×/day, up to 26 months | controls | 0/140 | | Ng *et al.* 1999 |
| | | 0.07–0.08 mg/kg | 12.9 % 7.1 % 5.0 % 0.7 % 1.4 % 5.0 % | tumours in gastrointestinal tract lungs liver spleen skin reproductive system (no other details) | |

**Table 7**. continued

| Species | Type of administration, duration | Exposure* | Tumour incidence | Type of tumour | References |
|---|---|---|---|---|---|
| **sodium arsenate** | | | | | |
| cynomolgus monkeys (*Macaca fascicularis*) | oral, 5 days/week, 17 years | controls | 1/130<br>1/130 | adenocarcinoma (kidneys)<br>adenocarcinoma (intestine) | Thorgeirsson *et al.* 1994 |
| | | 0.1 mg/kg body weight | 2/11 | adenoma (kidneys) | |
| **calcium arsenate** | | | | | |
| Syrian hamster ♂ | intratracheal, 1 ×/week, 15 weeks, lifelong observation | controls | 2/26 | adenomatous changes in the lungs[#] | Pershagen and Björklund 1985 |
| | | 3 mg/kg body weight | 3/35<br>4/35<br>10/35 | metaplasia (larynx, trachea)<br>adenomatous changes in the lungs<br>pulmonary adenomas[4] | |
| Syrian hamster ♀ | intratracheal, 1 ×/week, 15 weeks, lifelong observation | controls | 1/21<br>1/21<br>1/21 | adenocarcinoma (lungs)<br>adenoma (adrenal gland)<br>adenocarcinoma (adrenal gland) | Yamamoto *et al.* 1987 |
| | | 0.25 mg | 7/25<br>6/25<br>1/25<br>2/25<br>1/25 | hyperplasia (lungs)<br>adenoma (lungs)<br>adenocarcinoma (lungs)<br>adenocarcinoma (adrenal gland)<br>leukaemia | |
| **DMA** | | | | | |
| A/J mice ♂ | oral, 50 weeks | controls | 6/14<br>1/14 | adenomas (lungs)[#]<br>adenocarcinomas (lungs)<br>lung tumours per animal: 0.50 | Hayashi *et al.* 1998 |
| | | 0.4 mg/l drinking water° (about 0.1 mg/kg body weight) | 4/14<br>10/14<br>3/14<br>2/14 | hyperplasia (lungs)<br>adenomas (lungs)<br>adenocarcinomas (lungs)<br>tumours, not characterized (lungs)<br>lung tumours per animal: 1.39[5] | |

**Table 7**. continued

| Species | Type of administration, duration | Exposure* | Tumour incidence | Type of tumour | References |
|---|---|---|---|---|---|
| **DMA** | | | | | |
| F344 rats ♂ | oral 104 weeks | controls | 0/28 | | Wei *et al.* 1999 |
| | | 12.5 mg/l drinking water (about 0.6 mg DMA/kg body weight) | 0/33 | | |
| | | 50 mg/l drinking water (about 2.5 mg DMA/kg body weight) | 2/31 6/31[4] | papillomas (bladder)[#] carcinomas (bladder) | |
| | | 200 mg/l drinking water (about 10 mg DMA/kg body weight) | 2/31 12/31[6] | papillomas (bladder) carcinomas (bladder) | |

* exposure to arsenic if not otherwise stated, [#] other organs were not investigated, [°] only data for the highest concentration group are listed in the table, [1] $p = 0.244$, [2] $p = 0.052$, [3] $p = 0.07$, [4] $p < 0.05$, [5] $p = 0.002$, [6] $p < 0.001$, [7] significant in the Fisher exact test

## 5.7.2.3 Other routes of absorption

Groups of 25 male and 25 female Swiss mice were given intravenous doses of sodium arsenate of 0.5 mg/kg body weight (0.2 mg arsenic per kg body weight) once a week for 20 weeks. After an observation period of a further 76 weeks, a significant increase in interstitial cell hyperplasia was observed in the testes of the male mice. In the female animals, the incidence and severity of cystic hyperplasia in the uterus were significantly increased. In one animal a rare adenocarcinoma was found in the uterus. Also the incidence of preneoplastic foci in the liver was significantly increased (see Table 7 and Section 5.2.4) (Waalkes *et al.* 2000).

# 6 Manifesto (MAK value/classification)

Arsenic and inorganic arsenic compounds, with the exception of arsine, are carcinogenic in man. After inhalation of the substance, carcinogenic effects are observed in the lungs and after ingestion, in the bladder, kidneys, skin and lungs. However, neither the studies with inhalation exposure to arsenic nor those with oral administration of the substance can be used to derive a NOAEL (no observed adverse effect level). As a NOAEL for carcinogenicity cannot be derived from the epidemiological studies, no MAK value can be established for arsenic and inorganic arsenic compounds. Arsenic and arsenic compounds therefore remain in Carcinogenicity category 1.

Inorganic arsenic compounds are clearly genotoxic in soma cells *in vitro* and in rodents and man.

No dominant lethal mutations were induced in mice after intraperitoneal administration of the substance in the only available study with germ cells. The negative results in this dominant lethal test cannot, however, dispel the suspicion that arsenic and inorganic arsenic compounds cause germ cell mutagenicity, as the test was carried out with only one single intraperitoneal treatment. As a result of the mutagenic effects in soma cells, the formation of genotoxic and systemically effective metabolites, the bioavailability of the substance in the gonads and the inadequate investigation in germ cells, arsenic and inorganic arsenic compounds are classified in Category 3A for germ cell mutagens.

Possible contact allergy after exposure to inorganic arsenic compounds cannot be deduced with certainty from the available data as no clear separation from unspecific or toxic effects is possible, despite evidence indicating such effects above all in the earlier literature. In addition, a study with animals yielded negative results. There are no data available for immunological effects on the respiratory passages. Despite the suspected contact allergy the substance is therefore not designated with an "S".

Arsenic acid penetrates the skin *in vivo* and *in vitro* in amounts of less than 10 % and sodium arsenate from aqueous solution or as the solid *in vitro* in amounts of about 30 % to 60 %. Several studies have shown that the internal exposure to arsenic correlates well with the external exposure. Dermal absorption under workplace conditions does not, therefore, seem to be relevant, and arsenic and inorganic arsenic compounds are not designated with an "H".

# 7 References


ACGIH (American Conference of Governmental Industrial Hygienists) (1999) Arsenic and inorganic compounds. in: *Documentation of TLVs and BEIs*, ACGIH, Cincinnati, OH, USA

Ahmad AS, Sayed MHSU, Barua S, Khan MH, Faruquee MH, Jalil A, Hadi S, Talukter K (2001) Arsenic in drinking water and pregnancy outcomes. *Environ Health Perspect 109*: 629–630

Aranyi C, Bradof JN, O'Shea WJ, Graham JA, Miller FJ (1985) Effects of arsenic trioxide inhalation exposure on pulmonary antibacterial defenses in mice. *J Toxicol Environ Health 15*: 163–172

Aschengrau A, Zierler S, Cohen A (1989) Quality of community drinking water and the occurrence of spontaneous abortion. *Arch Environ Health 44*: 283–290

ATSDR (Agency for Toxic Substances and Disease Registry) (2000) *Toxicological profile for arsenic*, US Department of Health and Human Services, Atlanta, Georgia

Banu BS, Danadevi K, Jamil K, Ahuja YR, Rao KV, Ishaq M (2001) *In vivo* genotoxic effect of arsenic trioxide in mice using comet assay. *Toxicol 162*: 171–177

Barbaud A, Mougeolle JM, Schmutz JL (1995) Contact hypersensitivity to arsenic in a crystal factory worker. *Contact Dermatitis 33*: 272–273

Bates MN, Smith AH, Hopenhayn-Rich C (1992) Arsenic ingestion and internal cancers: a review. *Am J Epidemiol 135*: 462–476

Bau DT, Gurr JR, Jan KY (2001) Nitric oxide is involved in arsenite inhibition of pyrimidine dimer excision. *Carcinogenesis 22*: 709–716

Bazarbachi A, El-Sabban ME, Nasr R, Quignon F, Awaraji C, Kersual J, Dianoux L, Zermati Y, Haidar JH, Hermine O, de Thé H (1999) Arsenic trioxide and interferon-α synergize to induce cell cycle arrest and apoptosis in human T-cell lymphotropic virus type I-transformed cells. *Blood 93*: 278–283

Beckett WS, Moore JL, Keogh JP, Bleecker ML (1986) Acute encephalopathy due to occupational exposure to arsenic. *Br J Ind Med 43*: 66–67

Beckman L, Nordstrom S (1982) Occupational and environmental risks in and around a smelter in northern Sweden. IX. Fetal mortality among wives of smelter workers. *Hereditas 97*: 1–7

Beckman G, Beckman L, Nordenson I (1977) Chromosome aberrations in workers exposed to arsenic. *Environ Health Perspect 19*: 145–146

Bekemeier H, Hirschelmann R (1989) Reactivity of resistance blood vessels *ex vivo* after administration of toxic chemicals to laboratory animals: arteriolotoxicity. *Toxicol Lett 49*: 49–54

Bencko V, Rossner P, Havrankova H, Puznova A, Tucek M (1978) Effects of the combined action of selenium and arsenic on mice versus suspension culture of mice fibroblasts. in: Fours JR, Gut I (Eds) *Industrial and environmental xenobiotics: in vitro versus in vivo biotransformation and toxicity*, Proceedings of an international conference held in Prague, Czechoslovakia, 13th–15th September 1977, 312–316

Bencko V, Wagner V, Wagnerova M, Batora J (1988) Immunological profiles in workers of a power plant burning coal rich in arsenic content. *J Hyg Epidemiol Microbiol Immunol 32*: 137–146

Bernstam L, Nriagu J (2000) Molecular aspects of arsenic stress. *J Toxicol Environ Health, B3*: 293–322

Bertolero F, Pozzi G, Sabbioni E, Saffiotti U (1987) Cellular uptake and metabolic reduction of pentavalent to trivalent arsenic as determinants of cytotoxicity and morphological transformation. *Carcinogenesis 8*: 803–808

Birmingham DJ, Key MM, Holaday DA, Perone VB (1965) An outbreak of arsenical dermatoses in a mining community. *Arch Dermatol 91*: 457–464

Blakley BR (1987) Alterations in urethan-induced adenoma formation in mice exposed to selenium and arsenic. *Drug Nutr Interact 5*: 97–102

Bolla-Wilson K, Bleecker ML (1987) Neuropsychological impairment following inorganic arsenic exposure. *J Occup Med 29*: 500–503

Borzsonyi M, Bereczky A, Rudnai P, Csanady M, Horvath A (1992) Epidemiological studies on human subjects exposed to arsenic in drinking water in southeast Hungary. *Arch Toxicol 66*: 77–78

Bourrain JL, Morin C, Beani JC, Amblard P (1998) Airborne contact dermatitis from cacodylic acid. *Contact Dermatitis 38*: 364–365

Brouwer OF, Onkenhout W, Edelbroek PM, de Kom JF, de Wolff FA, Peters AC (1992) Increased neurotoxicity of arsenic in methylenetetrahydrofolate reductase deficiency. *Clin Neurol Neurosurg 94*: 307–310

Burgdorf W, Kuvink K, Cerevenka J (1977) Elevated sister chromatid exchange rate in lymphocytes of subjects treated with arsenic. *Hum Genet 36*: 69–72

Byron WR, Bierbower GW, Brouwer JB, Hansen WH (1967) Pathologic changes in rats and dogs from two-year feeding of sodium arsenite or sodium arsenate. *Toxicol Appl Pharmacol 10*: 132–147

Carmignani M, Boscolo P, Iannaccone A (1983) Effects of chronic exposure to arsenate on the cardiovascular function of rats. *Br J Ind Med 40*: 280–284

Cavigelli M, Li WW, Lin A, Su B, Yoshioka K, Karin M (1996) The tumor promoter arsenite stimulates AP-1 activity by inhibiting a JNK phosphatase. *EMBO J 15*: 6269–6279

Chen CJ, Wu MM, Lee SS, Wang JD, Cheng SH, Wu HY (1988) Atherogenicity and carcinogenicity of high-arsenic artesian well water. Multiple risk factors and related malignant neoplasms of blackfoot disease. *Arteriosclerosis 8*: 452–460

Chen G-Q, Zhu J, Shi X-G, Ni J-H, Zhong H-J, Si G-Y, Jin X-L, Tang W, Li X-S, Xong S-M, Shen Z-X, Sun G-L, Ma J, Zhang T-D, Gazin C, Naoe T, Chen S-J, Wang A-Y, Chen Z (1996) *In vitro* studies on cellular and molecular mechanisms of arsenic trioxide ($As_2O_3$) in the treatment of acute promyelocytic leukemia: $As_2O_3$ induces $NB_4$ cell apoptosis with downregulation of Bcl-2 expression and modulation of PML-RARα/PML proteins. *Blood 88*: 1052–1061

Costa M, Zhitkovich A, Harris M, Paustenbach D, Gargas M (1997) DNA-protein cross-links produced by various chemicals in cultured human lymphoma cells. *J Toxicol Environ Health 50*: 433–449

Dai J, Weinberg RS, Waxman S, Jing Y (1999) Malignant cells can be sensitized to undergo growth inhibition and apoptosis by arsenic trioxide through modulation of the glutathione redox system. *Blood 93*: 268–277

Das T, Roychoudhury A, Sharma A, Talukder G (1993) Modification of clastogenicity of three known clastogens by garlic extract in mice *in vivo*. *Environ Mol Mutagen 21*: 383–388

Deknudt G, Leonard A, Arany J, Jenar-Du Buisson G, Delavignette E (1986) *In vivo* studies in male mice on the mutagenic effects of inorganic arsenic. *Mutagenesis 1*: 33–34

DeSesso JM (2001) Teratogen update: inorganic arsenic. *Teratology 63*: 170–173

Dieke SH, Richter CP (1946) Comparative assays of rodenticides on wild Norway rats. *Public Health Rep 61*: 672–679

Done AK, Peart AJ (1971) Acute toxicities of arsenical herbicides. *Clin Toxicol 4*: 343–355

Dong JT, Luo XM (1993) Arsenic-induced DNA-strand breaks associated with DNA-protein crosslinks in human fetal lung fibroblasts. *Mutat Res 302*: 97–102

Dulout FN, Grillo CA, Seoane AI, Maderna CR, Nilsson R, Vahter M, Darroudi F, Natarajan AT (1996) Chromosomal aberrations in peripheral blood lymphocytes from native Andean women and children from northwestern Argentina exposed to arsenic in drinking water. *Mutat Res 370*: 151–158

Eberhartinger C, Ebner H, Klotz L (1969) Zur Kenntnis und Interpretation follikulärer, papulopustulöser Reaktionen im Epikutantest (Knowledge and interpretation of follicular papulopustular reactions in epicutaneous tests) (German). *Berufs-Dermatosen 17*: 241–252

EC (European Commission DG Environment) (2000) *Ambient air pollution by As, Cd, and Ni compounds*. Working Group on Arsenic, Cadmium and Nickel Compounds, Brussels, Belgium

Enterline PE, Henderson VL, Marsh GM (1987a) Exposure to arsenic and respiratory cancer: a reanalysis. *Am J Epidemiol 125*: 929–938

Enterline PE, Marsh GM, Esmen NA, Henderson VL, Callahan CM, Paik M (1987b) Some effects of cigarette smoking, arsenic, and $SO_2$ on mortality among US copper smelter workers. *J Occup Med 29*: 831–838

Enterline PE, Day R, Marsh GM (1995) Cancers related to exposure to arsenic at a copper smelter. *Occup Environ Med 52*: 28–32

EPA (Environmental Protection Agency) (1997) *Report on the expert panel on arsenic carcinogenicity: review and workshop.* National Center for Environmental Assessment, Washington DC, 1–26

EPA (Environmental Protection Agency) (2001) *Arsenic in drinking water.* National Academic Press, Washington DC

Feldman RG, Niles CA, Kelly-Hayes M, Sax DS, Dixon WJ, Thompson DJ, Landau E (1979) Peripheral neuropathy in arsenic smelter workers. *Neurology 29*: 939–944

Fowler BA, Woods JS, Schiller CM (1977) Ultrastructural and biochemical effects of prolonged oral arsenic exposure on liver mitochondria of rats. *Environ Health Perspect 19*: 197–204

Gaines TB (1960) The acute toxicity of pesticides to rats. *Toxicol Appl Pharm 2*: 88–99

Gebel T, Christensen S, Dunkelberg H (1997) Comparative and environmental genotoxicity of antimony and arsenic. *Anticancer Res 17*: 2603–2607

Germolec DR, Spalding J, Boorman GA, Wilmer JL, Yoshida T, Simeonova PP, Bruccoleri A, Kayama F, Gaido K, Tennant R, Burleson F, Dong W, Lang RW, Luster MI (1997) Arsenic can mediate skin neoplasia by chronic stimulation of keratinocyte-derived growth factors. *Mutat Res 386*: 209–218

Germolec DR, Spalding J, Yu H-S, Chen GS, Simeonova PP, Humble MC, Bruccoleri A, Boorman GA, Foley JF, Yoshida T, Luster MI (1998) Arsenic enhancement of skin neoplasia by chronic stimulation of growth factors. *Am J Pathol 153*: 1775–1785

Gibson DP, Brauninger R, Shaffi HS, Kerchaert GA, LeBoeuf RA, Isfort RJ, Aarderma MJ (1997) Induction of micronuclei in Syrian hamster embryo cells: comparison of results in the SHE cell transformation assay for national toxicology program test chemicals. *Mutat Res 392*: 61–70

Goebel HH, Schmidt PF, Bohl J, Tettenborn B, Kramer G, Gutmann L (1990) Polyneuropathy due to acute arsenic intoxication: biopsy studies. *J Neuropathol Exp Neurol 49*: 137–149

Goering PL, Aposhian HV, Mass MJ, Cebrián M, Beck BD, Waalkes MP (1999) The enigma of arsenic carcinogenesis: role of metabolism. *Toxicol Sci 49*: 5–14

Goncalo S, Silva MS, Goncalo M, Baptista AP (1989) Occupational contact dermatitis to arsenic trioxide. in: Forsch PJ, Dooms-Goossens A (Eds) *Current topics in Contact Dermatitis* 333–336

Gonsebatt ME, Vega L, Salazar AM, Montero R, Guzman P, Bias J, Del Razo LM, Garcia-Vargas G, Albores A, Cebrian ME, Kelsh M, Ostrosky-Wegman P (1997) Cytogenetic effects in human exposure to arsenic. *Mutat Res 386*: 219–228

Greim H, Lehnert G (Eds) (1994). Arsentrioxid. in: *Biologische Arbeitsstoff-Toleranz-Werte (BAT-Werte) und Expositionsäquivalente für krebserzeugende Arbeitsstoffe (EKA)*, 7th issue, Wiley-VCH, Weinheim

Hantson P, Verellen-Dumoulin C, Libouton JM, Leonard A, Leonard ED, Mahieu P (1996) Sister chromatid exchanges in human peripheral blood lymphocytes after ingestion of high doses of arsenicals. *Int Arch Occup Environ Health 68*: 342–344

Harrington-Brock K, Cabrera M, Collard DD, Doerr CL, McConnell R, Moore MM, Sandoval H, Fuscoe JC (1999) Effects of arsenic exposure on the frequency of HPRT-mutant lymphocytes in a population of copper roasters in Antofagasta, Chile: a pilot study. *Mutat Res 431*: 247–257

Harrison JWE, Packman EW, Abbott DD (1958) Acute oral toxicity and chemical and physical properties of arsenic trioxides. *AMA Arch Ind Health 17*: 118–123

Hartwig A, Groeblinghoff UD, Beyersmann D, Natarajan AT, Filon R, Mullenders LH (1997) Interaction of arsenic(III) with nucleotide excision repair in UV-irradiated human fibroblasts. *Carcinogenesis 18*: 399–405

Hayashi H, Kanisawa M, Yamanaka K, Ito T, Udaka N, Ohji H, Okudela K, Okada S, Kitamara H (1998) Dimethylarsinic acid, a main metabolite of inorganic arsenics, has tumorigenicity and progression effects in the pulmonary tumors of A/J mice. *Cancer Lett 125*: 83–88

Hei TK, Liu SX, Waldren C (1998) Mutagenicity of arsenic in mammalian cells: role of reactive oxygen species. *Proc Natl Acad Sci USA 95*: 8103–8107

Heinrich-Ramm R, Mindt-Prüfert S, Szadkowski D (2001) Arsenic species excretion in a group of persons in northern Germany – contribution to the evaluation of reference values. *Int J Hyg Environ Health 203*: 475–477

Hertz-Picciotto I, Smith AH (1993) Observations on the dose-response curve for arsenic exposure and lung cancer. *Scand J Work Environ Health 19*: 217–226

Heywood R, Sortwell RJ (1979) Arsenic intoxication in the Rhesus monkey. *Toxicol Lett 3*: 137–144

Holmqvist I (1951) Occupational arsenical dermatitis – A study among employees at a copper ore smelting work including investigations of skin reactions to contact with arsenic compounds. *Acta Dermatol Venereol 31 (Suppl 26)*: 1–214

Holson JF, Stump DG, Ulrich CE, Farr CH (1999) Absence of prenatal developmental toxicity from inhaled arsenic trioxide in rats. *Toxicol Sci 51*: 87–97

Holson JF, Stump DG, Clevidence KJ, Knapp JF, Farr CH (2000) Evaluation of the prenatal toxicity of orally administered arsenic trioxide in rats. *Food Chem Toxicol 38*: 459–466

Hopenhayn-Rich C, Browning SR, Hertz-Picciotto I, Ferreccio C, Peralta C, Gibb H (2000) Chronic arsenic exposure and risk of infant mortality in two areas of Chile. *Environ Health Perspect 108*: 667–673

Hsu C-H, Li S-J, Chiou H-Y, Yeh P-M, Liou J-C, Hsueh Y-M, Chang S-H, Chen C-J (1997) Spontaneous and induced sister chromatid exchanges and delayed cell proliferation in peripher lymphocytes of Bowen's disease patients and matched controls of arseniasis-hyperendemic villages in Taiwan. *Mutat Res 386*: 241–251

Hsu C-H, Yang S-A, Wang YY, Yu H-S, Lin S-R (1999) Mutational spectrum of p53 gene in arsenic-related skin cancer from the blackfoot disease endemic area of Taiwan. *Br J Cancer 80*: 1080–1086

Hu Y, Su L, Snow ET (1998) Arsenic toxicity is enzyme specific and its effects on ligation are not caused by the direct inhibition of DNA repair enzymes. *Mutat Res 408*: 203–218

Huang S-C, Lee T-C (1998) Arsenite inhibits mitotic division and perturbs spindle dynamics in HeLa S3 cells. *Carcinogenesis 19*: 889–896

Huang S-Y, Chang C-S, Tang J-L, Tien H-F, Kuo T-L, Huang S-F, Yao Y-T, Chou WC, Chung C-Y, Wang C-H, Shen M-C, Chen Y-C (1998) Acute and chronic arsenic poisoning associated with treatment of acute promyelocytic leukaemia. *Br J Haematol 103*: 1092–1095

Huff J, Chan P, Nyska A (2000) Is the human carcinogen arsenic carcinogenic to laboratory animals? *Toxicol Sci 55*: 17–23

Hughes MF, Thompson DJ (1996) Subchronic dispositional and toxicological effects of arsenate administered in drinking water to mice. J Toxicol Environ Health 49: 177–196

IARC (International Agency for Research on Cancer) (1980) *Arsenic and arsenic compounds.* IARC monographs on the evaluation of the carcinogenic risk of chemicals to humans, Vol. 23, IARC Lyon, 39–141

IARC (International Agency for Research on Cancer) (1987) *Arsenic and arsenic compounds (Group 1).* IARC monographs, Suppl 7: 100–107

Ihrig MM, Shalat SL, Baynes C (1998) A hospital-based case-control study of stillbirths and environmental exposure to arsenic using an atmospheric dispersion model linked to a geographical information system. *Epidemiology 9*: 290–294

Ishinishi N, Yamamoto A, Hisanaga A, Inamasu T (1983) Tumorigenicity of arsenic trioxide to the lung in Syrian golden hamsters by intermittent instillations. *Cancer Lett 21*: 141–147

Ishitsuka K, Hanada S, Suzuki S, Utsunomiya A, Chyuman Y, Takeuchi S, Takeshita T, Shimotakahara S, Uozumi K, Makino T, Arima T (1998) Arsenic trioxide inhibits growth of human T-cell leukaemia virus type I infected T-cell lines more effectively than retinoic acids. *Br J Haematol 103*: 721–728

Järup L, Pershagen G (1991) Arsenic exposure, smoking and lung cancer in smelter workers – a case-control study. *Am J Epidemiol 134*: 545–551

Järup L, Pershagen G, Wall S (1989) Cumulative arsenic exposure and lung cancer in smelter workers – a dose-response study. *Am J Ind Med 15*: 31–41

Jha AN, Noditi M, Nilsson R, Natarajan AT (1992) Genotoxic effects of sodium arsenite on human cells. *Mutat Res 284*: 215–221

Kaise T, Watanabe S, Itoh K (1985) The acute toxicity of arsenobetaine. *Chemosphere 14*: 1327–1332

Kashiwada E, Kuroda K, Endo G (1998) Aneuploidy induced by dimethylarsinic acid in mouse bone marrow cells. *Mutat Res 413*: 33–38

Kenyon EM, Hughes MF (2001) A concise review of the toxicity and carcinogenicity of dimethyl-arsenic acid. *Toxicol 160*: 227–236

Kerkvliet NI, Steppan LB, Koller LD, Exon JH (1980) Immunotoxicology studies of sodium arsenate – effects of exposure on tumor growth and cell-mediated tumor immunity. *J Environ Pathol Toxicol 4*: 65–79

Kitchin KT (2001) Recent advances in arsenic carcinogenesis: modes of action, animal model systems, and methylated arsenic metabolites. *Toxicol Appl Pharmacol 172*: 249–261

Kroes R, van Logten MJ, Berkvens JM, deVries T, van Esch GJ (1974) Study on the carcino-genicity of lead arsenate and sodium arsenate and on the possible synergistic effect of diethyl-nitrosamine. *Food Cosmet Toxicol 12*: 671–679

Lagerkvist BE, Linderholm H, Nordberg GF (1988) Arsenic and Raynaud's phenomenon. Vaso-spastic tendency and excretion of arsenic in smelter workers before and after the summer vaca-tion. *Int Arch Occup Environ Health 60*: 361–364

Lai MS, Hsueh YM, Chen CJ, Shyu MP, Chen SY, Kuo TL, Wu MM, Tai TY (1994) Ingested inorganic arsenic and prevalence of diabetes mellitus. *Am J Epidemiol 139*: 484–492

Lantzsch H, Gebel T (1997) Genotoxicity of selected metal compounds in the SOS chromotest. *Mutat Res 389*: 191–197

Lee T-C, Oshimura M, Barrett JC (1985) Comparison of arsenic-induced cell transformation, cytotoxicity, mutation and cytogenetic effects in Syrian hamster embryo cells in culture. *Carcinogenesis 6*: 1421–1426

Lee-Feldstein A (1986) Cumulative exposure to arsenic and its relationship to respiratory cancer among copper smelter employees. *J Occup Med 28*: 296–302

Lerda D (1994) Sister-chromatid exchange (SCE) among individuals chronically exposed to arsenic in drinking water. *Mutat Res 312*: 111–120

Li JH, Rossman TG (1989) Inhibition of DNA ligase activity by arsenite: a possible mechanism of its comutagenesis. *Mol Toxicol 2*: 1–9

Li W, Wanibuchi H, Salim EI, Yamamoto S, Yoshida K, Endo G, Fukushima S (1998) Promotion of NCI-Black-Reiter male rat bladder carcinogenesis by dimethylarsenic acid an organic arsenic compound. *Cancer Lett 134*: 29–36

Li YM, Broome JD (1999) Arsenic targets tubulins to induce apoptosis in myeloid leukemia cells. *Cancer Res 59*: 776–780

Lindgren A, Vahter M, Dencker L (1982) Autoradiographic studies on the distribution of arsenic in mice and hamsters administered [74]As-arsenite or -arsenate. *Acta Pharmacol Toxicol 51*: 253–265

Liu J, Kadiiska MB, Liu Y, Lu T, Qu W, Waalkes MP (2001a) Stress-related gene expression in mice treated with inorganic arsenicals. *Toxicol Sci 61*: 314–320

Liu SX, Athar M, Lippai I, Waldren C, Hei TK (2001b) Induction of oxyradicals by arsenic: implication for mechanism of genotoxicity. *Proc Natl Acad Sci USA 98*: 1643–1648

Liu YC, Huang H (1997) Involvement of calcium-dependent protein kinase c in arsenite-induced genotoxicity in Chinese hamster ovary cells. *J Cell Biochem 64*: 423–433

Lubin J, Pottern LM, Blot WJ, Tokudome S, Stone BJ, Fraumeni JF (1981) Respiratory cancer among copper smelter workers: recent mortality statistics. *J Occup Med 23*: 779–784

Lubin JH, Pottern LM, Stone BJ, Fraumeni JF (2000) Respiratory cancer in a cohort of copper smelter workers: results from more than 50 years of follow-up. *Am J Epidemiol 151*: 554–565

Lugo G, Cassady G, Palmisano P (1969) Acute and maternal arsenic intoxication with neonatal death. *Am J Dis Child 117*: 173–182

Lynn S, Gurr JR, Lai HAT, Jan KY (2000) NADH oxidase activation is involved in arsenite-induced oxidative DNA damage in human vascular smooth muscle cells. *Circ Res 86*: 514–519

Lynn S, Lai HT, Gurr JR, Jan KY (1997) Arsenite retards DNA break rejoining by inhibiting DNA ligation. *Mutagenesis 12*: 353–358

Maki-Paakkanen J, Kurttio P, Paldy A, Pekkanen J (1998) Association between the clastogenic effect in peripheral lymphocytes and human exposure to arsenic through drinking water. *Environ Mol Mutagen 32*: 301–313

Mass MJ, Tennat A, Roop BC, Cullen WR, Styblo M, Thomas DJ, Kligerman AD (2001) Methylated trivalent arsenic species are genotoxic. *Chem Res Toxicol 14*: 355–361

Mass MJ, Wang L (1997) Arsenic alters cytosine methylation patterns of the promoter of the tumor suppressor gene p53 in human lung cells: a model for a mechanism of carcinogenesis. *Mutat Res 386*: 263–277

Matsui M, Nishigori C, Toyokuni S, Takada J, Akaboshi M, Ishikawa M, Imamura S, Miyachi Y (1999) The role of oxidative DNA damage in human arsenic carcinogenesis: detection of 8-hydroxy-2'-deoxyguanosine in arsenic-related Bowen's disease. *J Invest Dermatol 113*: 26–31

Menendez D, Mora G, Salazar AM, Ostrosky-Wegman P (2001) ATM status confers sensitivity to arsenic cytotoxic effects. *Mutagenesis 16*: 443–448

Mohamed KB (1998) Occupational contact dermatitis from arsenic in a tin-smelting factory. *Contact Dermatitis 38*: 224–225

Moore LE, Smith AH, Hopenhayn-Rich C, Biggs ML, Kalman DA, Smith MT (1997a) Micronuclei in exfoliated bladder cells among individuals chronically exposed to arsenic in drinking water. *Cancer Epidemiol Biomarkers Prev 6*: 31–36

Moore MM, Harrington-Brock K, Doerr CL (1997b) Relative genotoxic potency of arsenic and its methylated metabolites. *Mutat Res 386*: 279–290

Morton WE, Caron GA (1989) Encephalopathy: an uncommon manifestation of workplace arsenic poisoning? *Am J Ind Med 15*: 1–5

Mouron SA, Golijow CD, Dulout FN (2001) DNA damage by cadmium and arsenic salts assessed by the single cell gel electrophoresis assay. *Mutat Res 498*: 47–55

Murphy MJ, Lyon LW, Taylor JW (1981) Subacute arsenic neuropathy: clinical and electro-physiological observations. *J Neurol Neurosurg Psychiatry 44*: 896–900

Nagymajtenyi L, Selypes A, Berencsi G (1985) Chromosomal aberrations and fetotoxic effects of atmospheric arsenic exposure in mice. *J Appl Toxicol 5*: 61–63

Neiger RD, Osweiler GD (1989) Effect of subacute level dietary sodium arsenite on dogs. *Fundam Appl Toxicol 13*: 439–451

Nemec MD, Holson JF, Farr CH, Hood RD (1998) Developmental toxicity assessment of arsenic acid in mice and rabbits. *Reprod Toxicol 12*: 647–658

Ng JC, Seawright AA, Qi L, Garnett CM, Chiswell B, Moore MR (1999) Tumours in mice induced by exposure to sodium arsenate in drinking water. in: Abernathy C, Caldron R, Chapell W (Eds) *Arsenic exposure and health effects*, Oxford Elsevier Science, 217–223

Nordenson I, Beckman G, Beckman L (1978) Occupational and environmental risks in and around a smelter in northern Sweden: II. Chromosomal aberrations in workers exposed to arsenic. *Hereditas 88*: 47–50

Nordenson I, Beckman L (1982) Occupational and environmental risks in and around a smelter in northern Sweden. VII. Reanalysis and follow-up of chromosomal aberrations in workers exposed to arsenic. *Hereditas 96*: 175–181

Nordenson I, Beckman L (1991) Is the genotoxic effect of arsenic mediated by oxygen free radicals? *Hum Hered 41*: 71–73

Nordström S, Beckman L, Nordenson I (1978) Occupational and environmental risks in and around a smelter in northern Sweden. I. Variations in birth weight. *Hereditas 88*: 43–46

Nordström S, Beckman L, Nordenson I (1979a) Occupational and environmental risks in and around a smelter in northern Sweden. V. Spontaneous abortion among female employees and decreased birth weight in their offspring. *Hereditas 90*: 291–296

Nordström S, Beckman L, Nordenson I (1979b) Occupational and environmental risks in and around a smelter in northern Sweden. VI. Congenital malformations. *Hereditas 90*: 297–302

Oh SJ (1991) Electrophysiological profile in arsenic neuropathy. *J Neurol Neurosurg Psychiatry 54*: 1103–1105

Ostrosky-Wegman P, Gonsebatt ME, Montero R, Vega L, Barba H, Espinosa J, Palao A, Cortinas C, Garcia-Vargas G, del Razo LM (1991) Lymphocyte proliferation kinetics and genotoxic findings in a pilot study on individuals chronically exposed to arsenic in Mexico. *Mutat Res 250*: 477–482

Ott MG, Holder BB, Gordon HL (1974) Respiratory cancer and occupational exposure to arsenicals. *Arch Environ Health 29*: 250–255

Paschoud JM (1964) Notes cliniques au sujet des eczémas de contact professionnels par l'arsenic et l'antimoine (Clinical notes on occupational contact eczemas due to arsenic and antimony) (French). *Dermatologica 129*: 410–415

Pershagen G, Björklund NE (1985) On the pulmonary tumorigenicity of arsenic trisulfide and calcium arsenate in hamsters. *Cancer Lett 27*: 99–104

Pershagen G, Lind B, Björklund NE (1982) Lung retention and toxicity of some inorganic arsenic compounds. *Environ Res 29*: 425–434

Pershagen G, Nordberg G, Björklund NE (1984) Carcinomas of the respiratory tract in hamsters given arsenic trioxide and/or benzo[a]pyrene by the pulmonary route. *Environ Res 34*: 227–241

Poma K, Degraeve N, Kirsch-Volders M, Susanne C (1981) Cytogenetic analysis of bone marrow cells and spermatogonia of male mice after *in vivo* treatment with arsenic. *Experientia 37*: 129–130

Poma K, Degraeve N, Susanne C (1987) Cytogenetic effects in mice after chronic exposure to arsenic followed by a single dose of ethylmethane sulfonate. *Cytologia 52*: 445–449

Prukop JA, Savage NL (1986) Some effects of multiple sublethal doses of monosodium methane-arsonate (MSMA) herbicide on hematology, growth and reproduction of laboratory mice. *Bull Environ Contam Toxicol 36*: 337–341

Rahman M, Axelson O (1995) Diabetes mellitus and arsenic exposure: a second look at case-control data from a Swedish copper smelter. *Occup Environ Med 52*: 773–774

Rahman M, Tondel M, Ahmad SA, Axelson O (1998) Diabetes mellitus associated with arsenic exposure in Bangladesh. *Am J Epidemiol 148*: 198–203

Ramirez P, Eastmond DA, Laclette JP, Ostrosky-Wegman P (1997) Disruption of microtubule assembly and spindle formation as a mechanism for the induction of aneuploid cells by sodium arsenite and vanadium pentoxide. *Mutat Res 38*: 291–298

Ramirez P, Del Razo LM, Gutierrez-Ruiz MC, Gonseblatt ME (2000) Arsenite induces DNA-protein crosslinks and cytokeratin expression in the WRL-68 human hepatic cell line. *Carcinogenesis 21*: 701–706

Rasmussen RE, Menzel DB (1997) Variation in arsenic-induced sister chromatid exchange in human lymphocytes and lymphoblastoid cell lines. *Mutat Res 386*: 299–306

Reichl FX, Szinicz L, Kreppel H, Fichtl B, Forth W (1990) Effect of glucose in mice after acute experimental poisoning with arsenic trioxide ($As_2O_3$). *Arch Toxicol 64*: 336–338

Reichl FX, Szinicz L, Kreppel H, Forth W (1988) Effect of arsenic on carbohydrate metabolism after single or repeated injection in guinea pigs. *Arch Toxicol 62*: 473–475

Rossberg J (1967) Berufsbedingte Arsen-Schädigungen der Haut (Skin lesions caused by occupation related to arsenic) (German). *Dermatol Wochenschr 153*: 977–987

RoyChoudhury A, Das T, Sharma A, Takukder G (1996) Dietary garlic extracts in modifying clastogenic effects of inorganic arsenic in mice: two-generation studies. *Mutat Res 359*: 165–170

Rupa DS, Schuler M, Eastmond DA (1997) Detection of hyperploidy and breakage affecting the 1cen-1q12 region of cultured interphase human lymphocytes treated with various genotoxic agents. *Environ Molec Mutagen 2*: 161–167

Saffiotti U, Bertolero F (1989) Neoplastic transformation of BALB/3T3 cells by metals and the quest for induction of a metastatic phenotype. *Biol Trace Elem Res 21*: 475–482

Salnikow K, Cohen MD (2002) Backing into cancer: effects of arsenic on cell differentiation. *Toxicol Sci 65*: 161–163

Schaumlöffel N, Gebel T (1998) Heterogeneity of the DNA damage provoked by antimony and arsenic. *Mutagenesis 13*: 281–286

Schrauzer GN, Ishmael D (1974) Effects of selenium and arsenic on the genesis of spontaneous mammary tumors in inbred C3H mice. *Ann Clin Lab Sci 4*: 441–447

Schrauzer GN, White DA, Schneider CJ (1976) Inhibition of the genesis of sponaneous mammary tumours in C3H mice: Effects of selenium-antagonistic elements and their possible role in human breast cancer. *Bioinorg Chem 6*: 265–270

Schroeder HA, Balassa JJ (1966) Abnormal trace metals in man: arsenic. *J Chron Dis 19*: 85–106

Schroeder HA, Balassa JJ (1967) Arsenic, germanium, tin and vanadium in mice: Effects on growth, survival and tissue levels. *J Nutrition 92*: 245–252

Schulz KH (1962) *Chemische Struktur und allergene Wirkung–unter besonderer Berücksichtigung von Kontaktallergenen* (Chemical structure and allergenic effect—particularly in consideration of contact allergens) (German). Editio Cantor, Aulendorf, 43–45

Schwerdtle T, Mackiw I, Bayerl A, Hartwig A (2002) Arsenite and its methylated metabolites MMAA and DMAA induce oxidative DNA damage in HeLa S3 cells. *Arch Pharmacol 365, Suppl 1*: 526

Seiwert M, Becker K, Friedrich C, Helm D, Hoffmann K, Krause C, Nöllke P, Schulz C, Seifert B (1999) *Umwelt-Survey 1990/92 Arsen-Zusammenhangsanalyse* (Environmental survey 1990/92, Arsenic multivariate statistical analysis) (German). WaBoLu 3/99, Umweltbundesamt, Berlin

Shen Z-X, Chen G-Q, Ni J-H, Li X-S, Xion S-M, Qui Q-Y, Zhu J, Tang W, Sun G-L, Yang K-Q, Chen Y, Zhou L, Fan Z-W, Wang Y-T, Ma J, Zhang P, Zhang T-D, Chen S-J, Chen Z, Wang Z-Y (1997) Use of arsenic trioxide ($As_2O_3$) in the treatment of acute promyelocytic leukemia (APL): II. Clinical efficacy and pharmacokinetics in relapsed patients. *Blood 89*: 3354–3360

Shirachi DY, Johansen MG, McGowan JP, Tu SH (1983) Tumorigenic effect of sodium arsenite in rat kidney. *Proc West Pharmacol Soc 26*: 413–415

Sikorski EE, McCay JA, White KL Jr, Bradley SG, Munson AE (1989) Immunotoxicity of the semiconductor gallium arsenide in female B6C3F$_1$ mice. *Fundam Appl Toxicol 13*: 843–858

Soignet SL, Maslak P, Wang Z-G, Jhanwar S, Calleja E, Dardashti LJ, Corso D, DeBlasio A, Gabrilove J, Scheinberg DA, Pandolfi PP, Warell RP (1998) Complete remission after treatment of acute promyelocytic leukemia with arsenic trioxide. *N Engl J Med 339*: 1341–1348

Sordo M, Herrera LA, Ostrosky-Wegman P, Rojas E (2001) Cytotoxic and genotoxic effects of As, MMA, and DMA on leukocytes and stimulated human lymphocytes. *Teratogen Carcinogen Mutagen 21*: 249–260

Stump DG, Holson JF, Fleeman TL, Nemec MD, Farr CH (1999) Comparative effects of single intraperitoneal or oral doses of sodium arsenate or arsenic trioxide during *in utero* development. *Teratology 60*: 283–291

Szinicz L, Forth W (1988) Effects of $As_2O_3$ on gluconeogenesis. *Arch Toxicol 61*: 444–449

Tabacova S, Hunter ES (1995) Pathogenic role of peroxidation in prenatal toxicity of arsenic. in: Zelikoff J, Bertin J, Burbacher T, Hunter S, Miller R, Silbergeld E, Tabacova S, Roger J: Health risks associated with prenatal exposure. *Experimental and Applied Toxicology 25*: 161–170

Taylor PR, Qiao YL, Schatzkin A, Yao SX, Lubin J, Mao BL, Rao JY, McAdams M, Xuan XZ, Li JY (1989) Relation of arsenic exposure to lung cancer among tin miners in Yunnan Province, China. *Br J Ind Med 46*: 881–886

Thomas DJ, Styblo M, Lin S (2001) The cellular metabolism and systemic toxicity of arsenic. *Toxicol Appl Pharmacol 176*: 127–144

Thorgeirsson UP, Dalgard DW, Reeves J, Adamson RH (1994) Tumor incidence in a chemical carcinogenesis study of nonhuman primates. *Regul Toxicol Pharm 19*: 130–151

Tian D, Ma H, Feng Z, Xia Y, Le XC, Ni Z, Allen J, Collins B, Schreinemachers D, Mumford JL (2001) Analyses of micronuclei in exfoliated epithelial cells from individuals chronically exposed to arsenic via drinking water in Inner Mongolia, China. *J Toxicol Environ Health, A 64*: 472–484

Tice RR, Yager JW, Andrews P, Crecelius E (1997) Effect of hepatic methyl donor status on urinary excretion and DNA damage in B6C3F$_1$ mice treated with sodium arsenite. *Mutat Res 386*: 315–334

Tinwell H, Stephens SC, Ashby J (1991) Arsenite as the probable active species in the human carcinogenicity of arsenic: mouse micronucleus assays on Na and K arsenite, orpiment, and Fowler's solution. *Environ Health Perspect 95*: 205–210

Trouba KJ, Wauson EM, Vorce RL (2000) Sodium arsenite-induced dysregulation of proteins involved in proliferative signaling. *Toxicol Appl Pharmacol 164*: 161–170

Vahter M (2000) Genetic polymorphism in the biotransformation of inorganic arsenic and its role in toxicity. *Toxicol Lett 112–113*: 209–217

Vig BK, Figueroa ML, Cornforth MN, Jenkins SH (1984) Chromosome studies in human subjects chronically exposed to arsenic in drinking water. *Am J Ind Med 6*: 325–338

Vogt BL, Rossman TG (2001) Effects of p53, p21 and cyclin D expression in normal human fibroblasts—a possible mechanism for arsenite's comutagenicity. *Mut Res 478*: 159–168

Waalkes MP, Keefer LK, Diwan BA (2000) Induction of proliferative lesions of the uterus, testes, and liver in Swiss mice given repeated injections of sodium arsenate: possible estrogenic mode of action. *Toxicol Appl Pharmacol 166*: 24–35

Wahlberg JE, Boman A (1986) Contact sensitivity to arsenical compounds—clinical and experimental studies. *Dermatosen Beruf Umwelt 34*: 10–12

Wang TS, Huang H (1994) Active oxygen species are involved in the induction of micronuclei by arsenite in XRS-5 cells. *Mutagenesis 9*: 253–257

Wang T-S, Hsu T-Y, Chung C-H, Wang ASS, Bau T-D, Jan K-Y (2001) Arsenite induces oxidative DNA adducts and DNA-protein cross-links in mammalian cells. *Free Radic Biol Med 31*: 321–330

Wanibuchi H, Yamamoto S, Chen H, Yoshida K, Endo G, Hori T, Fukushima S (1996) Promoting effects of dimethylarsinic acid on *N*-butyl-*N*-(4-hydroxybutyl)nitrosamine-induced urinary bladder carcinogenesis in rats. *Carcinogenesis 17*: 2435–2439

Warner ML, Moore LE, Smith MT, Kalman DA, Fanning E, Smith AH (1994) Increased micronuclei in exfoliated bladder cells of individuals who chronically ingest arsenic-contaminated water in Nevada. *Cancer Epidemiol Biomarkers Prev 3*: 583–590

Wauson EM, Langan AS, Vorce RL (2002) Sodium arsenite inhibits and reverses expression of adipogenic and fat cell-specific genes during *in vitro* adipogenesis. *Toxicol Sci 65*: 211–219

Wei M, Wanibuchi H, Yamamoto S, Li W, Fukushima S (1999) Urinary bladder carcinogenicity of dimethylarsinic acid in male F344 rats. *Carcinogenesis 20*: 1873–1876

Welch K, Higgins I, Oh M, Burchfiel C (1982) Arsenic exposure, smoking and respiratory cancer in copper smelter workers. *Arch Environ Health 37*: 325–335

WHO (World Health Organization) (2001) *Arsenic and arsenic compounds*. IPCS (International Programme on Chemical Safety) Environmental Health Criteria 224, WHO, Geneva

Wiencke JK, Yager JW, Varkonyi A, Hultner M, Lutze LH (1997) Study of arsenic mutagenesis using the plasmid shuttle vector pz189 propagated in DNA repair proficient human cells. *Mutat Res 386*: 335–344

Yager JW, Wiencke JK (1997) Inhibition of poly(ADP-ribose) polymerase by arsenite. *Mutat Res 386*: 345–351

Yamamoto A, Hisanaga A, Ishinishi N (1987) Tumorigenicity of inorganic arsenic compounds following intratracheal instillations to the lungs of hamsters. *Int J Cancer 40*: 220–223

Yamamoto S, Konishi Y, Matsuda T, Murai T, Shibata MA, Matsui-Yuasa I, Otani S, Kuroda K, Endo G, Fukushima S (1995) Cancer induction by an organic arsenic compound, dimethylarsinic acid (cacodylic acid), in GF344/DuCrj rats after pretreatment with five carcinogens. *Cancer Res 55*: 1271–1276

Yamamoto S, Wanibuchi H, Hori T, Yano Y, Matsui-Yuasa I, Otani S, Chen H, Yoshida K, Kuroda K, Endo G, Fukushima S (1997) Possible carcinogenic potential of dimethylarsinic acid as assessed in rat *in vivo* models: a review. *Mutat Res 386*: 353–361

Yamanaka K, Okada S (1994) Induction of lung-specific DNA damage by metabolically methylated arsenics via the production of free radicals. *Environ Health Perspect 102*: 37–40

Yamanaka K, Hasegawa A, Sawamura R, Okada S (1989) Dimethylated arsenics induce DNA strand breaks in lung via the production of active oxygen in mice. *Biochem Biophys Res Commun 165*: 43–50

Yih LH, Ho IC, Lee TC (1997) Sodium arsenite disturbs mitosis and induces chromosome loss in human fibroblasts. *Cancer Res 57*: 5051–5059

Zhang W, Ohnishi K, Shigeno K, Fujisawa S, Naito K, Nakamura S, Takeshita K, Takeshita A, Ohno R (1998) The induction of apoptosis and cell cycle arrest by arsenic trioxide in lymphoid neoplasms. *Leukemia 12*: 1383–1391

Zhao CQ, Young MR, Diwan B A, Coogan TP, Waalkes MP (1997) Association of arsenic-induced malignant transformation with DNA hypomethylation and aberrant gene expression. *Proc Natl Acad Sci USA 94:* 10907–10912

Zierler S, Theodore M, Cohen A, Rothman K (1988) Chemical quality of maternal drinking water and congenital heart disease. *Int J Epidemiol 17*: 589–594

completed 10.04.2002

# 2-Butoxyethanol        H

| | |
|---|---|
| **Classification/MAK value:** | **20 ml/m$^3$ (ppm)** <br> **100 mg/m$^3$** |
| **MAK value dates from:** | **1983** |
| Synonyms: | *n*-butoxyethanol <br> Butyl Cellosolve® <br> O-butyl ethylene glycol <br> butyl glycol <br> ethylene glycol *n*-butyl ether <br> ethylene glycol monobutyl ether <br> glycol butyl ether <br> monobutyl glycol ether <br> 3-oxa-1-hepatanol |
| Chemical name (CAS): | 2-butoxyethanol |
| CAS number: | 111-76-2 |
| Structural formula: | $H_2C-OH$ <br> $H_2C-O-CH_2-CH_2-CH_2-CH_3$ |
| Molecular formula: | $C_6H_{14}O_2$ |
| Molecular weight: | 118.17 |
| Melting point: | $-70.4\,°C$ |
| Boiling point: | $171.2\,°C$ |
| Vapour pressure at 20 °C: | $c$ 1 hPa |

**1 ml/m$^3$ (ppm) = 4.90 mg/m$^3$**      **1 mg/m$^3$ = 0.203 ml/m$^3$ (ppm)**

## 1 Toxic Effects and Modes of Action

The toxicological profile of 2-butoxyethanol is characterized by the following properties of the substance: it is readily absorbed through the skin, it has adverse effects on the blood count (however, this is less a cytotoxic effect on the stem cells such as is seen with the lower homologues, 2-methoxyethanol and 2-ethoxyethanol, and more a result of haemolytic effects in the peripheral blood) and it causes kidney and liver damage. Significant leukopenia or lymphopenia are not found. Very high oral doses of 2-butoxyethanol can have toxic effects on the seminiferous epithelium of the testes.

*Essential MAK Value Documentations.* DFG, Deutsche Forschungsgemeinschaft
Copyright © 2006 WILEY-VCH Verlag GmbH & Co. KGaA, Weinheim
ISBN: 3-527-31394-X

The haemolytic effects are the result of an increase in osmotic fragility of the erythrocytes. They are particularly marked in the rat, mouse and rabbit, less so in the primate, dog and guinea pig [1]. The secondary effects include a reduction in the blood haemoglobin level, increase in the reticulocyte number, haemosiderosis and severe kidney damage with increased kidney weights and cloudy swelling around the tubules and Henle's loop [1, 2].

The kidney damage does not appear to result solely from haemolysis. In the acutely toxic dose range, 2-butoxyethanol also has effects on the central nervous system in rats and rabbits, resulting in apathy, stupor and narcotic symptoms [3]. However, a specific effect on behaviour, such as is observed with 2-methoxyethanol, has not been established [4].

2-Butoxyethanol is oxidized in the human organism to butoxyacetic acid/butoxy-acetate, presumably by the hepatic alcohol dehydrogenase [1, 2]. This metabolite, which can be detected in human [1] and rat [5] urine after inhalation of 2-bu-toxyethanol, increased the osmotic fragility of erythrocytes when tested *in vitro* to a much greater extent than did 2-butoxyethanol itself [1, 6].

Butoxyacetate has been detected in blood only after intravenous administration of 2-butoxyethanol to rabbits [2].

## 2 Effects in Man

Irritation of the eyes and nasal epithelium [1, 2, 7] and headaches [1] were recorded during occupational exposure and during experimental inhalation of about 200 ml/ $m^3$ for 8 hours and 100 ml/$m^3$ for 4 or 8 hours by volunteers. Butoxyacetate was detected in the urine. Signs of haemolysis, such as were observed in 3 rats exposed under the same conditions, did not develop [1].

One case has been described in which transient haematuria occurred on two occasions 5 months apart after imprecisely specified exposure to 2-butoxyethanol and butyl diglycol [2].

In the eye, the solvent causes painful irritation, sometimes with corneal clouding; the symptoms generally regress within a few days [6, 7].

## 3 Effects on Animals

### 3.1 Acute and subacute toxicity

The oral $LD_{50}$ for 2-butoxyethanol in the rat is 550 mg/kg for old animals and up to 3000 mg/kg for young animals. The value is markedly dependent on age [1, 6]. The oral $LD_{50}$ in the mouse is given as 1230 mg/kg, in the rabbit as 320 mg/kg, and guinea pig as 1200 mg/kg [1]. Cats survived oral 2-butoxyethanol doses of 1.0 or 2.0 ml/kg; haemolysis was not observed [3].

That the substance is readily absorbed through the skin may be deduced from the values for the dermal $LD_{50}$ in the rabbit, 0.45 – 0.56 ml/kg [1] or 0.68 ml/kg [8], and guinea pig, 0.23 ml/kg [8]. After application of the substance to rabbit skin under non-occlusive conditions, the $LD_{50}$ was about 2 ml/kg [1].

After administration of single oral or dermal 2-butoxyethanol doses of about 0.5 to 1 ml/kg, deaths were observed. Pathological anatomical examination revealed fatty degeneration of the heart muscle and liver, and pulmonary oedema [3].

The values determined for the intravenous $LD_{50}$ were 0.28 ml/kg to 0.5 g/kg in the rabbit, 0.34 ml/kg to 0.38 g/kg in the rat and 1.13 g/kg in the mouse. After intraperitoneal administration, the $LD_{50}$ in the rat was 0.55 ml/kg [1].

The 7-hour $LC_{50}$ for 2-butoxyethanol in the mouse was given as 700 ml/m$^3$ [9].

Attention is drawn to an early study [7] which defines the approximate limits of the acutely toxic dose range and provides indications of subacute toxic effects. Single **oral** doses of 0.1 and 0.5 ml/kg body weight were not lethal for rabbits. However, 1.0 and 2.0 ml/kg killed the animals within 30 and 22 hours, respectively. Autopsy revealed severe haemoglobinuria [7]. Single **subcutaneous** injections of 0.05 and 0.1 ml/kg were tolerated by rabbits without symptoms, 0.2 ml/kg caused transient haematuria and 0.4 ml/kg was lethal with massive haemolysis and haemoglobinuria. No symptoms were seen in two rabbits which were given subcutaneous injections of 0.05 or 0.1 ml/kg body weight, 7 times within one week. Cats, however, tolerated a single subcutaneous injection of 1.0 ml/kg but 2.0 ml/kg produced symptoms of kidney damage and death [7]. **Inhalation** of concentrations up to practically saturated atmospheres (about 10 mg/l) for 1 hour had no effects on rabbits, cats, guinea pigs or mice. Daily 8-hour exposures to a 2-butoxyethanol concentration of 2.5 mg/l was lethal for cats after 9 days; one of two rabbits died 16 days after a 12-day exposure period. Guinea pigs died after 8 and 9 exposure days; mice survived the 12-day exposure period without conspicuous effects [7].

## 3.2 Subchronic toxicity

In a 90-day feeding study with doses of 1540, 310, 76, 18 and 0 mg/kg/day in the rat, no substance-related deaths occurred; in the highest dose group the daily dose was about $^3/_5$ of the acute $LD_{50}$. This indicates that administration of doses below the acutely toxic range, that is, division of the acutely toxic dose into several smaller doses, permits rapid detoxication and that cumulative effects are not to be expected. The animals which died during this study had pneumonia. Reduced body weight gain and kidney and liver damage with increased organ weights were recorded in the high dose groups but no haematuria. The no adverse effect level was between 76 and 310 mg/kg/day [1].

## 3.3 Effects on the blood

In rats exposed to a 2-butoxyethanol concentration of 320 ml/m$^3$, 7 hours daily, 5 days per week for 5 weeks, the erythrocyte count and blood haemoglobin level were reduced after the first week [10] and the number of reticulocytes and young granulocytes was increased. The values gradually returned to normal during the rest of the exposure period and after the end of exposure. The same research group reported haemoglobinuria in mice after a single exposure to toxic concentrations [9].

Haematuria was observed in rats exposed once or twice to a 2-butoxyethanol concentration of 50, 100, 200 or 400 ml/m$^3$. It regressed, however, during the remainder of the exposure period (4 hours daily, 5 days per week for 2 weeks) [4].

In a more recent study with male mice, a reduction in the erythrocyte count but not in the leukocyte count was seen after administration of 2-butoxyethanol doses of 500 or 1000 mg/kg by gavage, 5 times weekly for 5 weeks; a dose of 2000 mg/kg was lethal for the animals [11].

There is evidence that the haemolytic effects of 2-butoxyethanol are less marked in dogs and primates than in rats and mice.

– In dogs which inhaled a 2-butoxyethanol concentration of 145 ml/m$^3$, 7 hours daily, 5 days per week for 12 weeks, retention of urea was observed but the changes in the erythrocyte count remained marginal [12].

– Exposure of 3 test persons to 195 ml/m$^3$ for 8 hours did not produce signs of haemolytic effects nor of an increased osmotic fragility of the erythrocytes [1].

– The erythrocytes of guinea pigs, monkeys and man have a greater osmotic resistance *in vivo* than do those of mice, rats and rabbits under the same conditions. With butoxyacetate *in vitro* the same differences between species were found [1].

During a 2-week inhalation study [13] with rats exposed 9 times for 6 hours per day within 11 days to 2-butoxyethanol concentrations of 20, 86 or 245 ml/m$^3$, after 2 days transient haematuria was detected in the highest dose group and at 86 ml/m$^3$ a reduced haemoglobin concentration and increased erythrocyte cell volume.

Dermal exposure of rabbits according to the same exposure timetable caused haemoglobinuria when 1 ml of an undiluted or a 50% aqueous solution of 2-butoxyethanol was applied occlusively to each animal. No effects were seen with 1 ml of a 5% or 25% solution. In females treated with the undiluted substance, erythrocyte count and haemoglobin level were reduced; slight leukopenia developed in males but was considered by the authors not to be clearly substance-related [13].

In a subsequent 90-day inhalation study in rats, exposure to a 2-butoxyethanol concentration of 77 ml/m$^3$ caused reduction in erythrocyte count and Hb level and an increase in erythrocyte cell volume. Exposure to 25 or 5 ml/m$^3$ was without effects [13, 14].

## 3.4 Effects on the testes

Oral administration of 2-butoxyethanol doses of 2000, 1000 or 500 mg/kg, 5 times weekly for 5 weeks, to groups of 5 male mice caused testis atrophy in 1 of 5 animals in the 1000 mg/kg group [11]. The animals in the highest dose group died.

This shows that 2-butoxyethanol, like its lower homologues 2-ethoxyethanol and 2-methoxyethanol, has toxic effects on the seminiferous epithelium; they are, however, much less marked than those of the lower homologues. In parallel, the effect of 2-butoxyethanol on the development of the red and white blood cells in the bone marrow is also weak compared with that of the lower homologues.

In a 90-day feeding study, 2-butoxyethanol concentrations of 1.25% and 0.25% in the diet led to testis damage. With 0.05% and 0.01%, such effects were not detected [13]. Inhalation of 77 ml/m$^3$ for 90 days did not produce testis damage [13].

## 4 Reproductive and Developmental Toxicity

The teratogenicity of 2-butoxyethanol has been studied in three inhalation studies in the rat and one in the rabbit.

In the first study [15] with rats exposed to a concentration of 200 ml/m$^3$, 6 hours daily from day 7 to day 15 of gestation, there was no evidence of foetotoxicity or teratogenicity; haematuria was observed in the dams.

In another study [16], however, exposure of rats to 100, 200 or 300 ml/m$^3$ for 6 hours daily from day 6 to day 15 of gestation produced toxic effects in the dams of all dose groups. Food consumption and body weight gain were reduced even in the 100 ml/m$^3$ group; in the two higher dose groups there were also reductions in water consumption and the erythrocyte count (caused by haemolysis). In the 300 ml/m$^3$ group, all embryos or foetuses died in 16/32 dams. Other findings in this group included ossification disorders, enlarged cerebral ventricles and arterial malformations (shortened in comparison with the control values).

In a subsequent study by the same authors [16], rats and rabbits were exposed to 25, 50, 100 or 200 ml/m$^3$ for 6 hours daily from day 6 to day 15 or day 18 of gestation. The 200 ml/m$^3$ exposures caused haemolytic effects in the dams; in the uterus there were increased resorptions and delayed ossification. At 100 ml/m$^3$, these effects were only slight. An increase in the incidence of malformations above the control value was not seen either at 100 or at 200 ml/m$^3$. The 50 and 25 ml/m$^3$ exposures had no effects on foetuses or dams. In the rabbits there were slight maternal and embryonal toxic effects at 200 ml/m$^3$ but not at 100 ml/m$^3$. Here too, the incidence of malformations was not increased.

In a dermal study with rats, $4 \times 0.12$ ml (0.9 mmole) 2-butoxyethanol per animal was applied non-occlusively from day 7 to day 16 of gestation. Malformations, resorptions or delays in development were not observed [3].

In summary, embryotoxic effects were seen only in maternally toxic dose ranges (200 ml/m$^3$ in the rabbit and 300, 200 and 100 ml/m$^3$ in the rat). In one study at 300 ml/m$^3$ and with massive maternal toxicity, some malformations were seen. Under exposure conditions which did not lead to maternal toxicity, no embryotoxic effects were observed.

## 5 Manifesto (MAK value, classification)

The toxic effects typical of 2-methoxyethanol and 2-ethoxyethanol (pancytopenia, testis atrophy and teratogenic effects) are not found with 2-butoxyethanol. Testis damage was not seen in rats after 90-day exposure either to inhaled concentrations of 77 ml/m$^3$ or to 0.05% 2-butoxyethanol in the diet (about 25 mg/kg and day). Haemolytic effects were seen at 77 ml/m$^3$ but not at 25 ml/m$^3$. Assuming complete absorption, a 7-hour exposure to a 2-butoxyethanol concentration of 25 ml/m$^3$ (120 mg/m$^3$) is equivalent to a dose of about 30 mg/kg for the rat. Taking into account the fact that the haemolytic effects of the substance are weaker in man, it may be concluded that exposure to 20 ml/m$^3$ would not be expected to have adverse effects on health and includes an adequate safety margin as long as skin contact with the substance is avoided. This MAK value requires substantiation in appropriate animal studies and in systematic occupational medical investigations.

The substance is classified in category II,1 for the limitation of exposure peaks (peak level twice the MAK value, duration 30 minutes, four times per shift) because 2-butoxyethanol causes systemic effects which are expected to appear rapidly.

Because of the danger associated with absorption of the substance through the skin, 2-butoxyethanol is designated with an "H".

In reproductive toxicity studies, no significant teratogenicity was observed and no embryotoxic effects in the absence of maternal toxicity. The substance is therefore classified in pregnancy group C.

## 6 References


1. Carpenter, C. P., U. C. Pozzani, C. S. Weil, J. H. Nair III, G. A. Keck, H. F. Smyth, jr.: *Arch industr. Hlth 14*, 114 (1956)
2. Browning, E.: *Toxicity and Metabolism of Industrial Solvents*, p 608, Elsevier Publ. Co., Amsterdam, London, New York, 1965
3. BASF: Unpublished data from internal reports V/420, XV/329 and XVIII/354, BASF Aktiengesellschaft, D-6700 Ludwigshafen, 1959–1967
4. Goldberg, M. E., H. E. Johnson, U. C. Pozzani, H. F. Smyth, jr.: *Industr. Hyg. J. 25*, 369 (1964)
5. Jönsson, A. K., G. Steen: *Acta Pharmacol. (Kbh.) 42*, 354 (1978)
6. Rowe, V. K.: in Patty, F. A.: *Industrial Hygiene and Toxicology*, Vol. II, p 1552, Interscience Publishers, John Wiley & Sons, New York/London, 1962
7. Gross, E.: in Lehmann, K. B., F. Flury: *Toxikologie und Hygiene der technischen Lösungsmittel*, p 210, Springer Verlag, Berlin, 1938
8. Roudabush, R. L., C. J. Terhaar, D. W. Fassett, S. P. Dziuba: *Toxicol. appl. Pharmacol. 7*, 559 (1965)
9. Werner, H. W., J. L. Mitchell, J. W. Miller, W. F. von Oettingen: *J. industr. Hyg. 25*, 157 (1943)
10. Werner, H. W., C. Z. Nawrocki, J. L. Mitchell, J. W. Miller, W. F. von Oettingen: *J. industr. Hyg. 25*, 374 (1943)
11. Nagano, K., E. Nakayama, M. Koyano, H. Oobayashi, H. Adachi, T. Yamada: Jap. *J. industr. Hlth 21*, 29 (1979)
12. Werner, H. W., J. L. Mitchell, J. W. Miller, W. F. von Oettingen: *J. industr. Hyg. 25*, 409 (1943)
13. Tyler, T.: *Review of ethylene glycol monobutyl ether toxicity testing*, cited in ECETOC, Techn. Rep. No. 4, Avenue Louise 250, B. 63, B-1050 Brussels, Belgium, 1982
14. Dodd, D. E., W. M. Snellings, R. R. Maronpot, B. Ballantyne: *Toxicol. appl. Pharmacol. 68*, 405 (1983)
15. Nelson, B. K., J. V. Setzer, W. S. Brightwell, P. R. Mathinos, M. H. Kuczuk, T. E. Weaver: *Teratology 25*, 64A (1982)
16. Tyl, R. W., G. Millicovsky, D. E. Dodd, I. M. Pritts, K. A. France, L. C. Fischer: *Environm. Hlth Perspect. 57*, 47 (1984)
17. Hardin, B. D., P. T. Goad, J. R. Burg: *Environm. Hlth Perspect. 57*, 69 (1984)


MAK value completed 8. 12. 1982
Pregnancy group completed 1. 10. 1986

# Carbon disulfide

| | |
|---|---|
| **MAK value (1997)** | **5 ml/m$^3$ (ppm) $\triangleq$ 16 mg/m$^3$** |
| **Peak limitation (2001)** | **Category II, excursion factor 2** |
| **Absorption through the skin (1980)** | **H** |
| **Sensitization** | **–** |
| **Carcinogenicity** | **–** |
| **Prenatal toxicity (1985)** | **Pregnancy risk group B** |
| **Germ cell mutagenicity** | **–** |
| **BAT (1997)** | **4 mg 2-thio-thiazolidin-4-carboxyl acid (TTCA)/g creatinine** |
| Synonyms | carbon bisulfide<br>carbon sulfide<br>dithiocarbonic anhydride |
| Chemical name (CAS) | carbon disulfide |
| CAS number | 75-15-0 |
| Structural formula | S=C=S |
| Molecular formula | CS$_2$ |
| Molecular weight | 76.14 |
| Melting point | –112°C |
| Boiling point | 46°C |
| Vapour pressure at 20°C | 400 hPa |
| **1 ml/m$^3$ (ppm) $\triangleq$ 3.16 mg/m$^3$** | **1 mg/m$^3$ $\triangleq$ 0.317 ml/m$^3$ (ppm)** |

The toxicity of carbon disulfide was evaluated by the Commission in 1975 and 1997 (see Henschler 1975 and the 1997 documentation for Carbon disulfide in Volume 12 of the present series). The toxic effects discussed in these MAK documentations and the report from the Beratergremium für Altstoffe (BUA 1991) are summarised.

The long-known toxic effects of carbon disulfide on the central and peripheral nervous system have been confirmed in many studies of workers in the viscose industry. The peripheral neurotoxicity of carbon disulfide manifests itself in a characteristic polyneuropathy with an irreversible slowing of the sensorimotor nerve conduction velocity. It is dependent on the cumulative exposure (product of concentration and

*Essential MAK Value Documentations.* DFG, Deutsche Forschungsgemeinschaft
Copyright © 2006 WILEY-VCH Verlag GmbH & Co. KGaA, Weinheim
ISBN: 3-527-31394-X

length of exposure in years) (Henschler 1975, the 1997 documentation for Carbon disulfide in volume 12 of the present series). On this basis a NOEC (no observed effect concentration) for 40 years' exposure to carbon disulfide (8 hours/day, 5 days/week) of 4 ml/m$^3$ (about 12 mg/m$^3$) was calculated (Ruijten *et al.* 1990), which is beset with considerable uncertainties, however, since the earlier higher concentrations were not taken properly into account (see the 1997 documentation for Carbon disulfide in Volume 12 of the present series). In a cross-sectional study of 247 male workers with a median carbon disulfide concentration of 4 ml/m$^3$ (about 12 mg/m$^3$), a maximum concentration of 65.7 ml/m$^3$ (208 mg/m$^3$) and a median length of exposure of 66 months (4–220 months), no indications of neurotoxic effects compared with a group of 222 controls were found (Reinhardt *et al.* 1997a, 1997b, the 1997 documentation for Carbon disulfide in Volume 12 of the present series).

The vestibular damage, such as nystagmus and impaired hearing, observed in viscose workers after exposure to carbon disulfide concentrations of more than 30 mg/m$^3$ (about 10 ml/m$^3$) and balance disturbances are attributed to the neurotoxicity of carbon disulfide (BUA 1991). In a study of 74 workers exposed to carbon disulfide, effects of this kind were reported in workers with more than 20 years' exposure. Current exposure concentrations were 15 mg/m$^3$ (about 5 ml/m$^3$) on average, while earlier concentrations were higher (no further details) (Hirata *et al.* 1992, the 1997 documentation for Carbon disulfide in Volume 12 of the present series).

In many studies of viscose workers, vascular changes in the retina of the eye with microaneurisms, haemorrhages, vascular scleroses and pigment changes were described after several years' exposure to carbon disulfide concentrations of more than 30 mg/m$^3$ (10 ml/m$^3$) (BUA 1991, Vanhoorne *et al.* 1995, 1996). The studies carried out do not show a uniform picture in the different countries, so that ethnic differences have been postulated (BUA 1991). In the authors' opinion (Vanhoorne *et al.* 1995) it is still not clear, however, what effects on the eye are caused by carbon disulfide, by hydrogen sulfide and by a combination of the two (see the 1997 documentation for Carbon disulfide in Volume 12 of the present series).

After many years' exposure to carbon disulfide, increased blood pressure, angina pectoris, accelerated arteriosclerosis and higher mortality from myocardial infarct can be observed in viscose workers. There are no indications that hydrogen sulfide, which is always present at the same time during exposure in the viscose industry, plays any part. A NOEC for cardiotoxicity with long-term exposure cannot be derived with any certainty from the available studies, but it is probably around 12 mg/m$^3$ (about 4 ml/m$^3$) (see the 1997 documentation for Carbon disulfide in Volume 12 of the present series).

Various studies on workers exposed to carbon disulfide have been published since the 1997 documentation, necessitating a re-evaluation of the MAK value.

# 1 Effects in Man

## 1.1 Case reports

In four male patients, polyneuropathy occurred following 6–21 years' exposure to carbon disulfide (Huang *et al.* 2001). No data were given on the level of exposure.

Six male workers aged between 43 and 50 exposed to carbon disulfide for 4 to 23 years showed polyneuropathies and reduced nerve conduction velocities. The highest measured air concentrations were between 100 and 200 ml/m$^3$ (316 and 632 mg/m$^3$). Two years after the diagnosis, a biopsy of the *Nervus suralis* was carried out in one patient. This showed degeneration of axon und myelin, a loss of large myelinized fibres, but also remyelinization. Three years after the diagnosis, the carbon disulfide concentrations were only 10 to 20 ml/m$^3$ (32 to 63 mg/m$^3$). The reduced nerve conduction velocities showed a slight but insignificant improvement (Huang *et al.* 2002).

Four male workers aged between 50 and 58 exposed to carbon disulfide for 15 to 23 years (no data on level of exposure) showed signs of high blood pressure, coronary heart diseases, hypercholesterolaemia, hyperlipidaemia, atrial fibrillation and colour sense disturbances (Valic *et al.* 2001).

Thirteen workers aged between 21 and 63 were exposed to current carbon disulfide concentrations of between 20 and 50 ml/m$^3$ (63 and 158 mg/m$^3$) and peak concentrations of up to 100 or 200 ml/m$^3$ (316 or 632 mg/m$^3$). Five workers showed ECG changes (four of these workers were the most highly exposed), four workers increased cholesterol levels and a further four workers hearing impairments (Guidotti and Hoffman 1999).

Neuropsychological tests (Wechsler Adult Intelligence Scale) and magnetic resonance imaging (MRI) were carried out on patients with diagnosed carbon disulfide poisoning from a viscose factory which closed down in 1993. A total of 37 highly exposed patients (high level and long duration of exposure) and 37 slightly exposed patients were examined. There were no differences in the intelligence quotient between the groups. However, in 5 of 12 highly exposed persons (42 %), but in only 1 of 19 slightly exposed persons (5.3 %), an increased number of cerebral lacunae could be seen on the MRI images (Cho *et al.* 2002).

In another study, 91 patients with carbon disulfide poisoning were examined by MRI. In T2-weighted MRI images 70 patients (77 %) showed hyperintensities in the white cerebellar region and 27 patients lacunar infarcts (Cha *et al.* 2002).

After 27 years' exposure to carbon disulfide, a 56-year-old man developed headache, limb tremors, gait disturbance, dysarthria, memory impairment, and emotional lability. The brain MRI showed diffuse hyperintensity lesions in T2-weighted images in the subcortical white matter, basal ganglia, and brain stem. The brain computed tomography perfusion study revealed a diffusely decreased regional cerebral blood flow and prolonged regional mean transit time in the subcortical white matter and basal ganglion (Ku *et al.* 2003). Data on exposure levels are not given.

A man who worked in a viscose plant for 27 years showed, seven years after he had stopped working, coordination disturbances, dysmetria (wrong estimation of distance in target movements), speech disorders and adiadochokinesis (inability to perform antagonistic movements in rapid alternation). These were subsequently accompanied by cognitive impairment, memory disturbances, disorientation, emotional lability and paranoid-obsessive disturbances. The brain showed advanced atrophies of the cerebellum and less severe atrophies of the cortex (Fonte *et al.* 2003).

## 1.2 Cross-sectional studies

### 1.2.1 Effects on the nervous system

In a cross-sectional study carried out from 1992 to 1993 on a total of 432 Japanese viscose workers (on average 35.5 years old and exposed for 13.4 years) and 402 controls (average age 35.8), the prevalence of subjective symptoms (e.g. "feeling of heaviness" in the head, tremor in the limbs, numbness in the limbs, increased sensitivity of the skin, reduced strength of grip, lower sex drive, rough skin) was significantly higher than in the controls. The authors regard the increase as marginal and the significance for health to be unclear. In 30 workers from the sector "spinning/refining" (on average 34.9 years old and exposed for 13.8 years) the nerve conduction velocity was slightly but statistically significantly reduced compared with 123 workers from other sectors (on average 36.9 years old and exposed for 12.6 years) or with 402 controls (Takebayashi *et al.* 1998). In 1992, median carbon disulfide concentrations were 4.1 ml/m$^3$ (13 mg/m$^3$) and maximum concentrations up to 39.7 ml/m$^3$ (125 mg/m$^3$) (Omae *et al.* 1998). Prior to 1992 there were no data on exposure concentrations.

In a Chinese study, which is available as an English abstract, it was reported that a dose-dependent increase in electroneuromyographic anomalies in keeping with axonal polyneuropathy was observed among 175 workers exposed to carbon disulfide (Jiang *et al.* 1998).

Neuropsychological tests were carried out on 120 workers exposed to carbon disulfide and hydrogen sulfide and 67 controls. Depending on the activity, the carbon disulfide concentrations were between 3 mg/m$^3$ (1 ml/m$^3$) and 147 mg/m$^3$ (47 ml/m$^3$). Hydrogen sulfide exposure was lower than 14 mg/m$^3$. Only in the group of workers exposed to carbon disulfide concentrations of more than 90 mg/m$^3$ (about 28.5 ml/m$^3$) did psychomotor tests show significant impairments in the speed and quality of execution (De Fruyt *et al.* 1998).

There are also studies on heart rate variability, recording parameters of the vegetative nervous system, especially the sympathetic-parasympathetic equilibrium. Of 830 workers in a viscose factory with diagnosed carbon disulfide poisoning, 71 males were examined for heart rate variability and compared with 127 non-exposed persons. From the changes in the parameters the authors conclude that carbon disulfide can affect heart rate variability (Jhun *et al.* 2003).

152 male workers (24 to 66 years old and exposed for between 5 and 38 years) were exposed to a current carbon disulfide concentration of $18.1 \pm 15.8$ mg/m$^3$ ($5.7 \pm 5.0$ ml/m$^3$) and a maximum concentration of 109 mg/m$^3$ (34 ml/m$^3$). Compared with 93

controls, heart rate variability parameters were changed, indicating an impairment of neurovegetative regulation. In the group with the lowest carbon disulfide exposure of 1 to 10 mg/m$^3$ (0.3 to 3.2 ml/m$^3$), only one parameter (VLF power spectrum 0.0167–0.05 Hz) was reduced. In the intermediate exposure group of 10 to 18 mg/m$^3$ (3.2 to 5.7 ml/m$^3$) and the high exposure group (> 18 mg/m$^3$ or > 5.7 ml/m$^3$), several parameters were changed. However, individual parameters did not correlate with the current exposure. A correlation could be observed only if current exposure, length of exposure and cumulative lifetime exposure were taken into account together. The authors conclude that in addition to carbon disulfide other parameters may also be of significance (Bortkiewicz *et al.* 1997).

## Neurological effects on the eye

In 90 workers from three sectors (preparation, twisting, spinning) exposed to carbon disulfide for between 8.6 and 13.7 years on average, there was an increase in colour vision deficiency measured as the colour confusion index (CCI). The median CCI was 1.15 among viscose preparation workers, 1.11 among twisting workers and 1.24 among spinning workers. Average carbon disulfide concentrations in the air were 3.0 ml/m$^3$ (9.5 mg/m$^3$) in the preparation sector, 4.16 ml/m$^3$ (12.8 mg/m$^3$) in the twisting sector and 4.73 ml/m$^3$ (15.0 mg/m$^3$) in the spinning sector (Valic *et al.* 2001). It should be noted that a median CCI of 1.24 does not necessarily signify a clinical peculiarity. It is known from experience with persons exposed to toluene (HVBG 2001) that the Lanthony-D15d test used is influenced not only by the age of the persons examined but also by their level of qualification (CCI > 1.15 among workers) and that vocational practice in assessing colours has a training effect. Experience also shows that subjects with little knowledge of the language used in the test have difficulties in performing it. However, the study by Valic *et al.* (2001) does not give any information about the qualifications and the nationalities of persons examined.

191 male and 80 female workers at a viscose factory in China were exposed to current carbon disulfide concentrations of 13.7 to 20.05 mg/m$^3$ (4.3 to 6.4 ml/m$^3$). According to the authors, the exposure conditions had apparently not changed from the construction of the plant in the 1970s up to the time of the study (1997–1999). Male workers in particular reported increased burning of the eyes and hazy sight. The FM 100-Hue test results showed that the total error score of the exposed group, whether male or female, was higher than that of the control. Discrimination of the green and blue zones was also impaired significantly. The differences were significant in the case of men for the whole spectrum and for the green and blue zones and in the case of women only for the green zone; the lack of statistical significance might be attributable to the smaller number of women tested. Differentiated analyses, which took account of length of service, showed for controls that with increasing length of employment more errors in discriminating colours can be observed in both men and women. This would seem to indicate that colour discrimination deteriorates with age. This deterioration was most marked in the workers exposed to carbon disulfide compared with the controls in the group with the shortest length of employment (up to five years). With increasing length of employment (or age) the differences between controls and exposed persons

diminished. However, the poorest colour discrimination was observed with men exposed to carbon disulfide for 15–20 years (Wang *et al.* 2002). The interpretation of the results is made more difficult by the differing degrees of impairment of individual colour zones, by the fact that with increasing length of employment (or age) initial differences in colour discrimination between exposed persons and controls become smaller, and in particular by the failure to take account of possible confounders (especially age) in the analyses.

## Neurological effects on hearing

40 exposed workers with symptoms of chronic carbon disulfide poisoning (group I, on average 56.0 years old and exposed for 28.4 years), 40 exposed workers without carbon disulfide-specific symptoms (group II, on average 52.3 years old and exposed for 25.3 years) and 40 controls (on average 52.0 years old and employed elsewhere for 26.7 years) were tested for hearing ability. Current carbon disulfide concentrations were reported at 25.6 mg/m$^3$ (8.1 ml/m$^3$), in a range between 10 and 35 mg/m$^3$ (3.2 to 11 ml/m$^3$). The exposure to noise was above the permissible levels for both the workers exposed to carbon disulfide, at 88 to 92 dB(A), and the controls, at 86 to 93 dB(A). In group I the proportion of retrocochlear hearing reduction (hearing components lying behind the inner ear) was 97.5 %. In group II 45% of the workers showed retrocochlear hearing reduction, 32.5 % cochlear hearing reduction and 22.5 % no hearing reduction. In the control group there was a predominance of hearing reduction caused by cochlear damage (Kowalska *et al.* 2000). The authors do not give any precise data on the hearing ability of the control group. Nor does the study contain a statistical comparison between the measured hearing thresholds of group II and the controls. The authors interpret the results rather qualitatively and from the location of the damage conclude for group II that carbon disulfide exposure has a predominantly toxic effect. This study makes it clear that exposure periods of over 25 years and current carbon disulfide exposures of up to 35 mg/m$^3$ (11 ml/m$^3$) are associated with hearing impairments for which a toxic cause seems physiologically plausible. However, earlier carbon disulfide concentrations may have been higher.

Hearing loss was investigated in 131 men with exposure to noise (80–91 dB(A)) and carbon disulfide (1.6–20.1 ml/m$^3$) in a viscose rayon plant. The prevalence of > 25 dB hearing loss (dBHL) in rayon workers (67.9 %) was much higher than that in 110 administrative workers exposed to low noise (23.6 %) and in workers exposed to noise only (32.4 %). Hearing loss occurred mainly for speech frequencies of 0.5, 1, and 2 kHz. When the carbon disulfide exposure was measured by the product of carbon disulfide exposure level and employment years, the adjusted odds ratios (OR) of hearing loss of > 25 dBHL in rayon workers, compared with administrative workers, were 3.8 (95 % confidence interval (CI) 1.5–9.4) for those with the exposure of 37–214 year-ppm, 14.2 (95 % CI 4.4–45.9) for those with exposure of 215–453 year-ppm, and 70.3 (95 % CI 8.7–569.7) for those with exposure of > 453 year-ppm. The study suggests that carbon disulfide exposure enhances human hearing loss in a noisy environment and mainly affects hearing in lower frequencies (Chang *et al.* 2003).

## 1.2.2 Effects on the cardiovascular system

Various cross-sectional studies on cardiovascular effects in workers exposed to carbon disulfide have been published since the MAK documentation of 1997.

In 118 workers at a viscose factory in Taiwan exposed to current carbon disulfide concentrations of 0.1 to 54.6 ml/m$^3$ (0.3–173 mg/m$^3$), there were ECG changes compared with 44 controls (Kuo *et al.* 1997).

22 women (average age 42.5 years) chronically exposed (no further details) to carbon disulfide concentrations between 9.36 and 23.4 mg/m$^3$ (3.0–7.4 ml/m$^3$) showed in comparison with 20 women (average age 44.5 years) significant effects on the plasma lipid fraction and changes in the coagulation system (Stanosz *et al.* 1998).

In a group of 33 workers exposed to high levels of carbon disulfide of 50.6 ± 25.6 ml/m$^3$ (159.8 ± 76.8 mg/m$^3$) from two viscose factories in Taiwan, increased triglyceride and reduced HDL concentrations (after adjustment) were detected. In 64 moderately exposed workers (12.9 ± 5 ml/m$^3$; 40.7 ± 15.8 mg/m$^3$) and 35 low-exposed workers (3.5 ± 1.2 ml/m$^3$; 11.6 ± 3.7 mg/m$^3$), no increases were detected. The average length of exposure for all the groups was more than 21 years (Luo *et al.* 2003).

In order to test the hypothesis whether carbon disulfide has a weak negative inotropic effect on the heart muscle (Drexler *et al.* 1996), the diameter of the left ventricle of 325 workers exposed to carbon disulfide (on average 37 years old and exposed for 10 years) and 179 controls (average age 35 years) was determined by echocardiograph. The current carbon disulfide concentration in the air was given as 6 ml/m$^3$ (19 mg/m$^3$), with a range of 0.03–91.08 ml/m$^3$ (0.09–288 mg/m$^3$). The diameter of the left ventricle tended to be larger in the exposed workers than in the controls, but not to a significant extent. An evaluation by multiple linear regression showed that the diameter of the left ventricle correlates with the body mass index, body fat, alcohol consumption and physical fitness, but not with internal or external carbon disulfide exposure (Korinth *et al.* 2003).

In a group of 177 workers (on average 44 years old and exposed for between 5 and 38 years) and 93 controls (average age 41 years), the at-rest or 24-hour ECG, blood pressure and heart rate variability were recorded. Changes in the at-rest or 24-hour ECG were particularly apparent in workers with more than 20 years' exposure (Bortkiewicz *et al.* 2001). A further differentiation of the workers due to the level of exposure was not made.

252 workers in a Bulgarian viscose factory (111 men and 141 women, average age 42.4 years and exposed for at least 1 year) were compared with 252 controls of the same age and sex. Current carbon disulfide concentrations were between 10 and 64 mg/m$^3$ (3.2–20.3 ml/m$^3$). By multiplying the workplace-specific exposure concentration (in mg/m$^3$) by the length of exposure (in years) the authors calculated cumulative exposure indices. The odds ratios for elevated cholesterol were significantly higher with both a cumulative exposure index below 300 mg/m$^3$ at 2.19 (95 % CI 1.79–4.75; n=134) and a cumulative exposure index over 300 mg/m$^3$ at 3.05 (95 % CI 1.84–5.09; n=118). In the high-exposure group the odds ratio for coronary heart diseases was also significantly higher at 4.32 (95 % CI 1.84–10.11) (Kotseva and De Bacquer 2000).

In a later publication 141 workers (64 men and 77 women, on average 42.6 years old and exposed for at least 1 year) were compared with 141 similar controls. This is presumably a subgroup of the workers described by Kotseva and De Bacquer (2000). Current carbon disulfide concentrations were given as 1 to 30 mg/m$^3$ (about 0.3–9.5 ml/m$^3$) for this group. With a cumulative exposure index over 100 mg/m$^3$ the odds ratio for elevated cholesterol was significantly higher at 5.52 (95 % CI 2.81–10.83; n=70), but not with a cumulative exposure index under 100 mg/m$^3$. The author concludes that carbon disulfide concentrations under 31 mg/m$^3$ (about 10 ml/m$^3$) and cumulative exposure indices over 100 mg/m$^3$ can lead to higher cholesterol levels (Kotseva 2001). There are no data on earlier carbon disulfide concentrations.

Several studies were carried out in a Belgian viscose factory. One of them was published in 1992 (Vanhoorne *et al.* 1992, the 1997 documentation for Carbon disulfide in Volume 12 of the present series). For workers exposed to current carbon disulfide concentrations of 4 to 113 mg/m$^3$ (1.2–36 ml/m$^3$) increased blood pressure, reduced HDL cholesterol and higher apolipoproteins A1 and B were detected.

In more recent studies the current carbon disulfide concentrations in viscose processing were 1.3 to 20.8 mg/m$^3$ (0.4–6.6 ml/m$^3$) and in spinning 3.9 to 32.4 mg/m$^3$ (1.2–10.2 ml/m$^3$). In breathing masks the carbon disulfide concentrations in viscose processing were 6.0 to 10.09 mg/m$^3$ (about 1.9–3.2 ml/m$^3$) and in spinning 2.2 to 13.02 mg/m$^3$ (0.7–4.1 ml/m$^3$). The results for 91 exposed male workers (average age 41.1 years) were compared with those for 81 male controls (average age 39.5 years). There were no cardiovascular effects in workers with cumulative exposure indices under 150 mg/m$^3$ (n=47), but with cumulative exposure indices over 150 mg/m$^3$ (n=43) there were higher adjusted odds ratios for coronary heart diseases (OR$_{adj}$: 18.20; 95 % CI 2.44–135), ischaemic ECG (OR$_{adj}$: 8.48; 95 % CI 1.25–57.26) and ischaemia (OR$_{adj}$: 4.65; 95 % CI 1.60–13.54) (Kotseva *et al.* 2001a).

In order to determine earlier changes in the elasticity of arteries and functional changes in the vessel walls, 85 male workers (average age 37.1 years) and 37 controls (average age 40.5 years) from the same viscose plant were examined. In the exposed workers the distensibility of the vessel walls and the heart rate were increased (Braeckman *et al.* 2001, Kotseva *et al.* 2001b).

In 367 workers exposed to carbon disulfide concentrations of 15.47 ± 2.34 mg/m$^3$ (4.9 ± 0.74 ml/m$^3$) no significant increase in clinical complaints and abnormal electrocardiograms was noted compared to 125 reference workers. Nor were significant effects of carbon disulfide on blood pressure, cholesterol, HDL and LDL cholesterol or triglycerides detected (Tan et al. 2004).

In a study of a total of 432 Japanese viscose workers the prevalence of microaneurisms of the retinal artery at 8.1 % was significantly higher than for 402 controls at 3.4 %. The prevalence of arteriosclerosis was not increased (Takebayashi *et al.* 1998). The median carbon disulfide concentration was 4.1 ml/m$^3$ (13 mg/m$^3$) and the maximum concentration 39.7 ml/m$^3$ (125 mg/m$^3$) (Omae *et al.* 1998).

In an overview study the published literature on cardiovascular effects of carbon disulfide following occupational exposure was critically assessed (Sulsky *et al.* 2002). The evaluation covered 37 publications; 12 publications were excluded on account of methodological shortcomings (unsatisfactory presentation of exposure or the results, manifest selection bias without correction, no consideration of important confounders). Following a critical appraisal of the individual studies, specific hypotheses were tested on the basis of the evidence and consistency of the findings. A meta-analysis was not carried out. An important objective of this work was to describe the risks of cardiovascular diseases at various exposure concentrations. Concentrations over 20 ml/m$^3$ (63 mg/m$^3$) were regarded as "high-level" exposure, concentrations below that as "low-level" exposure. In 15 studies risk factors for cardiovascular diseases were examined at concentrations below 20 ml/m$^3$, with 11 of them showing at least one weak point. The four most significant studies were identified as Drexler *et al.* (1995), Kotseva (2001), Kotseva and De Bacquer (2000) and Kotseva *et al.* (2001a). In these studies a multivariant adjustment for important confounders (such as sex, smoking, alcohol, stress, shift work, noise, training), a dose-effect analysis and a classification of exposure were carried out. The authors conclude that there are no strong or consistent associations between carbon disulfide exposure and coronary heart disease mortality or relevant risk factors. Even at higher exposure concentrations the associations with coronary heart diseases or other clinical indicators were different, not very consistent and to some extent contradictory. At low exposure concentrations the only relatively consistent finding, which was also backed up by findings at high concentrations, was an increase in total or LDL cholesterol. As this effect was shown in only half of the studies (of comparable quality), it would seem that other factors than carbon disulfide exposure alone may play a part. It is unclear whether a slight increase in these parameters, even if they were associated with exposure, contributes to a disease or whether it is a physiological response to the exposure (Sulsky *et al.* 2002).

### 1.2.3 Other effects

In 29 men exposed to carbon disulfide for more than 20 years and in 24 patients with peripheral arteriosclerosis, but not in 30 non-exposed healthy subjects, there were indications of lipid peroxidation in the plasma (increased concentration of thiobarbituric reactive substances, reduced concentrations of α-tocopherol, glutathione-peroxidase, catalase). The carbon disulfide concentrations were between 30 and 80 mg/m$^3$ (9.5 to 25.4 ml/m$^3$) in the 1960s, between 22 and 45 mg/m$^3$ (7.0 to 14.3 ml/m$^3$) in the 1970s and over 18 mg/m$^3$ (5.7 ml/m$^3$) in later years (Wronska-Nofer *et al.* 2002). The increased incidence of coronary heart diseases observed in workers exposed to carbon disulfide is attributed by the authors to oxidative stress, leading to arteriosclerosis.

In a group of 67 workers and 88 controls biochemical indications of oxidative stress were obtained both in the case of workers exposed to carbon disulfide concentrations below 10 mg/m$^3$ (3.2 ml/m$^3$) (CuZn superoxide-dismutase increased, malondialdehyde reduced) and at concentrations over 10 mg/m$^3$ (3.2 ml/m$^3$) (CuZn superoxide-dismutase and malondialdehyde increased) (Jian and Hu 2000).

In workers exposed to carbon disulfide, the NO concentration in the plasma was significantly lower at 43.28 µmol/l (high-exposure group) and 50.00 µmol/l (low-exposure group) compared with controls at 70.66 µmol/l. The superoxide-dismutase activities in the erythrocytes were significantly higher in the high and low-exposure groups at 4832 U/g Hb and 3520 U/g Hb, respectively, than in the control group at 2425 U/g Hb. The concentrations of lipid peroxides in the plasma were also significantly higher in the high and low-exposure groups at 19.38 and 17.09 µmol/l, respectively, compared with the controls (4.37 µmol/l). The authors conclude that occupational long-term exposure to carbon disulfide reduces NO concentrations and increases those of the superoxide radicals (Jiang *et al.* 2000).

50 workers exposed to carbon disulfide at a concentration of $14.4 \pm 4.62$ mg/m$^3$ ($4.6 \pm 1.5$ ml/m$^3$) in a viscose factory showed in comparison with the control group reduced serum FSH ($7.50 \pm 7.07$ IU/l to $10.4 \pm 7.35$ IU/l) and prolactin ($75.72 \pm 4.18$ ng/l to $6.89 \pm 4.64$ ng/l). With increasing length of exposure serum LH decreased, as did serum FSH with increasing level of exposure (measured as TTCA elimination in urine) (Wang *et al.* 1999).

## 1.3 Cohort studies

There is a mortality study of workers in a Polish viscose factory. Male workers were engaged in production or the maintenance of equipment for at least one year in the period between 1950 and 1985 and lived in the vicinity of the factory. 43.7 % of the workers were employed for more than 21 years, while the average length of employment was 17.3 years. Over 50 % of the workers were considered to be severely exposed. Of 2878 workers in the cohort, 2762 were monitored, giving a total of 76465 man-years. The male population of Poland was taken as the reference. For the cohort there was a significantly increased SMR of 108 (95 % CI 101–115) for all fatalities. There was a significant increase in the risks of mortality from cardiovascular (SMR 114, 95 % CI 104–125) or cerebrovascular (SMR 208, 95 % CI 170–252) diseases. For the groups of highly exposed workers and workers in the spinning plant who were first employed before 1974, the risks of dying from a disease of the circulatory system were significantly increased, with SMRs of 209 (95 % CI 157–273) and 204 (95 % CI 138–291). There was also a statistically significant trend in mortality from all cardiovascular diseases in relation to the level of exposure. No clear relationship was apparent between the length of exposure and increased mortality (Peplońska *et al.* 2001).

A number of publications are available on a cohort of workers from 11 Japanese viscose rayon factories. In an initial survey in 1992 and 1993, a total of 432 viscose workers and 402 control workers were covered (see cross-sectional studies by Omae *et al.* (1998) and Takebayashi *et al.* (1998)). The median carbon disulfide concentration in 1992 was 4.1 ml/m$^3$ (13 mg/m$^3$) and the maximum concentration 39.7 ml/m$^3$ (125 mg/m$^3$) (Omae *et al.* 1998). Following these cross-sectional studies, longitudinal studies (cohort studies) were carried out. In 1994 and 1995 some viscose factories were shut down, so that 140 viscose workers were no longer exposed (ex-exposed workers). The other workers in the cohort study were exposed until the end of the study in 1998 or 1999 (exposed workers).

Cerebrovascular effects of carbon disulfide exposures were studied by brain magnetic resonance imaging to determine hyperintense spots (HIS, "silent cerebral infarctions"). The six-year follow-up study comprised 217 exposed workers, 125 ex-exposed workers and 324 referent workers. Mean duration of exposure was 19.6 years. The geometric mean exposure to carbon disulfide was 4.9 ml/m$^3$ (15.5 mg/m$^3$). The exposure level was

2.5 ml/m$^3$ (7.9 mg/m$^3$) in the lowest quartile and 8.1 ml/m$^3$ (25.6 mg/m$^3$) in the highest quartile. No significant difference in the prevalence of HIS was observed in the exposed or ex-exposed workers compared to reference workers. The proportion of increases in HIS in the follow-up period was 24, 15, and 12 % in the exposed, ex-exposed and referent workers, respectively. The multivariate adjusted odds ratio (95 % CL) was significantly increased in exposed (2.27 (1.37–3.76)), but not in the ex-exposed workers (1.33 (0.70–2.54)) (Nishiwaki *et al.* 2004). The authors of the study conclude that the results should be interpreted cautiously due to lack of an exposure-response relation and due to several potential limitations of the study. In addition, the relative newness of the application of MRI scans in the identification of toxic responses with the lack of criteria for the definition of adverse effects indicates that the results cannot be taken into account for setting an OEL at present.

Cardiovascular effects of carbon disulfide exposures were examined in 251 exposed, 140 ex-exposed and 359 referent workers. The incidence of coronary artery ischaemia, defined as Minnesota electrocardiographic codes I, IV$_{1-3}$, V$_{1-3}$ (tested at rest and after step test), or treatment received for ischaemia, was 6.4 % in referent workers and 11.8 % in exposed workers. When exposures were divided into quartiles based on the urinary TTCA concentration, the confounder-adjusted odds ratio for ischaemic findings was significantly increased in the highest quartile (4.2 (1.8–9.7)) corresponding to an exposure level of 8.7 ml/m$^3$ (27.5 mg/m$^3$). In the age restricted groups (< 35 years), the incidence of ischaemic findings was 3.8 % in the non-exposed group and 12.4 % in the exposed group. This indicates that recent exposure contributed to the development of the ischaemic findings. However, with more rigorous criteria for defining ischaemia (ST depression ≥ 2 mm) both incidence and prevalence of ischaemic findings were comparable among the exposed and non-exposed groups. In addition to ischaemic findings, the incidence of retinal microaneurysm was increased in exposed workers (9.2 %) compared to referent workers (5.0 %). In the highest quartile the confounder adjusted odds ratio for microaneurysms was increased significantly (3.8 (1.2–11.4)) (Takebayashi *et al.* 2004). The value of exposure-related findings with the Minnesota code is questioned, as with the application of generally accepted rigorous ECG criteria (ST depression ≥ 2 mm) no significant effects of carbon disulfide exposure were found. The incidence of retinal microaneurysm was increased with marginal significance in the high exposure group (8.7 ml/m$^3$) and has been reduced over time. In 1998 peak exposure was up to 30 ml/m$^3$ and may be responsible for the effects observed. In addition, the clinical relevance of this marginally statistically significant increase seems to be low.

In order to ascertain endocrinal changes, 259 workers exposed to carbon disulfide, 133 ex-exposed workers, and 352 controls were examined. The parameters measured with regard to glucose metabolism (glucose after fasting, insulin, HbA), the pituitary hormones (luteinising hormone, LH; follicle-stimulating hormone, FSH; adrenocorticotropic hormone, ACTH), the male sex organs (testosterone), the thyroid (triiodothyronine, T$_3$; tetraiodothyronine, T$_4$; thyroid-stimulating hormone, TSH; thyroxine-binding α-globulin, TBG) and subjective data on reduced sexual activity showed no biologically relevant changes between the groups (Takebayashi *et al.* 2003).

The cohort studies of workers in the Japanese viscose rayon industry thus show only in the group of workers exposed to carbon disulfide concentrations of more than 8 ml/m$^3$

($25\ mg/m^3$) initial indications of possible carbon disulfide-induced subclinical changes such as increased ischaemic findings in the ECG using the Minnesota code, retinal microaneurisms and hyperintense spots in T2-weighted MRI images of the brain. The study indicates no clinically relevant effects at mean exposure concentrations of $5\ ml/m^3$ ($16\ ml/m^3$) carbon disulfide.

A meta-analysis of cohort studies on cardiovascular effects (Tan *et al.* 2002) considers only older studies, which are already described in the earlier MAK documentation.


# 2 Manifesto (MAK value/classification)

In 1997 the MAK value was lowered from 10 to $5\ ml/m^3$ (30 to $16\ mg/m^3$), as there had been indications that with several decades' exposure at the workplace neurotoxic and cardiotoxic effects can occur in a concentration range below $10\ ml/m^3$ ($30\ mg/m^3$). For 40 years' exposure one working group estimated a NOEC for neurotoxicity of $4\ ml/m^3$ ($12\ mg/m^3$) (Ruijten *et al.* 1990). Another working group found at $4\ ml/m^3$ ($12\ mg/m^3$) no increase in blood lipids, no indication of an increase in the prevalence of coronary heart disease or arteriosclerotic findings, no difference in the distribution of the performance of ergometric tests (Drexler *et al.* 1995, 1996) and also no signs of effects on the nervous system (Reinhardt *et al.* 1997a, 1997b). A provisional MAK value was therefore set at $5\ ml/m^3$ ($16\ mg/m^3$). The main problem in setting a NOEC was and still is that, although the current concentrations are relatively well documented at the time of each study, earlier exposure concentrations at the workplace, which are of significance on account of the cumulative effect of carbon disulfide, are not available and can only be estimated. Previously, they were appreciably higher for the most part. Moreover, the usual method of measuring carbon disulfide in the air has considerable shortcomings, and it used to be quite common that the values measured were lower than actual exposure.

In cohort studies of workers in 11 Japanese viscose rayon factories, carried out from 1992 to 1999, signs of possible carbon disulfide-induced subclinical changes, such as ischaemic ECG findings, retinal microaneurisms (Takebayashi *et al.* 2004) and hyperintense spots in T2-weighted MRT brain images (Nishiwaki *et al.* 2004) were observed in workers of the highest exposure category of more than $8\ ml/m^3$ ($25\ mg/m^3$). No effects were observed in groups of workers exposed to $5\ ml/m^3$ ($16\ mg/m^3$) of carbon disulfide. These cohort studies support the MAK value established in 1997 of $5\ ml/m^3$ ($16\ mg/m^3$).

Well-performed cross-sectional studies did not show relevant cardiovascular effects at carbon disulfide concentrations of $5\ ml/m^3$ ($16\ mg/m^3$) (Tan *et al.* 2004) or $6\ ml/m^3$ ($19\ mg/m^3$) (Korinth *et al.* 2003). Cardiovascular or neurotoxic effects were mostly observed after long-term exposure to carbon disulfide at higher concentrations. However, some cardiovascular or neurological effects were reported at carbon disulfide concentrations of 3 to $10\ ml/m^3$ (9.5 to $32\ mg/m^3$). Due to methodological limitations, questionable clinical relevance of the findings, or higher exposure concentrations in previous years, these results cannot be used for lowering the MAK value below 5 ml/m³ (16 mg/m³). The provisional MAK value of 5 ml/m³ (16 mg/m³) will therefore be retained. Further research is required into effects at concentrations in the range of the MAK value.

For the limitation of exposure peaks carbon disulfide is classified in Peak limitation category II with an excursion factor of 2.

As prenatal toxic effects cannot be excluded in the concentration range of 5 ml/m³ (16 mg/m³), classification of carbon disulfide in Pregnancy risk group B is retained.

Designation with "H" has been retained because of the good dermal absorption of carbon disulfide.

There are no reports of sensitization of the skin or respiratory tract, nor have such results of animal experiments been published. Therefore carbon disulfide has not been designated with either "Sh" or "Sa".

# 3 References


Bortkiewicz A, Gadzicka E, Szymczak W (1997) Heart rate variability in workers exposed to carbon disulfide. *J Auton Nerv Syst 66*: 62–68

Bortkiewicz A, Gadzicka E, Szymczak W (2001) Cardiovascular disturbances in workers exposed to carbon disulfide. *Appl Occup Environ Hyg 16*: 455–463

Braeckman L, Kotseva K, Duprez D, De Bacquer D, De Buyzere M, Van De Veire N, Vanhoorne M (2001) Vascular changes in workers exposed to carbon disulfide. *Ann Acad Med Singapore 30*: 475–480

BUA (Beratergremium für umweltrelevante Altstoffe der Gesellschaft deutscher Chemiker) (1991) *Carbon disulfide, BUA report 83*. S. Hirzel, Wissenschaftliche Verlagsgesellschaft, Stuttgart

Cha JH, Kim SS, Han H, Kim RH, Yim SH, Kim MJ (2002) Brain MRI findings of carbon disulfide poisoning. *Korean J Radiol 3*: 158–162

Chang SJ, Shih TS, Chou TC, Chen CJ, Chang HY, Sung FC (2003) Hearing loss in workers exposed to carbon disulfide and noise. *Environ Health Perspect 111*: 1620–1624

Cho SK, Kim RH, Yim SH, Tak SW, Lee YK, Son MA (2002) Long-term neuropsychological effects and MRI findings in patients with $CS_2$ poisoning. *Acta Neurol Scand 106*: 269–275

De Fruyt F, Thiery E, De Bacquer D, Vanhoorne M (1998) Neuropsychological effects of occupational exposures to carbon disulfide and hydrogen sulfide. *Int J Occup Environ Health 4*: 139–146

Drexler H, Ulm K, Hubmann M, Hardt R, Göen T, Mondorf W, Lang E, Angerer J, Lehnert G (1995) Carbon disulphide. III. Risk factors for coronary heart diseases in workers in the viscose industry. *Int Arch Occup Environ Health 67*: 243–252

Drexler H, Ulm K, Hardt R, Hubmann M, Göen T, Lang E, Angerer J, Lehnert G (1996) Carbon disulphide. IV. Cardiovascular function in workers in the viscose industry. *Int Arch Occup Environ Health 69*: 27–32

Fonte R, Edallo A, Candura SM (2003) Cerebellar atrophy as a delayed manifestation of chronic carbon disulfide poisoning. *Ind Health 41*: 43–47

Guidotti TL, Hoffman H (1999) Indicators of cardiovascular risk among workers exposed to high intermittent levels of carbon disulphide. *Occup Med 49*: 507–515

Henschler D (Ed.) (1975) Schwefelkohlenstoffe (Carbon disulfide) (German). in: *Toxikologisch-arbeitsmedizinische Begründungen von MAK-Werten*, 4th issue. VCH-Verlagsgesellschaft, Weinheim

Hirata M, Ogawa Y, Okayama A, Goto S (1992) A cross-sectional study on the brainstem auditory evoked potential among workers exposed to carbon disulfide. *Int Arch Occup Environ Health* 64: 321–324

Huang CC, Chu CC, Chu NS, Wu TN (2001) Carbon disulfide vasculopathy: a small vessel disease. *Cerebrovasc Dis 11*: 245–250

Huang CC, Chu CC, Wu TN, Shih TS, Chu NS (2002) Clinical course in patients with chronic carbon disulfide polyneuropathy. *Clin Neurol Neurosurg 104*: 115–120

HVBG (Hauptverband der gewerblichen Berufsgenossenschaften) (2001) *Toluol in Tiefdruckereien. Abschlussbericht zu einem Forschungsprojekt.* (Toluene in rotogravure printing operations. Final report on a research project) (German). HVBG, Sankt Augustin

Jhun H-J, Yim S-H, Kim R, Paek D (2003) Heart-rate variability of carbon disulfide-poisoned subjects in Korea. *Int Arch Occup Environ Health 76*: 156–160

Jian L, Hu D (2000) Antioxidative stress response in workers exposed to carbon disulfide. *Int Arch Occup Environ Health 73*: 503–506

Jiang B, Zhang S, Huang J, Lu J (1998) Study on electroneuromyography of 175 workers exposed to carbon disulfide (Chinese). *Wei Sheng Yan Jiu 27*: 84–86

Jiang X, Jiann L, Hu D (2000) A preliminary study on body level of nitric oxide in workers exposed to carbon disulfide (Chinese). *Zhonghua Yu Fang Yi Xue Za Zhi 34*: 348–350

Korinth G, Göen T, Ulm K, Hardt R, Hubmann M, Drexler H (2003) Cardiovascular function of workers exposed to carbon disulphide. *Int Arch Occup Environ Health 76*: 81–85

Kotseva K (2001) Occupational exposure to low concentrations of carbon disulfide as a risk factor for hypercholesterolaemia. *Int Arch Occup Environ Health 74*: 38–42

Kotseva KP, De Bacquer D (2000) Cardiovascular effects of occupational exposure to carbon disulphide. Occup Med 50: 43–47

Kotseva K, Braeckman L, De Bacquer D, Bulat P, Vanhoorne M (2001a) Cardiovascular effects in viscose rayon workers exposed to carbon disulfide. *Int J Occup Environ Health 7*: 7–13

Kotseva K, Braeckman L, Duprez D, De Bacquer D, De Buyzere M, Van De Veire N, Vanhoorne M (2001b) Decreased carotid artery distensibility as a sign of early atherosclerosis in viscose rayon workers. *Occup Med 51*: 223–229

Kowalska S, Sulkowski W, Sinczuk-Walczak H (2000) Assessment of the hearing system in workers chronically exposed to carbon disulfide and noise (Polish). *Med Pra 51*: 123–138

Ku MC, Huang CC, Kuo HC, Yen TC, Chen CJ, Shih TS, Chang HY (2003) Diffuse white matter lesions in carbon disulfide intoxication: microangiopathy or demyelination. *Eur Neurol 50*: 220–224

Kuo HW, Lai JS, Lin M, Su ES (1997) Effects of exposure to carbon disulfide ($CS_2$) on electrocardiographic features of ischemic heart disease among viscose rayon factory workers. *Int Arch Occup Environ Health 70*: 61–66

Luo J-CJ, Chang H-Y, Chang S-J, Chou T-C, Chen C-J, Shih T-S, Huang C-C (2003) Elevated triglyceride and decreased high density lipoprotein level in carbon disulfide workers in Taiwan. *J Ocuup Environ Med 45*: 73–78

Nishiwaki Y, Takebayashi T, O'Uchi T, Nomiyama T, Uemura T, Sakurai H, Omae K (2004) Six year observational cohort study of the effect of carbon disulphide on brain MRI in rayon manufacturing workers. *Occup Environ Med 61*: 225–232

Omae K, Takebayashi T, Nomiyama T, Ishizuka C, Nakashima H, Uemura T, Tanaka S, Yamauchi T, O'Uchi T, Horichi Y, Sakurai H (1998) Cross sectional observation of the effects of carbon disulphide on arteriosclerosis in rayon manufacturing workers. *Occup Environ Med 55*: 468–472

Peplońska B, Sobala W, Szeszenia-Dabrowska N (2001) Mortality pattern in the cohort of workers exposed to carbon disulfide. *Int J Occup Med Environ Health 14*: 267–274

Reinhardt F, Drexler H, Bickel A, Claus D, Angerer J, Ulm K, Lehnert G, Neundorfer B (1997a) Neurotoxicity of long-term low-level exposure to carbon disulphide: results of questionnaire, clinical neurological examination and neuropsychological testing. *Int Arch Occup Environ Health 69*: 332–338

Reinhardt F, Drexler H, Bickel A, Claus D, Ulm K, Angerer J, Lehnert G, Neundorfer B (1997b) Electrophysiological investigation of central, peripheral and autonomic nerve function in workers with long-term low-level exposure to carbon disulphide in the viscose industry. *Int Arch Occup Environ Health 70*: 249–256

Ruijten MWMM, Sallé HJA, Verbek MM, Muijser H (1990) Special nerve function and colour discrimination in workers with long term low level exposure to carbon disulphide. *Br J Ind Med 47*: 589–595

Stanosz S, Kuligowska E, Kuligowski D (1998) Coefficient of linear correlation between levels of fibrinogen, antithrombin III, thrombin-antithrombin complex and lipid fractions in women exposed chronically to carbon disulfide (Polish). *Med Pra 49*: 51–57

Sulsky SI, Hooven FH, Burch MT, Mundt KA (2002) Critical review of the epidemiological literature on the potential cardiovascular effects of occupational carbon disulfide exposure. *Int Arch Occup Environ Health 75*: 365–380

Takebayashi T, Omae K, Ishizuka C, Nomiyama T, Sakurai H (1998) Cross sectional observation of the effects of carbon disulphide on the nervous system, endocrine system, and subjective symptoms in rayon manufacturing workers. *Occup Environ Med 55*: 473–479

Takebayashi T, Nishiwaki Y, Nomiyama T, Uemura T, Yamauchi T, Tanaka S, Sakurai H, Omae K (2003) Lack of relationship between occupational exposure to carbon disulfide and endocrine dysfunction: a six-year cohort study of the Japanese rayon workers. *J Occup Health 45*: 111–118

Takebayashi T, Nishiwaki Y, Uemura T, Nakashima H, Nomiyama T, Sakurai H, Omae K (2004) A six year follow up study of the subclinical effects of carbon disulphide exposure on the cardiovascular system. *Occup Environ Med 61*: 127–134

Tan X, Peng X, Wang F, Joyeux M, Hartemann P (2002) Cardiovascular effects of carbon disulfide: meta-analysis of cohort studies. *Int J Hyg Environ Health 205*: 473–477

Tan X, Chen G, Peng X, Wang F, Bi Y, Tao N, Wang C, Yan J, Ma S, Cao Z, He J, Yi P, Braeckman L, Vanhoorne M (2004) Cross-sectional study of cardiovascular effects of carbon disulfide among Chinese workers of a viscose factory. *Int J Hyg Environ Health 207*: 217–225

Valic E, Pilger A, Pirsch P, Pospischil E, Waldhör T, Rüdiger HW, Wolf C (2001) Nachweis erworbener Farbsinnstörungen bei CS$_2$ exponierten Arbeitern in der Viskoseproduktion (The detection of acquired colour vision deficiency in workers exposed to CS$_2$ during the production of viscose) (German). *Arbeitsmed Sozialmed Umweltmed 36*: 59–63

Vanhoorne M, De Bacquer D, De Backer G (1992) Epidemiological study of the cardiovascular effects of carbon disulphide. *Int J Epidemiol 21*: 745–752

Vanhoorne M, De Rouck A, De Bacquer D (1995) Epidemiological study of eye irritation by hydrogen sulphide and/or carbon disulphide exposure in viscose rayon workers. *Ann Occup Hyg 39*: 307–315

Vanhoorne M, De Rouck A, Bacquer D (1996) Epidemiological study of the systemic ophthalmological effects of carbon disulfide. *Arch Environ Health 51*: 181–188

Wang C, Bi Y, Tan X, Yan J (1999) Serum sex hormone and urinary metabolites of male workers exposed to carbon disulfide (Chinese). *Wei Sheng Yan Jiu 30*: 132–133

Wang C, Tan X, Bi Y, Su Y, Yan J, Ma S, He J, Braeckman L, De Bacquer D, Wang F, Vanhoorne D (2002) Cross-sectional study of the ophthalmological effects of carbon disulfide in Chinese viscose workers. *Int J Hyg Environ Health 205*: 367–372

Wronska-Nofer T, Chojnowska-Jezierska J, Nofer JR, Halatek T, Wisniewska-Knypl J (2002) Increased oxidative stress in subjects exposed to carbon disulfide (CS$_2$) - an occupational coronary risk factor. *Arch Toxicol 76*: 152–157

completed 28.11.2002; updated 10.11.2004

# Cereal flour dusts

# Rye and Wheat

| | |
|---|---|
| **MAK value** | – |
| **Peak limitation** | – |
| **Absorption through the skin** | – |
| **Sensitization (1995)** | **Sa** |
| **Carcinogenicity** | – |
| **Prenatal toxicity** | – |
| **Germ cell mutagenicity** | – |
| **BAT value** | – |

## 1 Allergenic Effects

Disease symptoms caused by cereal flour dusts have been known for hundreds of years. In 1713 Ramazzini used bakers' asthma as an example when he described this obstructive respiratory disorder for the first time as a typical occupational disease. As early as 1909 it was possible to demonstrate immediate type reactions to wheat in skin tests carried out on bakers with respiratory disorders. Bakers' asthma has been defined as an occupational disease in its own right since 1929 (O'Hollaren 1992).

The symptoms induced by various cereal flour dusts (conjunctivitis, rhinitis, bronchial asthma, protein dermatitis) are almost entirely IgE-mediated. Obstructive respiratory disorders resulting from allergic reactions to wheat flour and rye flour are currently the respiratory allergies most frequently notified as occupational diseases (HVGB 1985). Persons with atopic diathesis are most at risk (Drexler *et al.* 1992). Other factors which affect the incidence of sensitizations are workplace hygiene and duration of exposure.

In Germany, bakers are most frequently shown to be sensitized to allergic components of rye flour and wheat flour. In a study of 261 persons with symptoms who were employed in bakeries and cake shops, IgE-mediated sensitization to rye flour or to wheat flour was demonstrated in both cases in 57 % of the tested persons with the radioallergosorbent test (RAST) (Bauer 1991). Rye flour components are the more potent

*Essential MAK Value Documentations.* DFG, Deutsche Forschungsgemeinschaft
Copyright © 2006 WILEY-VCH Verlag GmbH & Co. KGaA, Weinheim
ISBN: 3-527-31394-X

allergens. The allergens in wheat flour and rye flour produce frequent allergic cross-reactions, but to date little is known of the structures involved (König 1991). Clinically relevant sensitizations to all four main protein fractions of wheat (albumin, globulin, gliadin, glutenin) have been demonstrated (Walsh *et al.* 1985). An alpha-amylase inhibitor has also been identified as an allergen in wheat flour (Fränken *et al.* 1994).

## 2 Manifesto (sensitization)

Wheat flour dust and rye flour dust cause respiratory allergies in man and are therefore designated with an "Sa" (airway sensitization).

## 3 References

Baur X (1991) Klinisch relevante Allergene bei Bäckerasthma. in: Berufsgenossenschaft Nahrungsmittel und Gaststätten (Ed.) *Bäckerasthma*, Knopf-Druck, Erdingen-Neckarhausen

Drexler H, Escher S, Schmelz M, Weltle D, Lehnert G (1992) Atopie und berufsbedingte Atemwegsallergien vom Soforttyp. *Arbeitsmed Sozialmed Präventivmed 27*: 368–376

Fränken J, Stephan U, Meyer HE, König W (1994) Identification of alpha-amylase inhibitor as a major allergen of wheat flour. *Int Arch Allergy Immunol 104*: 171–174

HVGB (Hauptverband der gewerblichen Berufsgenossenschaften) (1985) *Das Berufskrankheitengeschehen*, Schriftenreihe des Hauptverbandes der gewerblichen Berufgenossenschaften, St. Augustin

König W (1991) Immunologische Grundlagen zur IgE-Regulation. in: Berufsgenossenschaft Nahrungsmittel und Gaststätten (Ed.) *Bäckerasthma*, Knopf-Druck, Erdingen-Neckarhausen

O'Hollaren MT (1992) Bakers' asthma and reactions secondary to soybean and grain dusts. in: Bardana EJ, Montanaro A, O'Hollaren MT (Eds) *Occupational Asthma*, Hanley & Belfus, Philadelphia

Walsh BJ, Wrigley CW, Musk AW, Baldo BA (1985) A comparison of the binding of the IgE in sera of patients with bakers' asthma to soluble and insoluble wheat-grain proteins. *J Allergy Clin Immunol 76*: 23–28

completed 23.03.1995

# Chloroform

| | |
|---|---|
| **MAK value (1999)** | **0.5 ml/m$^3$ $\hat{=}$ 2.5 mg/m$^3$** |
| **Peak limitation (1983)** | **II,1** |
| **Absorption through the skin (1999)** | **H** |
| **Sensitization** | **–** |
| **Carcinogenicity (1999)** | **Category 4** |
| **Prenatal toxicity (1999)** | **C** |
| **Germ cell mutagenicity** | **–** |
| **BAT value** | **–** |
| Synonyms | trichloromethane |
| Chemical name (CAS) | trichloromethane |
| CAS number | 67-66-3 |
| Structural formula | $CHCl_3$ |
| Molecular weight | 119.38 |
| Melting point | –63.5°C |
| Boiling point | 61.7°C |
| Density at 20°C | 1.48 g/cm$^3$ |
| Vapour pressure at 20°C | 211 hPa |
| **1 ml/m$^3$ (ppm) $\hat{=}$ 4.962 mg/m$^3$** | **1 mg/m$^3$ $\hat{=}$ 0.202 ml/m$^3$ (ppm)** |

Since the appearance of the last documentation of a MAK value for chloroform in 1976 and the supplement about risks during pregnancy in 1989, more data for metabolism and mechanism of action of the substance, for chronic toxicity, genotoxicity, reproductive toxicity and carcinogenic activity of chloroform have appeared in numerous publications. Therefore the MAK value for chloroform required re-evaluation and the carcinogenic potential had to be re-assessed.

*Essential MAK Value Documentations.* DFG, Deutsche Forschungsgemeinschaft
Copyright © 2006 WILEY-VCH Verlag GmbH & Co. KGaA, Weinheim
ISBN: 3-527-31394-X

# 1 Toxic Effects and Mode of Action

Chloroform is the classic narcotic with a characteristic odour. It was first used as an analgesic in medical practice in 1847. Because of its narrow therapeutic range and the severe respiratory depression, negative inotropic effects and liver toxicity which it caused, however, as new methods were developed it was replaced by other anaesthetics. In animal studies, chloroform has been shown to be carcinogenic for the rat and mouse.

Uptake of chloroform is generally by inhalation of the vapour, rarely by ingestion. Depending on the concentration and the exposure time, initially local irritation is observed followed by central nervous excitation or inebriation, then relaxation and paralysis (tolerance phase and narcosis). After an overdose of chloroform, death is caused by heart failure and accompanied by characteristic damage to the parenchymal organs. To achieve a state of inebriation, concentrations of 0.5 % to 0.7 % v/v (5000–7000 ppm) in the inhaled air are required, for the tolerance phase about 1.0 % v/v (10000 ppm), for deep anaesthesia about 1.4 % v/v (14000 ppm). The asphyxial stage is reached at concentrations above 1.6 % v/v (16000 ppm) (Rosenfeld 1896, Spenzer 1954). Chronic intoxications are rare as is chloroformismus, habitual misuse as an inebriating drug. In both cases, the symptoms and clinical findings are like those of chronic alcoholism.

# 2 Mechanism of Action

During the metabolism of chloroform, reactive intermediates such as phosgene and dichloromethyl radicals are formed and can react with cellular components such as fatty acids and phospholipids. The resulting lipid peroxidation accounts at least in part for the hepatotoxic and nephrotoxic effects of chloroform. After incubation of radioactively labelled chloroform with liver microsomes from the mouse and rat, covalent binding to proteins and lipids was shown to be a function of the partial pressure of oxygen (Testai *et al.* 1990, 1995, 1996). *In vitro* chloroform was not mutagenic. The existence of weak genotoxicity *in vivo* cannot be excluded. DNA binding has been demonstrated only to purified calf thymus DNA (DiRenzo *et al.* 1982). After oral administration of chloroform to the rat and mouse, DNA binding could not be detected in the target organs, liver and kidney (Diaz-Gomez and Castro 1980, Pereira *et al.* 1982, Reitz *et al.* 1982). There is no evidence that the substance acts by a genotoxic mechanism. Tumours were only observed in animals with simultaneous hepatotoxicity and nephrotoxicity; this indicates that the carcinogenic activity of the substance requires regenerative hyperplasia.

# 3 Toxicokinetics and Metabolism

## 3.1 Absorption, distribution

Chloroform is taken up readily and distributed in the organism like other substances which are highly soluble in lipids (von Oettingen 1964).

In a study of the effects of the vehicle on the absorption and distribution of chloroform, male Wistar rats were given oral doses of 75 mg/kg body weight in aqueous solution or in corn oil. Six minutes after dosing, the concentration of chloroform in blood was 39.3 μg/ml when the substance had been administered in water but only 6 μg/ml when it had been administered in corn oil (Withey *et al.* 1983). Four F344 rats were given chloroform doses of 15, 30, 90 or 180 mg/kg body weight and four B6C3F$_1$ mice doses of 70, 238 or 477 mg/kg body weight by gavage in corn oil, in water or as an aqueous emulsion (with 2 % Emulphor® for doses of 30 mg/kg body weight or more for rats and 238 mg/kg body weight or more for mice). Rate and level of absorption determined in the blood, liver and kidneys of the rats were shown to be the same for a given vehicle. The rate of absorption was slightly lower after administration in corn oil. Unlike in the rat, the concentrations in mouse tissues were higher after administration of the substance in water than after administration in corn oil (Dix *et al.* 1997). After administration of chloroform doses of 19, 38, 81 or 160 mg/kg body weight to rats in the drinking water for a period of 18 months, the chloroform levels in the blood were shown to be 0.7, 7.5, 22.75 and 124 μg/l; after 24 months levels of 0.4, 18, 14, 81 and 135 μg/l were found (Jorgenson *et al.* 1985).

Groups of 6 to 10 male F344 rats were treated by epicutaneous application of 2 ml undiluted chloroform or aqueous chloroform solutions for a period of 24 hours. The maximum concentration in blood (51 μg/ml) was achieved 8 h after application of pure chloroform and remained constant for up to 24 hours. After application of the aqueous solutions (about 9.6, 22, 37.6 mg/kg body weight), maximum concentrations of 150 to 1000 ng/ml were found after 2 hours; after 24 hours the levels did not differ from the control values (Morgan *et al.* 1991). After exposure of the entire skin of female hairless guinea pigs to an aqueous solution of [$^{14}$C]-chloroform (19–52 ng/ml) for 70 minutes, a permeability coefficient of 0.13 cm/hour was determined. The reduction in radioactivity in the solution was measured (Bogen *et al.* 1992).

Mice inhaled [$^{14}$C]-chloroform for a period of 10 minutes. After 2 hours, radioactivity was shown to have accumulated in adipose tissue, in the liver, kidneys, lungs and blood. High concentrations of bound radioactivity were detected in the testes, preputial glands and epididymis (WHO 1994).

## 3.2 Elimination

The substance is eliminated mostly by exhalation; 18 hours after intraduodenal administration of radioactively labelled chloroform to rats, 70 % of the dose has been exhaled unchanged, 4 % as $^{14}CO_2$. The formation of $CO_2$ from chloroform takes place at four times the rate of its formation from carbon tetrachloride (Paul and Rubinstein 1963). There are marked species differences: after oral administration of labelled chloroform, mice exhale 80 % of the dose as $CO_2$ within 48 hours, rats 66 % and monkeys 18 % (Brown *et al.* 1974).

Male B6C3F$_1$ mice and Osborne-Mendel rats inhaled [$^{14}$C]-chloroform for 6 hours (mouse: 10, 89 or 366 ml/m$^3$; rat: 93, 356 or 1041 ml/m$^3$). Within 48 hours after the end of exposure, the mice in the highest concentration group exhaled 23 mg equivalent/kg body weight as unchanged chloroform and 52 mg eq/kg body weight as $CO_2$. In the faeces and urine, respectively, about 3.8 and 14.2 mg eq/kg body weight of the administered radioactivity was recovered. In the rats of the highest concentration group, 78.3 mg eq/kg body weight unchanged chloroform and 64 mg eq/kg body weight $CO_2$ was detected in the exhaled air. Of the administered radioactivity, 1.1 mg eq/kg body weight and 11 mg eq/kg body weight were recovered in faeces and urine, respectively. The lower the exposure concentration, the higher the proportion of absorbed substance which was metabolized and the lower the amount exhaled unchanged. Furthermore, because the rates of metabolism and respiration in the mouse are higher than in the rat, a higher level of binding of radioactivity to macromolecules was observed in the target organs of the mouse (Andersen *et al.* 1993, Corley *et al.* 1990).

The half time for elimination of radioactivity from the adipose tissue of male Wistar rats after intravenous injection of doses of 3, 6, 9, 12 or 15 mg/kg was 106 minutes. The rates of elimination from various tissues were comparable with that from the blood and were dependent on the solvent used: 46 minutes with water and 39 minutes with corn oil (WHO 1994).

After oral administration of chloroform doses of 500 mg to volunteers, 50 % of the dose was exhaled as $CO_2$ and 40 % eliminated unchanged within 8 hours. After 1.5 hours, maximum blood chloroform concentrations of 1 to 5 ng/l were obtained and the amount exhaled reached 18 % to 66 % of the dose (WHO 1994). After inhalation of a single dose of 5 mg [$^{38}$Cl]-chloroform by volunteers, retention of 80 % was measured (Morgan *et al.* 1970). A blood level of 100 mg/l was found in a person who had inhaled chloroform at a concentration of 50000 mg/m$^3$ (narcosis) (WHO 1994).

## 3.3 Metabolism

### 3.3.1 Rat and mouse

Chloroform is oxidized by cytochrome P450 to trichloromethanol, which is converted by dehydrochlorination to phosgene. Phosgene reacts with water to yield HCl and $CO_2$ or forms unstable adducts with biological macromolecules. However, phosgene also reacts directly with cysteine to form 2-oxothiazolidine-4-carboxylic acid or is conjugated with glutathione to form *S*-chlorocarbonyl glutathione, which in its turn can react with glutathione to yield diglutathionyl dithiocarbonate or glutathione disulfide and carbon monoxide. Chloroform is also metabolized by the cytochrome P450 enzyme system under reducing conditions; the product is the dichloromethyl radical which reacts preferentially with fatty acids or phospholipids or is converted to dichloromethane (Testai *et al.* 1995). In *in vitro* studies of liver microsomes from Osborne-Mendel rats and from B6C3F$_1$ mice, both the oxidative and the reductive metabolism of chloroform was demonstrated. At low chloroform concentrations (< 0.1 mM) chloroform is oxidized preferentially by cytochrome P450 2E1, at higher concentrations (5 mM) also by the cytochrome P450 enzymes 2B1/2 and 2C11 (Guengerich *et al.* 1991). Reductive metabolism of chloroform first takes place *in vitro* at higher concentrations (5 mM) (Testai *et al.* 1996). Whether chloroform is metabolized by the oxidative or the reductive pathway and with what metabolic capacity depends also on the species of animal, the strain and the level of enzyme induction. The mouse liver has a higher capacity for metabolism of chloroform than that of the rat. In incubations of microsomes from mouse liver and kidney, a much higher level of oxidative and reductive metabolism is observed than in those from the rat (Gemma *et al.* 1996). This higher metabolic capacity is thought to account for the fact that the mouse is more sensitive than the rat to liver toxins (Corley *et al.* 1990, Mink *et al.* 1986). Chloroform metabolism is both organ-specific and sex-specific. Chloroform is metabolized by microsomes from the renal cortex of male mice but not by those from females (Smith *et al.* 1983, Smith and Hook 1984). This can explain the fact that chloroform produces much less nephrotoxicity in female mice than in the males.

Covalent binding to protein and lipids of liver microsomes from the mouse and rat was shown to depend on the partial pressure of oxygen and was demonstrated under both hypoxic and anoxic conditions; this indicates that the reductive metabolism of chloroform is of significance (Testai *et al.* 1990, 1992, 1995, 1996). Metabolism of chloroform by microsomes from rat colon mucosa was not detected (Vittozzi *et al.* 1991).

Chloroform is metabolized in the vagina *in situ* by cytochrome P450. In C57Bl mice and F344 rats *in vivo* and *in vitro*, chloroform was shown to bind to the vaginal epithelium but not to the epithelium of the uterus (Brittebo et al. 1987). After male knock-out mice which were deficient in cytochrome P450 2E1 had inhaled chloroform at a concentration of 90 ml/m$^3$, no severe effects on the kidney or liver were detected (Butterworth *et al.* 1998, Constan *et al.* 1998). This result indicates that the production of reactive metabolites by cytochrome P450 2E1 plays an decisive role in the hepatotoxicity and nephrotoxicity of chloroform.

### 3.3.2 Man

Chloroform is metabolized oxidatively by the cytochrome P450 2E1 isozyme which has to date been detected only in the liver and not in the kidney in man (Amet *et al.* 1997, Guengerich *et al.* 1991). The rate of metabolism of chloroform by liver and kidney microsomes is highest in the mouse, followed by the rat and man. *In vivo*, V$_{max}$ values of 1.89 nmol/g/hour were found for the mouse and 2.42 nmol/g/hour for the rat. The corresponding K$_m$ values were 0.352 and 0.543 mg/l. From the average K$_m$ values for the mouse and rat, the V$_{max}$ for a 70 kg man was calculated to be 307 mg/h. With a physiologically based pharmacokinetic (PBPK) model and the above data, it was predicted that, given the same exposure concentrations, the binding of chloroform metabolites to macromolecules in the liver would be highest for the mouse and lowest for man (Corley *et al.* 1990).

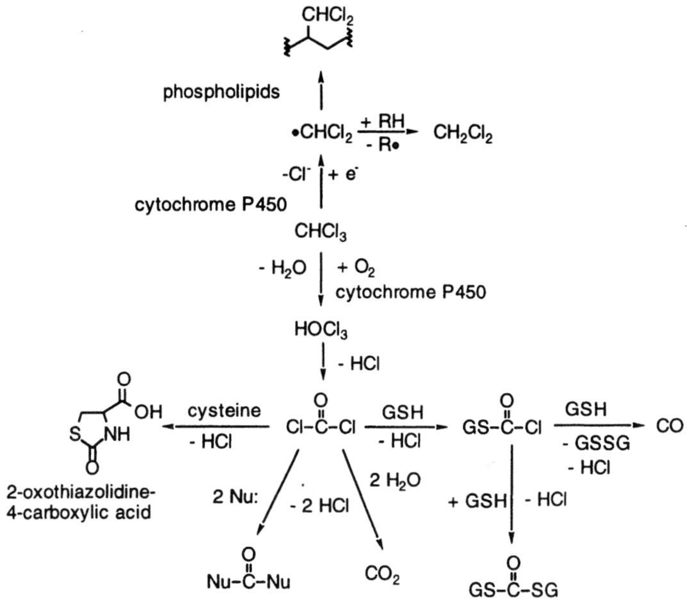

**Figure 1.** Metabolism of chloroform

# 4 Effects in Man

## 4.1 Single exposures

A severe acute intoxication developed after injection of 0.5 ml chloroform and ingestion of 120 ml (180 g) on the subsequent day. The woman was found unconscious but became responsive a few minutes later. Serum biomarkers indicated the presence of reversible liver cell necrosis and liver damage (Rao *et al.* 1993). The average lethal dose for man is given as 45 g but inter-individual sensitivity varies widely. The animals survived doses of up to 270 g chloroform (Winslow and Gerstner 1978). Chloroform was used as an anaesthetic in concentrations of 12000 to 48000 mg/m$^3$. Because of the chronic kidney and liver damage it caused and the acute respiratory disorders and cardiac arrhythmia which developed during the narcosis, the medical use of chloroform has been discontinued. The concentration-dependent effects of inhalation of chloroform include deeper breathing, increased respiration rate, hyperthermia, adrenaline depletion, hypotension, reduced gastrointestinal motility, respiratory acidosis, hyperglycaemia, congestion in the spleen, increased leukocyte count, reduced coagulation time and increased prothrombin time (WHO 1994).

## 4.2 Repeated exposures

Blood concentrations of 1 to 2.9 mg/l were determined in 5 of 13 persons exposed to chloroform concentrations of about 1950 mg/m$^3$ for up to 6 months. Jaundice was diagnosed in all of the persons and considered to be an effect of the exposure to chloroform (Phoon *et al.* 1975). In one factory 18 cases of jaundice were described in persons who had been exposed for less than 4 months to chloroform concentrations of 80 to 160 mg/m$^3$. Infection with hepatitis B virus was excluded (no other details) (Phoon *et al.* 1983).

## 4.3 Local effects on skin and mucous membranes

Chloroform is irritating in the eyes and causes reddening of the conjunctiva. Corneal damage was reversible within a few days. Dermatitis developed after skin contact with chloroform (Winslow and Gerstner 1978).

## 4.4 Reproductive and developmental toxicity

Chloroform can pass the placenta in man. The chloroform concentrations determined in umbilical cord blood from new-born babies were the same as those in the blood of the mother. It has been reported that during the course of pregnancy, eclampsia developed in two women who were exposed to a large number of organic compounds including chloroform in concentrations between 300 and 1000 ml/m$^3$ (Barlow and Sullivan 1982). Further human data are not available.

# 5 Animal Experiments and *in vitro* Studies

## 5.1 Acute toxicity

### 5.1.1 Inhalation

Exposure of male mice to chloroform vapour (5000 mg/m$^3$) resulted in necrosis in the proximal and distal renal tubules and calcification of the renal cortex. Anaesthesia was not observed (Deringer *et al.* 1953). In rats exposed to a chloroform concentration of 49000 mg/m$^3$, respiratory acidosis and liver toxicity developed (WHO 1994). After exposure of male rats to chloroform concentrations up to 5250 mg/m$^3$ for four hours, increased activities of glutamate dehydrogenase, sorbitol dehydrogenase and aspartate aminotransferase were detected in serum (Brondeau *et al.* 1983). Female mice were exposed to chloroform concentrations up to 980 mg/m³. Liver necrosis and an increase in serum ornithine carbamoyl transferase were observed. Exposure of mice, rabbits, guinea pigs and cats to chloroform concentrations of 10000 to 100000 mg/m³ caused anaesthesia which lasted for 30 minutes to several hours. These concentrations could also be lethal (WHO 1994).

### 5.1.2 Ingestion

After administration of single oral chloroform doses of 546, 765, 1071, 1500 or 2100 mg/kg body weight, all animals given the highest dose died within 7 days. The effects included sleepiness, reduced muscle tonus, ataxia, piloerection and prostration, occasionally also lacrimation. Liver and kidneys were congested and enlarged (Chu *et al.* 1980) (Table 1). Male F344 rats were given chloroform doses of 15, 22.4, 30, 59.7, 89.5, 119.4 or 179.1 mg/kg body weight by gavage. After 24 hours in the animals which had been given doses of 59.7 mg/kg body weight or more, body weights were reduced and the activities of alanine aminotransferase, aspartate aminotransferase and sorbitol dehydrogenase were increased. The authors concluded that the no observed adverse effect level (NOAEL) for acute toxic effects in the male rat is 30 mg/kg body weight (Keegan *et al.* 1998).

As little as one day after administration of chloroform doses of 34, 180 or 477 mg/kg body weight by gavage to male F344 rats and 34, 238 or 477 mg/kg body weight to female B6C3F₁ mice, necrosis was found in the proximal tubules and scattered centrilobular focus formation in the livers of the low dose group animals. In rats given doses of 180 mg/kg body weight or more the labelling index was increased, and in the group treated with 477 mg/kg body weight there was extensive necrosis in the kidneys. In addition, in the high dose group, slight necrosis, infiltration of inflammatory cells and an increase in the labelling index were found in the liver. Sorbitol dehydrogenase, alanine aminotransferase and aspartate aminotransferase activities in the plasma were increased. In the mice, foci and infiltration of inflammatory cells were detected in the liver after doses of 238 mg/kg or more. In the high dose group, centrilobular necrosis developed and the labelling index was significantly increased after 2 days. A no observed effect level (NOEL) for the mouse of 34 mg/kg body weight can be derived from this study (Larson *et al.* 1993).

In a comparative study, male Osborne-Mendel and F344 rats were given chloroform doses of 10, 34, 90, 180 or 477 mg/kg body weight by gavage in corn oil. In both strains of rat, dose-dependent peripheral nasal lesions were seen in animals given 10 mg/kg body weight or more and central nasal lesions from doses of 90 mg/kg body weight; the labelling index in the kidney was also increased from doses of 10 mg/kg body weight. From 180 mg/kg body weight (Osborne-Mendel) and at 477 mg/kg body weight (F344), minimal to slight vacuolation of the epithelial cells in the proximal renal tubules was seen. Only in the F344 rats were midzonal hepatocellular vacuolation with fatty degeneration of the liver and an increase in the labelling index observed. There were no differences between the two strains in the effects on the kidneys and nose (Templin *et al.* 1996a).

**Table 1.** Acute toxicity of chloroform (WHO 1994)

| Species | Vehicle Administration route | Observation period (days) | LD$_{50}$ (mg/kg body weight) ♂/♀ | References |
|---|---|---|---|---|
| **mouse** | | | | |
| C3H/tif | sesame oil oral | 15 | 36/353 | Pericin and Thomann 1979 |
| DBA/2/j | sesame oil oral | 15 | 101/679 | Pericin and Thomann 1979 |
| Tif:MAGf | sesame oil oral | 15 | 213/1366 | Pericin and Thomann 1979 |
| A/J | sesame oil oral | 15 | 253/774 | Pericin and Thomann 1979 |
| Tif:MF2f | sesame oil oral | 15 | 336/1126 | Pericin and Thomann 1979 |
| C57Bl/6j | sesame oil oral | 15 | 460/820 | Pericin and Thomann 1979 |
| **rat** | | | | |
| Sprague-Dawley | no vehicle oral | 14 | 908/1117 | Chu *et al.* 1980 |
| Sprague-Dawley | no vehicle oral | 14 | 2000/n.d. | Torkelson *et al.* 1976 |
| Sprague-Dawley | arachis oil intraperitoneal | 24 h 14 | n.d./1379 n.d./894 | Lundberg *et al.* 1986 |

n.d.  not determined

## 5.1.3 Dermal absorption

Two weeks after occlusive application of single dermal chloroform doses of 1000 or 4000 mg/kg body weight to two rabbits for a period of 24 hours, dose-dependent severe degenerative changes were observed in the renal tubules. Effects were not seen in the liver (Torkelson *et al.* 1976).

## 5.1.4 Intraperitoneal injection

After intraperitoneal injection of chloroform doses of 75 to 1500 mg/kg body weight to ICR mice of both sexes, liver toxicity, increased body weights and increased levels of alanine aminotransferase were seen in animals given doses of 375 mg/kg body weight and more. In the males, lesions and necrosis were seen in the proximal renal tubules (Smith *et al.* 1983); such changes were also observed in an earlier study in male Swiss mice given doses as low as 48 mg/kg body weight (WHO 1994).

## 5.2 Subacute, subchronic and chronic toxicity

### 5.2.1 Inhalation

Groups of 5 male F344 rats and 5 female B6C3F$_1$ mice were exposed to chloroform concentrations of 1, 3, 10, 30, 100, or 300 ml/m$^3$, 6 hours daily for 7 days. In the mice, there was a concentration-dependent increase in cell proliferation in the liver and vacuolation in the hepatocytes of animals exposed to 10 ml/m$^3$ or more, hepatocellular necrosis from 100 ml/m$^3$ and increased cell proliferation and histopathological changes (regeneration of the tubulus epithelium) also in the kidneys in the animals exposed to 300 ml/m$^3$. In the rat, increased cell proliferation was seen in the kidney from 30 ml/m$^3$ and in the liver from 100 ml/m$^3$. Histopathological changes (hepatocellular necrosis, regeneration of the tubular epithelium) were not seen in these organs unless the animals had been exposed to the high concentration of 300 ml/m$^3$. In addition in the rats exposed to 10 ml/m$^3$ or more, dose-dependent histopathological changes developed in the nasal cavity (atrophy of Bowman's glands and bone regeneration). In the mice, cell proliferation (without bone hyperplasia) was only seen in the nasal region in the group of animals exposed to 300 ml/m$^3$. From this study a NOEL of 3 ml/m$^3$ can be deduced for both species (Larson *et al.* 1994b; Mery *et al.* 1994).

Groups of 12 male Wistar rats were exposed for 4 weeks to a chloroform concentration of 32 ml/m$^3$ (160 mg/m$^3$). After continuous exposure (24 hours/day, 7 days/week), the liver damage (fatty degeneration of the hepatocytes, focal necrosis) was more severe than after intermittent exposure (6 hours/day, 5 days/week). The liver toxicity was not increased by pretreatment of the animals with phenobarbital or 1,3-butanediol. Clearance was significantly increased after pretreatment with 1,3-butanediol but unaffected by phenobarbital. Other organs were not investigated (Plummer *et al.* 1990).

Male F344 rats were exposed to chloroform concentrations of 2, 10, 30, 90 or 300 ml/m$^3$, 6 hours/day, 7 days/week for 4 days or 3, 6 or 13 weeks and female animals only for 3 or 13 weeks. Other groups of animals were exposed 5 days/week for 13 weeks or 7 days/week for 6 weeks and then observed for another 7 weeks. The level of cell proliferation was measured by determining the amount of bromodeoxyuridine (BrdU) incorporated into the DNA of various organs. The results obtained after the 13 week exposures are summarized in Table 2. Body weight gains were reduced in a concentration-dependent manner at all times in all animals exposed to concentrations of 10 ml/m$^3$ or more. In animals exposed to 90 ml/m$^3$ or more, changes in organ weights were found. In the kidneys of male animals exposed to concentrations of 30 ml/m$^3$ or more, regenerative cell proliferation was increased reversibly. In the females, kidney lesions were seen only in the highest concentration group. Degenerative changes and vacuolation of liver cells developed in males and females exposed to 90 ml/m$^3$ or more. A significant increase in regenerative cell proliferation in the liver was found only in the 300 ml/m$^3$ group. In the group with a 7-week recovery period after exposure to 300 ml/m$^3$, necrosis and inflammatory cells were found in the liver. The effects on the nose were not reversible within the 7-week recovery period in animals exposed for 7 days/week to concentrations of 90 ml/m$^3$ or more. In animals exposed for 5 days per week, the effects were generally less marked than in those exposed for 7 days per week. From these results a NOEL of 10 ml/m$^3$ can be derived for the target organs kidney and liver, but effects on the nose were still seen at 2 ml/m$^3$ (Templin *et al.* 1996b).

Groups of 5–15 female B6C3F$_1$ mice were exposed to chloroform concentrations of 0.3, 2, 10, 30 or 90 ml/m$^3$, 6 hours/day, 5 or 7 days/week for 4 days or 3, 6 or 13 weeks and male animals only for 3 or 13 weeks. Effects on the nose were seen in mice exposed for 4 days to 10 to 90 ml/m$^3$, but not in those exposed for longer periods. Liver lesions characterized by swelling of nuclei and vacuolation of hepatocytes developed at concentrations of 10 ml/m$^3$ or more. Nephropathy, regenerative cell proliferation and mineralization of the renal cortex were also recorded at concentrations of 10 ml/m$^3$ and more. In female mice there was no kidney damage even at 90 ml/m$^3$. The results obtained after the 13 week exposures are summarized in Table 2. From this study, a NOEL of 2 ml/m$^3$ can be derived for the mouse exposed to chloroform (Larson *et al.* 1996).

The results of Larson *et al.* (1996) were confirmed in another study (Templin *et al.* 1998). Groups of 5–8 male mice were exposed to chloroform concentrations of 1, 5, 30 or 90 ml/m$^3$ and female mice to 5, 30, 90 ml/m$^3$ for 3, 7 or 13 weeks. The nose was not examined. The authors derived a NOAEL of 5 ml/m$^3$ for cell proliferation and tumour development in the target organs liver and kidney in the mouse, with reference to a carcinogenicity study by Matsushima (1994) in which no increase in tumour incidence was found in animals exposed to 5 ml/m$^3$ (Table 7).

Groups of 8–15 male and female F344 rats were exposed to chloroform concentrations of 2, 10, 30, 90 or 300 ml/m$^3$, 6 hours/day, 7 or 5 days/week for 3, 6 or 13 weeks to study, in particular, liver lesions and the development of cholangiofibromas. In the animals exposed to 300 ml/m$^3$ 'intestinal crypt-like ducts with periductal fibrosis' were observed and shown to be quite different from true cholangiofibrosis. The lesions were most severe in the right lobe of the liver. They were associated with liver necrosis and increased regenerative hepatocyte proliferation (determined as labelling index) which could not be demonstrated in bile duct cells. The authors concluded that the cholangiofibrosis was probably not true bile duct fibrosis (Jamison *et al.* 1996).

## 5.2.2 Ingestion

In studies of B6C3F$_1$ mice and F344 rats given chloroform by gavage or in the drinking water, degenerative changes were found in the hepatocytes or in the epithelial cells of the proximal renal tubules after 3 weeks of treatment. The results of studies of subacute, subchronic and chronic toxicity of ingested chloroform are summarized in Table 3. The NOEL and LOEL (lowest observed effect level) values are shown in Table 4. Administration in the drinking water produced less severe toxic effects than did the administration by gavage (bolus effect). After four days treatment the effects were more severe than after treatment for three weeks; this is evidence of adaptation. It is noteworthy that the nasal lesions develop in rats even when the substance is administered orally; distribution of the substance with the bloodstream must be responsible for these effects. This is also of significance for interpretation of the results of the inhalation studies (Section 4.1.2).

**Table 2.** Results of the 13-week inhalation studies (Larson *et al.* 1996, Templin *et al.* 1996b)

| Concentration (ml/m$^3$) | Rat | Mouse |
|---|---|---|
| 2* | ♂, ♀ **nose:** minimal atrophy of the olfactory epithelium in the ethmoid turbinates, cell proliferation unaffected | ♂, ♀ **liver, kidneys:** NOEL |
| 10* | ♂, ♀: reduced body weight gains, <br>**nose:** oedema in the *lamina propria*, loss of Bowman's glands, minimal atrophy of the olfactory epithelium in the ethmoid turbinates, cell proliferation increased by a factor of 4 <br>**kidneys:** NOEL <br>**liver:** NOEL | ♂ **liver:** mild swelling and vacuolation of centrilobular hepatocytes (4/13, 5 days/week; 5/15, 7 days/week) <br>**kidneys:** nuclei of the tubulus epithelial cells enlarged (5 and 7 days/week), cell proliferation unaffected (7 days/week) or increased by a factor of 5 (5 days/week), focal regeneration (5 days/week) <br>♀: **liver** as in ♂ (4/14, 7 days/week), unaffected (5 days/week) |
| 30 | ♂: **kidneys:** cell proliferation in the cortex increased by a factor of 2 (7 days/week) or unaffected (5 days/week), no microscopic findings (5 and 7 days/week) <br>**nose:** generalized atrophy of the ethmoid turbinates, severity the same for 5 days/week and 7 days/week exposures, cell proliferation increased by a factor of 4 <br>**liver:** NOEL <br>♀: **kidneys:** vacuolation, cell proliferation increased by a factor of 2–3 <br>**liver:** no effects <br>**nose:** as in ♂ | ♂: **liver:** organ weights increased, swelling of centrilobular hepatocytes, enlarged nuclei, vacuolation (12/12) <br>**kidneys:** as in the 10 ml/m$^3$ group and, in addition, focal regeneration in the cortex (11/12), not reversible <br>♀: **liver:** as in the 10 ml/m$^3$ group (10/15, 7 days/week), mild effects (5 days/week), reversible, cell proliferation unaffected |
| 90 | ♂: **kidneys:** relative organ weights increased by 10 %, cell proliferation increased by a factor of 4 (7 days/week) or 2 (5 days/week), reversible; vacuolation of the epithelial cells in the proximal tubules (7 days/week) or cells normal (5 days/week); **liver:** occasional vacuolated hepatocytes, necrosis, effects less marked in the 5 day/week group, reversible <br>**nose:** as in the 30 ml/m$^3$ group, not reversible | ♂: **liver:** as in the 30 ml/m$^3$ group (14/14) and, in addition, necrosis (2/14, 7 days/week) swelling and vacuolation (10/12, 5 days/week), cell proliferation increased by a factor of 20 (7 days/week) or 6 (5 days/week), reversible <br>**kidneys:** as in the 30 ml/m$^3$ group (14/14) and, in addition, mineralization in the cortex (14/14, 7 days/week; 8/12, 5 days/week), not reversible, cell proliferation increased by a factor of 2 (7 days/week) or 4 (5 days/week), reversible <br>**nose:** NOEL |

**Table 2.** continued

| Concentration (ml/m³) | Rat | Mouse |
| --- | --- | --- |
| 90 | ♀: **liver:** relative organ weights increased by 10 %, vacuolation of midzonal hepatocytes (7 days/week) or cells unaffected (5 days/week), reversible; cell proliferation unaffected. **kidneys:** no effects, cell proliferation increased by a factor of 7. **nose:** as in ♂ | ♀: **liver:** organ weights increased, moderate swelling and vacuolation (15/15, 7 days/week), mild effects (12/12, 5 days/week), after observation period only nuclei enlarged, cell proliferation increased by a factor of 14 (7 days/week) or 7 (5 days/week), reversible. **kidneys:** NOEL. **nose:** NOEL |
| 300 | ♂: **kidneys:** relative organ weights increased by 30 %, cell proliferation increased by a factor of 9 (5 and 7 days/week), reversible, microscopic kidney changes, also necrosis of the tubulus epithelial cells (5 and 7 days/week). **liver:** as for the 90 ml/m³ group; relative organ weights increased by 30 %, cell proliferation increased by a factor of 25–29, irreversible. **nose:** as for the 30 ml/m³ group. ♀: **kidneys:** relative organ weights increased by 50 %, vacuolation without necrosis, reversible; cell proliferation increased by a factor of 10–12. **liver:** relative organ weights increased by 50 %, degeneration of centrilobular to midzonal hepatocytes (5 and 7 days/week), reversible; cell proliferation increased by a factor of 40 (5 and 7 days/week), reversible. **nose:** as in ♂ | not exposed |

* only 7 days/week exposures for the rats

**Table 3**. Subacute and subchronic toxicity of ingested chloroform

| Number Species Sex | Duration | Dose | Effects | References |
|---|---|---|---|---|
| 24 per group F344 rat ♂ | 4 days; 3 weeks, 5 days/week | 3, 10, 34, 90, 180 mg/kg body weight (gavage, corn oil); 60, 200, 400, 900, 1800 mg/l drinking water (about 3.5, 12, 24, 53, 106 mg/kg body weight) | **gavage:**<br>**4 days: from 10 mg/kg body weight:** increased relative liver weights<br>**from 34 mg/kg body weight:**<br>**kidneys:** degenerative changes in the proximal tubules,<br>**liver:** mild centrilobular sinusoidal leukostasis, centrilobular necrosis<br>**from 90 mg/kg body weight:** increased alanine aminotransferase and sorbitol dehydrogenase<br>**liver:** increased cell proliferation<br>**180 mg/kg body weight:** decreased body weights<br>**kidneys:** increased cell proliferation<br>**liver:** more severe necrosis<br>**3 weeks: from 90 mg/kg body weight:** decreased body weights, increased relative liver weights<br>**180 mg/kg body weight:**<br>**kidneys:** increased organ weights, progressive colourless degeneration of the proximal tubules, cell proliferation unaffected<br>**liver:** effects like those after 4 days, increased alanine aminotransferase and sorbitol dehydrogenase<br>**drinking water:**<br>**4 days: from 53 mg/kg body weight:** degenerative changes in the liver<br>**106 mg/kg body weight:** decreased body weights, kidneys unaffected<br>**3 weeks: from 12 mg/kg body weight:** focal areas of regenerating epithelium and cell proliferation in the kidneys<br>**106 mg/kg body weight:**<br>**liver:** vacuolation of hepatocytes, focal inflammation, cell proliferation unaffected<br>**kidneys:** increased relative organ weights, cell proliferation unaffected | Larson *et al.* 1995a |

**Table 3.** continued

| Number<br>Species<br>Sex | Duration | Dose | Effects | References |
|---|---|---|---|---|
| 10 per group<br>F344 rat<br>♀ | 4 days;<br>3 weeks,<br>5 days/week | 34, 100, 200,<br>400 mg/kg body<br>weight (gavage,<br>corn oil) | **4 days: see effects after 3 weeks**<br>**3 weeks: from 34 mg/kg body weight:**<br>**nose:** minimal peripheral lesions<br>**from 100 mg/kg body weight:**<br>**liver:** degenerative changes, increased regenerative cell proliferation<br>**nose:** central lesions, damaged olfactory epithelium, increased regenerative cell proliferation<br>**kidneys:** increased regenerative cell proliferation<br>**from 200 mg/kg body weight:**<br>**kidneys:** degeneration and necrosis in the proximal tubules | Larson *et al.*<br>1995b |
| 10 per group<br>B6C3F₁ mouse<br>♂ | 4 days;<br>3 weeks<br>5 days/week | 34, 90, 138, 277<br>mg/kg body<br>weight (gavage,<br>corn oil) | **4 days: from 34 mg/kg body weight:**<br>**liver:** vacuolation and slight swelling of centrilobular hepatocytes, increased cell proliferation<br>**kidneys:** tubule necrosis, increased cell proliferation<br>**277 mg/kg body weight:**<br>**liver:** aggregations of inflammatory cells<br>**3 weeks: from 34 mg/kg body weight:**<br>**kidneys:** formation of subcapsular foci<br>**from 90 mg/kg body weight:** swelling of hepatocytes, increased eosinophil count, increased cell proliferation<br>**from 138 mg/kg body weight:**<br>**liver:** degeneration, necrosis, increased cell proliferation<br>**kidneys:** increased cell proliferation<br>**277 mg/kg body weight:** severe nephropathy | Larson *et al.*<br>1994a |

**Table 3.** continued

| Number Species Sex | Duration | Dose | Effects | References |
|---|---|---|---|---|
| 28 per group B6C3F$_1$ mouse ♀ | 4 days; 3 weeks 5 days/week | 3, 10, 34, 90, 238, 477 mg/kg body weight (gavage, corn oil); 60, 200, 400, 900, 1800 mg/l drinking water (16, 43, 82, 184, 329 mg/kg body weight) | **4 days: see effects after 3 weeks** **3 weeks: gavage:** **from 34 mg/kg body weight:** **liver:** histopathological changes **from 90 mg/kg body weight:** **liver:** increased cell proliferation **477 mg/kg body weight:** **kidneys:** increased cell proliferation **drinking water:** **from 82 mg/kg body weight:** minimal histopathological changes in the liver | Larson *et al.* 1994d |
| 10 per group B6C3F$_1$ mouse ♀ | 3 weeks | 55, 110, 238, 477 mg/kg body weight (gavage, corn oil); | **from 55 mg/kg body weight:** decreased body weights, increased relative liver weights, increased alanine aminotransferase and sorbitol dehydrogenase **liver:** degeneration, necrosis, mineralization **from 110 mg/kg body weight:** **liver:** significantly increased cell proliferation | Melnick *et al.* 1998 |
| 4–5 per group B6C3F$_1$ mouse ♀ | 33 days 31 days | 300, 1800 mg/l drinking water (about 70, 400 mg/kg body weight); 120, 240, 480 mg/l drinking water (about 30, 60, 120 mg/kg body weight) | from day 30 or 28, an additional chloroform dose of 263 mg/kg body weight and day was administered by gavage in corn oil for 3 days. Other animals were given chloroform *per os* only for 3 days. In these animals the centrilobular necrosis and liver cell proliferation were more severe than in animals which had also been given chloroform previously in the drinking water. After administration of chloroform only in the drinking water, hepatotoxicity was not seen. | Pereira and Grothaus 1997 |

**Table 3.** continued

| Number Species Sex | Duration | Dose | Effects | References |
|---|---|---|---|---|
| 8–12 per group CD-1 mouse ♂, ♀ | 14 or 90 days | 50, 125, 150 mg/kg body weight (gavage) | **14 days: from 50 mg/kg body weight:** increased spleen weights, **from 125 mg/kg body weight:** increased liver weights, increased aspartate and alanine aminotransferase in serum **250 mg/kg body weight:** decreased body weights in ♂ **90 days: from 50 mg/kg body weight:** depression of humoral immunity **kidneys:** intertubular aggregations of inflammatory cells **liver:** degeneration of hepatocytes and focal aggregations of lymphocytes; increased glutathione levels in ♀ **from 125 mg/kg body weight:** increased serum glucose levels **250 mg/kg body weight:** increased liver weights | Munson *et al.* 1982 |
| 10 per group B6C3F$_1$ mouse ♂, ♀ | 90 days | 60, 130, 270 mg/kg body weight (gavage) | the hepatotoxic effects of chloroform were more severe after administration in corn oil than in aqueous solution ♂: **from 130 mg/kg body weight (corn oil):** increased liver weights, **from 270 mg/kg body weight:** decreased body weight gain ♀: **from 60 mg/kg body weight:** increased liver weights, hepatocellular vacuolation **from 130 mg/kg body weight:** very mild focal liver necrosis ♂, ♀: **270 mg/kg body weight:** increased serum aspartate aminotransferase, decreased triglycerides | Bull *et al.* 1986 |

**Table 4.** NOEL and LOEL values after administration of chloroform for three weeks

| Species Sex | | Dose (mg/kg body weight) liver | kidneys | nose | References |
|---|---|---|---|---|---|
| **gavage (5 days/week)** | | | | | |
| rat | | | | | |
| ♂ | NOEL | 34 | 90 | no data | Larson *et al.* 1995a |
|  | LOEL | 90 | 180 | no data | |
| ♀ | NOEL | 34 | 100 | < 34 | Larson *et al.* 1995b |
|  | LOEL | 100 | 200 | 34 | |
| mouse | | | | | |
| ♂ | NOEL | 34 | – | no data | Larson *et al.* 1994a |
|  | LOEL | 138 | 277 | no data | |
| ♀ | NOEL | 10 | 238 | no data | Larson *et al.* 1994d |
|  | LOEL | 34 | 477 | no data | |
| ♀ | NOEL | < 55 | no data | no data | Melnick *et al.* 1998 |
|  | LOEL | 55 | no data | no data | |
| **drinking water (5 days/week)** | | | | | |
| rat | | | | | |
| ♂ | NOEL | 53 | 3.5 | no data | Larson *et al.* 1995a |
|  | LOEL | 106 | 12 [1] | no data | |
| mouse | | | | | |
| ♀ | NOEL | 43 | 329 | no data | Larson *et al.* 1994d |
|  | LOEL | > 329 | > 329 | no data | |

[1] not dose-dependent, therefore questionable

# 5.3 Local effects on skin and mucous membranes

## 5.3.1 Skin

After the first of four applications of chloroform to the rabbit ear, mild hyperaemia and desquamation were observed. After application of chloroform for 24 hours to the abdominal skin of rabbits, mild hyperaemia, moderate necrosis and scab formation were recorded (no other details) (Torkelson *et al.* 1976). Application of undiluted chloroform to the skin of six rabbits caused severe skin irritation (Duprat *et al.* 1976).

## 5.3.2 Eyes

In the rabbit eye, chloroform caused slight irritation of the conjunctiva and corneal damage which had regressed within a week. After 2 days and more, purulent exudate was formed (Torkelson *et al.* 1976). Application of undiluted chloroform to the eyes of rabbits resulted in severe eye irritation with mydriasis and keratitis, purulent haemorrhage and corneal damage. Apart from corneal clouding in one rabbit, all effects were completely reversible within 3 weeks (Duprat *et al.* 1976).

## 5.4 Reproductive and developmental toxicity

Groups of 25 pregnant Sprague-Dawley rats were given daily chloroform doses of 20, 50 or 126 mg/kg body weight administered by gavage in two portions per day. The 50 and 126 mg/kg doses were toxic for the dams. Body weight gains were reduced in animals given 50 mg/kg body weight or more; in those given 126 mg/kg body weight, feed consumption was also reduced and alopecia and poor general condition were observed. Histopathogical examination of liver, kidneys and heart in two dams from each dose group which were killed on day 15 of gestation revealed fatty degeneration of the liver in the two animals from the 126 mg/kg group and in one of the two from the 50 mg/kg group. The maternally toxic dose of 126 mg/kg body weight also led to reduced birth weights of the progeny. No other embryotoxic and no teratogenic effects were observed (Thomson *et al.* 1974).

Groups of 15 pregnant Dutch Belted rabbits were treated once daily with chloroform doses of 20, 35 or 50 mg/kg body weight by gavage on days 6 to 18 of gestation. A caesarean section was carried out on day 29. The 50 mg/kg body weight dose was toxic for the dams. Of the 7 animals which died during the course of the study (2 control animals, 1 from the 20 mg/kg group and 4 from the 50 mg/kg group), those from the 50 mg/kg group died as a result of hepatotoxicity. Histopathological examination of liver, kidneys and heart in the animals which survived until they were killed on day 29 revealed no substance-related changes. In the group given the maternally toxic dose of 50 mg/kg body weight, the birth weights of the pups were reduced significantly relative to the control values. No other embryotoxic and no teratogenic effects were observed (Thomson *et al.* 1974).

Groups of 15 inseminated Sprague-Dawley rats were given chloroform doses of 100, 200 or 400 mg/kg body weight daily from day 6 to day 15 of gestation by gavage. All doses were maternally toxic (particularly delayed body weight gains, increased liver weights, reduced haemoglobin and haematocrit values). Whereas in this dose range neither embryotoxic or teratogenic effects were seen, in the foetuses of the group treated with 400 mg/kg there were signs of beginning embryotoxicity (delayed development) (Ruddick *et al.* 1983).

In a behavioural teratology study, male and female albino mice (strain ICR) were treated for 21 days by gavage of chloroform doses of 31.1 mg/kg body weight and day and then mated while the treatment was continued for another 21 days or until a vaginal plug was recognizable. The treatment of the dams was continued during gestation and lactation. The pups were then given the same dose from age 7 days. For the study, 5 litters of treated animals and 5 from the vehicle (Emulphor®) controls were selected. From each of the total of 10 litters, 3 pups were chosen at random every day for 15 days and subjected to a series of behavioural teratological experiments (e.g. various reflexes).

In a total of 15 pups on day 17 after birth the motor performance was tested and on days 22 and 23 passive avoidance learning was studied. The results revealed no effects of the chloroform treatment (Burkhalter and Balster 1979).

Groups of 31, 28 and 20 inseminated Sprague-Dawley rats were exposed to chloroform concentrations of 30, 100 or 300 ml/m$^3$ for 7 hours daily from day 6 to day 15 of gestation. Caesarean section was carried out on day 21. All the concentrations were maternally toxic (feed consumption and body weight gains reduced in a dose-dependent manner). Tests for hepatotoxic effects in the dams (alanine aminotransferase, gross liver pathology, liver weights) revealed that in animals which had inhaled a chloroform concentration of 100 ml/m$^3$ absolute and relative liver weights were increased whereas at 300 ml/m$^3$ the absolute liver weights were decreased. In the rats exposed to 300 ml/m$^3$ the number of implantations was markedly reduced (15 % compared to 88 % in the controls), resorptions were frequent and foetal weights were reduced. Corresponding effects were not seen in the animals exposed to 30 or 100 ml/m$^3$. In the 30 ml/m$^3$ group, but not in the 100 ml/m$^3$ group, the crown-rump lengths of the foetuses were reduced significantly below the control values. In addition, in the foetuses of the 30 ml/m$^3$ group, the ossification of the skull was delayed and nodular enlargements developed on the ribs. In the foetuses exposed to chloroform concentrations of 100 ml/m$^3$, signs of delayed development (missing ribs and delayed ossification of the sternebrae) were seen as well but also subcutaneous oedema and in three cases also genuine terata – namely shortened or missing tails and absent anus openings. Thus, in this study there was evidence of teratogenic potential of chloroform. The absence of similar symptoms in the (few) foetuses in the 300 ml/m$^3$ group is not in disagreement with this finding because in this dose range chloroform had mainly embryolethal effects and potential teratogenic effects would therefore not be manifested (Schwetz *et al.* 1974).

Pregnant rats (not specified more exactly in the abstract) were exposed for 1 hour daily to a chloroform concentration of 4100 ml/m$^3$ (20.1 ± 1.2 g/kg body weight) from day 7 to day 14 of gestation. This concentration caused increased foetal mortality and delayed body weight gains. Teratogenic effects were not seen (Dilley *et al.* 1977).

Groups of pregnant CF1 mice (numbers not specified) were exposed to chloroform concentrations of 100 ml/m$^3$ for 7 hours daily during various phases of gestation (days 1 to 7, 6 to 15, and 8 to 15). This concentration was toxic for the dams (delayed body weight gains, reduced feed and water consumption) and was lethal for one animal (autopsy revealed ulceration of the stomach). Hepatotoxic effects were manifest in significantly increased liver weights in the dams exposed on days 6 to 15 and 8 to 15 and in significantly increased serum glutamate pyruvate transaminase levels. The gestation index was reduced in all groups (but the reduction was statistically significant only in the groups exposed on days 1 to 7 and 6 to 15 of gestation). An increased incidence of resorptions was seen especially in animals exposed on days 1 to 7. The body weights and crown-rump lengths of the foetuses were reduced in the animals exposed on days 1 to 7 and 8 to 15 of gestation. Cleft palate was the main malformation, seen only in the foetuses of animals exposed on days 8 to 15 of gestation and most common in foetuses with delayed growth, which could be evidence of an indirect effect (Murray *et al.* 1979).

In a multi-generation study, groups of 10 male and 30 female ICR mice were given chloroform in the drinking water in concentrations of 0.1, 1 or 5 mg/ml (about 20, 200 and 1000 mg/kg body weight, assuming 6 ml water consumption per day and 30 g body weight). The treatment began 5 weeks before mating of the $F_0$ animals and ended when the $F_{2b}$ progeny were killed. In the highest concentration group the mortality was increased and the body weight gain reduced in both sexes. In the middle concentration group the body weights of the female $F_{1b}$ animals were reduced. In the $F_0$ and $F_{1b}$ animals there were dose-dependent toxic effects on the liver (slight yellow-grey to grey-black discoloration with nodules of 3 mm diameter or more). In all the $F_1$ and $F_2$ animals in the highest concentration groups there were significant effects on reproduction (reduced fertility, litter sizes, gestation and viability parameters). There was no evidence of teratogenic effects (WHO 1994).

Pregnant Wistar rats (20 per group) were exposed by inhalation to chloroform concentrations of 30, 100 or 300 ml/m$^3$ for 6 hours daily on days 7 to 16 of gestation. Concentration-dependent reduction in food consumption and delayed body weight gains were observed. In the 30 ml/m$^3$ group in two of the dams, in the 100 ml/m$^3$ group in three, and in the 300 ml/m$^3$ group in eight of the dams, all embryos died shortly after implantation. Since this effect was very rare in historical controls and in the control group there were no post-implantation losses in any of the dams, the death of the embryos must be considered to be an effect of the exposure to chloroform. In all dose groups the growth of the foetuses was retarded, resulting in lower crown-rump lengths or lower body weights. In this study, chloroform proved to be both maternally toxic and embryotoxic. Evidence of teratogenicity was not seen (Hoechst AG 1988b). In another study, pregnant Wistar rats (20 per group) were exposed by inhalation to chloroform concentrations of 3, 10 or 30 ml/m$^3$ for 6 hours daily on days 7 to 16 of gestation. Food consumption was reduced and body weight gain delayed in animals exposed to 10 ml/m$^3$ or more. In the highest concentration group, growth of the foetuses was slightly delayed and ossification delayed as well; in one dam there were early resorptions. On the basis of these results, the authors derived a NOAEL of 10 ml/m$^3$ for embryotoxic effects and of 3 ml/m$^3$ for maternal toxicity (Hoechst AG 1990a, 1990b).

## 5.5 Genotoxicity

### 5.5.1 *In vitro*

In the *Salmonella* mutagenicity test, chloroform proved not to be mutagenic. In two of three HPRT tests carried out under identical conditions, chloroform was mutagenic; the authors do not explain the different results. Two SCE tests yielded positive results. In one of these the authors draw attention to the high test concentration of 50 mM which could account for the positive results. None of the studies reveal a clastogenic potential of chloroform. There is evidence that chloroform binds to purified DNA *in vitro* and causes DNA double strand breaks in hepatocytes (Table 5).

**Table 5.** Genotoxicity of chloroform *in vitro*

| Test system | Concentration | S9 mix | Result | References |
|---|---|---|---|---|
| *Salmonella typhimurium* TA98, TA100, TA1535, TA1537, TA1538 | 10–1000 µg/plate | ± | negative | Daniel *et al.* 1980 |
|  | 10–10000 µg/plate | ± | negative | Van Abbe *et al.* 1982 |
|  | up to 15000 µg/plate | ± | negative | Gocke *et al.* 1981, Nestmann *et al.* 1980 |
| *S. typhimurium* TA1535, TA1538, *Escherichia coli* K12 | 5 mM | ± | negative | Greim *et al.* 1977 |
| *S. typhimurium* TA100 | 30–10000 µg/ml | ± | negative | Le Curieux *et al.* 1995 |
| *S. typhimurium* TA98, TA1535, TA1537 | 0.02–0.3 % (v/v) | ± | negative | San Augustin and Lim-Sylianco 1978 |
| *S. typhimurium* TA98, TA100, TA1537 | 50–5000 µg/plate (DMSO) | ± | negative | MacDonald 1981 |
| *E. coli* WP2p, WP2uvrA⁻p | 0.1–1000 µg/plate | ± | negative | Kirkland *et al.* 1981 |
| *E. coli* PQ37 (SOS chromotest) | 10–10000 µg/ml | ± | negative | Le Curieux *et al.* 1995 |
| *E. coli* WP2$_s$ ($\gamma$) (microscreen prophage-induction assay) | 225–3750 µg/ml | ± | negative | DeMarini *et al.* 1991 |
| *B. subtilis* H17; M45 (rec assay) | 20 µl/plate | + | negative | Kada 1981 |
| *Saccharomyces cerevisiae* D7 (mitotic gene conversion) | 21–54 mM | – | positive (high cytotoxicity) | Callen *et al.* 1980 |
| *Saccharomyces cerevisiae* D4 (mitotic gene conversion) | 0.33–333.3 µg/plate | – | negative | Jagannath *et al.* 1981 |
| V79 (HPRT test) | 100-1500 µg/ml | ± | positive results in 2/3 tests | Hoechst AG 1987 |
|  | 1–2.5% | – | negative | Sturrock 1977 |
| L5178Y TK$^{+/-}$ (mouse lymphoma test) | 0.39–1.5 µl/ml | – | negative | Mitchell *et al.* 1988 |
|  | 0.007–0.06 µl/ml | + / – | positive / negative | Caspary *et al.* 1988 |
| human lymphocytes (chromosomal aberrations) | 50–400 µg/ml | ± | negative | Kirkland *et al.* 1981 |

**Table 5.** continued

| Test system | Concentration | S9 mix | Result | References |
|---|---|---|---|---|
| human lymphocytes (sister chromatid exchange) | 25–400 µg/ml | ± | negative | Kirkland *et al.* 1981 |
| | 0.016–50 mM | no data | positive | Morimoto and Koizumi 1983 |
| CHO (sister chromatid exchange) | 7000 ml/m$^3$ | ± | negative | White *et al.* 1979 |
| K$_3$D (sister chromatid exchange) | 0.02–2 mM | + − | positive negative | Fujie *et al.* 1993 |
| SHE (sister chromatid exchange) | 0.1–10 mM | ± | positive | Suzuki 1987 |
| rat hepatocytes (unscheduled DNA synthesis) | 0.8 µM–8.4 mM | − | negative | Althaus *et al.* 1982 |
| human lymphocytes (unscheduled DNA synthesis) | 2.5–10 µl/ml | ± | negative | Perocco and Prodi 1981 |
| SHE (unscheduled DNA synthesis) | 0.3–20 mM | ± | negative | Suzuki 1987 |
| mouse hepatocytes (unscheduled DNA synthesis) | 0.01–10 mM | − | negative | Larson *et al.* 1994c |
| rat hepatocytes (alkaline elution) | 0.03–3 mM | − | negative | Sina *et al.* 1983 |
| hepatocytes F344 rat, B6C3F$_1$ mouse (DNA double strand breaks) | 0.1–5.0 mM | − | positive (no cytotoxicity) | Ammann and Kedderis 1997 |
| calf thymus (DNA binding) | 0.3 nmol/mg (radioactivity count) | − | positive | DiRenzo *et al.* 1982 |
| BHK-21 CL3/HRC 1 (cell transformation) | no data | ± | negative | Daniel *et al.* 1980 |
| BHK cells (cell transformation) | 0.025–250 µg/ml | ± | negative | Styles 1981 |
| SHE, SA7 adenovirus (viral transformation) | 0.12–2.0 ml per exposure chamber | − | positive | Hatch *et al.* 1983 |

### 4.5.2 *In vivo*

The studies of *in vivo* genotoxicity of chloroform are reviewed in Table 6.

Both positive and negative results have been obtained. A test for chromosomal aberrations yielded positive results in the rat after single intraperitoneal injections of 1.2 mg/kg body weight but negative results in the mouse after three intraperitoneal injections of 200–1000 mg/kg body weight. Seen in this light, the positive results for chromosomal aberrations obtained in the rat after intraperitoneal injection of the low dose are not explicable and are therefore not included in the present evaluation.

**Table 6.** Genotoxicity of chloroform *in vivo*

| Species | Dose | Result | References |
|---|---|---|---|
| mouse, *S. typhimurium* TA1535, TA1537 (host-mediated assay) | no data | weak positive | San Agustin and Lim-Sylianco 1978 |
| *Drosophila melanogaster* (test for sex-linked recessive lethal mutations) | 25 mM (diet) 0.1 % or 0.2 % (diet) | negative negative | Gocke *et al.* 1981, Vogel *et al.* 1981 |
| *Drosophila melanogaster* (eye spot test) | 2000–16000 ml/m$^3$ (inhalation) | negative | Vogel and Nivard 1993 |

**Tests for chromosome damage**

| Species | Dose | Result | References |
|---|---|---|---|
| hamster (chromosomal aberrations) | 40, 120, 400 mg/kg bw* (p.o.) | positive (at 400 mg/kg bw) | Hoechst AG 1988a |
| rat (chromosomal aberrations) | 1.2–119.4 mg/kg bw, i.p. or p.o., once daily for 5 days (aqueous solution) | positive (from 1.2 mg/kg i.p. or 119.4 mg/kg bw p.o.) | Fujie *et al.* 1990 |
| mouse (chromosomal aberrations) | 200, 400, 800, 1000 mg/kg bw (i.p., corn oil) | negative | Shelby and Witt 1995 |
| mouse (chromosomal aberrations) | 100, 200 mg/kg bw (s.c.) | positive | Sharma and Anand 1984 |
| mouse (sister chromatid exchange) | 25, 50, 100, 200 mg/kg bw, once daily for 4 days (p.o.) in olive oil | positive (from 50 mg/kg bw) | Morimoto and Koizumi 1983 |
| | 300 ml/m$^3$, 3 or 6 h, inhalation | positive | Iijima *et al.* 1982 |
| mouse (micronucleus test) | 238, 476, 952 mg/kg bw, i.p. once daily for 2 days in olive oil | negative | Gocke *et al.* 1981 |
| mouse (micronucleus test) | about 80 % of the LD$_{50}$ (i.p.) twice | negative | Salamone *et al.* 1981 |
| | 100–900 mg/kg bw (administration route not specified) | positive from 700 mg/kg bw, not dose-dependent | San Agustin and Lim-Sylianco 1978 |
| | 200, 400, 600, 800 mg/kg bw (i.p., corn oil), once daily for 3 days | weak positive | Shelby and Witt 1995 |
| | 0.0225–0.09 mg/kg bw, i.p. once daily for 2 days in DMSO | negative | Tsuchimoto and Matter 1981 |
| rat (micronucleus test) (kidney) | 480 mg/kg bw (corn oil, p.o.) | positive | Robbiano *et al.* 1998 |

**Table 6.** continued

| Species | Dose | Result | References |
|---|---|---|---|
| **Tests for DNA repair** | | | |
| rat (autoradiography) (liver) | 40, 400 mg/kg bw (corn oil, p.o.) | negative | Mirsalis *et al.* 1982 |
| rat (autoradiography) (liver) | 200, 500 mg/kg bw (corn oil, p.o.) | ambiguous | Mirsalis *et al.* 1989 |
| mouse (autoradiography) (liver) | 238, 477 mg/kg bw (corn oil, p.o.) | negative | Larson *et al.* 1994c |
| **Tests for DNA single strand breaks** | | | |
| rat (alkaline elution) (kidney) | 1.5 mmol/kg bw (about 180 mg/kg bw ) (p.o.) (4 % Emulphor$^®$) | negative | Potter *et al.* 1996 |
| **Tests for DNA binding** | | | |
| mouse (liver) | 15 mg/kg bw (i.p.) | negative | Diaz-Gomez and Castro 1980 |
| mouse (liver, kidney) (radioactivity) | 119 mg/kg bw (corn oil, p.o.) | negative | Pereira *et al.* 1982 |
| rat (liver, kidney) (radioactivity) | 48 mg/kg bw (p.o.) | weak positive | Pereira *et al.* 1982 |
| mouse (liver, kidney) (radioactivity) | 15, 60, 240 mg/kg bw (p.o.) | negative | Reitz *et al.* 1982 |
| mouse (sperm abnormality test) | 0.025–0.25 mg/kg bw, (i.p.) 5 days (corn oil) | negative | Topham 1980 |
| | 400, 800 ml/m$^3$, 4 h/d, 5 days, inhalation | positive | Land *et al.* 1981 |

* bw  body weight, p.o. *per os*

Positive results were obtained in a test for chromosomal aberrations in the mouse after subcutaneous injection of 100 or 200 mg/kg body weight (Sharma and Ananad 1984). After oral administration of doses of 119 mg/kg body weight to rats, chromosome damage was found. When assessing these results it must be taken into account that the NOEL for the target organ liver in the rat is 34 mg/kg body weight after gavage and 53 mg/kg body weight after administration of the substance with the drinking water. A NOEL of 10 mg/kg body weight for the target organ liver in the mouse was derived after administration of the substance by gavage.

A micronucleus test in which chloroform was administered to mice by intraperitoneal injection yielded weak positive results, but negative results were obtained in another similar test. The other available micronucleus tests also yielded negative results, apart from one which is not considered to be meaningful because the control values were too high (San Augustin and Lim-Sylianco 1978).

Tests for unscheduled DNA synthesis (UDS) in hepatocytes from the rat and mouse yielded negative results (Larson *et al.* 1994c, Mirsalis *et al.* 1982) but cell proliferation was increased by chloroform (Mirsalis *et al.* 1989). Negative results were obtained in a test for sex-linked recessive lethal (SLRL) mutations in the germ cells of *Drosophila melanogaster* and in a test for somatic mutations (eye spot test) after administration of chloroform in the diet or by inhalation. In several DNA binding studies, the effects of chloroform were in the range of the detection limit of the test.

The data provide evidence of a clastogenic potential of chloroform *in vivo* when administered in doses which are cytotoxic or which have been associated with an increased tumour incidence in carcinogenicity studies.

# 5.6 Carcinogenicity

## 5.6.1 Initiation promotion studies

15-day-old Swiss mice were given ethylnitrosourea doses of 5 or 20 mg/kg body weight by intraperitoneal injection. From the age of 5 weeks until they were 51 weeks old, the mice were given chloroform (1800 mg/l) in the drinking water. The female mice developed no liver tumours either with or without chloroform. In the male mice the number of liver adenomas induced by ethylnitrosourea was reduced by about one half by the subsequent treatment with chloroform. Chloroform alone induced neither liver nor lung tumours (Pereira *et al.* 1985).

Male B6C3F$_1$ mice were treated with the initiator diethylnitrosamine (10 mg/l drinking water, 4 weeks) and then given chloroform in concentrations of 600 or 1800 mg/l in the drinking water every day for 52 weeks. In the high concentration group chloroform reduced the incidence of liver and lung tumours in the animals pretreated with diethylnitrosamine. Without initiation with diethylnitrosamine, the incidence of tumours in the animals treated with chloroform was not higher than in the control group (Klaunig *et al.* 1986).

Groups of 4–6 female Sprague-Dawley rats were treated once for initiation with an oral diethylnitrosamine dose of 8 mg/kg body weight and then, from one week later, twice weekly for 11 weeks with chloroform doses of 25, 100, 200 or 400 mg/kg body weight. An additional four groups of female rats were given the chloroform doses without previous initiation with diethylnitrosamine. Chloroform on its own had no effect on the liver. In the rats treated with diethylnitrosamine, chloroform doses of 100 mg/kg body weight or more caused a dose-dependent increase in the number and size of preneoplastic liver foci (Deml and Oesterle 1987).

Thus, the data provide no evidence of an initiating effect of chloroform but suggest a tumour-promoting effect.

## 5.6.2 Long-term studies

For an assessment of the carcinogenic potential of chloroform, three appropriate long-term studies are available (Jorgenson *et al.* 1985, Matsushima 1994, NCI 1976; Table 7).

The other studies which have been published are not appropriate for evaluation because only one exposure concentration was used and because of low survival of the animals (Palmer *et al.* 1979, Tumasonis *et al.* 1985).

Administration of chloroform doses of 90 or 180 mg/kg body weight by gavage in corn oil to male Osborne-Mendel rats resulted in a dose-dependent increase in the incidence of renal adenomas and carcinomas. In female rats, the incidence of thyroid tumours was increased in a dose-dependent manner. These tumours were considered not biologically relevant by the authors because their spontaneous incidence in this strain of rat is high and they were of various origins so that statistical significance was attained only in the female rats. The incidence of liver tumours was increased in both male and female B6C3F$_1$ mice (NCI 1976). The results of other long-term studies (Bull *et al.* 1986, Larson *et al.* 1994a, 1994d, 1995a, 1995b) in which chloroform was administered orally indicate that cytotoxic effects on kidney and liver may be expected after doses of 90 mg/kg body weight.

Re-evaluation of the NCI study (1976) revealed an increased incidence of cholangiofibromas and cholangiocarcinomas in the female rats. The female rats were more sensitive than the male rats to induction of tumours in parenchyma and bile duct cells. Liver necrosis was also detected frequently but no cirrhosis. In addition, it was established that the renal carcinomas in the male rats only developed in the presence of chronic inflammation of the kidneys. The other tumour incidences were confirmed. The re-evaluation of the mouse findings confirmed the increased incidence of liver tumours. In addition, a significant increase in the incidence of lymphomas was found in the low dose group. In 23 % of the high dose group female animals cardiac thrombosis was also detected, a very rare finding which was therefore considered to be substance-related (Reuber 1979).

After long-term administration of chloroform in the drinking water, a significantly increased incidence of tubulus cell adenomas, adenocarcinomas and renal tumours was seen in the male rats of the high dose group (160 mg/kg body weight); the incidence of nephropathy was 92 %. The incidence of thyroid tumours was significantly reduced in the exposed animals. In the female mice, no increases in tumour incidence were observed. The authors pointed out that they were unable to confirm the increased incidence of liver tumours in mice which was found in the NCI study. They ascribed the different results to the different administration routes or to the vehicle (corn oil) used in the NCI study (Jorgenson *et al.* 1985). As shown in the study by Bull *et al.* (1986) (see Table 3), administration of chloroform in corn oil causes much more severe toxic effects on the liver than does administration of the substance in aqueous suspension. The liver tumours which developed during the NCI study were considered to be a result of the hepatotoxic effects.

The results of the drinking water study (Jorgenson *et al.* 1985) were re-evaluated with respect to cytotoxicity and regenerative cell proliferation in the kidneys of the Osborne-Mendel rats. In the highest dose group, an increase in the basophil count, vacuolation, karyomegaly, nuclear polymorphism and slight hyperplasia were seen in the tubulus cells after 6, 12, 18 or 24 months. These changes were also seen in the 81 mg/kg group after 18 and 24 months. This re-evaluation demonstrates that the renal tumours only developed in the presence of cytotoxicity and regenerative hyperplasia (Hard and Wolf 1999).

**Table 7.** Carcinogenicity studies with chloroform

| Author: | NCI 1976 |
|---|---|
| Substance: | chloroform (99 %, 0.5 %–1 % ethanol) |
| Species: | rat (Osborne-Mendel), mouse (B6C3F$_1$), groups of 50 ♂ and 50 ♀, control animals: 20 per sex and species, vehicle: corn oil, positive control: carbon tetrachloride, rat: ♂ 47, 94 mg/kg body weight, ♀ 80, 160 mg/kg body weight, mouse: 1250–2500 mg/kg body weight: |
| Administration route: | gavage, vehicle corn oil |
| Dose: | rat: ♂ 90, 180 mg/kg body weight, ♀ 100, 200 mg/kg body weight (MTD), mouse: ♂ 138, 277 mg/kg body weight, ♀ 238, 477 mg/kg body weight |
| Duration: | 78 weeks, 5 days/week, recovery period for the rat until week 111, for the mouse until week 92 or 93 |
| Toxicity: | after week 10: reduced feed consumption in the treated groups, urine discoloured, eyelids reddened, reduced body weight gains<br>in first year: discoloured urine, hunched posture, wheezing<br>in second year: symptoms more severe, piloerection, hair loss on the extremities and rump, wounds on the body, head and tail |

| | | | controls | low dose | high dose |
|---|---|---|---|---|---|
| Survivors in week 80 | rat | ♂ | c 95 % | c 76 % | c 52 % |
| | | ♀ | c 90 % | c 58 % | c 45 % |
| | mouse | ♂ | c 75 % | c 85 % | c 82 % |
| | | ♀ | c 88 % | c 88 % | c 72 % |

| **Results: rat** | | controls | low dose | high dose |
|---|---|---|---|---|
| histiocytomas | ♂ | 2/19 | 4/50 | 1/50 |
| skin tumours | ♂ | 0/19 | 2/50 | 0/50 |
| liver: neoplastic nodules | ♂ | 0/19 | 1/50 | 2/50 |
| | ♀ | 2/20 | 4/49 | 3/48 |
| liver carcinomas | ♂ | 0/19 | 0/50 | 1/50 |
| | ♀ | 0/20 | 1/49 | 0/48 |
| bile duct haematoma | ♀ | 0/20 | 0/49 | 1/48 |
| kidney adenomas and adenocarcinomas | ♂ | 0/19 | 6/50 | 13/50 |
| | ♀ | 0/20 | 1/49 | 2/48 |
| renal pelvis carcinoma | ♀ | 0/20 | 0/49 | 1/48 |
| thyroid carcinomas and adenomas | ♂ | 4/19 | 3/50 | 4/50 |
| | ♀ | 1/19 | 8/49 | 10/46 |
| pituitary adenomas | ♂ | 0/16 | 4/44 | 1/47 |
| | ♀ | 6/20 | 3/45 | 10/45 |
| mammary adenomas | ♂ | 1/19 | 0/50 | 0/49 |
| | ♀ | 7/20 | 13/48 | 10/46 |
| mammary adenocarcinoma | ♀ | 0/20 | 0/48 | 1/46 |

**Table 7.** continue

**Results: mouse**

| | | controls | low dose | high dose |
|---|---|---|---|---|
| alveolar cell adenomas and sarcomas | ♂ | 1/18 | 3/50 | 3/44 |
| | ♀ | 0/20 | 3/46 | 0/41 |
| liver tumours | ♂ | 2/18 | 19/50 | 44/45 |
| | ♀ | 0/20 | 37/45 | 39/41 |
| kidney adenomas and adenocarcinomas | ♂ | 1/18 | 2/50 | 3/45 |
| adrenal haemangiosarcomas | ♂ | 0/18 | 0/50 | 2/44 |
| | ♀ | 0/20 | 0/43 | 1/41 |
| lymph node tumours | ♂ | 0/18 | 1/50 | 3/45 |
| spleen haemangiosarcomas and haemangiomas | ♂ | 0/18 | 1/49 | 2/45 |
| | ♀ | 0/19 | 1/46 | 0/41 |

Author: Reuber 1979 (re-evaluation of the data from NCI 1976)

| **Results: rat** | | controls | vehicle | low dose | high dose |
|---|---|---|---|---|---|
| cholangiofibromas | ♀ | 0/20 | 0/20 | 1/39 | 3/39 |
| cholangiocarcinomas | ♀ | 0/20 | 0/20 | 2/39 | 8/39 |
| hyperplastic nodules | ♂ | 0/20 | 2/20 | 5/50 | 8/49 |
| | ♀ | 1/20 | | 7/39 | 12/39 |
| hepatocellular carcinomas | ♂ | 0/20 | 0/20 | 0/50 | 2/49 |
| | ♀ | 0/20 | | 2/39 | 2/39 |
| thyroid | | | | | |
| adenomas | ♂ | 0/20 | 2/19 | 3/50 | 4/49 |
| | ♀ | 1/20 | 0/20 | 8/39 | 7/39 |
| carcinomas | ♂ | 1/25 | 2/19 | 0/50 | 3/49 |
| | ♀ | 2/20 | 1/20 | 3/39 | 5/39 |
| **Results: mouse** | | | | | |
| lymphomas | ♂ | 0/17 | 0/17 | 14/46 | 10/44 |
| | ♀ | 0/20 | 0/19 | 9/45 | 4/40 |

| | |
|---|---|
| Author: | Jorgenson *et al.* 1985 |
| Substance: | chloroform |
| Species: | rat (Osborne-Mendel) 630 ♂, mouse (B6C3F$_1$) 730 ♀, control animals: 50 per sex and species |
| Administration route: | drinking water |
| Concentration: | rat (number of animals): 200 (330), 400 (150), 900 (50), 1800 (50) mg/l (19, 38, 81, 160 mg/kg body weight); mouse: 200 (439), 400 (150), 900 (50), 1800 (50) mg/l (34, 65, 130, 263 mg/kg body weight) |
| Duration: | 104 weeks |
| Toxicity: | reduced body weight gains at concentrations of 400 mg/l or more with reduced feed consumption, explanation of the dose-dependent increase in survival |

**Table 7.** continued

| Survivors in week 104 | controls | matched controls | 200 mg/l | 400 mg/l | 900 mg/l | 1800 mg/l |
|---|---|---|---|---|---|---|
| rat | 12 % | 54 % | 25 % | 29 % | 60 % | 66 % |
| mouse | c 90 % | c 90 % | c 90 % | c 85 % | c 75 % | c 62 % |

**Results: rat**

| | controls | matched controls | 200 mg/l | 400 mg/l | 900 mg/l | 1800 mg/l |
|---|---|---|---|---|---|---|
| neurofibromas | 2/303 (1)[1] | 1/50 (2) | 2/316 (1) | 1/148 (1) | 0/48 (0) | 3/50 (6) |
| lymphomas leukaemia | 5/303 (2) | 1/50 (2) | 19/316 (6)[3] | 5/148 (3) | 2/48 (4) | 3/50 (6)[3] |
| tumours of the cardiovascular system | 5/303 (2) | 0/50 (0) | 6/316 (2) | 3/148 (2) | 3/48 (6)[3] | 3/50 (6)[2] |
| kidney tumours | 5/301 (2) | 1/50 (2) | 6/313 (2) | 7/148 (5) | 4/48 (6) | 7/50 (14)[3] |
| tubulus cell adenomas | 4/301 (1) | 0/50 (0) | 2/313 (1) | 3/148 (2) | 2/48 (4) | 5/50 (10)[3] |
| tubulus cell adenomas and adenocarcinomas | 4/301 (1) | 1/50 (2) | 4/313 (1) | 4/148 (3) | 3/48 (6) | 7/50 (14)[3] |
| adrenal cortex carcinomas | 91/298 (31) | 16/50 (32) | 86/311 (28)[2] | 36/144 (25)[2] | 17/48 (35) | 11/50 (22)[2] |
| phaeochromocytomas | 76/298 (26) | 8/50 (16) | 71/311 (23) | 25/144 (17) | 12/48 (25) | 5/50 (10) |
| thyroid adenomas | 44/294 (15) | 9/49 (18) | 33/303 (11) | 18/148 (12) | 6/48 (13) | 3/50 (6) |
| thyroid adenomas and carcinomas | 47/294 (16) | 12/49 (24) | 49/303 (16) | 27/148 (18) | 7/48 (15) | 7/50 (14) |

**Results: mouse**

| | controls | matched controls | 200 mg/l | 400 mg/l | 900 mg/l | 1800 mg/l |
|---|---|---|---|---|---|---|
| hepatocellular adenomas | 19/415 | 0/47 | 8/410 | 8/142 | 0/47 | 0/44 |
| hepatocellular carcinomas | 2/415 | 0/47 | 7/410 | 1/142 | 0/47 | 1/44 |

| | |
|---|---|
| Author: | Tumasonis *et al.* 1985 |
| Substance: | chloroform |
| Species: | rat (Wistar) 32 ♂, 45 ♀, control animals: 26 ♂, 22 ♀ |
| Administration route: | drinking water |
| Concentration: | 2900 mg/l (♂ 180 mg/kg body weight, ♀ 240 mg/kg body weight) |
| Duration: | exposure for life |
| Toxicity: | survival and body weight gains were reduced relative to the control values, severe hepatic adenofibrosis |

| Results | | 0 | 2900 mg/l |
|---|---|---|---|
| liver: hyperplastic nodules | ♀ | 0/18 | 10/40 |
| | ♂ | 5/22 | 5/28 |
| hepatocarcinomas | ♀ | 0/18 | 1/40 |
| | ♂ | 0/22 | 1/28 |

**Table 7.** continued

| | | controls | 60 mg/kg body weight: |
|---|---|---|---|
| lymphosarcomas | ♂ | 14/22 | 6/28 |
| | ♀ | 2/18 | 4/40 |
| pituitary tumours | ♂ | 1/22 | 2/28 |
| | ♀ | 6/18 | 1/40 |
| haemangiomas and | ♀ | 0/18 | 2/40 |
| haemangiosarcomas | ♂ | 3/22 | 1/28 |
| kidney adenoma | ♂ | 0/22 | 1/28 |
| kidney carcinoma | ♂ | 0/22 | 1/28 |
| rhabdomyosarcoma | ♂ | 0/22 | 1/28 |
| mesotheliomas | ♂ | 1/22 | 2/28 |

| | |
|---|---|
| Author: | Palmer *et al.* 1979 |
| Substance: | chloroform |
| Species: | rat (Sprague-Dawley) 50 ♂, 50 ♀, control animals: 75 ♂, 75 ♀ |
| Administration route: | gavage, vehicle: toothpaste with ethereal oils |
| Dose: | 60 mg/kg body weight |
| Duration: | 6 days/week, 80 weeks (recovery period 15 weeks) |
| Toxicity: | respiratory diseases caused by infections developed during weeks 42 and 56, relative liver weights significantly reduced in ♀ |
| Results: | no evidence of substance-related increases in the incidence of kidney or liver tumours, incidence of mammary tumours increased in female rats but not significantly |

| | | controls | 60 mg/kg body weight: |
|---|---|---|---|
| Survivors in week 95 | ♂ | 26 % | 32 % |
| | ♀ | 14 % | 28 % |

**Results:**

| | | | |
|---|---|---|---|
| benign mammary tumours | ♀ | 15/50 | 15/49 |
| malignant mammary tumours | ♀ | 1/50 | 6/49 |
| chromophobic pituitary adenomas | ♂ | 0/48 | 3/49 |
| | ♀ | 7/50 | 9/49 |

| | |
|---|---|
| Author: | Matsushima 1994 |
| Substance: | chloroform |
| Species: | 50 rats (F344), 50 mice (BDF$_1$) |
| Administration route: | inhalation |
| Concentration: | rat: 0, 10, 30, 90 ml/m$^3$; mouse: 0, 5, 30, 90 ml/m$^3$ |
| Duration: | 6 h/day, 5 days/week, 2 years |
| Toxicity: | survival unaffected, mouse: ♂ nephrotoxicity after exposure to concentrations of 30 ml/m$^3$ or more |

**Table 7.** continued

| Results: mouse | | 0 | 5 ml/m$^3$ | 30 ml/m$^3$ | 90 ml/m$^3$ |
|---|---|---|---|---|---|
| kidney adenomas and carcinomas | ♂ | 0/50 | 1/50 | 7/50[4] | 12/48[4] |
| liver adenomas and carcinomas | ♂ | 14/50 | 7/50 | 12/50 | 17/48 |
| | ♀ | 2/50 | 2/49 | 4/50 | 6/48 |

[1] incidence (%); [2] p < 0.05; [3] p < 0.01; [4] p < 0.0001

Oral administration of chloroform doses of 15 or 30 mg/kg body weight and day to 8 male and 8 female beagle dogs for a period of 7.5 years caused an increased incidence of liver hyperplasia and fatty degeneration in the livers of the high dose group females. Tumour incidences were not increased (Heywood *et al.* 1979).

Exposure of F344 rats and BDF$_1$ mice by inhalation of chloroform concentrations of 10, 30 or 90 ml/m$^3$ (rat) or 5, 30 or 90 ml/m$^3$ (mouse) for 2 years (6 hours/day, 5 days per week) did not cause tumour development in the rats but did result in a significant increase in the incidence of kidney tumours in male mice exposed to 30 ml/m$^3$ or more. Nephrotoxicity developed at concentrations of 30 ml/m$^3$ or more. In the female animals the incidence of liver tumours was increased from 30 ml/m$^3$; the increase was significant at 90 ml/m$^3$ (no other details) (Matsushima 1994).

### 5.6.3 Dose-effect relationships

In male mice exposed for 13 weeks (Larson et al. 1996) by inhalation of chloroform concentrations of 10 ml/m$^3$ or more, cell proliferation was increased in the kidneys in a concentration-dependent manner; in the 2-year inhalation study (Matsushima 1994) kidney tumours developed in animals exposed to 30 ml/m$^3$ or more. The kidney tumours in the male mouse developed only in the presence of chronic inflammation of the kidneys. It is conceivable that the lack of inflammation of the kidneys and therefore the lack of tumours in female mice is a result of the much lower level in the female kidney of the cytochrome P450 2E1 isozyme which determines the level of reactive metabolites formed from chloroform. For the development of kidney tumours and for the regenerative cell proliferation in the kidney, a NOEL of 5 ml/m$^3$ can be derived.

A statistically significant increase in cell proliferation in the livers of male and female mice was seen after 13 weeks inhalation of a chloroform concentration of 90 ml/m$^3$; exposure to 30 ml/m$^3$ had no effect. A significantly increased incidence of liver tumours was seen in the 2-year study only in the female mice exposed to 90 ml/m$^3$. The correlation between the increase in cell proliferation and the development of tumours is not as clear in the liver as in the kidney; however, only at the highest concentration tested in the inhalation study was an increased incidence of both cell proliferation and tumours observed in the liver (Butterworth *et al.* 1998, Larson *et al.* 1996).

Thus both kidney and liver tumours develop only as a result of cytotoxic effects.

# 6 Manifesto (MAK value/classification)

Chloroform administered orally by gavage causes kidney tumours in male rats, thyroid tumours in female rats and liver tumours in male and female mice. Also after administration in the drinking water, chloroform produced an increase in the incidence of kidney tumours in the male rat. Inhalation of chloroform resulted in an increased incidence of kidney tumours only in male mice; the incidence of tumours in rats was increased neither in the kidney nor in the liver. The results document a carcinogenic potential of chloroform in experimental animals. The findings listed below suggest that the mechanisms of tumour development are largely non-genotoxic. Chloroform has been shown not to be mutagenic in numerous mutagenicity tests *in vitro*. A clastogenic potential of cytotoxic concentrations *in vivo* cannot be excluded. Clastogenic effects are seen only with toxic doses. Binding to the DNA of liver or kidney has not been detected. On the other hand, cytotoxic effects are well documented in animal studies and in man. These cytotoxic effects can be put down to the production of phosgene and the damage to proteins and lipids that it causes. Tumours develop in animal studies only when there are also cytotoxic effects. Thus the main effect of chloroform is to induce regenerative hyperplasia. In an initiation-promotion study, chloroform had tumour-promoting activity in the rat liver. Therefore chloroform is classified in Carcinogen category 4 of Section III of the *List of MAK and BAT Values*.

A MAK value can be established on the basis of the results of animal studies. These are supported by the results of calculations carried out with a PBPK model which indicate that, because of the different cytochrome P450 2E1 enzyme activities in the species mouse, rat and man, exposure of the three species to similar concentrations of chloroform will result in lowest levels of conversion to reactive metabolites in the kidney and liver in man. The NOEL for an increase in cell proliferation in the liver and kidney after inhalation of chloroform for 13 weeks is 5 ml/m$^3$ for the rat and mouse; cytotoxic effects are seen from concentrations of 10 ml/m$^3$. In the rat but not in the mouse, exposure to a concentration of 2 ml/m$^3$ caused minimal effects on the olfactory epithelium. The MAK value is established provisionally at 0.5 ml/m$^3$. The effects on the rat nose require further clarification.

In 1989 on the basis of the information then available (slight foetotoxicity and embryotoxicity in rats exposed to 30 ml/m$^3$), chloroform was classified in Pregnancy risk group B. Since then more studies of exposures at lower concentrations have been published in which exposure to 10 ml/m$^3$ was shown to be neither embryotoxic nor teratogenic for the rat. Thus adequate data are now available for two animal species and indicate that, given observance of the reduced MAK value of 0.5 ml/m$^3$, prenatal toxicity is not to be expected. Chloroform is therefore classified in Pregnancy risk group C.

From the data of Bogen *et al.* (1992) it can be estimated that, during dermal exposure of 2000 cm$^2$ of skin to a saturated aqueous solution of chloroform (8.1 mg/ml) for 1 hour, 2106 mg of chloroform is absorbed. The formulae of Fiserova-Bergerova *et al.* (1990) and Guy and Potts (1993) predict the amounts absorbed to be 2288 and 137 mg, respectively. If the MAK value is observed, the maximum amount of chloroform taken up via the respiratory tract, assuming 100 % retention, is only 25 mg in 8 hours. Therefore, under some conditions, the dermal uptake could be much higher than the uptake by inhalation. After application of chloroform to the skin of rabbits, degenerative changes were seen in the kidneys. This documents systemic toxic effects of dermally applied chloroform. Chloroform is therefore designated with an "H". Since the sensitizing effects of chloroform have not been studied, it cannot be decided whether the substance should be designated with an "S".

# 7 References


Althaus FR, Lawrence SD, Sattler GL, Longfellow DG, Pitot HC (1982) Chemical quantification of unscheduled DNA synthesis in cultured hepatocytes as an assay for the rapid screening of potential chemical carcinogens. *Cancer Res 42*: 3010–3015

Amet Y, Berthou F, Fournier G, Dréano Y, Bardou L, Clèdes J, Ménez JF (1997) Cytochrome P450 4A and 2E1 expression in human kidney microsomes. *Biochem Pharmacol 53*: 765–771

Ammann P, Kedderis GL (1997) Chloroform-induced DNA double-strand breaks in freshly isolated male B6C3F$_1$ mouse and F-344 rat hepatocytes. *Toxicologist 36*: 223

Andersen ME, Krewski D, Withey JR (1993) Physiological pharmacokinetics and cancer risk assessment. *Cancer Lett 69*: 1–14

Barlow SM, Sullivan FM (Eds) (1982) *Reproductive Hazards of Industrial Chemicals.* p 230, Academic Press, London

Bogen KT, Colston Jr BW, Machiacao LK (1992) Dermal absorption of dilute aqueous chloroform, trichlorethylene and tetrachloroethylene in hairless guinea pigs. *Fundam Appl Toxicol 18*: 30–39

Brittebo EB, Kowalski B, Brandt I (1987) Binding of the aliphatic halides 1,2-dibromoethane and chloroform in the rodent vaginal epithelium. *Pharmacol Toxicol 60*: 294–298

Brondeau MT, Bonnet P, Guenier JP, De Ceaurriz J (1983) Short-term inhalation test for evaluating industrial hepatotoxicants in rats. *Toxicol Lett 19*: 139–146

Brown DM, Langley PF, Smith D, Taylor DC (1974) Metabolism of chloroform. I. The metabolism of (14C)chloroform by different species. *Xenobiotica 4*: 151–163.

Bull RJ, Brown JM, Meierhenry EA, Jorgenson TA, Robinson M, Stober JA (1986) Enhancement of the hepatotoxicity of chloroform in B6C3F$_1$ mice by corn oil: implications for chloroform carcinogenesis. *Environ Health Perspect 69*: 49–58

Burkhalter JE, Balster RL (1979) Behavioral teratology evaluation of trichloromethane in mice. *Neurobehav Toxicol 1*: 199205

Butterworth BE, Kedderis GL, Conolly RB (1998) The chloroform cancer risk assessment: a mirror of scientific understanding. *CIIT Act 18*: 1–12

Callen DF, Wolf CR, Philpot RM (1980) Cytochrome P-450 mediated genetic activity and cytotoxicity of seven halogenated aliphatic hydrocarbons in *Saccharomyces cerevisiae. Mutat Res 77*: 55–63

Caspary WJ, Daston DS, Myhr BC, Mitchell AD, Rudd CJ, Lee PS (1988) Evaluation of the L5178Y mouse lymphoma cell mutagenesis assay: interlaboratory reproducibility and assessment. *Environ Mol Mutagen 12, Suppl 13*: 1–18

Chu I, Secours V, Marino I, Villeneuve DC (1980) The acute toxicity of four trihalomethanes in male and female rats. *Toxicol Appl Pharmacol 52*: 351–353

Constan AA, Sprankle CS, Peters J, Kedderis GL, Everitt JI, Wong BA, Gonzalez FL, Butterworth BE (1998) Chloroform metabolism by cytochrome P450 2E1 is required for induction of toxicity in the liver, kidney, and nose of male mice. *Toxicol Sci 42, Suppl 1*: 17 (Abstract)

Corley RA, Mendrala AL, Smith FA, Staats DA, Gargas ML, Conolly RB, Andersen ME, Reitz RH (1990) Development of a physiologically based pharmacokinetic model for chloroform. *Toxicol Appl Pharmacol 103*: 512–527

Daniel MR, Richold M, Allen J, Jones E, Roe FJC, Uttley M, Van Abbe NJ (1980) Bacterial mutagenicity and cell transformation studies with chloroform. *Toxicol Lett 6*: 247

DeMarini DM, Lawrence BK, Brooks HG, Houk VS (1991) Compatibility of organic solvents with the microscreen prophage-induction assay: solvent-mutagen interactions. *Mutat Res 263*: 107–113

Deml E, Oesterle D (1987) Dose-response of promotion by polychlorinated biphenyls and chloroform in rat liver foci bioassay. *Arch Toxicol 60*: 209–211

Deringer MK, Dunn TB, Heston WE (1953) Results of exposure of strain C3H mice to chloroform. *Proc Soc Exp Biol Med 83*: 474–479

Diaz-Gomez MI, Castro JA (1980) Covalent binding of chloroform metabolites to nuclear proteins—no evidence for binding to nucleic acids. *Cancer Lett 9*: 213–218

Dilley JV, Chernoff N, Kay D, Winslow N, Newell GW (1977) Inhalation teratology studies of five chemicals in rats. *Toxicol Appl Pharmacol 41*: 196

DiRenzo AB, Gandolfi AJ, Sipes IG (1982) Microsomal bioactivation and covalent binding of aliphatic halides to DNA. *Toxicol Lett 11*: 243–252

Duprat P, Delsaut L, Gradiski D (1976) Pouvoir irritant des principaux chlorés aliphatiques sur la peau et les musqueuses oculaires du lapin. *Eur J Toxicol Environ Hyg 9*: 171–177

Dix KJ, Kedderis GL, Borghoff SJ (1997) Vehicle-dependent oral absorption and target tissue dosimetry of chloroform in male rats and female mice. *Toxicol Lett 91*: 197–209

Fiserova-Bergerova V, Pierce JT, Droz PO (1990) Dermal absorption potential of industrial chemicals: criteria for skin notation. *Am J Ind Med 17*: 617–635

Fujie K, Aoki T, Wada M (1990) Acute and subacute cytogenetic effects of the trihalomethanes on rat bone marrow cells *in vivo*. *Mutat Res 242*: 111–119

Fujie K, Aoki T, Ito Y, Maeda S (1993) Sister-chromatid exchanges induced by trihalomethanes in rat erythroblastic cells and their suppression by crude catechin extracted from green tea. *Mutat Res 300*: 241–246

Gemma S, Ade P, Sbraccia M, Testai E, Vittozzi L (1996) *In vitro* quantitative determination of phospholipid adducts of chloroform intermediates in hepatic and renal microsomes from different rodent strains. *Environ Toxicol Pharmacol 2*: 233–242

Gocke E, King MT, Eckhardt K, Wild D (1981) Mutagenicity of cosmetic ingredients licensed by the European communities. *Mutat Res 90*: 91–109

Greim H, Bimboes D, Egert G, Göggelmann W, Krämer M (1977) Mutagenicity and chromosomal aberrations as an analytical tool for *in vitro* detection of mammalian enzyme-mediated formation of reactive metabolites. *Arch Toxicol 39*: 159–169

Guengerich FP, Kim D-H, Iwasaki M (1991) Role of human cytochrome P-450 IIE1 in the oxidation of many low molecular weight cancer suspects. *Chem Res Toxicol 4*: 168–179

Guy RH, Potts RO (1993) Penetration of industrial chemicals across the skin: a predictive model. *Am J Ind Med 23*: 711–719

Hard GC, Wolf DC (1999) Re-evaluation of the chloroform 2-year drinking water bioassay in Osborne-Mendel rats indicates that sustained renal tubule injury is associated with renal tumor development. *Toxicologist 48*: 140

Hatch GG, Mamay PD, Ayer MC, Castro BC, Nesnow S (1983) Chemical enhancement of viral transformation in Syrian hamster embryo cells by gaseous and volatile chlorinated methanes and ethanes. *Cancer Res 43*: 1945–1950

Heywood R, Sortwell RJ, Noel PRB, Street AE, Prentice DE, Roe FJC, Wadsworth PF, Worden AN (1979) Safety evaluation of toothpaste containing chloroform III. Long-term study in beagle dogs. *J Environ Pathol Toxicol 2*: 835–851

Hoechst AG (1987) *Chloroform: Detection of gene mutations in somatic cells in culture HGPRT-test with V79 cells*. Report No 870692, unpublished study

Hoechst AG (1988a) *Chromosome aberrations in chinese hamster bone marrow cells*. Report No 880445, unpublished study

Hoechst AG (1988b) *Chloroform: Prüfung auf embryotoxische Wirkung an Wistar-Ratten bei inhalativer Verabreichung.* Report No 880609, unpublished study

Hoechst AG (1990a) *Chloroform: Supplementary inhalation embryotoxicity study in Wistar rats.* Report No 910902, unpublished study

Hoechst AG (1990b) *Chloroform: Supplementary inhalation embryotoxicity study in Wistar rats.* Report No 921047, Amendment No 1 to Report No 910902, unpublished study

Iijima S, Morimoto K, Koizumi A (1982) Induction of sister chromatid exchanges in mouse bone marrow cells by inhaled chloroform. *Igaku No Ayumi 122*: 978–980

Jagannath G, Vultaggio DM, Brusick DJ (1981) Genetic activity of 42 coded compounds in the mitotic gene conversion assay using *Saccharomyces cerevisiae* strain D4. in: Ashby J, de Serres (Eds) (1981) *Progress in mutation research, Vol. 1, Evaluation of short-term tests for carcinogens,* Elsevier, New York, 456–467

Jamison KC, Larson JL, Butterworth BE, Harden, Skinner BL, Wolf DC (1996) A non-bile duct origin for intestinal crypt-like ducts with periductular fibrosis induced in livers of F344 rats by chloroform inhalation. *Carcinogenesis 17*: 675–682

Jorgenson TA, Meierhenry EF, Rushbrook CJ, Bull RJ, Robinson M (1985) Carcinogenicity of chloroform in drinking water to male Osborne-Mendel rats and female B6C3F$_1$ mice. *Fundam Appl Toxicol 5*: 760–769

Kada T (1981) The DNA-damaging activity of 42 coded compounds in the rec-assay. in: Ashby J, de Serres (Eds) (1981) *Progress in mutation research, Vol. 1, Evaluation of short-term tests for carcinogens,* Elsevier, New York, 175–182

Keegan TE, Simmons JE, Pegram RA (1998) NOAEL and LOAEL determinations of acute hepatotoxicity for chloroform and bromodichloromethan delivered in an aqueous vehicle to F344 rats. *J Toxicol Environ Health, Part A 55*: 65–75

Kirkland DJ, Smith KC, Van Abbe NJ (1981) Failure of chloroform to induce chromosome damage or sister chromatid exchanges in cultured human lymphocytes and failure to induce reversion in *E. coli. Food Cosmet Toxicol 19*: 651–656

Klaunig JE, Ruch RJ, Pereira MA (1986) Carcinogenicity of chlorinated methane and ethane compounds administered in drinking water to mice. *Environ Health Perspect 69*: 89–95

Land PC, Owen EL, Linde HW (1981) Morphologic changes in mouse spermatozoa after exposure to inhalational anesthetics during early spermatogenesis. *Anesthesiology 54*: 53–56

Larson JL, Wolf DC, Butterworth BE (1993) Acute hepatotoxic and nephrotoxic effects of chloroform in male F-344 rats and female B6C3F$_1$ mice. *Fundam Appl Toxicol 20*: 302–315

Larson JL, Wolf DC, Butterworth BE (1994a) Induced cytolethality and regenerative cell proliferation in the livers and kidneys of male B6C3F$_1$ mice given chloroform by gavage. *Fundam Appl Toxicol 23*: 537–543

Larson JL, Wolf DC, Morgan KT, Mery S, Butterworth BE (1994b) The toxicity of 1-week exposures to inhaled chloroform in female B6C3F$_1$ mice and male F-344 rats. *Fundam Appl Toxicol 22*: 431–446

Larson JL, Sprankle CS, Butterworth BE (1994c) Lack of chloroform-induced DNA repair *in vitro* and *in vivo* in hepatocytes of female B6C3F$_1$ mice. *Environ Mol Mutagen 23*: 132–136

Larson JL, Wolf DC, Butterworth BE (1994d) Induced cytotoxicity and cell proliferation in the hepatocarcinogenicity of chloroform in female B6C3F$_1$ mice: comparison of administration by gavage in corn oil vs *ad libitum* in drinking water. *Fundam Appl Toxicol 22*: 90–102

Larson JL, Wolf DC, Butterworth BE (1995a) Induced regenerative cell proliferation in livers and kidneys of male F-344 rats given chloroform in corn oil by gavage or *ad libitum* in drinking water. *Toxicology 95*: 73–86

Larson JL, Wolf DC, Mery S, Morgan KT, Butterworth BE (1995b) Toxicity and cell proliferation in the liver, kidneys and nasal passages of female F-344 rats induced by chloroform administered by gavage. *Food Chem Toxicol 33*: 443–456

Larson JL, Templin MV, Wolf DC, Jamison KC, Leininger JR, Mery S, Morgan KT, Wong BA, Conolly RB, Butterworth BE (1996) A 90-day chloroform inhalation study in female and male B6C3F$_1$ mice: implications for cancer risk assessment. *Fundam Appl Toxicol 30*: 118–137

Le Curieux F, Gauthier L, Erb F, Marzin D (1995) Use of the SOS chromotest the Ames-fluctuation test and the new micronucleus test to study the genotoxicity of four trihalo-methanes. *Mutagenesis 10*: 333–341

MacDonald DJ (1981) *Salmonella*/microsome tests on 42 coded chemicals. in: Ashby J, de Serres FJ (Eds) (1981) *Progress in mutation research, Vol. 1, Evaluation of short-term tests for carcinogens*, Elsevier, New York, 285–297

Mavournin KH, Blakey DH, Cimino MC, Salamone MF, Heddle JA (1990) The *in vivo* micronucleus assay in mammalian bone marrow and peripheral blood. A report of the USEPA, Gene-Tox Program. *Mutat Res 239*: 29–80

Matsushima T (1994) *An inhalation carcinogenesis study of chloroform*. Initial Report. Japan Bioassay Laboratory, Japan Industrial Safety and Health Association, 2445 Ohshibahara, Hirasawa Hando Kanagawa 257, Japan

Melnick RL, Kohn MC, Dunnick JK, Leininger JR (1998) Regenerative hyperplasia is not required for liver tumor induction in female B6C3F$_1$ mice exposed to trihalomethanes. *Toxicol Appl Pharmacol 148*: 137–147

Mery S, Larson JL, Butterworth BE, Wolf DC, Harden R, Morgan KT (1994) Nasal toxicity of chloroform in male F-344 rats and female B6C3F$_1$ mice following a 1-week inhalation exposure. *Toxicol Appl Pharmacol 125*: 214–227

Mink FL, Brown TJ, Richabaugh J (1986) Absorption, distribution, and excretion of $^{14}$C-trihalomethanes in mice and rats. *Bull Environ Contam Toxicol 37*: 752–758

Mirsalis JC, Tyson CK, Butterworth BE (1982) Detection of genotoxic carcinogens in the *in vivo-in vitro* hepatocyte DNA repair assay. *Environ Mutagen 4*: 553–562

Mirsalis JC, Tyson CK, Steinmetz KL, Loh EK, Hamilton CM, Bakke JP, Spalding JW (1989) Measurement of unscheduled DNA synthesis and S-phase synthesis in rodent hepatocytes following *in vivo* treatment: testing of 24 compounds. *Environ Mol Mutagen 14*: 155–164

Mitchell AD, Myhr BC, Rudd CJ, Caspary WJ, Dunkel VC (1988) Evaluation of the L5178Y mouse lymphoma cell mutagenesis assay: methods used and chemicals evaluated. *Environ Mol Mutagen 12, Suppl 13*: 37–101

Morgan D, Black A, Belcher DR (1970) The excretion in breath of some aliphatic halogenated hydrocarbons following administration by inhalation. *Ann Occup Hyg 13*: 219–233

Morgan D, Cooper SW, Carlock DL, Sykora JJ, Sutton B, Mattie DR, McDougal JN (1991) Dermal absorption of neat and aqueous volatile organic chemicals in the Fischer 344 rat. *Environ Res 55*: 51–63

Morimoto K, Koizumi A (1983) Trihalomethanes induce SCE in human lymphocytes *in vitro* and mouse bone marrow cells *in vivo*. *Environ Res 32*: 72–79

Munson AE, Sain LE, SandersVM, Kauffmann BM, White KL, Page DG, Barnes DW, Borzelleca JF (1982) Toxicology of organic drinking water contaminants: trichloromethane, bromodichloromethane, dibromomethane and tribromomethane. *Environ Health Perspect 46*: 117–126

Murray FJ, Schwetz BA, McBride JG, Staples RE (1979) Toxicity of inhaled chloroform in pregnant mice and their offspring. *Toxicol Appl Pharmacol 50*: 515–522

NCI (National Cancer Institute) (1976) *Report on carcinogenesis bioassay of chloroform*. NIH 76-1279, NCI Bethesda, MD, NTIS PB264018, Springfield, VA, USA

Nestmann ER, Lee EGH, Matula TI, Douglas GR, Mueller JC (1980) Mutagenicity of constituents identified in pulp and paper mill effluents using the *Salmonella*/mammalian microsomes assay. *Mutat Res 79*: 203–212

von Oettingen WF (1964) *The Halogenated Hydrocarbons of Industrial and Toxicological Importance.* p 77, Elsevier, Amsterdam 1964

Palmer AK, Street AE, Roe FJC, Worden AN, Van Abbe NJ (1979) Safety evaluation of toothpaste containing chloroform. II. Long-term studies in rats. *J Environ Pathol Toxicol 2*: 821–833

Paul B B, Rubinstein D (1963) Metabolism of carbon tetrachloride and chloroform by the rat. *J Pharmacol exp Ther 141*: 141–148

Pereira MA, Grothaus M (1997) Chloroform in drinking water prevents hepatic cell proliferation induced by chloroform administered by gavage in corn oil to mice. *Fundam Appl Toxicol 37*: 82–87

Pereira MA, Lin LHC, Lippitt JM, Herren SL (1982) Trihalomethanes as initiators and promoters of carcinogenesis. *Environ Health Perspect 46*: 151–156

Pereira MA, Knutsen GL, Herren-Freund SL (1985) Effect of subsequent treatment of chloroform or phenobarbital on the incidence of liver and lung tumors initiated by ethylnitrosourea in 15 day old mice. *Carcinogenesis 6*: 203–207

Pericin C, Thomann P (1979) Comparison of the acute toxicity of clioquinol, histamine and chloroform in different strains of mice. *Arch Toxicol, Suppl 2*: 371–373

Perocco P, Prodi G (1981) DNA damage by haloalkanes in human lymphocytes cultured *in vitro*. *Cancer Lett 13*: 213–218

Phoon WH, Liang OK, Kee CP (1975) An epidemiological study of an outbreak of jaundice in a factory. *Ann Acad Med Singap 4*: 396–399

Phoon WH, Goh KT, Lee LT, Tan KT, Kwok SF (1983) Toxic jaundice from occupational exposure to chloroform. *Med J Malays 38*: 31–34

Plummer JL, Hall PM, Ilsley AH, Jenner MA, Cousins MJ (1990) Influence of enzyme induction and exposure profile on liver injury due to chlorinated hydrocarbon inhalation. *Pharmacol Toxicol 67*: 329–335

Potter CL, Chang LW, DeAngelo AB, Daniel FB (1996) Effects of four trihalomethanes on DNA strand breaks, renal hyaline droplet formation and serum testosterone in male F-344 rats. *Cancer Lett 106*: 235–242

Rao KN, Virji MA, Moraca MA, Diven WF, Martin TG, Schneider SM (1993) Role of serum markers for liver function and regeneration of chloroform poisoning. *J Anal Toxicol 17*: 99–102

Reitz RH, Fox TR, Quast JF (1982) Mechanistic considerations for carcinogenic risk estimation: chloroform. *Environ Health Perspect 46*: 163–168

Reuber MD (1979) Carcinogenicity of chloroform. *Environ Health Perspect 31*: 171–182

Robbiano L, Mereto E, Morando AM, Pastore P, Brambilla G (1998) Increased frequency of micronucleated kidney cells in rats exposed to halogenated anaesthetics. *Mutat Res 413*: 1–6

Roe FJC, Palmer AK, Worden AN, Van Abbe NJ (1979) Safety evaluation of toothpaste containing chloroform. I. Long-term studies in mice. *J Environ Pathol Toxicol 2*: 799–819

Rosenfeld M (1896) Über die Chloroformnarkose bei bestimmtem Gehalt der Inspirationsluft an Chloroformdämpfen. *Arch Exp Pathol Pharm 37*: 52

Ruddick JA, Villeneuve DC, Chu I, Valli VE (1983) A teratological assessment of four trihalomethanes in the rat. *J Environ Sci Health B 18*: 333–349

Salamone MF, Heddle JA, Katz M (1981) Mutagenic activity of 41 compounds in the *in vivo* micronucleus assay. in: Ashby J, de Serres FJ (Eds) (1981) *Progress in mutation research, Vol. 1, Evaluation of short-term tests for carcinogens*, Elsevier, New York, 686–697

San Agustin J, Lim-Sylianco CY (1978) Mutagenic and clastogenic effects of chloroform. *Bull Philipp Biochem Soc 1*: 17–23

Schwetz BA, Leong BKJ, Gehring PJ (1974) Embryo- and fetotoxicity of inhaled chloroform in rats. *Toxicol Appl Pharmacol 28*: 442–451

Sharma GP, Anand RK (1984) Two halogenated hydrocarbons as inducers of chromosome aberrations in rodents. *Proc Nat Acad Sci India 54*: 61–67

Shelby MD, Witt KL (1995) Comparison of results from mouse bone marrow chromosome aberration and micronucleus tests. *Environ Mol Mutagen 25*: 302–313

Sina JF, Bean CL, Dysart GR, Taylor VI, Bradley MO (1983) Evaluation of the alkaline elution/rat hepatocyte assay as a predictor of carcinogenic and mutagenic potential. *Mutat Res 113*: 357–391

Smith JH, Hook JB (1984) Mechanism of chloroform nephrotoxicity. III. Renal and hepatic microsomal metabolism of chloroform in mice. *Toxicol Appl Pharmacol 73*: 511–524

Smith JH, Maita K, Sleight SD, Hook JB (1983) Mechanism of nephrotoxicity. I. Time course of chloroform toxicity in male and female mice. *Toxicol Appl Pharmacol 70*: 467–479

Spenzer (1954) in: Killian H *Die Narkose*. p 328, Thieme Verlag, Stuttgart 1954

Sturrock JE (1977) Lack of mutagenic effect of halothane or chloroform on cultured cells using the azaguanine test system. *Br J Anaesth 49*: 207–210

Styles JA (1981) Activity of 42 coded compounds in the BHK-21 cell transformation test. in: Ashby J, de Serres (Eds) (1981) *Progress in mutation research, Vol. 1, Evaluation of short-term tests for carcinogens*, Elsevier, New York, 639–646

Suzuki H (1987) Assessment of the carcinogenic hazard of 6 substances used in dental practices. *Shigaku 74*: 1385–1403

Templin MV, Jamison KC, Wolf DC, Morgan KT, Butterworth BE (1996a) Comparison of chloroform-induced toxicity in the kidneys, liver and nasal passages of male Osborne-Mendel and F-344 rats. *Cancer Lett 104*: 71–78

Templin MV, Larson JL, Butterworth BE, Jamison KC, Leininger JR, Mery S, Morgan KT, Wong BA, Wolf DC (1996b) A 90-day chloroform inhalation study in F-344 rats: profile of toxicity and relevance to cancer studies. *Fundam Appl Toxicol 32*: 109–125

Templin MV, Constan AA, Wolf DC, Wong BA, Butterworth BE (1998) Patterns of chloroform-induced regenerative cell proliferation in BDF$_1$ mice correlate with organ specificity and dose-response of tumor formation. *Carcinogenesis 19*: 187–193

Testai E, Di Marzio S, Vittozzi L (1990) Multiple activation of chloroform in hepatic microsomes from uninduced B6C3F$_1$ mice. *Toxicol Appl Pharmacol 104*: 496–503

Testai E, Gemma S, Vittozzi L (1992) Bioactivation of chloroform in hepatic microsomes from rodent strains susceptible or resistant to CHCl$_3$ carcinogenicity. *Toxicol Appl Pharmacol 114*: 197–203

Testai E, Di Marzio S, Di Domenico A, Piccardi A, Vittozzi L (1995) An *in vitro* investigation of the reductive metabolism of chloroform. *Arch Toxicol 70*: 83–88

Testai E, De Curtis V, Gemma S, Fabrizi L, Gervasi P, Vittozzi L (1996) The role of cytochrome P450 isoforms in *in vitro* chloroform metabolism. *J Biochem Toxicol 11*: 305–312

Thompson DJ, Warner SD, Robinson VB (1974) Teratology studies on orally administered chloroform in the rat and rabbit. *Toxicol Appl Pharmacol 29*: 348–357

Topham JC (1980) Do induced sperm-head abnormalities in mice specifically identify mammalian mutagens rather than carcinogens? *Mutat Res 74*: 379–387

Torkelson TR, Oyen F, Rowe VK (1976) The toxicity of chloroform as determined by single and repeated exposure of laboratory animals. *Am Ind Hyg Assoc J 37*: 697–705

Tsuchimoto T, Matter BE (1981) Activity of coded compounds in the micronucleus test. in: Ashby J, de Serres FJ (Eds) (1981) *Progress in mutation research, Vol. 1, Evaluation of short-term tests for carcinogens*, Elsevier, New York, 705–711

Tumasonis CF, McMartin DN, Bush B (1985) Lifetime toxicity of chloroform and bromodi-chloromethane when administered over a lifetime in rats. *Ecotoxicol Environ Safety 9*: 233–240

Van Abbe NJ, Green TJ, Jones E, Richold M, Roe FJC (1982) Bacterial mutagenicity studies on chloroform *in vitro*. *Food Cosmet Toxicol 20*: 557–561

Vittozzi L, Testai E, De Biasi A (1991) Multiple bioactivation of chloroform: a comparison between man and experimental animals. *Adv Exp Med Biol 283*: 665–667

Vogel E, Blijleven WG, Kortselius MJH, Zijlstra JA (1981) Mutagenic activity of 17 coded compounds in the sex-linked recessive lethal test in *Drosophila melanogaster*. in: Ashby J, de Serres (Eds) (1981) *Progress in mutation research, Vol. 1, Evaluation of short-term tests for carcinogens*, Elsevier, New York, 660–665

Vogel EW, Nivard MJM (1993) Performance of 181 chemicals in a *Drosophila* assay predominantly monitoring interchromosomal mitotic recombination. *Mutagenesis 8*: 57–81

White AE, Takehisa S, Eger EI, Wolff S, Stevens WC (1979) Sister chromatid exchanges induced by inhaled anesthetics. *Anesthesiology 50*: 426–430

WHO (World Health Organization) (1994) *Chloroform. IPCS – Environmental health criteria 163*, WHO, Genf

Winslow SG, Gerstner HB (1978) Health aspects of chloroform—a review. *Drug Chem Toxicol 1*: 259–275

Withey JR, Collins BT, Collins PG (1983) Effect of vehicle on the pharmacokinetics and uptake of four halogenated hydrocarbons from the gastrointestinal tract of the rat. *J Appl Toxicol 3*: 249–253

completed 01.03.1999

# 1,4-Dioxane

| | |
|---|---|
| **MAK value (1996)** | **20 ml/m$^3$ (ppm) $\hat{=}$ 73 mg/m$^3$** |
| **Peak limitation (2000)** | **Category I, excursion factor 2** |
| **Absorption through the skin (1966)** | **H** |
| **Sensitization** | **–** |
| **Carcinogenicity (1998)** | **Category 4** |
| **Prenatal toxicity (1989)** | **Pregnancy risk group D** |
| **Germ cell mutagenicity** | **–** |
| **BAT value** | **–** |
| Synonyms | 1,4-diethylene dioxide<br>1,4-diethylene oxide<br>diethylene ether<br>1,4-dioxacyclohexane<br>dioxane<br>*p*-dioxane<br>dioxyethylene ether<br>tetrahydro-*p*-dioxin |
| Chemical name (CAS) | dioxane |
| CAS number | 123-91-1 |
| Structural formula | |
| Molecular formula | $C_4H_8O_2$ |
| Molecular weight | 88.11 |
| Melting point | 11.8°C |
| Boiling point | 101.3°C |
| Vapour pressure at 20°C | 41 hPa |

**1 ml/m$^3$ (ppm) $\hat{=}$ 3.662 mg/m$^3$**　　**1 mg/m$^3$ $\hat{=}$ 0.273 ml/m$^3$ (ppm)**

*Essential MAK Value Documentations.* DFG, Deutsche Forschungsgemeinschaft
Copyright © 2006 WILEY-VCH Verlag GmbH & Co. KGaA, Weinheim
ISBN: 3-527-31394-X

# 1 Toxic Effects and Mode of Action

1,4-Dioxane vapour causes mucosal irritation. Liquid 1,4-dioxane is also irritating in the eye but causes relatively little skin irritation.

1,4-Dioxane causes central nervous depression. In workers exposed to high concentrations of 1,4-dioxane, kidney failure and liver damage were diagnosed; in one case, dermal exposure was reported. In long-term experiments with mice, hepatocellular adenomas and carcinomas developed at the lowest dose tested of 0.05 % (about 50 mg/kg body weight and day). In rats, liver and kidney toxicity as well as a slight increase in the incidence of hepatocellular adenomas or carcinomas was observed in animals given 0.1 % 1,4-dioxane in the drinking water (about 100 mg/kg body weight and day) or more. Inflammation of the nasal turbinates with squamous cell carcinomas was seen in mice and rats given 1,4-dioxane in concentrations of 0.5 % (about 500 mg/kg body weight and day) or more. In an inhalation study, no carcinogenic effects were detected at the only 1,4-dioxane concentration tested of 111 ml/m$^3$, corresponding to a dose of about 100 mg/kg body weight and day.

Most genotoxicity studies have yielded negative results. In the rat liver treated *in vitro* and *in vivo* with cytotoxic concentrations of 1,4-dioxane, DNA strand breaks were observed.

Administered to pregnant rats in maternally toxic doses, 1,4-dioxane caused reductions in foetal body weights. Teratogenic effects were not observed. There are no studies of the reproductive toxicity in other species.

# 2 Mechanism of Action

1,4-Dioxane was shown to be carcinogenic in several drinking water studies in rats, mice and guinea pigs. The target organs were mainly liver and nasal cavities. 1,4-Dioxane is considered to be a typical cytotoxic, but not genotoxic carcinogen. Furthermore, the cytotoxicity requires exposure levels resulting in accumulation of 1,4-dioxane and enzyme induction. Without these, no carcinogenic effects are expected.

In the dose range up to about 10 mg/kg body weight and day, 1,4-dioxane is metabolized rapidly by the rat and by man to 1,4-dioxane-2-one (or β-hydroxyethoxyacetic acid) and excreted in the urine with an excretion half-time of about 1 hour (Young *et al.* 1977, 1978a, 1978b). In the dose range between 10 and 100 mg/kg body weight and day, the metabolism becomes saturated (Young *et al.* 1978a, 1978b) and the levels of 1,4-dioxane in blood and urine increase disproportionately with accumulation of the parent compound in the circulating blood (Young *et al.* 1978a, 1978b). For doses in this range and above, enhanced cell proliferation, toxic effects in liver and kidney and carcinogenic effects in liver and nasal epithelium could be established. It has been shown that if the eliminating metabolism becomes saturated, a further pathway is induced with some latency leading to formation of a further species with a low excretion rate, presumably dioxan-2-ol and its hemiacetal 2-hydroxyethoxyacetaldehyde, a potential reactive metabolite (Hecht and Young 1981, Woo *et al.* 1977a).

Several studies have investigated cell proliferating, peroxisomal proliferating and DNA damaging effects of 1,4-dioxane in the liver and the nasal epithelium. Since 1,4-dioxane (and even more so, dioxan-2-ol) have protein-denaturing effects, one would expect cytostatic as well as proliferating effects, the latter being caused by the replacement of necrotic cells.

## Liver

1,4-Dioxane was administered in the drinking water at approximately 10 or 1000 mg/kg body weight and day for 11 weeks. Hepatocytes were isolated by collagenase perfusion and labelled *in vitro* with 6-$^3$H-TdR. Relative liver weights and the labelling index were increased at 1000 mg/kg body weight and day, but not at 10 mg/kg body weight and day (Stott *et al.* 1981).

The hepatic labelling index was investigated in rats by administration of $^3$H-TdR (single injection or osmotic pump) and subsequent quantitative historadiography. Administration of 0.01 % 1,4-dioxane in the drinking water (about 10 mg/kg body weight and day) for up to 28 days did not increase the labelling index. Administration of 0.1 % for 9 weeks, 1 % for 2 weeks and 2 % for 1 week led to a dose-dependent increase in the labelling index which could be detected after more than 3 days of administration (BASF 1997). Also administration of 1 % 1,4-dioxane via the drinking water for 2 weeks (about 1000 mg/kg body weight and day) followed by a single 1,4-dioxane dose of 1000 mg/kg body weight resulted in an increase in the labelling index. No such effects were seen after a single dose. No activity was seen in the *in vivo* hepatocyte DNA repair assay (as an indicator of reactivity with DNA) in animals given a single 1,4-dioxane dose of up to 1000 mg/kg body weight or up to 2 % 1,4-dioxane in the drinking water for 1 week. Treatment of rats with 1 % 1,4-dioxane in the drinking water for 5 days yielded no increase in liver/body weight nor induction of palmitoyl CoA oxidase, indicating that dioxane does not belong to the class of peroxisomal proliferating carcinogens (Goldsworthy *et al.* 1991). The authors conclude that repair-inducing DNA adduct formation and peroxisomal proliferation in the liver do not appear to be involved in tumour formation by 1,4-dioxane but that there may be a role of 1,4-dioxane-induced cell proliferation in the formation of liver tumours, the quantitative relationships between induced cell proliferation and tumourigenic potential being still to be established (Goldsworthy *et al.* 1991).

The application of 1,4-dioxane doses of 1000, 1500 or 2000 mg/kg body weight by gavage increased replicative DNA synthesis detected by the [$^3$H]thymidine technique. At 2000 mg/kg body weight, also an increased replicative DNA synthesis detected by 5-bromo-2'-deoxyuridine incorporation was observed. According to the authors, the findings support the hypothesis that the capacity to induce cell proliferation may play a key role in 1,4-dioxane hepatocarcinogenesis (Miyagawa *et al.* 1999).

**Nose**

No DNA repair was seen in either nasoturbinate or maxilloturbinate nasal epithelial cells isolated from animals treated with 1 % 1,4-dioxane in the drinking water for 8 days (about 1000 mg/kg body weight and day) followed by a single 1,4-dioxane dose of up to 1000 mg/kg body weight by gavage 12 h before sacrifice (Goldsworthy *et al.* 1991).

Re-examination of the nasal passages of male rats in archived material from the NTP bioassay (NCI 1978), revealed that the primary site of tumour formation was the anterior third of the dorsal meatus. The location of these tumours supports the proposal that inhalation of 1,4-dioxane-containing drinking water may account for the site specificity of these nasal lesions. *In vivo* studies showed no increase relative to controls in cell proliferation at the site of tumour formation in the nose in response to 1 % 1,4-dioxane in the drinking water for 2 weeks (Goldsworthy *et al.* 1991).

Thus, short-term induction of cell proliferation in the nose does not appear to be involved in tumour formation by 1,4-dioxane (Goldsworthy *et al.* 1991).

# 3 Toxicokinetics and Metabolism

## 3.1 Uptake

Exposure of 4 rats via the head and nose for 6 hours to 1,4-dioxane in a concentration of 50 ml/m$^3$ resulted in uptake of doses of about 72 mg/kg body weight (Young *et al.* 1978a, 1978b). When 4 persons were exposed to 50 ml/m$^3$ for 6 hours, the doses of 1,4-dioxane were only about 5.4 mg/kg body weight (Young *et al.* 1977).

In an *in vivo* skin penetration study, $^{14}$C-labelled 1,4-dioxane in methanol or skin lotion (4 µg/cm$^2$) was applied on the forearm of rhesus monkeys without occlusion and removed after 24 hours with soap and water. Urine was collected over a 5-day period. The peak rate of urinary excretion of $^{14}$C was seen during the first 4 hours. Skin penetration was 2.3 % of the dose (methanol) and 3.4 % (skin lotion). Absorption rates were calculated from urinary excretion as 0.2 % (methanol) and 1.0 % (skin lotion). The authors conclude that the volatility of 1,4-dioxane would be expected initially to favour skin penetration, but loss of material by evaporation would reduce the amount which actually traversed the skin barriers (Marzulli *et al.* 1981).

The penetration of 1,4-dioxane through human skin is poor. *In vitro* studies show that 3.2 % of an applied dose passes through excised skin under occlusion and only 0.3 % without occlusion (IARC 1999).

## 3.2 Metabolism

Administered at low exposure concentrations (50 ml/m$^3$) and doses (10 mg/kg body weight), more than 90 % of the absorbed 1,4-dioxane is metabolized by man (Young *et al.* 1977) and the rat (Braun and Young 1977, Young *et al.* 1978a, 1978b) to yield 2-hydroxyethoxyacetic acid which is excreted in the urine (Figure 1). When a different method of isolation was used, dioxan-2-one was also reported as the main metabolite of 1,4-dioxane in the rat (Woo *et al.* 1977a); this substance exists in pH-dependent equilibrium with 2-hydroxyethoxyacetic acid.

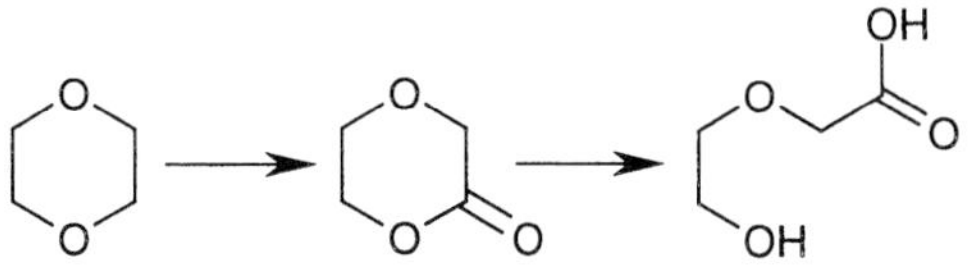

**Figure 1.** Conversion of 1,4-dioxane to dioxan-2-one and 2-hydroxyethoxyacetic acid

The metabolism involves cytochrome P450-dependent oxidation which may be induced by phenobarbital (Woo *et al.* 1977b) or 1,4-dioxane (Young *et al.* 1978a, 1978b).

It has been postulated that when the dioxane-eliminating metabolic pathway is saturated, a second pathway involving $\alpha$-hydroxylation can yield dioxan-2-ol (Hecht *et al.* 1983, Woo *et al.* 1977a). This cyclic hemiacetal is in equilibrium with 2-hydroxyethoxyacetaldehyde, a compound which is presumably reactive and cytotoxic (Figure 2; Young *et al.* 1978b). To date this substance has been detected, by derivatization with 2,4-dinitrophenylhydrazine to yield 2-hydroxyethoxyacetaldehyde-2,4-dinitrophenylhydrazone, only as a metabolite of *N*-nitrosomorpholine which is also metabolized to dioxan-2-ol and 2-hydroxyethoxyacetaldehyde (Hecht and Young 1981). It has not yet been identified as a metabolite of 1,4-dioxane.

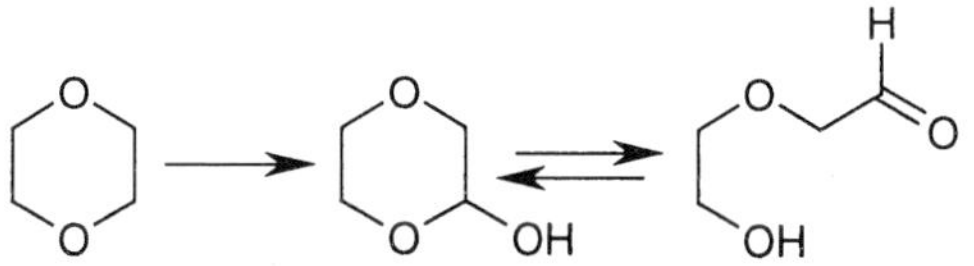

**Figure 2.** Conversion of 1,4-dioxane to dioxan-2-ol and 2-hydroxyethoxyacetaldehyde

## 3.3 Elimination

The elimination half time determined in persons exposed to 1,4-dioxane in a concentration of 50 ml/m$^3$ was $59 \pm 7$ minutes for plasma and $48 \pm 17$ minutes for urine. 90 % of the absorbed substance was excreted with the urine during the exposure period; 6 hours after the end of exposure the substance could no longer be detected (Young *et al.* 1977). The half-life of $^{14}$C-labelled 1,4-dioxane in rat plasma was also about 1 hour, both in animals which had inhaled the substance in a concentration of 50 ml/m$^3$ for 6 hours and in those given single oral doses of up to 10 mg/kg body weight. Of the administered radioactively labelled substance, 98.74 % was excreted with the urine and 3.5 % was exhaled, 3.07 % in the form of $CO_2$ and 0.43 % as unchanged 1,4-dioxane (Young *et al.* 1978a, 1978b; see also Table 1).

In rats given the substance orally, the proportion of the radioactivity excreted in the urine decreased with increasing dose (100, 1000 mg/kg body weight) and the proportion of exhaled 1,4-dioxane increased. The proportions excreted in the faeces and exhaled as $CO_2$, on the other hand, remained unchanged (Young *et al.* 1978a, 1978b; see also Table 1). In animals given 1,4-dioxane by intravenous injection, a dose-dependent increase in the half-life of the substance in plasma from 1.1 hours (3 and 10 mg/kg body weight) to 14.2 hours (1000 mg/kg body weight) was detected, whereby the plasma clearance was reduced from 3.33 to 0.25 ml/min and the metabolic clearance from 2.82 to 0.17 ml/min. The authors interpreted these results as demonstrating saturation of the oxidation of 1,4-dioxane to 2-hydroxyethoxyacetic acid (Kociba *et al.* 1975, Young *et al.* 1978a, 1978b).

When a high dose of 1,4-dioxane (1000 mg/kg body weight) was administered either in 17 aliquots or as a single dose, the animals given the 17 aliquots excreted a larger proportion of the dose in the urine (82.32 % and 75.74 %) and exhaled a larger proportion as $CO_2$ (6.95 % and 2.39 %) and a markedly smaller proportion as unchanged 1,4-dioxane (8.86 % and 25.25 %) than did those given the single dose (see Table 1). According to the authors, the results suggest the induction of metabolizing enzymes (determined *in vitro* as activities of aniline hydroxylase and aminopyridine-*N*-demethylase; no other details; Young *et al.* 1978a, 1978b).

**Table 1.** Elimination of single and repeated oral doses of [14]C-labelled 1,4-dioxane (from Young *et al.* 1978a, 1978b)

| Recovered radioactivity (%) | Single dose [1] | | | 17 aliquots [2] | |
| --- | --- | --- | --- | --- | --- |
| Dose (mg/kg body weight) | 10 | 100 | 1000 | 10 | 1000 |
| urine | 98.74 | 85.52 | 75.74 | 98.87 | 82.32 |
| faeces | 0.95 | 1.95 | 1.06 | 0.46 | 2.05 |
| exhaled 1,4-dioxane | 0.43 | 4.69 | 25.25 | 1.33 | 8.86 |
| exhaled $CO_2$ | 3.07 | 3.13 | 2.39 | 4.17 | 6.95 |
| body | 3.11 | 1.47 | 1.02 | 0.63 | 0.53 |
| total | 106.30 | 96.75 | 105.46 | 105.45 | 100.70 |

[1] mean value from determinations with 3 animals
[2] mean value from determinations with 2 animals

The difficulty in the evaluation of the non-linear kinetics is that the metabolism of 1,4-dioxane is still inadequately understood. 2-Hydroxyethoxyacetic acid is considered to be a detoxification product and main urinary metabolite, but it is in equilibrium with the lactone dioxan-2-one. A conceivable activation pathway would be the directly inducible *alpha*-oxidation to dioxan-2-ol, a hemiacetal in equilibrium with a hydroxyaldehyde (Woo *et al.* 1985). Clarification could be expected from a determination of the dose-dependence of the formation of a critical metabolite or one of its reaction products.

# 4 Effects in Man

No data are available for the reproductive toxicity of 1,4-dioxane in man.

## 4.1 Single exposures

Test persons exposed to a 1,4-dioxane concentration of 5500 ml/m$^3$ for one minute reported eye irritation, a burning feeling in the nose and throat and, in 3 of 5 cases, slight dizziness. Exposure to 1600 ml/m$^3$ for 10 minutes caused slight irritation of the eyes, nose and throat. This concentration was not intolerable but readily noticeable (Yant *et al.* 1930).

During exposure to 2000 ml/m$^3$ for 3 minutes, the test persons detected a marked odour but no mucosal irritation. During a 5-minute exposure to 1000 ml/m$^3$, the odour of the substance was detected immediately. One of four persons reported constriction in the throat and a desire to breathe more quickly (Fairley *et al.* 1934).

2800 ml/m$^3$ caused intensive mucosal irritation, 1400 ml/m$^3$ marked mucosal irritation and 280 ml/m$^3$ slight mucosal irritation but with rapid habituation. At 5.6 ml/m$^3$ the odour of the substance was readily identifiable and at 2.8 ml/m$^3$ still detectable but no longer recognizable as 1,4-dioxane (Wirth and Klimmer 1937).

According to another publication, persons found exposure for 8 hours to 200 ml/m$^3$ still tolerable but higher concentrations caused irritation of the nose, throat and eyes (Silverman *et al.* 1946).

In studies of the kinetics of 1,4-dioxane in persons exposed once for 6 hours by inhalation of 50 ml/m$^3$, eye irritation was reported frequently (Young *et al.* 1977).

## 4.2 Repeated exposures

Liver damage and more especially kidney damage has been described in man. Inhalation of high concentrations of 1,4-dioxane vapour for short periods has even caused deaths from kidney failure; in at least one case, high levels of percutaneous absorption appear to have played a role (Fairley *et al.* 1934, De Nevasquez 1935, Laug *et al.* 1939, Schrenk and Yant 1936, Barber 1934, Johnstone 1959).

There is a report of 5 persons exposed to 1,4-dioxane in a synthetic silk factory in England who became ill with symptoms of renal insufficiency and died within 5 to 8 days. The affected persons had worked in a process which had been used in the factory for a long time without any adverse effects on health. One to two weeks before the persons became ill, one of the two machines was readjusted for technical reasons and this resulted in an increase in the workplace 1,4-dioxane concentration. At the same time, the shift length was increased from 8 to 12 hours. All the persons who died had worked at the readjusted machine. After one to two weeks the first symptoms were gastrointestinal complaints (loss of appetite, vomiting, stomach-ache) followed by progressive renal insufficiency (oliguria, anuria, coma) and death. Autopsy revealed swollen kidneys with subcapsular bleeding and pale, enlarged livers. Microscopic examination of the kidneys revealed haemorrhage in and around the glomerulus, and cortical necrosis. In addition, central lobular necrosis was detected in the liver. 80 other persons working in the same factory had no clinical symptoms. However, only 11 of them were exposed to high levels of 1,4-dioxane and only 3 or 4 to the same levels as the persons who died. From this incident it may be concluded that the intoxication was not the result of long-term exposure to low concentrations but that it was a subacute intoxication caused by a few exposures to high levels of the substance. Analytical data are, however, not available (Barber 1934).

Another worker became ill after working with 1,4-dioxane for a week without a mask in a closed unventilated room. He too developed gastrointestinal symptoms first and also hypertonia and neurological symptoms. He died with renal insufficiency (oliguria, anuria) after one week in hospital. Autopsy and histological examination revealed necrosis in the renal cortex with severe interstitial haemorrhage, and erythrocytes in the tubule lumen. Central lobular necrosis was detected in the liver and in the brain demyelination and partial loss of nerve fibres together with malacia caused by anoxia and oedema. The 1,4-dioxane concentration at the workplace was determined only afterwards and was then between 208 and 650 ml/m$^3$. High-level percutaneous exposure was also said to have been involved. It is very probable that the intoxication was caused by 1,4-dioxane (Johnstone 1959).

An extensive occupational medical study of 74 workers who had been employed for 3 to 4 decades in a factory producing 1,4-dioxane revealed no evidence of adverse effects of the substance on health. The 1,4-dioxane concentrations in the air at various places in the factory were found to lie between 0 and 14.3 ml/m$^3$; simulation of earlier workplace conditions produced levels which were not much higher. Two cases of cancer (haematopoietic organs) diagnosed in persons between 65 and 75 years old was approximately the expected number; chromosome changes determined in 6 workers were in the normal range (Thiess *et al.* 1976).

## 4.3 Local effects on skin and mucosa, allergenic effects

Application of 0.05 ml 1,4-dioxane to human skin, three times within one day, caused dryness but no signs of inflammation (Wirth and Klimmer 1937).

In a 47-year-old female laboratory technician, several weeks of dermal exposure to 1,4-dioxane led to inflammatory skin changes in the upper extremities and, to a lesser extent, in the face. After having been excused work for four weeks, the patient resumed her duties with intact skin. However, just a few days of renewed exposure caused her to have a relapse. Histological examination of the stripy skin changes revealed the clinical symptoms of eczema. In the epicutaneous test the patient showed a positive reaction to undiluted 1,4-dioxane after 48 hours. Control tests on two other persons yielded negative results. The female patient was assumed to have increased dermal sensitivity to toxic allergic substances as a result of an isoprene burn only a few months previously (BUA 1991, Sonneck 1964).

In an inadequately documented additional case report the frequent use of a solvent containing 94 to 97 % 1,1,1-trichloroethane and 2.4 to 3 % 1,4-dioxane for cleaning metal parts has been made responsible for occupational contact dermatitis of the hand of a 52-year-old male worker. He showed a 2+ reaction to a 0.5 % solution of 1,4-dioxane in water; 22 control persons did not show a reaction. In the worker a 25 % solution of the cleaning agent in olive oil caused a 1+ reaction. No reaction was found to 5 % 1,1,1-trichloroethane (Fregert 1974).

## 4.4 Genotoxicity

In human lymphocytes obtained from six workers employed in 1,4-dioxane production and exposed to unspecified airborne levels of the compound for 6 to 15 years, no increase in the incidence of chromosomal aberrations was found relative to that observed in an equal number of controls (IARC 1999, Thiess *et al.* 1976).

## 4.5 Carcinogenicity

In a small prospective mortality study of 165 workers who had been exposed to low concentrations of 1,4-dioxane since 1954, seven deaths had occurred in the manufacturing department by 1975, two of which were from cancer. Expected numbers, based on Texas mortality rates, were 4.9 and 0.9, respectively. In the processing department, five deaths were observed versus 4.9 expected, of which one was from cancer (0.8 expected) (Buffler *et al.* 1978, IARC 1999).

# 5 Animal Experiments and *in vitro* Studies

## 5.1 Acute toxicity

The $LC_{50}$ for a 4-hour exposure of rats to 1,4-dioxane is 142600 ml/m$^3$ (Rowe and Wolf 1982). Data for the acute inhalation toxicity of 1,4-dioxane are shown in Table 2. At the higher concentrations, marked irritation of the mucous membranes was apparent. Deaths occurring during exposure or shortly afterwards were usually due to respiratory failure because of lung oedema, but the animals exhibited congestion of the brain. Delayed deaths were usually due to pneumonia. Liver and kidney injuries were almost always apparent upon microscopic examination of animals dying days after exposure as well as in those apparently recovering but killed several days after exposure (Rowe and Wolf 1981).

**Table 2.** Acute inhalation toxicity of 1,4-dioxane

| Species | Concentration ($ml/m^3$) | Exposure duration (h) | Symptoms | | References |
|---|---|---|---|---|---|
| mouse | 2800 | 8–9.5 | mucosal irritation, unsteady gait | | Wirth and Klimmer 1937 |
| | 8300–3900 | 3.5–1 | mucosal irritation, unsteady gait, death | | |
| rat | 37000 | 1 | death of 0/12 animals | dyspnoea, narcosis, irritation | BASF 1972 |
| | | 2 | death of 6/6 animals | | |
| | | 4 | death of 6/6 animals | | |
| rat | 43000 | 1 | death of 0/12 animals | dyspnoea, apathy, narcosis, irritation, | BASF 1981c |
| | | 3 | death of 6/12 animals | | |
| | | 7 | death of 4/18 animals | | |
| guinea pig | 1000 | 8 | transient nasal irritation | | Yant *et al.* 1930 |
| | 2000 | 8 | mucosal irritation | | |
| | 3000 | 8 | mucosal irritation | | |
| | 30000 | 3.6–9 | mucosal irritation, narcosis, death | | |
| cat | 12000 (technical grade 1,4-dioxane) | 7 | mucosal irritation, unsteady gait, survived | | Wirth and Klimmer 1937 |
| | 12000 (pure 1,4-dioxane) | 7 | mucosal irritation, unsteady gait, death | | |
| | 18000 | 4.3 | mucosal irritation, unsteady gait, lateral position, death of 2/2 animals after 4 and 5 days | | |
| | 24000 | 4 | mucosal irritation, unsteady gait, lateral position, death of 2/2 animals after 1 and 2 days | | |
| | 31000 | 3 | mucosal irritation, unsteady gait, lateral position, death of 1/2 animals after 2 min | | |

Oral toxicity of 1,4-dioxane was studied rather thoroughly by Laug and co-workers (1939). They found the $LD_{50}$ values for mice, rats, and guinea pigs to be 5660, 5170, and 3900 mg/kg body weight, respectively. In rabbits, cats and dogs, $LD_{50}$ values were about 2500, 3500, and 2000 mg/kg body weight (Laug *et al.* 1939). Other $LD_{50}$ values for rats have been reported as 5600, 6200, 7120, 7350 mg/kg body weight (Rowe and Wolf 1981) and 6.3 ml/kg body weight (about 6300 mg/kg body weight; BASF 1958). 1,4-Dioxane in doses of 2.0 ml/kg body weight (about 2000 mg/kg body weight) was lethal for rabbits and dogs, 4.0 ml/kg body weight (about 4000 mg/kg body weight) lethal for cats (BASF 1958).

The symptoms of intoxication were dose-dependent and included central nervous depression, liver damage and especially kidney damage. After lethal doses, the animals generally died within a week with progressive renal insufficiency. Autopsy revealed swollen kidneys with capsular congestion, histological examination revealed typical necrosis in the renal cortex with capsular exudate, haemorrhage and obstruction of the cortical tubules. In some species liver damage was also seen, but the severe kidney damage was predominant (BASF 1958, De Nevasquez 1935, Fairley *et al.* 1934, Laug *et al.* 1939, Schrenk and Yant 1936).

After dermal application of 1,4-dioxane to rabbits, an $LD_{50}$ value of 7600 mg/kg body weight was reported (Rowe and Wolfe 1981). No information about occlusion was given. Application of 1,4-dioxane in doses of 2100, 4200 or 8300 mg/kg body weight to the shaved back and flanks of Wistar rats without occlusion did not result in irritation or mortality (Clark *et al.* 1984).

An $LD_{50}$ value given for i.p. injection into rats was between 2640 and 4400 mg/kg body weight (Knoefel 1935). After subcutaneous injection of 1,4-dioxane into mice, the $LD_{50}$ was 4.21 ml/kg body weight (about 4210 mg/kg body weight) (BASF 1958). The $LD_{50}$ value for i.v. injection into rabbits was between 1.0 and 2.0 ml/kg (about 1000 to 2000 mg/kg body weight; De Nevasquez 1935).

## 5.2 Subacute, subchronic and chronic toxicity

### 5.2.1 Inhalation

Exposure to 1,4-dioxane in a concentration of 1400 ml/m$^3$, 8 hours daily for 17 days, was tolerated by mice without symptoms apart from slight mucosal irritation (Wirth and Klimmer 1937).

Groups of 3 to 5 mice, 3 to 6 rats, 3 to 6 guinea pigs and 2 to 4 rabbits were exposed 11 times weekly for 1.5 hours. At 10000 ml/m$^3$ the animals died after 2 to 7 exposures, at 5000 ml/m$^3$ the mice died after 2 to 34 exposures, the rats after 6 to 10, one guinea pig after 29 and one rabbit after 11 exposures; 3 guinea pigs and 3 rabbits were killed after 33 and 63 exposures, respectively. At 2000 ml/m$^3$ one of 4 rabbits died after 46 exposures; 3 rabbits, 4 guinea pigs, 6 rats and 5 mice were killed after 30 to 68 exposures. After exposure to a concentration of 1000 ml/m$^3$, no animals died; 4 mice, 3 rats, 3 guinea pigs and 2 rabbits were killed after 52 to 135 exposures. Autopsy of the animals that died revealed mainly degenerative and haemorrhagic changes in the renal cortex and in some animals also degenerative liver changes. In the animals which were killed similar changes were found in all concentration groups down to 1000 ml/m$^3$ (Fairley *et al.* 1934).

Inhalation of 1,4-dioxane in a concentration of 1400 ml/m$^3$ for 6½ hours daily by three cats caused sleepiness, occasional vomiting and increased drinking. In the blood, the haemoglobin level and erythrocyte count were increased and lymphocytosis was diagnosed. The animals survived (Wirth and Klimmer 1937).

During a medium-term inhalation study, 7 male and 8 female guinea pigs were exposed to 1,4-dioxane at a concentration of 50 ml/m$^3$ for 7 hours daily 82 times during 118 days, and under the same conditions groups of 24 male and 24 female rats, 3 male and 3 female rabbits and 2 female dogs 130 to 136 times during 180 to 185 days; 12

male and 12 female rats and 2 male and 2 female rabbits were exposed to 100 ml/m³ for 7 hours daily 133 to 136 times. No differences from the control animals were found in appearance, behaviour, growth, mortality, haematology, clinicochemical parameters, organ weights, or in the results of the gross pathological or histological examinations (Torkelson *et al.* 1974).

In a 2-year inhalation study with 288 male and 288 female Wistar rats exposed to 1,4-dioxane in a concentration of 111 ml/m³ (7 h/day, 5 days/week), no organ toxicity and no tumour formation was observed (see Section 5.6.2; Torkelson *et al.* 1974)

## 5.2.2 Oral application

When an 80 % 1,4-dioxane solution was added to the drinking water of rats in a concentration of 5 %, 5 of 6 animals died after 14 to 34 days. The 6th rat and 6 mice survived the treatment for 67 days. Autopsy of the rats revealed markedly enlarged kidneys, the histological examination cortical necrosis and liver cell degeneration. The effects were more marked in rats than in mice (Fairley *et al.* 1934).

Ingestion of up to 15 weekly doses of 0.2 ml/kg body weight was tolerated without symptoms by rabbits (De Nevasquez 1935).

Dioxane concentrations of 5 % in the drinking water were lethal for 2 dogs within 9 to 10 days. The total 1,4-dioxane doses were 3 and 3.5 ml/kg body weight. After a few days, reduced appetite, vomiting, weakness and inability to stand were observed, then increased respiration, increased rest nitrogen and unconsciousness. Autopsy revealed swollen kidneys and livers, and distension of the renal cortex; the stomachs contained a bloody fluid. A third dog survived a total dose of about 3.3 ml/kg body weight administered within 5 days (Schrenk and Yant 1936).

There are a number of long-term studies with application of 1,4-dioxane via drinking water to rats and mice (see Section 5.6.2 and Table 5).

In rats given 1,4-dioxane in the drinking water in a concentration of 0.1 %, toxicity in liver and kidney and slightly increased incidences of liver tumours were observed (Kociba *et al.* 1974, Yamazaki *et al.* 1994). With 1,4-dioxane concentrations of 0.5 % or 0.75 %, marked increases in the incidences of liver tumours can be observed. In addition, inflammation and tumours in the nasal turbinates were seen (Argus *et al.* 1973, NCI 1978, Yamazaki *et al.* 1994). With 0.01 % 1,4-dioxane, no effects could be detected (NOAEL) (Kociba *et al.* 1974).

In mice, effects comparable to those in rats, toxicity in the liver, kidney and nasal turbinates, are observed (NCI 1978, Yamazaki *et al.* 1994). At the lowest concentration tested (0.05 %) the incidences of hepatocellular adenomas and carcinomas were still increased significantly (Yamazaki *et al.* 1994). Therefore, no NOAEL could be derived for mice.

## 5.2.2 Dermal application

Local irritation developed on the shaved skin of rabbits and guinea pigs after application 11 times weekly of 10 drops of an 80 % aqueous 1,4-dioxane solution. Autopsy and histological examination of animals killed after 49, 66, 77 and 101 days revealed cell degeneration in the renal tubules and glomerulus, haemorrhage in the renal medulla and liver cell degeneration (Fairley *et al.* 1934).

## 5.3 Local effects on skin and mucous membranes, allergenic effects

Occlusive application of undiluted 1,4-dioxane on the dorsal skin of rabbits for 1, 5, or 15 minutes, caused relatively little primary skin irritation; oedema was observed after application for 20 hours (BASF 1972). Application of 1,4-dioxane in doses of 2100 to 8300 mg/kg body weight to rats without occlusion did not result in irritation (BUA 1991).

Severe eye irritation was observed in rabbits after instillation of 50 ml 1,4-dioxane (BASF 1972, BUA 1991).

## 5.4 Reproductive and developmental toxicity

In a 2-generation drinking water study with mice and a teratogenicity study (inhalation) with rats and mice 1,4-dioxane was used as a stabilizer for 1,1,1-trichloroethane. Neither in the 2-generation study (highest 1,4-dioxane dose 30 mg/kg body weight) nor in the inhalation study (32 ml/m$^3$) were reproductive toxic effects detected (BUA 1991). Because of the low levels of 1,4-dioxane used in these studies, they are not suitable for evaluation of the reproductive toxicity of the substance.

SD rats (17–20 per group) were given oral 1,4-dioxane doses of 0, 0.25, 0.5 or 1.0 ml/kg body weight from day 6 to day 15 *post coitum*. The highest dose caused slight maternal toxicity in the form of reduced body weight gains. As a result of the reduced maternal weight gains, the foetuses of this group were lighter than the controls. Teratogenic effects were not observed (Giavini *et al.* 1985).

## 5.5 Genotoxicity

The genotoxicity studies with 1,4-dioxane *in vitro* are shown in Table 3, *in vivo* in Table 4.

### 5.5.1 *In vitro*

The tests for mutagenicity in the form of *Salmonella* and *Escherichia coli* mutagenicity tests, thymidine kinase (TK) tests in mouse lymphoma cells, hypoxanthine guanine phosphoribosyl transferase (HPRT) test in CHO cells (a cell line from Chinese hamster ovary), clastogenicity (chromosomal aberrations, micronucleus formation and sister chromatid exchange (SCE) in CHO cells) and for DNA repair (unscheduled DNA synthesis (UDS) in rat hepatocytes) yielded negative results with 1,4-dioxane (see Table 3). A positive result was obtained in a test for DNA strand breaks in rat hepatocytes incubated with 1,4-dioxane in cytotoxic concentrations above 0.3 mM (survival < 57 %). At the non-toxic concentration of 0.03 mM (survival 98 %), the test yielded negative results (Sina *et al.* 1983). In a test for SCE, a weak positive result was obtained only without S9 at the highest tested concentration of 10.2 mg/ml (10.62 SCE/cell; negative control 8.34 SCE/cell; positive control 55.88 SCE/cell) (Galloway *et al.* 1987). The significance of this result is questionable because in this study the negative control value was relatively low and the "weak positive" result was like the control value in other studies.

No binding of radioactivity to calf thymus DNA could be detected after *in vitro* incubation with 1,4-dioxane and microsomes (Woo *et al.* 1977).

The tests which have been carried out with the 1,4-dioxane metabolite dioxan-2-one (*Salmonella* mutagenicity test, HPRT test in CHO cells, UDS test in rat hepatocytes) have yielded negative results (Table 3). Studies with dioxan-2-ol are not available.

**Table 3.** *In vitro* genotoxicity studies with 1,4-dioxane and dioxan-2-one

| Test | Test conditions | Results [1] | | References |
|---|---|---|---|---|
| | | – S9 | + S9 | |
| **1,4-dioxane** | | | | |
| *Salmonella mutagenicity test* TA98, TA100, TA1535, TA1537, TA1538 | up to 10000 µg/plate | – | – | BASF 1979a, Haworth *et al.* 1983, Khudoley *et al.* 1987, Morita and Hayashi 1998, Stott *et al.* 1981 |
| *Escherichia coli mutagenicity test*, WP2 uvrA, WP2 | 156–5000 µg/plate | – | – | Morita and Hayashi 1998 |
| *aneuploidy, Saccharomyces cerevisiae*, D61.M | 1.48–4.75 % | – | n. t. | Zimmermann *et al.* 1985 |
| *TK mutations*, mouse lymphoma cells | 300–5000 µg/ml | – | – | McGegror *et al.* 1991 |
| *TK mutations*, mouse lymphoma cells | 1250–5000 µg/ml | – | – | Morita and Hayashi 1998 |
| *HPRT test*, CHO cells | 50–10000 µg/ml | – | – | BASF 1988 |
| *chromosomal aberrations*, CHO cells | 1050–10520 µg/ml | – | – | Galloway *et al.* 1987 |
| *chromosomal aberrations*, CHO cells | 1250–5000 µg/ml | – | – | Morita and Hayashi 1998 |
| *micronucleus test*, CHO cells | 1250–5000 µg/ml | – | – | Morita and Hayashi 1998 |
| *sister chromatid exchange*, CHO cells | 1000, 3500 µg/ml 10500 µg/ml | – (+) | – – | Galloway *et al.* 1987 |
| *sister chromatid exchange*, CHO cells | 1250–5000 µg/ml | – | – | Morita and Hayashi 1998 |

**Table 3.** continued

| Test | Test conditions | Results[1] | | References |
|---|---|---|---|---|
| | | – S9 | + S9 | |
| *DNA strand breaks (alkaline elution)*, rat hepatocytes | 0.03 mM, 3 h (survival 98 %)<br>0.3–30 mM, 3 h (survival < 57 %) | –<br>+ | n. t. | Sina *et al.* 1983 |
| *DNA repair (UDS)*, rat hepatocytes | 0.00001–1000 mM | – | n. t. | Stott *et al.* 1981 |
| *DNA repair (UDS)*, rat hepatocytes | 0.001–1 mM (with and without pretreatment with 1 % 1,4-dioxane in the drinking water) | – | n. t. | Goldworthy *et al.* 1991 |
| *DNA binding*, isolated DNA from calf thymus | 82 µCi $^3$H-1,4-dioxane, 12 µCi $^{14}$C-1,4-dioxane | n. t. | – | Woo *et al.* 1977c |
| **1,4-dioxane-2-one** | | | | |
| *Salmonella mutagenicity test* TA98, TA100, TA1535, TA1537, TA1538 | 4–2500 µg/plate | – | – | BASF 1979b |
| *HPRT test,* CHO cells | 30–4600 µg/ml | – | – | BASF 1985 |
| *DNA repair (UDS)*, rat hepatocytes | 0.001–1 mM (with and without pretreatment with 1 % 1,4-dioxane in the drinking water) | – | n. t. | Goldworthy *et al.* 1991 |

[1] –   negative
 +   positive
(+)  weak positive
n. t.  not tested

### 5.5.2 *In vivo*

*In vivo* studies (dominant lethal test, DNA repair in rat hepatocytes and nasal epithelial cells, binding of 1,4-dioxane to rat liver DNA; Table 4) have yielded negative results. A positive result was obtained in one of three micronucleus tests in mouse bone marrow (Mirkova 1994) but could not be reproduced (Tinwell and Ashby 1994). In a further micronucleus test in mice, micronucleus formation was observed in liver but not in peripheral lymphoctyes (Morita and Hayashi 1998). Morita and Hayashi (1998) also cited micronucleus formation in rat liver at a 1,4-dioxane dose of 1000 mg/kg body weight (Suzuki *et al.* 1995). A weak positive result for DNA strand breaks in rat liver was obtained in animals given 1,4-dioxane doses of 2500 mg/kg body weight and above. These doses caused slight degenerative vacuolation in the periportal region of the liver but no necrosis (Kitchin and Brown 1990; Table 4).

Application of a single i.p. dose of $^3$H-1,4-dioxane of 5000 µCi/kg body weight to Sprague-Dawley rats resulted in increased radioactivity in the nuclear fraction of the liver (Woo *et al.* 1977c). However, in the liver of Sprague-Dawley rats given an oral $^{14}$C-1,4-dioxane dose of 1000 mg/kg body weight, no alkylation of nucleotides could be identified after isolation of DNA and analysis by high performance liquid chromatography (HPLC) (Stott *et al.* 1981).

**Table 4.** *In vivo* genotoxicity studies with 1,4-dioxane

| Test | Dose | Results[1] | References |
|---|---|---|---|
| *dominant lethal test*, mouse (NMRI) | 2500 µl/kg body weight, i.p. (about 2.58 mg/kg body weight; i.e. ½ LD$_{50}$) | – | BASF 1977 |
| *micronucleus test*, mouse (B6C3F$_1$), bone marrow | 500, 1000, 2000 mg/kg body weight, 3 × i.p.<br>2000, 3000, 4000 mg/kg body weight, 1 × i.p. | –<br>– | McFee *et al.* 1994 |
| *micronucleus test*, mouse (BALB/c), bone marrow | 5000 mg/kg body weight, p.o. | – | Mirkova 1994 |
| *micronucleus test*, mouse (C57BL6), bone marrow | 450 mg/kg body weight, p.o.,<br>900, 1800, 3600 mg/kg body weight, p.o. | –<br>+ | Mirkova 1994 |
| *micronucleus test*, mouse (CBA, C57BL6), bone marrow | 1800 mg/kg body weight (CBA), p.o.<br>3600 mg/kg body weight (C57BL6), p.o. | –<br>– | Tinwell and Ashby 1994 |
| *micronucleus test*, mouse (CD-1), peripheral lymphocytes | 3000 mg/kg body weight, 1 × p.o. | – | Morita and Hayashi 1998 |
| *micronucleus test*, mouse (CD-1), liver | 2000 mg/kg body weight, 1 × p.o. | + | Morita and Hayashi 1998 |
| *micronucleus test*, rat (F344), liver | 1000 mg/kg body weight, p.o. | + | Suzuki *et al.* 1995 |
| *DNA strand breaks*, rat (Sprague-Dawley), liver | 168 mg/kg body weight, 2 × p.o.<br>840 mg/kg body weight, 2 × p.o.<br>2550 mg/kg body weight, 2 × p.o.<br>4200 mg/kg body weight, 2 × p.o. | –<br>–<br>+<br>+ | Kitchin and Brown 1990 |
| *DNA repair*, rat (F344), nasal epithelial cells | 0, 10, 100, 1000 mg/kg body weight p.o. after pretreatment with 1 % 1,4-dioxane in the drinking water for 1 week | – | Goldsworthy *et al.* 1991 |
| *DNA repair*, rat (F344), hepatocytes | 1000 mg/kg body weight p.o.<br>1 % or 2 % 1,4-dioxane in the drinking water for 2 weeks or 1 week, respectively | – | Goldsworthy *et al.* 1991 |
| *DNA repair*, rat (Sprague-Dawley), liver | 1000 mg/kg body weight, p.o. (hydroxyurea 500 mg/kg body weight) | – | Stott *et al.* 1981 |
| *sex-linked recessive lethal test, Drosophila melanogaster* | 35000 ppm in the diet or injection | – | Yoon *et al.* 1985 |

[1] – negative
+ positive

# 5.6 Carcinogenicity

## 5.6.1 Short term studies

### Cell transformation test

In the cell transformation test, exposure of BALB/3T3 cells to 1,4-dioxane in concentrations of 0.1, 0.3 or 1.0 mg/ml for 20 to 24 hours did not cause any discernible increase in the transformation rate or any cytotoxic effects. Exposure time was 2 hours in the presence of S9 mix and 20 to 24 hours without metabolic activation (BASF 1980a, 1980b, BUA 1991).

Other authors recorded transforming activity at concentrations of 2 mg/ml or more, concurrent cytotoxicity following incubation with 1,4-dioxane concentrations of 0.25 to 4 mg/ml for 48 hours using no metabolic activation (Sheu *et al.* 1988). After incubation with 1,4-dioxane in concentrations of 0.25 to 2 mg/ml for 13 days, statistically significant transforming activity was registered at 0.5 and 2 mg/ml, without—according to the authors' data—marked cytotoxicity (Sheu *et al.* 1988). Cytotoxicity was, however, only assessed once—from the cell count—at the end of the 13-day incubation period; hence it is to be assumed that adaptation occurred in the culture (BUA 1991).

The metabolite of 1,4-dioxane, 1,4-dioxan-2-one, was also tested for transforming activity in BALB/3T3 cells, over the range 0.1 to 1.0 mg/ml. In the absence of metabolic activation (S9 mix, incubation for 24 hours) statistically significant transforming activity was detected at concentrations of 0.25 mg/ml or more, also without cytotoxic effects. When S9 mix was added and incubation performed for two hours, the effect was no longer statistically significant (BASF 1981a, 1981b, 1986, BUA 1991).

### Inhibition of metabolic cooperation

Inhibition of metabolic cooperation was observed in V79 cells (in the thioguanine nucleotide transfer test) after treatment with 1,4-dioxane in concentrations of 1 to 16 mg/ml. The effect correlated with the dose used and the severity of the cytotoxic effect. The authors suspect a connection between the two effects (BUA 1991, Chen *et al.* 1984).

In contrast, 1,4-dioxan-2-one did not inhibit metabolic cooperation, even in the cytotoxic range. The concentration range investigated was 0.01 to 1 mg/ml (BASF 1990, BUA 1991).

### Tumour initiation and promotion studies

SENCAR mice (20 to 40 animals per dose group) were given a single oral, subcutaneous or dermal 1,4-dioxane dose of up to 1000 mg/kg body weight and were treated then dermally thrice weekly for 20 weeks with 1.0 µg 12-*O*-tetradecanoylphorbol-13-acetate (TPA). After 24 weeks, the incidence of papillomas had not increased, i.e. 1,4-dioxane did not function as an initiator (promoting activity of 1,4-dioxane following administration of an initiator was not tested) (BUA 1991, Bull *et al.* 1986).

Tumour-promoting activity ($\gamma$-GT-foci) was observed in the liver of male Sprague-Dawley rats (9 animals per dose group). The animals were initiated with DEN, partially hepatectomized and treated with 1000 mg/kg body weight and day (gavage administration; 5 days per week) for 7 weeks. In animals treated with 100 mg/kg body weight and day, no promoting activity was detected (Lundberg *et al.* 1987).

## Short-term carcinogenicity studies

Groups of 30 male A/J mice, six to eight weeks of age, were given 1,4-dioxane (purity unspecified) by intraperitoneal injection three times per week for eight weeks for total doses of 0, 400, 1000 and 2000 mg /kg body weight. The high dose increased the number of lung tumours to 0.97 per mouse ($p < 0.05$) compared with 0.28 per mouse in controls given the vehicle alone (IARC 1999, Maronpot *et al.* 1986).

In a mouse-lung adenoma assay, 1,4-dioxane produced a significant increase in the incidence of lung tumours in males given an intermediate intraperitoneal dose (38 %); no such increase was noted in males given a lower or higher intraperitoneal dose or in females given three intraperitoneal doses or in either males or females given 1,4-dioxane orally (IARC 1999, Stoner *et al.* 1986). The significant increase in lung tumours in the intermediate dose group was, however, attributed to the very low incidence of 7 % tumours in the vehicle control (water); for comparison, the incidence in a second water control group was 33 % (BUA 1991).

## 5.6.2 Long term studies

### 5.6.2.1 Inhalation

In a long-term inhalation study, 288 male and 288 female Wistar rats were exposed in inhalation chambers for 7 hours per day, 5 days per week for 2 years to a 1,4-dioxane concentration of 111 ml/m$^3$ (400 mg/m$^3$). Growth, deaths, haematological and clinico-chemical parameters and the gross pathological and histological examinations revealed no differences from the control values. Tumours did not develop, nor was there liver or kidney damage (Torkelson *et al.* 1974). The daily dose of 1,4-dioxane which could be absorbed via the lungs was about 100 mg/kg body weight and thus of the same order of magnitude as that resulting from ingestion of 0.1 % 1,4-dioxane in the drinking water. Although the latter treatment also produced no tumours, it caused marked liver and kidney damage (see Table 5). The validity of the inhalation study is, however, limited by the fact that the maximum tolerated dose was not attained and that the typical organo-toxic effects of 1,4-dioxane on the liver and kidneys did not develop.

### 5.6.2.2 Oral application

**Rats**

In long-term drinking water studies with the rat, 1,4-dioxane concentrations of 0.5 % to 2.0 % in the drinking water (about 500–2000 mg/kg body weight and day) were hepatotoxic and nephrotoxic and produced tumours of the liver and the nasal cavity and also occasional renal carcinomas (see Table 5).

Tumours developed in 7 of 26 male Wistar rats given 1 % 1,4-dioxane in the drinking water for 63 weeks (about 840 mg/kg body weight and day); 6 of the 7 rats had malignant liver tumours like those seen with dimethylnitrosamine and diethylnitrosamine, one of these 6 animals also had a renal carcinoma and the seventh animal leukaemia (Argus *et al.* 1965; Table 5).

Groups of 30 male Sprague-Dawley rats were given 1,4-dioxane in the drinking water for 13 months in concentrations of 0 (control), 0.75 %, 1.0 %, 1.4 % and 1.8 % (0 and about 760, 1000, 1400 and 1800 mg/kg body weight and day). In the control group no tumours were found (Argus *et al.* 1973, Hoch-Ligeti *et al.* 1970). In each of the 0.75 % and 1 % groups one of 30 animals and in each of the 1.4 % and 1.8 % groups 2 of 30 animals developed keratinizing epithelial papillocarcinomas or adenocarcinomas in the nasal cavity (Hoch-Ligeti *et al.* 1970; Table 5). In animals of the 0.75 %, 1 %, 1.4 % and 1.8 % groups 4, 9, 13 and 11 "incipient" liver tumours were detected as well as 3 and 12 hepatomas in the two high dose groups. In this study the clearly dose-dependent histopathological development of precancerous hepatomas was followed by electron microscopy. The minimum total dose required to produce tumours ($TD_5$) was given as 72 g, the mean ($TD_{50}$) as 149 g and the maximum ($TD_{95}$) as 260 g per rat. In addition in all dose groups, marked kidney damage was observed (Argus *et al.* 1973; Table 5).

In another drinking water study which demonstrates clearly the dose-dependency of the carcinogenic effects of 1,4-dioxane, groups of 60 male and 60 female rats were given 1,4-dioxane in the drinking water for up to 716 days in concentrations of 0.01 %, 0.1 % and 1.0 %. The treatment with 1 % 1,4-dioxane in the drinking water resulted in reduced body weight gains, reduced fluid consumption, liver and kidney damage, reduced survival and hepatocellular carcinomas in 10 of 120 rats and nasal cavity carcinomas in 3 of 120 rats. In the animals given 0.1 % 1,4-dioxane in the drinking water, liver and kidney damage was also detected and—as in the control group—a hepatocellular carcinoma in one of 120 rats. 0.01 % 1,4-dioxane in the drinking water (daily dose for the male animals about 9.6 mg/kg body weight, females about 19 mg/kg body weight) did not cause systemic toxicity or tumour development. In these studies too, the carcinogenic effects of ingested 1,4-dioxane were shown to be clearly dose-dependent (Kociba *et al.* 1974; Table 5). From this study a NOAEL of 0.01 % can be derived.

In a drinking water study, 1,4-dioxane was administered to rats for 110 weeks in concentrations of 1 % and 0.5 %. Significantly increased incidences of squamous-cell carcinoma in the nasal turbinates of male and female animals and of liver adenoma in female animals were recognized in both dose groups (NCI 1978; Table 5).

The carcinogenicity of 1,4-dioxane administered in drinking water was recently confirmed in rats given the substance in concentrations of 0.02 %, 0.1 % or 0.5 % and in mice given 0.05 %, 0.2 % or 0.8 % in the drinking water (50 animals per sex and dose). Male rats given 0.5 % developed nasal cavity metaplasia, proliferation and malignant tumours and increases in the incidence of hepatocellular adenomas and carcinomas, peritoneal mesotheliomas, fibromas of the subcutis and fibromas of the mammary glands. In female rats given 0.5 %, similar effects in the nasal cavities including malignant tumours were observed, furthermore increases in the incidence of hepatocellular

124    *1,4-Dioxane*

adenomas and carcinomas and adenomas of the mammary gland. At 0.1 % no nasal effects at all were observed in male or female rats and no hepatocellular carcinomas. The incidences of hyperplasia of the liver and hepatocellular adenomas, however, were still increased in both sexes. At 0.02 % no effects were noted for the females; for males a slight increase in *spongiosis hepatis* and two hepatocellular adenomas appeared to be of borderline significance in this group (Yamazaki *et al.* 1994; Table 5).

**Table 5.** Drinking water studies with 1,4-dioxane in rats and mice

| Author: | **Argus *et al.* 1965** | | |
|---|---|---|---|
| Species: | rat, Wistar, 26 ♂ | | |
| Administration route: | drinking water | | |
| Dose: | 1 %, about 840 mg/kg body weight and day | | |
| Duration: | 63 weeks | | |
| Toxicity: | 1 %: toxic effects in liver and kidney | | |
| Tumours: | | | |
| 1,4-dioxane concentration in drinking water | | 0 % | 1 % |
| liver tumours | ♂ | 0/9 | 6/26 |
| renal carcinomas | ♂ | 0/9 | 1/26 |
| leukaemia | ♂ | 0/9 | 1/26 |

| Author: | **Argus *et al.* 1973, Hoch-Ligeti *et al.* 1970** | | | | |
|---|---|---|---|---|---|
| Species: | rat, Sprague-Dawley, 30 ♂ per group | | | | |
| Administration route: | drinking water | | | | |
| Dose: | 0.75 %, 1.0 %, 1.4 %, 1.8 % about 760, 1000, 1400, 1800 mg/kg body weight and day | | | | |
| Duration: | 13 months | | | | |
| Toxicity: | > 0.75 %: toxic effects in liver and kidney | | | | |
| Tumours: | | | | | |
| 1,4-dioxane concentration in drinking water | | 0 % | 0.75 % | 1.0 % | 1.4 % | 1.8 % |
| nasal turbinate | | | | | |
| – carcinomas | ♂ | 0/30 | 1/30 | 1/30 | 2/30 | 2/30 |
| liver | | | | | |
| – "incipient tumours" | ♂ | 0/30 | 4/30 | 9/30 ** | 13/30 ** | 11/30 ** |
| – hepatocellular carcinomas | ♂ | 0/30 | 0/30 | 0/30 | 3/30 | 12/30 ** |
| ** p < 0.01, Fisher's exact test | | | | | |

**Table 5.** continued

| Author: | **NCI 1978** | | | |
| --- | --- | --- | --- | --- |
| Species: | rat, Osborne-Mendel, 35 ♂/♀ per group | | | |
| Administration route: | drinking water | | | |
| Dose: | 0.5 %, 1.0 %<br>approx. dose ♂: 240, 530, ♀: 350, 640 mg/kg body weight and day | | | |
| Duration: | 110 weeks | | | |
| Toxicity: | > 0.5 % survival reduced (♂/♀), toxic effects in nose (nasal turbinate inflammation), liver (degeneration and necrosis), kidney (degeneration of the tubuli) | | | |

Tumours:

| 1,4-dioxane concentration in drinking water | | 0 % | 0.5 % | 1 % |
| --- | --- | --- | --- | --- |
| nasal turbinate | | | | |
| – squamous-cell carcinomas | ♂ | 0/33 | 12/33 * | 16/34 * |
| | ♀ | 0/34 | 10/35 * | 8/35 * |
| – adenocarcinomas | ♂ | 0/33 | 0/33 | 3/34 |
| liver | | | | |
| – adenoma or carcinomas | ♂ | 2/31 | 2/32 | 1/33 |
| – adenomas | ♀ | 0/31 | 10/33 * | 11/32 * |

*p < 0.05

| Author: | **Kociba *et al*. 1974** | | | | |
| --- | --- | --- | --- | --- | --- |
| Species: | rat, Sherman, 60 ♂/♀ per group | | | | |
| Administration route: | drinking water | | | | |
| Dose: | 0.01 %, 0.1 %, 1 %<br>approx. dose ♂: 10, 94, 1000 mg/kg body weight and day<br>♀: 19, 148, 1600 mg/kg body weight and day | | | | |
| Duration: | 102 weeks | | | | |
| Toxicity: | 0.01 %: no toxicity, 0.1 %: toxicity in liver and kidney<br>1.0 %: body weight gain and survival reduced (♂/♀), liver and kidney toxicity | | | | |

Tumours:

| 1,4-dioxane concentration in drinking water | | 0 % | 0.01 % | 0.1 % | 1 % |
| --- | --- | --- | --- | --- | --- |
| nasal turbinate | | | | | |
| – squamous cell carcinomas | ♂+♀ | 0/106 | 0/110 | 0/106 | 3/66 * |
| liver | | | | | |
| – hepatocellular carcinomas | ♂+♀ | 1/106 | 0/110 | 1/106 | 10/66 ** |
| – hepatic tumours all types | ♂+♀ | 2/106 | 0/110 | 1/106 | 12/66 ** |

* p = 0.0549, ** p < 0.01

**Table 5.** continued

| Author: | **Yamazaki *et al.* 1994** | | | |
|---|---|---|---|---|
| Species: | rat, F344, 50 ♂/♀ | | | |
| Administration route: | drinking water | | | |
| Dose: | 200, 1000, 5000 ppm (0.02 %, 0.1 %, 0.5 %) approx. dose not given | | | |
| Duration: | 104 weeks | | | |
| Toxicity: | > 5000 ppm: reduced survival (♂/♀) | | | |
| Tumours: | | | | |

| 1,4-dioxane concentration in drinking water | | 0 % | 0.02 % | 0.1 % | 0.5 % |
|---|---|---|---|---|---|
| nasal turbinate | | | | | |
| – proliferation nasal gland | ♂ | 0/50 | 0/50 | 0/50 | 5/50 |
| | ♀ | 0/50 | 0/50 | 0/50 | 11/50 |
| – squamous cell hyperplasia | ♂ | 0/50 | 0/50 | 0/50 | 2/50 |
| | ♀ | 0/50 | 0/50 | 0/50 | 5/50 |
| – squamous cell metaplasia | ♂ | 0/50 | 0/50 | 0/50 | 31/50 |
| | ♀ | 0/50 | 0/50 | 0/50 | 35/50 |
| – esthesioneuroepitheliomas | ♂ | 0/50 | 0/50 | 0/50 | 1/50 |
| | ♀ | 0/50 | 0/50 | 0/50 | 1/50 |
| – rhabdomyosarcomas | ♂ | 0/50 | 0/50 | 0/50 | 1/50 |
| – sarcomas | ♂ | 0/50 | 0/50 | 0/50 | 2/50 |
| – squamous cell carcinomas | ♂ | 0/50 | 0/50 | 0/50 | 3/50 |
| | ♀ | 0/50 | 0/50 | 0/50 | 7/50 * |
| liver | | | | | |
| – hyperplasia | ♂ | 3/50 | 2/50 | 10/50 | 24/50 |
| | ♀ | 3/50 | 2/50 | 11/50 | 47/50 |
| – *spongiosis hepatis* | ♂ | 12/50 | 20/50 | 25/50 | 40/50 |
| | ♀ | 0/50 | 0/50 | 1/50 | 20/50 |
| – hepatocellular adenomas | ♂ | 0/50 | 2/50 | 4/50 | 24/50 |
| | ♀ | 1/50 | 0/50 | 5/50 | 38/50 |
| – hepatocellular carcinomas | ♂ | 0/50 | 0/50 | 0/50 | 14/50 |
| | ♀ | 0/50 | 0/50 | 0/50 | 10/50 |
| – hepatocellular adenomas and carcinomas | ♂ | 0/50 | 2/50 | 4/50 | 38/50 ** |
| | ♀ | 1/50 | 0/50 | 5/50 | 48/50 ** |
| peritoneal mesotheliomas | ♂ | 2/50 | 2/50 | 5/50 | 28/50 ** |
| subcutaneous fibromas | ♂ | 5/50 | 3/50 | 5/50 | 12/50 |
| mammary fibroadenomas | ♂ | 1/50 | 1/50 | 0/50 | 4/50 |
| mammary adenomas | ♀ | 6/50 | 7/50 | 10/50 | 16/50 * |

* p < 0.05, **p < 0.01, Fisher's exact test

**Table 5**. continued

| Author: | **NCI 1978** |
|---|---|
| Species: | mouse, B6C3F₁, 50 ♂/♀ |
| Administration route: | drinking water |
| Dose: | 0.5 %, 1 % |
| | approx. dose ♂: 720, 830, ♀: 380, 860 mg/kg body weight and day |
| Duration: | 90 weeks |
| Toxicity: | > 0.5 % body weight gain and survival reduced (♀), toxic effects on nose (inflammation), lung (inflammation, pneumonia), liver (necrosis, hyperplasia) |
| Tumours: | |

| 1,4-dioxane concentration in drinking water | | 0 % | 0.5 % | 1.0 % |
|---|---|---|---|---|
| nasal turbinate | | | | |
| – adenocarcinomas | ♂ | 0/49 | 0/50 | 1/50 |
| | ♀ | 0/50 | 1/48 | 0/37 |
| liver | | | | |
| – hepatocellular carcinomas | ♂ | 2/49 | 18/50 * | 24/47 * |
| | ♀ | 0/50 | 12/48 * | 29/37 * |
| – hepatocellular adenoma or carcinomas | ♂ | 8/49 | 19/50 * | 28/47 * |
| | ♀ | 0/50 | 21/48 * | 35/37 * |

* p < 0.01

| Author: | **Yamazaki *et al.* 1994** |
|---|---|
| Species: | mouse, BDF₁, 50 ♂/♀ |
| Administration route: | drinking water |
| Dose: | 500, 2000, 8000 ppm (0.05 %, 0.2 %, 0.8 %), approx. dose not given |
| Duration: | 104 weeks |
| Toxicity: | > 2000 ppm: survival reduced (♀) |
| Tumours: | |

| 1,4-dioxane concentration in drinking water | | 0 % | 0.05 % | 0.2 % | 0.8 % |
|---|---|---|---|---|---|
| nasal turbinate | | | | | |
| – esthesioneuroepitheliomas | ♂ | 0/50 | 0/50 | 0/50 | 1/50 |
| – adenocarcinomas | ♀ | 0/50 | 0/50 | 0/50 | 1/50 |
| liver | | | | | |
| – hepatocellular adenomas | ♂ | 7/50 | 16/50 | 22/50 | 8/50 |
| | ♀ | 4/50 | 30/50 | 20/50 | 2/50 |
| – hepatocellular carcinomas | ♂ | 15/50 | 20/50 | 23/50 | 36/50 |
| | ♀ | 0/50 | 6/50 | 30/50 | 45/50 |
| – hepatocellular adenomas and carcinomas combined | ♂ | 22/50 | 36/50 | 45/50 * | 44/50 * |
| | ♀ | 4/50 | 36/50 ** | 50/50 ** | 47/50 ** |

* p < 0.05, ** p < 0.01, Fisher's exact test

### Mice

In mice, increased incidences of hepatocellular adenomas and carcinomas were detected from the lowest concentration tested of 0.05 %. Nasal tumours were detected at low incidences (2 %) in animals given concentrations of 0.5 % 1,4-dioxane or more.

In a drinking water study with B6C3F$_1$ mice, 1,4-dioxane was administered for 90 weeks in concentrations of 1 % and 0.5 %. One female animal of the low and one male animal of the high dose group developed nasal adenocarcinomas. Hepatocellular carcinomas developed in both dose groups (NCI 1978; Table 5).

In a recent study with application of 1,4-dioxane in concentrations of 0.05 %, 0.2 %, or 0.8 % in the drinking water to BDF$_1$ mice, the incidences of hepatocellular adenomas and carcinomas were increased in all dose groups in both sexes. Throughout all dose levels, with increasing dose there was an increasing tendency to malignancy in the liver. Nasal tumours were detected only in one male and one female animal of the high dose group. Thus, 0.05 % (66 mg/kg body weight and day) was the LOAEL in this study (Yamazaki *et al.* 1994; Table 5).

The overall pattern of tumorigenicity and organ toxicity is exactly the same as was found in earlier drinking water studies. Toxic tissue damage and cell proliferation appear to precede the tumour formation.

According to a less reliable study, 1,4-dioxane administered to guinea pigs in the drinking water in concentrations of 0.5 % to 2 % caused the development of lung tumours and occasional other tumours (King *et al.* 1970).

# 6 Manifesto (MAK value/classification)

In a long-term experiment with mice, hepatocellular adenomas and carcinomas developed at the lowest dose tested of 0.05 % (about 50 mg/kg body weight and day). In rats, liver and kidney toxicity as well as a slight increase in the incidence of hepatocellular adenomas or carcinomas were observed at concentrations of 0.1 % 1,4-dioxane in the drinking water (about 100 mg/kg body weight and day) or more. Inflammation of the nasal turbinates with squamous cell carcinomas developed in mice and rats given 1,4-dioxane in concentrations of 0.5 % (about 500 mg/kg body weight and day) or more. No toxic or carcinogenic effects in rats were seen at 0.01 % (10 mg/kg body weight and day). In an inhalation study, no carcinogenic effects were detected at 111 ml/m$^3$, the only 1,4-dioxane concentration tested and corresponding to a dose of about 100 mg/kg body weight and day.

It has been demonstrated in genotoxicity studies that both *in vitro* and *in vivo*, cytotoxic concentrations of 1,4-dioxane cause DNA strand breaks and micronucleus formation in rat liver *in vivo* but otherwise do not produce any mutagenic effects.

1,4-Dioxane was one of the first substances for which non-linear toxicokinetics and accumulation of the substance at high doses were demonstrated experimentally. The half-life of the substance in plasma increases from 1.1 hours in rats given doses of 3 or

10 mg/kg body weight to 14.2 hours after 1000 mg/kg body weight; this suggests saturation of metabolism. Since morphological and biochemical changes first appear after 1,4-dioxane doses in the range producing saturation of metabolism, it is assumed that other adverse effects such as tumours also only develop when metabolism is saturated. Repeated administration of high doses yielded evidence of enzyme induction and presumed appearance of a second metabolite (Young *et al.* 1978a, 1978b). During inhalation of 1,4-dioxane at a concentration of 50 ml/m$^3$ by rats and by volunteers, the pharmacokinetic constants were unchanged and the elimination of 1,4-dioxane was not inhibited. At this concentration, enzyme induction does not occur nor are there any organotoxic or cytotoxic effects. In this study, however, marked eye irritation developed in persons exposed to as little as 50 ml/m$^3$ under controlled conditions. Therefore in 1996, the MAK value was reduced provisionally to 20 ml/m$^3$. Reports of more observations or field studies are required to confirm this value. For the limitation of exposure peaks, 1,4-dioxane is classified in Category I with an excursion factor of 2.

In spite of its still inadequately understood metabolism, 1,4-dioxane may be seen as a well-studied substance which has carcinogenic potential when administered at high levels to experimental animals. The available data suggest that genotoxic properties play little or no role in the effects of the substance (Ashby 1994). At 1,4-dioxane concentrations over 50 ml/m$^3$, on the other hand, cytotoxic effects determine the observable processes and have been shown to obey non-linear toxicokinetics. Thus, 1,4-dioxane meets essential criteria for a classification in Carcinogen category 4 of the *List of MAK and BAT Values*. The MAK value of 20 ml/m$^3$ should prevent irritant effects on the human eye and so should also provide protection from cytotoxic effects.

The reproductive toxicity of 1,4-dioxane has been studied in only one species, the rat. Therefore until new studies (preferably with inhalation exposure) relevant for this classification have been published, 1,4-dioxane can only be classified in Pregnancy risk group D.

Because of the toxic effects seen in experimental animals treated with 1,4-dioxane by dermal application, the substance is designated with an "H".

Apart from two inadequately documented case reports on contact dermatitis no relevant studies on sensitizing effects of 1,4-dioxane on skin or respiratory passages are available. Therefore 1,4-dioxane is not designated with "Sh" or "Sa".


# 7 References

Argus MF, Arcos JC, Hoch-Ligeti C (1965) Studies on the carcinogenic activity of protein-denaturating agents: Hepatocarcinogenicity of dioxane. *J Nat Cancer Inst 35*: 949–958

Argus MF, Sohal RS, Bryant GM, Hoch-Ligeti C, Arcos JC (1973) Dose-response and ultrastructural alterations in dioxane carcinogenesis. Influence of methylcholanthrene on acute toxicity. *European J Cancer 9*: 237–243

Ashby J (1994) Current issues in mutagenesis and carcinogenesis, No. 45. The genotoxicity of 1,4-dioxane. *Mutat Res 322*: 141–150

Barber H (1934) Haemorrhagic nephritis and necrosis of the liver from dioxane poisoning. *Guy's Hospital Reports 84*: 267–280

BASF (1958) *Bericht über die eingehende toxikologische Prüfung von Tetrahydrofuran. Teil I: Akute Toxizität im Vergleich zu Diethyläther, Äthanol, Aceton und Dioxan* (Report of current toxicological testing of tetrahydrofuran. Part 1: Acute toxicity comparison with diethylether, ethanol, acetone and dioxane) (German). Oettel H, Hofmann HT, BASF, Gewerbehygienisch-Pharmakologisches Institut, unpublished report, 24.06.1958

BASF (1972) *Ergebnis der gewerbetoxikologischen Vorprüfung* (Results of preliminary toxicological tests) (German), Zeller H, Hofmann HT, BASF, Gewerbehygiene und Toxikologie, unpublished report, 28.12.1972

BASF (1977) *Bericht über die Prüfung von 1,4-Dioxan auf mutagene Wirkung an männlichen Mäusen nach einmaliger intraperitonealer Applikation; Dominanter Letaltest* (Testing of 1,4-dioxane for mutagenic effects in the male mouse after single intraperitoneal injections; dominant lethal test) (German), Zeller H, Engelhardt G, BASF, Gewerbehygiene und Toxikologie, unpublished report, 21.03.1977

BASF (1979a) *Bericht über die Prüfung von 1,4-Dioxan (peroxidfrei) im Ames-Test* (Results of the Ames test with 1,4-dioxane (peroxide-free)) (German), Zeller H, Engelhardt G, BASF, Gewerbehygiene und Toxikologie, unpublished report, 03.04.1979

BASF (1979b) *Bericht über die Prüfung von 1,4-Dioxan-2-on im Ames-Test* (Results of the Ames test with 1,4-dioxan-2-one) (German), Zeller H, Engelhardt G, BASF, Gewerbehygiene und Toxikologie, unpublished report, 26.07.1979

BASF (1980a) *Activity of T1561 in the in vitro mammalian cell transformation assay in the absence of exogenous metabolic activation,* Microbiological Associates, Bethesda, Maryland, unpublished report, 15.05.1980

BASF (1980b) *Activity of T1561 in the in vitro mammalian cell transformation assay in the presence of exogenous metabolic activation,* Microbiological Associates, Bethesda, Maryland, unpublished report, 11.06.1980

BASF (1981a) *Activity of T1681 in the in vitro mammalian cell transformation assay in the absence of exogenous metabolic activation,* Microbiological Associates, Bethesda, Maryland, unpublished report, 27.10.1981

BASF (1981b) *Activity of T1681 in the in vitro mammalian cell transformation assay in the presence of exogenous metabolic activation,* Microbiological Associates, Bethesda, Maryland, unpublished report, 28.10.1981

BASF (1981c) *Bestimmung des akuten Inhalationsrisikos an Ratten* (Determination of the acute inhalation risk in the rat) (German). Zeller H, Klimisch H-J, BASF Gewerbehygiene und Toxikologie, unpublished report, 23.02.1981

BASF (1985) *Bericht über die Durchführung eines Punktmutationstestes an CHO-Zellen (HGPRT-Locus) mit der Substanz Dioxan-2-on* (Test for point mutations in CHO cells (HGPRT locus) with dioxan-2-one) (German), Gelbke H-P, Jäckh R, BASF, Department of Toxicology, unpublished report, 20.03.1985

BASF (1986) *Activity of dioxan-2-one in the in vitro morphological transformation of BALB73T3 mouse embryo cells in the absence and presence of exogenous metabolic activation,* Microbiological Associates, Bethesda, Maryland, unpublished report, 30.10.1986

BASF (1987) *Dioxan; Untersuchungen am CIIT* (Dioxane; studies at the Chemical Industry Institute of Toxicology) (German), Jäckh R, BASF, Department of Toxicology, unpublished report, 19.10.1987

BASF (1988) *Report on a point mutation test carried out on CHO cells (HGPRT locus) with the test substance 1,4-dioxane,* Hoffmann HD, Jäckh R, BASF, Department of Toxicology, unpublished report, 30.11.1988

BASF (1990) *Investigation of the inhibition of metabolic cooperation in the V79 cell contact feeding assay with the test substance dioxan-2-one* (substance no.: 86/63), Hoffmann HD, Jäckh R, BASF, Department of Toxicology, unpublished report, 06.03.1990

Braun WH, Young JD (1977) Identification of $\beta$-hydroxyethoxyacetic acid as the major urinary metabolite of 1,4-dioxane in the rat. *Toxicol Appl Pharmacol 39*: 33–38

BUA (Beratergremium für umweltrelevante Altstoffe der Gesellschaft Deutscher Chemiker) (1991) *Stoffbericht 80: 1,4-Dioxan*, also available in English. VCH Verlagsgesellschaft, Weinheim

Buffler PA, Wood SM, Suarez L, Kilian DJ (1978) Mortality follow-up of workers exposed to 1,4-dioxane. *J Occup Med 20*: 255–259

Bull RJ, Robinson M, Laurie RD (1986) Association of carcinoma yield with early papilloma development in SENCAR mice. *Environ Health Perspect 68*: 11–17

Chen TH, Kavanagh TJ, Chang CC, Trosko JE (1984) Inhibition of metabolic cooperation in Chinese hamster V79 cells by various organic solvents and simple compounds. *Cell Biol Toxicol* 1: 155–171

Clark B, Furlong JW, Ladner A, Slovak AJM (1984) Dermal toxicity of dimethyl acetylene dicarboxylate, *N*-methyl pyrrolidone, triethylene glycol dimethyl ether, dioxane and tetralin in the rat. *IRCS Med Sci 12*: 296–297

De Nevasquez S (1935) Experimental tubular necrosis of the kidney accompanied by liver changes due to dioxane poisoning. *J Hyg (Lond) 35*: 540–548

Fairley A, Linton EC, Ford-Moore AM (1934) The toxicity to animals of 1:4 dioxan. *J Hyg (Lond) 34*: 486–501

Fregert (1974) Allergic contact dermatitis from dioxane in a solvent for cleaning metal parts. *Contact Dermatitis Newslett*: 438

Galloway SM, Armstrong MJ, Reuben C, Colman S, Brown B, Cannon C, Loorn AD, Nakamura F, Ahmed M, Duk S, Rimpo J, Margolin BH, Resnick MA, Anderson B, Zeiger E (1987) Chromosome aberrations and sister chromatid exchanges in Chinese hamster ovary cells: Evaluations of 108 chemicals. *Environ Mol Mutagen 10, Suppl 10*: 1–175

Giavini E, Vismara C, Broccia ML (1985) Teratogenesis study of dioxane in rats. *Toxicol Lett 26*: 85–88

Goldsworthy TL, Monticello TM, Morgan KT, Bermudez E, Wilson DM, Jäckh R, Butterworth BE (1991) Examination of potential mechanisms of carcinogenicity of 1,4-dioxane in rat nasal epithelial cells and hepatocytes. *Arch Toxicol 65*: 1–9

Haworth S, Lawlor T, Mortelmans K, Speck W, Zeiger E (1983) *Salmonella* mutagenicity test results for 250 chemicals. *Environ Mutagen, Suppl 1*: 3–142

Hecht S, Young R (1981) Metabolic α-hydroxylation of *N*-nitrosomorpholine and 3,3,5,5-tetra-deutero-*N*-nitrosomorpholine in the F344 rat. *Cancer Res 41*: 5039–5043

Hecht SS, Castonguay A, Hoffmann D (1983) Nasal cavity carcinogens: Possible routes of metabolic activation. in: Reznik G, Stinson SF (Eds) *Nasal Tumours in Animals and Man*, Vol 3, CRC Press Inc, Boca Raton, Florida, 201–232

Hoch-Ligeti C, Argus MF, Arcos JC (1970) Induction of carcinomas in the nasal cavity of rats by dioxane. *Brit J Cancer 24*: 164–167

IARC (1999) 1,4-Dioxane, *IARC Monographs on the evaluation of carcinogenic risks to humans, Re-evaluation of some organic chemicals, hydrazine and hydrogen peroxide,* Vol 71, Part 2: 589–602

Johnstone RT (1959) Death due to dioxane? *Arch Industr Hyg 20*: 445–447

Khudoley VV, Mizgireuv I, Pliss GB (1987) The study of mutagenic activity of carcinogens and other chemical agents with *Salmonella typhimurium* assay: testing of 126 compounds. *Arch Geschwulstforsch 57*: 453–462

King ME, Shefner AM, Bates RR (1970) Carcinogenesis bioassay of chlorinated dibenzodioxins and related chemicals. *Environm Health Perspect 5*: 163–170

Kitchin KT, Brown JL (1990) Is 1,4-dioxane a genotoxic carcinogen? *Cancer Lett 53*: 67–71

Knoefel PK (1935) *J Pharmacol Exp Ther 53*: 440

Kociba RJ, McCollister SB, Park C, Torkelsen TR, Gehring PJ (1974) 1,4-Dioxane. I: Results of a 2-year ingestion study in rats. *Toxicol Appl Pharmacol 30*: 275–286

Kociba RJ, Torkelson TR, Young JD, Gehring PJ (1975) *1,4-Dioxane: Correlation of the results of chronic ingestion and inhalation studies with its dose-dependent fate in rats.* Aerospace Medical Research Laboratories, Wright-Patterson Air Force Base, Ohio, Proceedings of the 6th Annual Conference on Environmental Toxicology, 21–23 October 1975, pp 345–353, NTIS AD/A-024899

Laug EP, Calvery MO, Morris HJ, Woodard GJ (1939) The toxicology of some glycols and derivatives. *J Industr HygToxicol 21*: 173–201

Lundberg I, Högberg J, Kronevi T, Holmberg B (1987) Three industrial solvents investigated for tumor promoting activity in the rat liver. *Cancer Lett 36*: 29–33

Maronpot RR, Shimkin MB, Witschi HP, Smith LH, Cline JM (1986) Strain A mouse pulmonary tumor test results for chemicals previously tested in the National Cancer Institute carcinogenicity tests. *J Natl Cancer Inst 76*: 1101–1112

Marzulli FN, Anjo DM, Maibach HI (1981) *In vivo* skin penetration studies of 2,4-toluenediamine, 2,4-diaminoanisole, 2-nitro-*p*-phenylenediamine, *p*-dioxane and *N*-nitrosodiethanolamine in cosmetics. *Food Cosmet Toxicol 19*: 743–747

McFee AF, Abbott MG, Gulati DK, Shelby MD (1994) Results of mouse bone marrow micronucleus studies on 1,4-dioxane. *Mutat Res 322*: 145–148

McGregor DB, Brown AG, Howgate S, McBride D, Riach C, Caspary WJ (1991) Responses of the L5178Y mouse lymphoma cell forward mutation assay. V: 27 coded chemicals. *Environm Mol Mutagen 17*: 196–219

Mirkova ET (1994) Activity of the rodent carcinogen 1,4-dioxane in the mouse bone marrow micronucleus assay. *Mutat Res 322*: 142–144

Miyagawa M, Shirotori T, Tsuchitani M, Yoshikawa K (1999) Repeat-assessment of 1,4-dioxane in a rat-hepatocyte replicative DNA synthesis (RDS) test: evidence for stimulus of hepatocyte proliferation. *Exp Toxicol Pathol 51*: 555–558

Morita T, Hayashi M (1998) 1,4-Dioxane is not mutagenic in five *in vitro* assays and mouse peripheral blood micronucleus assay, but is in mouse liver micronucleus assay. *Environ Mol Mutagen 32*: 269–280

NCI (National Cancer Institute) (1978) *Bioassay of 1,4-dioxane for possible carcinogenicity.* National Cancer Institute, Bethesda, NIH 78-1330, NTIS PB-285 711

Rowe VK, Wolf MA (1982) Derivatives of glycols, 19, dioxane, in: Clayton GD, Clayton FE (Eds) *Patty's Industrial Hygiene and Toxicology, Volume 2*, 3rd ed, John Wiley & Sons, New York, 3909–4052

Schrenk HH, Yant WP (1936) Toxicity of dioxane. *J Indust Hyg Toxicol 18*: 448–460

Sheu CW, Moreland FM, Lee JK, Dunkel VC (1988) *In vitro* BALB/3T3 cell transformation assay of nonoxynol-9 and 1,4-dioxane. *Environ Mol Mutagen 11*: 41–48

Silverman L, Schulte HF, First MW (1946) Further studies on sensory response to certain industrial solvent vapors. *J Indust Hyg Toxicol 28*: 262–266

Sina JF, Bean CL, Dysart GR, Taylor VI, Bradley MO (1983) Evaluation of the alkaline elution/rat hepatocyte assay as a predictor of carcinogenic/mutagenic potential. *Mutat Res 113*: 357–391

Sonneck HJ (1964) Kontaktekzem durch Dioxan in überwiegender lineärer Anordnung (Contact dermatitis caused by dioxane) (German). *Dermatol Wochenschr 1*: 24–27

Stoner GD, Conran PB, Greisiger EA, Stober J, Morgan M, Pereira MA (1986) Comparison of two routes of chemical administration on the lung adenoma response in strain A/J mice. *Toxicol Appl Pharmacol 82*: 19–31

Stott WT, Quast JF, Watanabe PG (1981) Differentiation of the mechanism of oncogenicity of 1,4-dioxane and 1,3-hexachlorobutadiene in the rat. *Toxicol Appl Pharmacol 60*: 287–300

Suzuki M, Noguchi T, Noda K, Fukuda K, Matsushima T (1995) Rat liver micronucleus test with organic solvents (Japanese). 24th JEMS, Osaka, Japan, p 131 (Abstract), cited in: Morita and Hayashi (1998)

Thiess AM, Tress E, Fleig I (1976) Arbeitsmedizinische Untersuchungen von Dioxan-exponierten Mitarbeitern (Occupational medical examinations of workers exposed to dioxane) (German). *Arbeitsmed Sozialmed Präventivmed 11*: 36–46

Tinwell H, Ashby J (1994) Activity of 1,4-dioxane in mouse bone marrow micronucleus assay. *Mutat Res 322*: 148–150

Torkelson TR, Leong BK, Kociba RJ, Richter WA, Gehring PJ (1974) 1,4-Dioxane. II: Results of a 2-year inhalation study in rats. *Toxicol Appl Pharmacol 30*: 287–298

Wirth W, Klimmer O (1937) Zur Toxizität der organischen Lösungsmittel. 1,4-Dioxan (Toxicity of organic solvents) (German). *Int Arch Gewerbepath Gewerbehyg 7*: 192–206

Woo YT, Arcos JC, Argus MF, Griffin GW, Nishiyama K (1977a) Structural identification of *p*-dioxane-2-one as the major urinary metabolite of *p*-dioxane. *Naunyn-Schmiedeberg's Arch Pharmacol 299*: 283–287

Woo YT, Argus MF, Arcos JC (1977b) Metabolism *in vivo* of dioxane: effect of inducers and inhibitors of hepatic mixed-function oxidases. *Biochem Pharmacol 25*: 1539–1542

Woo YT, Argis MF, Arcos JC (1977c) Tissue and subcellular distzribution of 3*H*-dioxane in the rat and apparent lack of microsome-catalyzed covalent binding in the target tissue. *Life Sci 21*: 1447–1456

Woo YT, Lai DY, Arcos JC, Argus MF (1985) Ethylene glycol, diethylene glycol, dioxanes, and related compounds. in: *Chemical Induction of Cancer, Vol IIIB, Aliphatic and polyhalogenated carcinogens*, Academic Press, Orlando, 275–290

Yamazaki K, Ohno H, Asakura M, Narumi A, Ohbayashi H, Fujita H, Ohnishi M, Katagiri T, Seno H, Yamanouchi K, Nakayama E, Yamamoto S, Noguchi T, Nagano K, Enomoto M, Sakabe H (1994) Two-year toxicological and carcinogenesis studies of 1,4-dioxane in F344 rats and BDF1 mice – drinking studies. Japan Bioassay Laboratory, *Proceedings – Second Asia-Pacific Symposium on Environmental and Occupational Health*, 22–24 July 1993, Kobe University, National University of Singapore, 193–198

Yant WP, Schrenk HH, Waite CP, Patty FA (1930) Acute response of guinea pigs to vapors of some new commercial organic compounds: VI. Dioxane. *Public Health Report (Wash) 45*: 2023–2033

Yoon JS, Mason JM, Valencia R, Woodruff RC, Zimmering S (1985) Chemical mutagenesis testing in *Drosophila*. IV. Results of 45 coded compounds tested for the National Toxicology Program. *Environ Mutagen 7*: 349–367

Young JD, Braun WH, Gehring PJ (1978a) The dose-dependent fate of 1,4-dioxane in rats. *J Environ Pathol Toxicol 2*: 263–282

Young JD, Braun WH, Gehring PJ (1978b) Dose-dependent fate of 1,4-dioxane in rats. *J Toxicol Environm Health 4*: 709–726

Young JD, Braun WH, Rampy LW, Chenoweth MB, Blau GE (1977) Pharmacokinetics of 1,4-dioxane in humans. *J Toxicol Environm Health 3*: 507–520

Zimmermann FK, Mayer VW, Scheel I, Resnick MA (1985) Acetone, methyl ethyl ketone, ethyl acetate, acetonitrile and other polar aprotic solvents are strong inducers of aneuploidy in *Saccharomyces cerevisiae*. Mutat Res 149: 339–351

completed 28.11.2002

# General Threshold Limit Value for Dust

**Classification/MAK value (1983)**     4 mg/m$^3$ I (inhalable fraction)
                                   **(1997)**     1.5 mg/m$^3$ R (respirable fraction)

Since the general threshold limit value for dust was last established in 1983 (see *Occupational Toxicants* Volume 2) new data which make a reexamination of this value necessary have been published. In the present document, a new approach has been applied to the evaluation of both the newly published data and the data of the DFG study "Chronische Bronchitis" (Lange *et al.* 1983). The data for the mining cohorts and the cohort exposed to cement dust were not included because coal mine dust (black coal) was being reviewed for a MAK value at that time (and has since been classified as a suspected carcinogen (MAK List Section III, Category 3) (Greim 1998)) and cement dust has its own MAK value (see 1993 MAK documentation for "Portland cement" in Volume 11 of the present series).

## Applicability

The general threshold limit value for dust which was valid until 1997 applied only to the dust fraction which could enter the alveolar space, the respirable fraction (previously called fine dust, F).

The new general threshold limit values for dust for the respirable fraction (R) (fine dust) and for the inhalable fraction (I) (total dust) are intended to prevent the unspecific effects which all insoluble dusts can produce in the respiratory organs, effects such as impairment of airway clearance by overloading, chronic inflammatory changes in the bronchial mucosa and obstructive ventilation disorders. Both values (for R and I) are to be used for poorly soluble and insoluble dusts which do not have other threshold values and for mixtures of dusts together with any MAK and TRK values which apply to components of the mixtures.

The purpose of the general threshold limit value for dust is the prevention of adverse effects on health. However, even when the general threshold limit value for dust is observed, it may only be assumed that exposure to the dust has no effects on health if it has been demonstrated in appropriate experimental studies that genotoxic, carcinogenic, fibrogenic, allergenic or other systemic toxic effects of the dust are not to be expected.

The general threshold limit values for dust (I and R) do not apply to soluble particles or to ultrafine particle fractions. The value for inhalable dust does not apply to dusts with an unusually high proportion of dispersed coarse particles (see "Aerosols", this volume). Whereas insoluble dusts and dispersed coarse particles are excluded because they are not

*Essential MAK Value Documentations.* DFG, Deutsche Forschungsgemeinschaft
Copyright © 2006 WILEY-VCH Verlag GmbH & Co. KGaA, Weinheim
ISBN: 3-527-31394-X

involved in causing the general dust effect, ultrafine particles are considered to have additional effects which are specific for particles in this size range.

**Soluble particles**: the general threshold limit value for dust applies only to dusts which are persistent enough in the lungs to impair the clearance mechanisms; therefore it does not apply to readily soluble particles such as the salts from rock salt and potash deposits.

**Dispersed coarse particles** can be present in airborne dust especially where wind speeds are high, for example, in mines. They then have a large effect on the gravimetrically determined concentration values without a corresponding effect on the human organism. In certain situations where the particle size distributions have been shown to be displaced towards large particles, the use of the general threshold limit value for inhalable dust can be dispensed with. However the general threshold limit value for respirable dust must still be observed in these situations. The reader is reminded that high wind speeds are not covered by the convention EN 481 (CEN 1993).

**Ultrafine particles** (< 0.1 μm) or aggregates of such particles must be considered separately because there is evidence that these fine dusts cause pathological changes and even tumours in the lungs at much lower concentrations than compact dust particles with diameters > 1 μm (Heinrich *et al.* 1995). That these small primary particles penetrate cell membranes more readily than do larger particles and that they possess a large specific surface area with corresponding adsorption and reaction capacities is considered to account for the specific toxic effects. Processes and work areas in which ultrafine particle fractions in relevant concentrations are to be expected in the workplace air and for which, therefore, the general threshold limit value for dust does not apply are listed, for example, in the BIA handbook (BIA-Handbuch 1985).


# 1 Toxic Effects and Modes of Action


Since the dusts to be discussed here are generally poorly soluble particles, the possibility of their accumulation in the lungs is of great importance. Such dust accumulation can mark the beginning of a slow progressive disorder, a chronic obstructive lung disease which manifests itself as coughing, sputum production and dyspnoea and mostly develops over years or decades.

Which of the inhaled dust fractions is responsible for the pathogenesis of the chronic unspecific reaction of the respiratory system is not yet fully understood. The currently available data include only workplace concentrations of inhalable dust (previously called total dust) and the respirable fraction (fine dust). The critical dust fractions are probably those deposited in the bronchi and alveoli, the latter because they are also eliminated in part from the lungs via the bronchial clearance mechanisms and so also contribute to the airways burden.

The particles deposited in the bronchi and bronchioli are not all eliminated from the lungs by the mucociliary clearance system within one day. Some of the particles are retained in the lungs for much longer periods, having elimination half-times of days to weeks and, because of these long retention times, can cause marked irritation in the bronchial region (Gore and Patrick 1982, Patrick and Sterling 1977, Scheuch *et al.* 1996, Stahlhofen *et al.* 1986, 1995). In obstructive lung disorders, the deposition in the tracheobronchial region is increased and, at the same time, the rate of clearance from this region is reduced. The coarser fractions of the inhaled dust which are deposited in the upper airways are not of significance for the deeper airways until they have caused marked changes in the region of the nasal sinuses and this has resulted in dissemination of inflammatory processes into the lung.

In a series of medium-term and long-term studies in which animals were exposed to dust concentrations in the range between 5 and 10 mg/m$^3$, a reduction in the alveolar clearance rate, chronic inflammatory reactions and interstitial fibrosis were observed (Muhle *et al.* 1991). The lung changes which develop in rats exposed to such concentrations are interpreted as the result of exceeding a dose which is critical for the lung clearance mechanisms and as a result of the subsequent dust overload (McClellan 1990) which leads to a reduction of the mobility of the macrophages caused by the poorly soluble phagocytosed particles, further functional changes in these cells and subsequently a chronic inflammatory reaction of the lungs (Morrow 1988, Morrow *et al.* 1991, Yu *et al.* 1989). The development of tumours as a result of overloading has been described only in rats and not in other rodents (mouse, hamster) or in man. In studies with rats exposed to very high concentrations (e.g. titanium dioxide at a concentration of 250 mg/m$^3$, Lee *et al.* 1985), the incidence of lung tumours was increased.


# 2 Exposure


Dusts are aerosols of solid particles in the air. The composition and particle size distribution of such aerosols at the workplace depends on the materials and machinery being used and varies widely.

The assay of dusts in the workplace air was carried out until 1996 on the basis of the concentrations of total dust ($C_G$) and fine dust ($C_F$) as defined in the definitions and conventions for assay of particles detailed in the *List of MAK and BAT Values* 1995 (DFG 1995). From 1996 onwards the conventions to be used for assay of airborne particles are those defined in EN 481 (CEN 1993) and the internationally agreed definitions of the "inhalable fraction" (abbreviation I) instead of total dust (abbreviation G) and "respirable fraction" (abbreviation R) instead of fine dust (abbreviation F). The data for dust exposure were obtained with the old methods, so that the assay results must still be discussed as fine dust and total dust. The new definitions are, however, largely equivalent to the definitions on which MAK values were based in the past and also the sampling and assay equipment used in the past fulfill the new requirements as a good approximation (BIA-Arbeitsmappe 1989).

As demonstrated below, it is not possible to derive the fractions from each other by application of constant conversion factors.

The data for the concentrations of total dust and fine dust at workplaces in the states of the former West Germany were obtained in the course of routine measurements of workplace concentrations by the professional trade associations. For the eastern states relatively extensive data are available from the former German Democratic Republic and they have been supplemented by occasional more recent assay results. Although the sampling and measuring equipment was based on different concepts in East and West, the results for total dust and fine dust in these two sets of data can, with some qualifications, be compared directly (conversion factor 1 for the total dust concentration, 1.3 for the fine dust concentration) (BIA-Report 1996). Most of the data were obtained with static area sampling equipment. Data published in the international literature were obtained in some cases on the basis of other conventions for total and fine dust and different characteristics of the deposition curves (see *List of MAK and BAT Values* 1998 (DFG 1998), pp 147–154).

The values for dust concentrations and particle sizes at individual workplaces differ very widely because of differences in the mechanism of dust formation, the kind of dust and the measures taken to reduce dust exposure. The particle size distribution determines the relationship between total dust ($C_G$) and fine dust ($C_F$) concentrations. In Figures 1 and 2 are plotted a number of total and fine dust concentration values found for various kinds of factories, work areas, occupations and dust types. The values for the ratio $C_G{:}C_F$ are very widely scattered between 1:1 and more than 50:1.

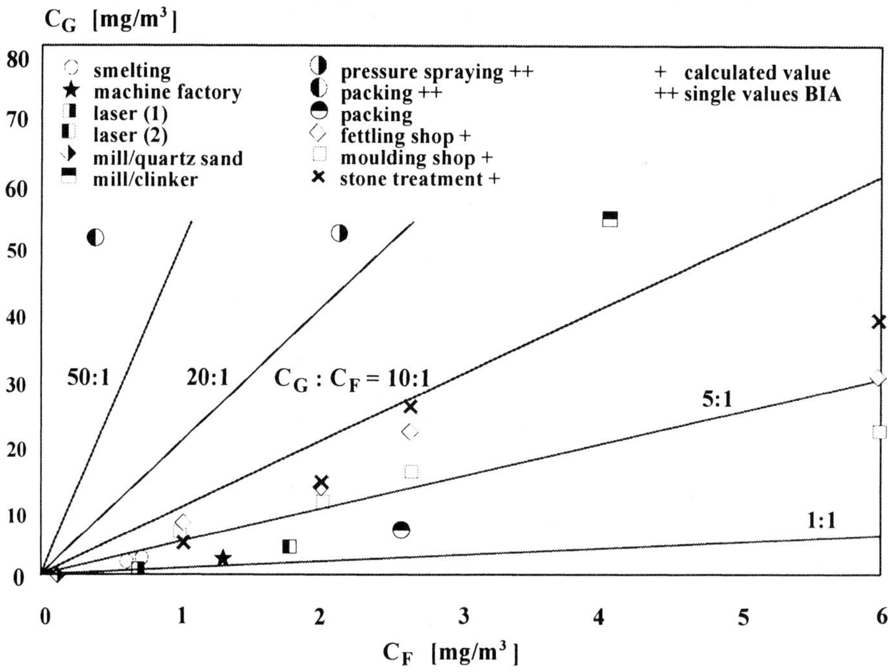


**Figure 1.** Ratio of total dust to fine dust concentrations for various workplaces and kinds of factories

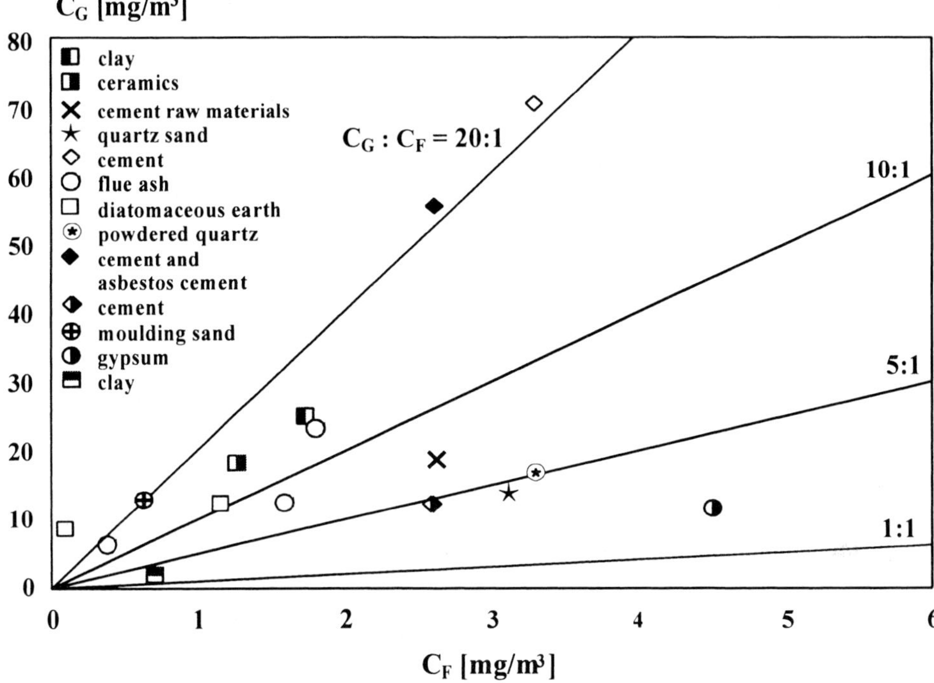


**Figure 2.** Ratio of total dust to fine dust concentrations for various kinds of dust

Evaluation of the assay data available from the BIA (Berufsgenossenschaftlichen Institut für Arbeitssicherheit) for total and fine dust concentrations in the years 1981 to 1993 (BIA-Report 5/96) yields the conclusions listed below.

1. The average value of the ratio of total dust to fine dust is about 3.5:1. (The data pool included measurements only in selected kinds of factories, workplaces and activities and includes much data for processes producing a relatively high proportion of fine dust (e.g., welding). Thus the average value of the ratio of total dust to fine dust could be higher for industry as a whole.)
2. The average total dust concentration (50 % value) varies in the region of 2.5 mg/m$^3$, the 90 % value is about 16 mg/m$^3$.
3. Whereas the higher fine dust concentrations (e.g., the 90 % value) decreased markedly in the 1970s and the 1980s, at the end of the 1980s the levels did not go on decreasing everywhere. However, in foundry fettling shops, for example, the measured fine dust concentrations decreased continually (BIA-Handbuch 1985).

The results of an analysis of the available data for total and fine dust concentrations in the former German Democratic Republic are shown in Table 1 (Thürmer 1995). In contrast with the results obtained by the BIA, the ratios of total dust to fine dust for the various kinds of dust varied between 5 and 11.

**Table 1.** Data for total dust ($C_G$) and fine dust ($C_F$) concentrations from the former German Democratic Republic

| Industrial exposure | Works | Pairs of values | Ratio $C_G$:$C_F$ average | minimum | maximum |
|---|---|---|---|---|---|
| ceramics | 13 | 52 | 5.25 | 1.1 | 22.4 |
| moulding | 8 | 364 | 5.5 | 1.35 | 42.7 |
| quarrying | 9 | 81 | 6.17 | 1.3 | 43.7 |
| furnace | 21 | 79 | 6.46 | 1.95 | 19.5 |
| fettling shop | 3 | 89 | 6.61 | 2.2 | 30.2 |
| glass production | 7 | 65 | 7.24 | 1.9 | 58.9 |
| flue ash | 1 | 141 | 9.33 | 5.6 | 18.2 |
| wood | 17 | 57 | 10.72 | 2.0 | 58.9 |

The data were obtained with SPG 210 or SPG P2 (two step gravimetric analysis with simultaneous measurement of total dust and fine dust)

It may be concluded that there is no constant relationship between the amounts of total dust and fine dust and that even for specific kinds of industries, processes and dusts the ratio must be expected to vary widely. Therefore, separate threshold limit values must be established for the respirable and inhalable dust fractions.

In all the studies which have been assessed for the establishment of this general threshold limit value for dust, the data were obtained by static area dust sampling. Experience has shown that the actual concentrations in the air inhaled by the exposed persons, concentrations which can be determined accurately after personal sampling, are much higher (Figure 3).

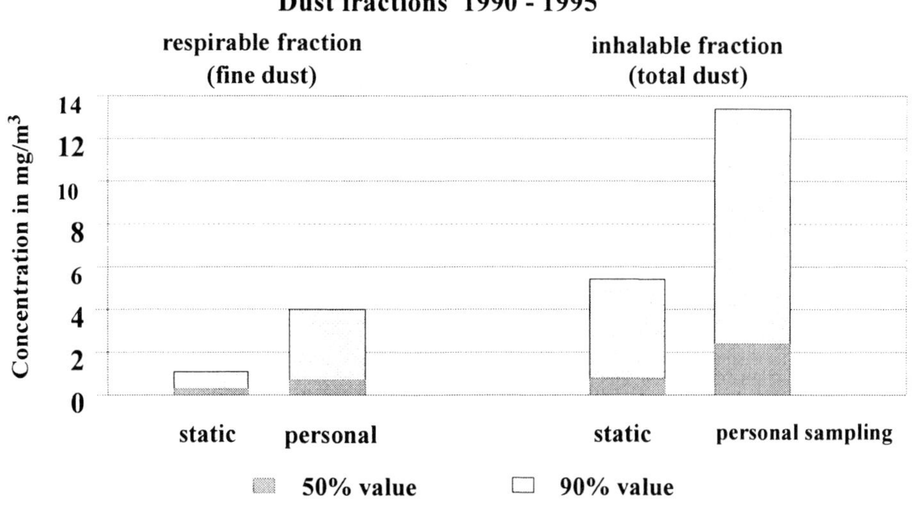



**Figure 3.** Dust fractions determined by area and personal sampling (BGMG, Berufsgenossenschaftliches Meßsystem Gefahrstoffe)

Analysis of the data obtained for the BGMG (Berufsgenossenschaftliches Meßsystem Gefahrstoffe) for the years 1990 to 1995 revealed the following average values for the ratio of the dust concentrations determined after personal sampling ($C_P$) and area sampling ($C_{st}$) (cf. Figure 3):

**inhalable fraction (I)**
50 % values: $C_P:C_{st} = 3.05$
90 % values: $C_P:C_{st} = 2.47$

**respirable fraction (R)**
50 % values: $C_P:C_{st} = 2.18$
90 % values: $C_P:C_{st} = 3.77$

# 3 Toxicokinetics and Effects

## 3.1 Particle deposition

In which of the various compartments of the respiratory tract (nasopharynx, tracheo-bronchiolar region, pulmonary region—the respiratory bronchioles and alveoli) and to what extent inhaled particles are deposited is a function of the aerodynamic diameter of the particles. In Figure 4 the deposition ratios for particles of different diameters in the lungs of man and rat are shown, and for man the deposition curves for both oral and nasal breathing. The percentage of particles deposited in the lungs is much higher for man than for the rat. There are five fundamental mechanisms which can play a role in the deposition of particles in the respiratory tract: impaction, sedimentation, Brownian movement, interception (only with elongated particles/fibres) and electrostatic precipitation (Schlesinger 1989).

The effects of anatomy and respiratory physiology result in very marked variability in the proportion of dust deposited, both in experimental animals and in man. Cuddihy *et al.* (1979) estimated that 2 % of experimental animals inhale a lung-dose which is three times higher than the average value for the group of animals in the study. For man, the authors estimate that deposition is as much as 5 times higher in a few percent of the population and that this results in very markedly higher lung doses in these persons.

The deposition ratios depend on various factors, e.g., nature of the particles and geometry of the respiratory tract (see Table 1 in Schlesinger 1989); in several of these factors individual differences can play a role (ICRP 1966).

In the kinetics of transport of particles out of the lungs, the alveolar clearance of poorly soluble or insoluble particles is the slowest process (Schlesinger 1989). For this reason in the present document most emphasis is placed on the alveolar clearance.

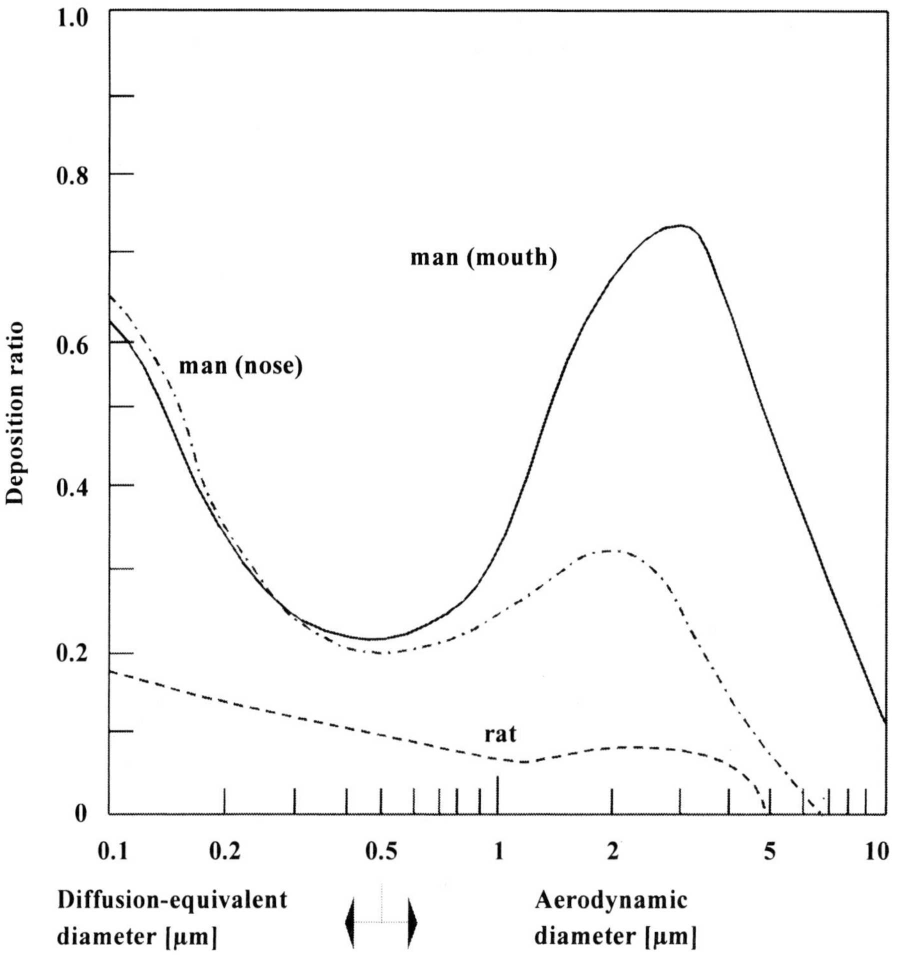


**Figure 4**. Alveolar particle deposition in man and rat (WHO 1988)

## 3.2 Retained mass

The dose is defined as the integral over time of the effective concentration of a material (retained weight of dust per unit weight of tissue) which is present at a specific site in the organism (Morrow and Mermelstein 1988). In practice the dose is frequently difficult to determine.

In the assessment of inhaled poorly soluble material it should be remembered, on the one hand, that in some cases only part of the retained material contributes to the effective dose because the rest is not in a dispersed molecular form. On the other hand, if exposure is continual both deposition and clearance must be taken into account. These relationships can be described in a simplified form, ignoring dust overload, for poorly soluble particles in a model assuming first order kinetics (Raabe 1967).

$$\frac{dm}{dt} = D - k \cdot m$$

where
m   is the retained weight of material in the lungs
t   is time
D   is the weight of material deposited per unit time
k   is the clearance factor.

The weight of material deposited in the lungs per unit time is given by

$$D = f \cdot c \cdot \dot{V}$$

where
f   is the fraction of the inhaled weight of material which is deposited
c   is the exposure concentration (w/v)
$\dot{V}$   is the inhaled volume per unit time.

According to this model, after exposure for a time t the retained weight of material m is given by

$$m = \frac{D}{k(1 - e^{-kt})}$$

At equilibrium the retained weight of material is given by

$$m = \frac{D}{k} = \frac{D \cdot t_{1/2}}{\ln 2}$$

where $t_{1/2}$ is the half time for the lung clearance. Thus the relationship between k and $t_{1/2}$ is

$$k = \frac{\ln 2}{t_{1/2}} \, .$$

## 3.3 Particle clearance in experimental animals

Alveolar clearance was studied in Fischer 344 rats by means of radioactively labelled particles (gamma emitters) which were administered briefly during medium-term or long-term exposure to dust and the clearance followed over a period of about 100 days (Figure 5). The results shown in Figure 5 are assembled from studies with various kinds of particles (Bellmann *et al.* 1991, 1994; Muhle *et al.* 1990, 1991). The clearance factor k is plotted against the volume of dust retained in the rat lung. Since the clearance of insoluble particles does not obey first order kinetics exactly and becomes progressively slower during an observation period of, for example, one year, for the results shown in Figure 5, in each case only the time period between day 15 and day 90 after the exposure to the radioactively labelled particles was evaluated (Muhle *et al.* 1990). The clearance factor $k = 0.012$ corresponds to a clearance half-time of 58 days; $k = 0.001$ is equivalent to a half-time of 693 days. Figure 5 demonstrates that the alveolar clearance becomes slower with increasing lung burden of particles.

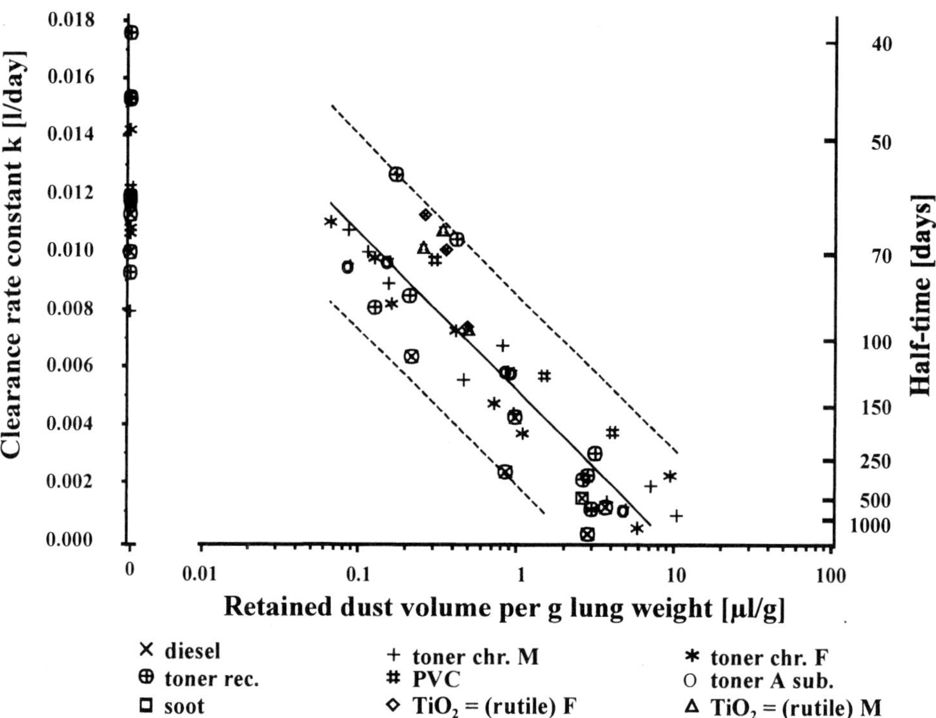



**Figure 5.** Clearance factors for γ-labelled particles ($^{85}$Sr-polystyrene) or toner particles in rat lungs as a function of the retained volume of various test materials. Area between the dotted lines: 95 % confidence interval (Muhle *et al.* 1990)

As a result of the large number of experimental observations of animals with lungs overloaded with dust, the term "dust overloading" was coined (McClellan 1990, Morrow 1988, 1992).

The physiological consequences of this overloading include changes in macrophage function, increased lung dust burden and impairment of the lung clearance mechanisms. Morrow (1988) suggested that the reduced rate of particle clearance resulting from overload is rather a case of volume overloading than of weight overloading. There are a number of experimental results which support this hypothesis (Oberdörster 1994).

In Fischer 344 rats, dust overloading of the lungs resulting in reduced rates of particle clearance was observed in a dose range between 0.5 and 1.5 µl/g lung. This range was found for particles with low specific toxicity. For cytotoxic particles such as quartz, this range does not apply. It may be seen in Figure 5 that a lung burden of 1 µl/g lung increases the half-time for the alveolar clearance by a factor of about two. Since the wet weight of the rat lung is about 1.5 g, a dust burden of 1 µl/g lung is equivalent to 1500 nl dust per lung.

Impairment of particle clearance from the lungs—progressive reduction in the rate of lung clearance—as a result of loading the alveolar macrophages with poorly soluble dust particles has been demonstrated in all species which have been studied such as the rat, mouse, hamster and dog (Bolton *et al.* 1983, Muhle *et al.* 1988, 1990, Oberdörster 1994, Snipes and Clem 1981, Snipes *et al.* 1984). Therefore it may be assumed that overloading of macrophages with insoluble dust particles and subsequent impairment of particle clearance from the lungs can occur in man too, especially during long-term exposure (see Section 3.4).

However, it should be pointed out that in the experimental studies the animals were frequently exposed to relatively higher concentrations of dust than are exposed persons. The usual justification for using this procedure is that it is intended to reveal potential effects with a minimum number of experimental animals. But such an approach involves the danger of overloading organ-specific defence and clearance mechanisms and this increases the probability of producing unspecific effects.

Various hypotheses have been developed to account for the mechanism of action of dust (Morrow 1988, Oberdörster 1988, Yu *et al.* 1989). They are based essentially on the significance of the transport function of the alveolar macrophages and its impairment and may be summarized as follows: a reduction in the mobility of macrophages by a factor of more than 2 resulting from phagocytosis of poorly soluble particles at a retained dust volume of 1 µl/g lung causes dysfunction of these cells and leads subsequently to a chronic inflammatory reaction in the lung; exceeding of a critical dose can also result in lung fibrosis.

## 3.4 Particle clearance in man

The lung clearance of inert, practically insoluble dusts can be shown to take place in two phases in healthy non-smokers. The first phase involves clearance of particles deposited in the ciliated region of the respiratory tract and has a half-time in the range of hours or a few days. The half-time of the second phase is in the range of hundreds of days (Schlesinger 1989). In man the average half-time for alveolar clearance was found to be 400 days (Bailey *et al.* 1985).

As has been demonstrated especially in animal studies, the rate of macrophage-mediated alveolar clearance of poorly soluble particles decreases with increasing volume of particles retained in the lungs. The effect was also observed in people working in black coal mines who were exposed for long periods to high concentrations of airborne particles and accumulated high weights of dust in their lungs. With magneto-pneumographic methods, it was demonstrated that the alveolar clearance in miners in black coal mines was significantly slower than that in a control cohort of non-smokers (Freedman *et al.* 1988). In an earlier study, the kinetics of accumulation of coal mine dust in the lungs of dead miners were estimated; it was demonstrated that the clearance half-time for this dust in the miners must be in the range of about 5 years (Stöber *et al.*

1967). This value is very much higher than the listed values of 33 to 602 days for the slow phase of alveolar clearance in non-smokers exposed to dust (see Table 3 in Schlesinger 1989). In another study, after exposure of smokers and non-smokers to magnetic dust of $Fe_2O_3$, it was demonstrated that smoking reduces the alveolar clearance rate (Cohen *et al.* 1979).

The retained weight of dust in the lungs of miners was given as 5–40 g per lung; the mean value was about 15 g per lung (Stöber *et al.* 1967).

An indication of how unusually high these retained weights are is provided, in the absence of genuine normal values, by the data of the DMM study (diffuse malignant mesothelioma study) (Rödelsperger 1996, Woitowitz *et al.* 1993) which show that the average weight of ash (without water-soluble salts) in the lungs of 124 patients (DMM and controls) was about 0.02 g per gram dry weight of lung tissue. As a lung wet weight of about 1 kg corresponds to a lung dry weight of about 100 g, that was about 2 g (maximum 23 g) per lung.

## 3.5 Pathophysiology of the chronic unspecific effects of dust in man

The most frequent cause of chronic obstructive lung disease in man is the inhalation of dust and the resulting overloading of the pulmonary clearance mechanisms (Fruhmann and Woitowitz 1997, Valentin and Woitowitz 1967).

Overloading of the clearance mechanisms of the airways can result in chronic inflammation of the airways in which alveolar and haematogenic macrophages, the epithelia of the respiratory passages and other cellular components of the mucous membranes as well as granulocytes and lymphocytes play a role. The interactions of these cells involve numerous mediators which are still only partly understood. As a result of the inflammatory processes, characteristic lung function disorders may develop. They include especially obstructive ventilation disorders, ventilation perfusion mismatches, unspecific bronchial hypersensitivity and finally gas exchange disorders leading to insufficient partial pressure of oxygen in the arterial blood. In the advanced stages of the chronic obstructive lung disease, pulmonary hypertonia can result in *cor pulmonale* and premature death.

The initial clinical signs of clearance insufficiency resulting from overloading of the airways with dust are coughing and sputum production. A person is said to have chronic bronchitis (as defined by the WHO) when in the previous three years he or she has suffered from coughing or sputum production on most days of the week during the cold season and for at least 3 months per year. The adverse effects on lung function are seen in respiratory distress, initially only during physical work. The person is said to suffer from the obstructive form of the chronic obstructive lung disease, which has a poor prognosis, when the parameters characteristic for obstruction (airway resistance, $R_{aw}$, forced expiratory volume in one second, $FEV_1$, maximum expiratory flow) are no longer in the normal range.

The term chronic obstructive lung disease (COLD) includes as separate clinical diagnoses both chronic bronchitis and pulmonary emphysema. The diagnosis should differentiate as far as possible between these syndromes, although satisfactory differential diagnosis is not always possible because of the similarity of the symptoms and interrelation of the disorders. The target criteria which have been used frequently to date for the demonstration of unspecific effects of dust in epidemiological studies include:

– standardized recording of individual symptoms such as chronic coughing or sputum production as stipulated in the WHO definition of chronic bronchitis
– records of medical diagnoses such as that of chronic unspecific/obstructive lung disease (ICD 496), simple chronic bronchitis (ICD 491.0), mucopurulent chronic bronchitis (ICD 491.1), obstructive chronic bronchitis (ICD 491.2) or the formation of variable combinations of symptoms or functional diagnostic findings (see DFG 1981), which fit these diagnoses
– deviations in lung function parameters characteristic for obstruction (airway resistance, $R_{aw}$, forced expiratory volume in one second, $FEV_1$) from the normal range or the observation of the course of changes in lung function parameters in single individuals (rate of change per year in forced expiratory volume, $FEV_1$, and forced expiratory vital capacity, FVC).

In epidemiological analyses, known confounders such as cigarette smoking and age must be taken into account. The individual symptoms of chronic obstructive lung disease provide high sensitivity. When, however, coughing and sputum production are observed as symptoms on their own, there is no agreement on how increased mucociliary clearance is to be differentiated from a genuine disorder. Therefore in the derivation of the general threshold limit value for dust, studies of diagnostic functional variables, i.e. the evaluation of obstructive forms of chronic bronchitis, are preferred. Especially for ventilation parameters, evaluations of changes in the parameters for single individuals are more sensitive than comparison of the measured values with normal values. Since the normal values can be scattered over very wide ranges, persons with high initial values can suffer the functional losses associated with a genuine disorder without being classified as having an obstructive disorder when the data are evaluated by comparison with a mean value.

# 4 Effects in man

The literature on chronic bronchitis and chronic obstructive lung disease (COLD) is extensive (e.g. Ulmer 1979, Valentin and Woitowitz 1967, Woitowitz 1972).

On the effects of dust, with the exception of dust in mines, there are only a few publications in which data for the individual exposure levels are given for sufficiently long exposure periods and the effects described with sufficiently sensitive parameters (see Section 4.1).

# 4.1 DFG study "Chronische Bronchitis"

## 4.1.1 Concept for the evaluation

The data from the DFG chronic bronchitis study, especially the data for mines, were used to establish the MAK value of 6 mg/m$^3$ for fine dust which was valid until 1996 (see the 1983 documentation "General Threshold Limit Value for Dust" in Volume 2 of the present series). At that time, on the basis of a model without a threshold value, a concentration above which an effect of dust could be considered certain was determined from the regression coefficient and confidence interval in the regression analysis.

In 1993 in a similar way, for Portland cement dust a specific MAK value of 5 mg/m$^3$ averaged over the shift was established; a specific value for coal mine dust was also planned at that time. Therefore, for the establishment of a new general threshold limit value for dust, only the three cohorts Moers, München and Saarbrücken were considered. These are two cohorts from foundries (Moers and Saarbrücken) and one cohort from a machine factory with a foundry (München).

Although the data do not reveal a dust concentration without any effects, by means of statistical model calculations a threshold value for the occurrence of additional dust effects was determined. For the statistical analysis, logistic and isotonic regression were used.

These statistical methods and a comparison of the methods for just one part of the cohort (smokers from the cohort München) have been described (Küchenhoff and Ulm 1997) and are summarized below.

Logistic regression is a standard method for the evaluation of studies in which the outcome, such as the chronic bronchial reaction in the DFG study, is a "yes/no" parameter. Isotonic regression is also suitable for the analysis of this kind of data.

Each method has its own advantages. With the logistic model, threshold values can be estimated statistically and at the same time tested for significance. With isotonic regression, a model for the relationship between the independent parameters and the outcome can be established without any assumptions about their effects. Isotonic regression is an optimum method for describing associations. The logistic model is particularly suited for the statistical testing of hypotheses and for testing for the existence of a threshold. With the logistic model, the optimum value of the threshold can also be estimated. With isotonic regression this estimate can be checked. To date it is not possible to estimate a threshold level by means of isotonic regression. Therefore the result of the isotonic regression is not expressed as a threshold value but as the concentration at which the background risk (at a dust level < 0.5 mg/m$^3$) is exceeded by at least 5 % in at least two age groups. The relative increase in risk depends on the size of the background risk. For example, if the background risk is 5 %, an absolute increase in risk of 5 % would double the risk. The choice of a 5 % increase would make more sense, however, if the background risk were 25 %, as it is for at least some of the parameters being considered here. Therefore an increase of 5 % was used as a basis for threshold determination.

## 4.1.2 Data and methods

The DFG study has been described in detail in the literature (DFG 1975, 1981, Lange and Ulm 1983; see also 1983 MAK documentation "General Threshold Limit Value for Dust") and is therefore described only briefly here.

In the years 1966–1970, the DFG chronic bronchitis study was carried out initially as an epidemiological cross-sectional study. The data collection was continued in the years 1972–1977 in the form of a longitudinal study. Each of the persons included in the cross-sectional study was examined again with the same methods about five years later.

Recorded during the medical examination were anamnesis, thorax X-ray, ECG and detailed lung function parameters including vital capacity, forced expiratory volume in one second, airway resistance, thoracic gas volume and arterial partial pressure of oxygen (see 1983 MAK documentation "General Threshold Limit Value for Dust").

The results of determinations of fine dust and total dust which were carried out between 1974 and 1976 after area sampling at representative workplaces of the three cohorts are shown in Table 2. The ratio of total:fine dust in Moers and Saarbrücken averaged about 4:1, in München, however, only 1.7:1.

For the diagnosis of CBR (chronic bronchial reaction) various combinations of high grade anamnestic, clinical and functional diagnostic evidence of an obstructive ventilation disorder were assembled; grades 2 and 3 of these combined variables were considered indicative of CBR (see Figures 2b and 2c, in the 1983 MAK documentation "General Threshold Limit Value for Dust"). This procedure yielded relatively high background incidences in the non-exposed groups because, for example, anamnestic-clinical evidence was also accepted in the absence of impairment of lung function parameters (17.4 % and 12.4 % of cases according to Tables 2a and 2b in the 1983 MAK documentation). The inclusion of whole body plethysmographic determinations of airway resistance and the arterial partial pressure of oxygen (the latter with a threshold of 85 % of the normal value, equivalent to about one standard deviation) probably also resulted in higher background incidences of CBR compared with those found in studies in which only spirometry was used.

In all, data are available for 5518 employees. The prevalence of CBR was 31.8 % with considerable variation between the cohorts (Moers, n = 2562: 45.6 %; München, n = 1246: 23.4 %; Saarbrücken, n = 1705: 17.3 %).

The proportion of smokers was 73.5 %, with little variation between the cohorts (70.6 %–75.5 %). The time since the beginning of exposure was on average 27 years (1–66 years). All three cohorts were subdivided before analysis according to smoking habits (non-smokers and ex-smokers, and smokers).

The median value of the average concentration of respirable dust (fine dust) was between 0.14 mg/m$^3$ and 0.3 mg/m$^3$ and for inhalable dust (total dust) between 0.3 mg/m$^3$ and 1.4 mg/m$^3$. The majority of the persons (> 90 %) were exposed during the whole of their working lives to an average concentration of less than 10 mg/m$^3$ total dust, and for 50 % of the persons the average fine dust concentration was less than 0.5 mg/m$^3$ (DFG 1981).

**Table 2.** Average fine and total dust concentrations (w/v) to which the persons in the cohorts München, Saarbrücken and Moers were exposed (Ulm *et al.* 1996)

| Cohort | Kind of dust | Non-smokers and ex-smokers | | | | Smokers | | | |
|---|---|---|---|---|---|---|---|---|---|
| | | No. of persons | Concentration (mg/m$^3$) | | | No. of persons | Concentration (mg/m$^3$) | | |
| | | | median | min | max | | median | min | max |
| München | fine | 326 | 0.3 | 0.0 | 5.9 | 920 | 0.3 | 0.0 | 6.0 |
| | total | | 1.4 | 0.0 | 15.0 | | 1.4 | 0.2 | 15.0 |
| Saarbrücken | fine | 501 | 0.14 | 0.0 | 3.8 | 1204 | 0.14 | 0.0 | 3.8 |
| | total | | 0.3 | 0.0 | 14.3 | | 0.8 | 0.0 | 16.0 |
| Moers | fine | 629 | 0.2 | 0.0 | 4.7 | 1933 | 0.24 | 0.0 | 4.5 |
| | total | | 0.8 | 0.0 | 21.6 | | 0.8 | 0.0 | 21.0 |

## 4.1.3 Results

The results of these analyses for the cohorts München, Moers and Saarbrücken have been described in detail (Ulm *et al.* 1996) and are summarized below.

### a) Total dust – inhalable dust fraction

In Table 3 the threshold values—suitable exposure limits in mg/m$^3$ for the inhalable dust fraction—found by logistic regression for the cohorts Moers, München and Saarbrücken are listed. For comparison, the concentrations obtained by isotonic regression for which the background risk determined for total dust concentrations less than 0.5 mg/m$^3$ is exceeded by 5 % (5 % value) are given.

**Table 3.** Threshold values for the inhalable dust fraction (in mg/m$^3$)

| | Non-smokers and ex-smokers | | Smokers | |
|---|---|---|---|---|
| | logistic model | isotonic regression | logistic model | isotonic regression |
| Moers | 20.6 | 2.5 | 18.0* | 4.5 |
| München | 8.0 | 6.0 | 3.8* | 5.0 |
| Saarbrücken | 7.5 | – | 13.8 | 3.5 |

* $p < 0.05$

Significant thresholds (3.8 and 18 mg/m$^3$) could be found only for smokers in Moers and München. In all the other analyses, the model was not significantly different from one with no threshold value, that is, in these cases a threshold of 0 cannot be excluded. It is conspicuous that the two significant threshold values differ by a factor of 5. These results do not support the use of a model with a threshold value for the effects of total dust. In contrast with the threshold values obtained by logistic regression, the 5 % values found by isotonic regression are relatively low, between 2.5 and 6.0 mg/m$^3$. It should be pointed out that the background prevalence at concentrations below 0.5 mg/m$^3$ is considerably higher in the Moers cohort than in those from München and Saarbrücken (Ulm *et al.* 1996). Therefore the 5 % increase in prevalence results in a much smaller increase above the background for Moers than for München and Saarbrücken.

Further conclusions which may be drawn from the results of Ulm *et al.* (1996) are listed below.

– Overall a marked, statistically significant effect of the inhalable dust fraction on the risk of disease in smokers can be seen (cohorts München, Moers).
– For the cohort Saarbrücken, however, the effect of the inhalable dust fraction on the risk of disease in smokers is only small.
– For non-smokers and ex-smokers from all three cohorts, only a slight effect of the inhalable dust fraction on the risk of disease can be detected.

The differences in the results obtained by the logistic and isotonic regression analyses are mainly a consequence of the definitions used. For the isotonic regression an increase in the risk of disease by more than 5 % in at least two age groups is sufficient whereas in the logistic regression analysis the whole cohort is always analyzed.

In summary, for the inhalable fraction (total dust) in the logistic model, threshold values between 3.8 and 20.6 mg/m$^3$ and significant thresholds in two cohorts at 3.8 and 18 mg/m$^3$ were found.

With isotonic regression an increase in risk of 5 % was found at concentrations between 2.5 and 6.0 mg/m$^3$.

## b) Fine dust – respirable dust fraction

The threshold values—suitable exposure limits for the respirable dust fraction in mg/m$^3$ —found by logistic regression for the three cohorts are listed in Table 4. The concentrations obtained by isotonic regression for which the background risk determined for fine dust concentrations below 0.5 mg/m$^3$ is exceeded by 5 % are given for comparison.

**Table 4.** Threshold values for the respirable dust fraction (in mg/m$^3$)

| | Non-smokers and ex-smokers | | Smokers | |
|---|---|---|---|---|
| | logistic regression | isotonic regression | logistic regression | isotonic regression |
| Moers | 4.3* | 4.5 | 4.1* | 2 |
| München | 5.4 | – | 5.0* | 3.5 |
| Saarbrücken | 1.7* | 2 | 4.0* | 2.0 |

* $p < 0.05$

Only for non-smokers from the cohort München is it not possible to demonstrate the existence of a significant threshold value. However, the groups of non-smokers in München and Saarbrücken were relatively small, with only 51 and 48 persons with CBR. In all other cohorts at least 200 patients with CBR were observed. In addition, in both the München and Saarbrücken cohorts, half of the patients with CBR were allocated to the lowest exposure concentration groups, $\leq 0.4$ and $\leq 0.15$ mg/m$^3$, respectively, so that there were only small numbers of patients in the groups exposed to levels above the thresholds observed at 5.4 and 1.7 mg/m$^3$. The results of the analysis of the two non-smoking subgroups are therefore associated with considerable uncertainty. On the other hand, for the other 4 cohorts relatively good agreement in the threshold values (between 4.0 and 5.0 mg/m$^3$) is seen. The 5 % values from the isotonic regression, which lie between 2.0 and 4.5 mg/m$^3$, are relatively lower.

Further conclusions which may be drawn from the results of Ulm *et al.* (1996) are listed below.

- Overall a marked, and in the case of the logistic regression statistically significant, effect of the respirable dust fraction on the risk of disease in smokers can be seen.
- For the cohorts Moers and Saarbrücken, a marked, statistically significant effect of the respirable dust fraction on the risk of disease in non-smokers and ex-smokers is also seen whereas for the cohort München this effect is only small.

For the respirable fraction (fine dust), the logistic model reveals threshold values between 1.7 and 5.4 mg/m$^3$ and significant thresholds between 1.7 and 5.0 mg/m$^3$. The isotonic regression reveals a 5 % increase in risk at concentrations between 2.0 and 4.5 mg/m$^3$.

## 4.2 Longitudinal study of foundry workers

In a longitudinal study, employees of two foundries were examined four times at five year intervals (standardized questionnaire, clinical examination, spirometric lung function test, flow-volume curve and oscilloresistometry and, if indicated in the individual case, also thorax X-ray). The cohort comprised initially 207 foundry workers (manual moulder, mould maker, manual fettler, polisher, machine moulder, core inserter and machine fettler) and 50 control persons from two factories (Schneider *et al.* 1986). Analysis of random samples indicated that effects of toxic chemicals could be largely excluded. The measured levels of metal oxide fumes, carbon monoxide and mineral oil mist were below the MAK values valid at that time in the German Democratic Republic. With the moulding systems used (e.g. waterglass/clay-bound moulding materials) the development of high levels of pyrolysis products is not to be expected.

Unifactorial and multifactorial statistical analysis made use of the pairwise multiple t-test or Welch test and the Chi-square test of Pearson. The data were standardized by regression analysis for the effects of age, size, weight and smoking habits. Thresholds were calculated by the analysis of mixture distributions as described by Dietz (1992). This study produced the first evidence for a dependence of the impairment of lung function on the dust load.

By the third examination of the foundry workers (Karsten *et al.* 1992, Schneider *et al.* 1994) 36 persons, i.e. 17 %, were for various reasons no longer available for observation. The initial data for the persons eliminated from the study were not significantly different from those for the rest of the group either for lung function or for frequency of symptoms. The average age of the persons at the time of the third examination was 42 years, the average period of exposure 23 years.

For the unifactorial analysis the cohort was divided into 4 groups exposed to different levels of total dust. The groups were defined as follows: level of total dust averaged over the total exposure period $E_1$ = below 5 mg/m$^3$, $E_2$ = 5 mg/m$^3$ to below 10 mg/m$^3$, $E_3$ = 10 mg/m$^3$ to below 15 mg/m$^3$, $E_4$ = 15 mg/m$^3$ and above.

For the parameter changes/year in FVC (forced expiratory vital capacity) the group exposed to the highest dust concentration (> 15 mg/m$^3$) differs significantly from all 3 other groups. For the parameter changes/year in FEV$_1$ (forced expiratory volume in one second) there is also a difference between the group exposed to < 5 mg/m$^3$ and that exposed to < 10 mg/m$^3$ (Figure 6).

For analysis according to the average fine dust exposure levels, the persons were divided into three groups ($F_1$ = below 1.0 mg/m$^3$, $F_2$ = 1.0 mg/m$^3$ to below 2.5 mg/m$^3$, $F_3$ = 2.5 mg/m$^3$ and above). The group $F_3$ differs from both the other groups for both parameters (Figure 7).

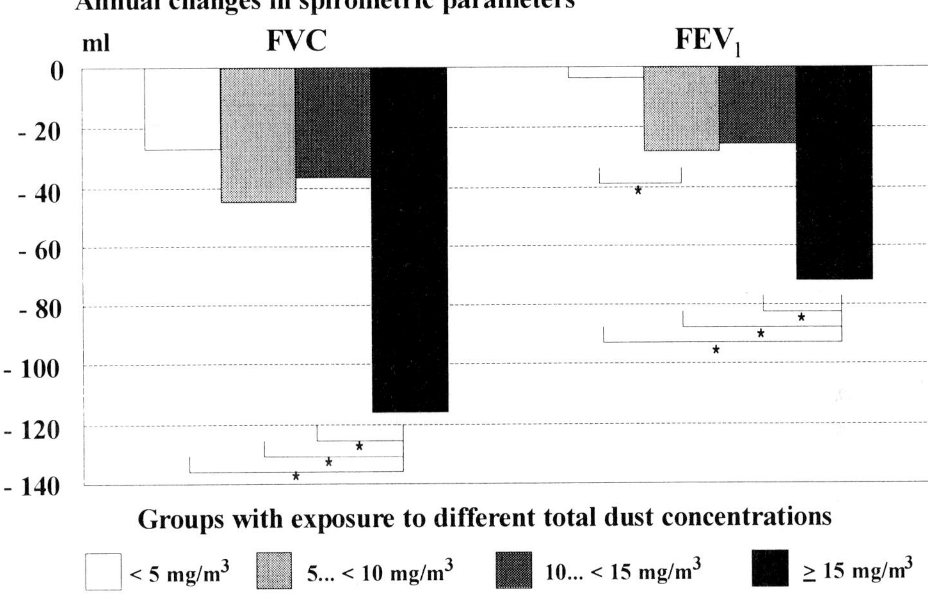


**Figure 6.** Reductions (ml/year) in ventilation values in groups exposed to different concentrations of total dust (* = significant at the 5 % error level)

In analogy to the methods used in the DFG study, threshold values were estimated for the two exposure parameters total dust and fine dust. In the same way, combinations of variables reflecting the anamnestic and clinical evidence for a bronchial reaction, for functional evidence of an obstructive ventilation disorder (FOV, forced obstructive ventilation disorder) and for the association of these two (COV, combined obstructive ventilation disorder) were established. These variables differ from those used in the DFG study in the diagnostic program (different questionnaire, oscilloresistometry instead of whole body plethysmography, no blood gas determinations) and reflect the target CBR with a different, probably reduced sensitivity. For the variable COV the number of cases was so small that it did not make sense to carry out an analysis. The functional diagnostic combination variable FOV attained a relative maximum at a total dust concentration of 4.4 mg/m$^3$.

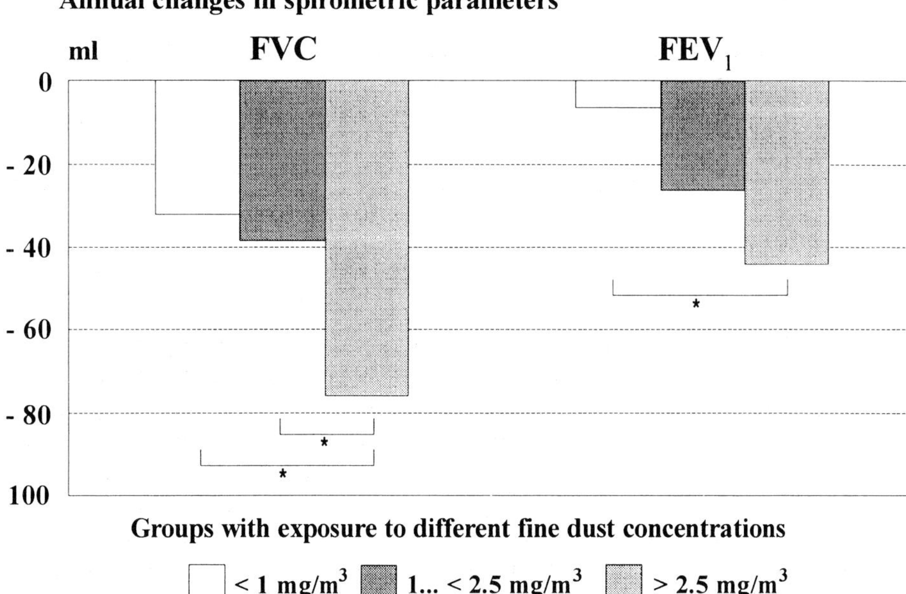



**Figure 7.** Reductions (ml/year) in ventilation values in groups exposed to different concentrations of fine dust (* = significant at the 5 % error level)

In addition, thresholds for total dust were calculated directly from the lung function parameters (FEV$_1$, FVC). The results revealed for the various parameters of lung function a relatively narrow threshold range of 4.0 to 5.9 mg/m$^3$ total dust for the time-weighted average dust concentration, averaged over the whole period of exposure.

The analysis of the functional diagnostic variable FOV for exposure to fine dust revealed a threshold of 1.6 mg/m$^3$. The thresholds for fine dust calculated directly from the data for the lung function parameters were in the range from 1.5 mg/m$^3$ (FEV$_1$) to 2.3 mg/m$^3$ (FVC).

At the fourth examination in 1992, because of changes in the economic situation, only 100 persons from the original cohorts could still be reached in the foundries (Schneider *et al.* 1995). Given the limitations applying to interpretation of the data as a result of this "selection" of workers, for inhalable dust there was a trend towards slightly lower threshold values between 3.25 and 5.89 mg/m$^3$. For respirable dust significant thresholds could no longer be derived because considerable loss of function was present even at low exposure concentrations. The average, time-weighted total dust concentration to which the subgroup with a significantly increased incidence of impaired lung function parameters had been exposed was $7.2 \pm 1.1$ mg/m$^3$. The range of shift average values to which these foundry workers had been exposed was between 2.3 and 21.5 mg/m$^3$ total dust.

## 4.3 Cross-sectional study of agricultural workers

In a study (Hofmann 1992, Kuthe and Pernack 1994) of 364 tractor drivers (field cultivators) in agriculture the lung function parameters FEV$_1$ and FVC determined by spirometry were significantly lower in the group with the highest dust exposure (more than 300 g/m$^3$·h) than in workers exposed to dust levels below 100 g/m$^3$·h. Since the average age of the persons in the group exposed to the high levels was $50.3 \pm 7.9$ years, a lower limit for the average dust concentration can be calculated to be 5 mg/m$^3$, assuming a 30-year working life and 2000 working hours per year (actually 1096 to 1805 hours/year exposure period and 1986 to 2623 hours/year total working time). The measured shift-average dust concentrations varied widely depending on the machinery being used and the wetness of the soil. In addition it must be pointed out that although the persons examined in this study were only occupied in cultivation of fields (that is, not in livestock breeding), they were exposed during the course of the year to very different kinds of dust (harvesting with combine harvesters, spreading of fertilizer, etc.) so that the exposures must be described as exposures to dust mixtures. The components of these mixtures which could conceivably have specific effects include not only low levels of quartz but also especially plant fragments including fibrous components and also diesel soot from the tractor engines.

## 4.4 Results of occupational medical check-ups

In the German Democratic Republic in the years 1982 to 1990 all results of occupational medical examinations were obtained and documented with standardized methods, and recorded centrally. The objectives were diagnoses according to the International Classification of Diseases (ICD 9). For chronic bronchitis the WHO definition applied. The severity was classified according to functional diagnostics, clinical appearance and effectiveness of therapy (Bräunlich 1993). Severity grades 2–4 of chronic bronchitis corresponded with the obstructive form, that is, the parameter FEV$_1$ in such cases was in the pathological range (less than the reference value minus 1.6 s).

Occupational exposures were classified in five grades (reference numbers), which were allocated by occupational hygienists on the basis of measured concentration values, catalogue values, or evaluation of analogous situations. The general threshold limit value for dust was a total dust concentration of 10 mg/m$^3$ expressed as a shift average value.

Employees whose respiratory tract was exposed to dusts and chemicals were examined at two-year to four-year intervals. Chronic obstructive lung disease was diagnosed in 2.58 % of male employees (n = 367968) and in 1.53 % of female employees (n = 135468). As is usual, the incidence increased steeply with age.

Interpretation of the results is limited by the relatively coarse grades of the reference number classes used to characterize dust exposure. Therefore it was only possible to differentiate between exposures equivalent to one to two times the then valid threshold (10–20 mg/m$^3$) and more than twice the threshold (> 20 mg/m$^3$). There was no requirement for examination of persons exposed to levels below the threshold. Among the persons (n = 1892) exposed to non-fibrogenic dusts (> 10 mg/m$^3$ total dust, 0 % to < 5 % quartz), the incidence of obstructive lung disorders increased in the 45-year old to 59-year old men from 2.68 % in non-exposed persons to 4.25 % (non-smokers) or from 4.90 % to 7.48 % (smokers). Both differences are statistically significant. The control group in this case comprised employees who were not exposed to inhaled dust but who had to be subjected to medical examination for other reasons (noise, shift work, etc.).

The incidence of obstructive lung disease in non-smokers exposed to non-fibrogenic dust concentrations above 20 mg/m$^3$ was twice that found in non-smokers exposed to 10–20 mg/m$^3$. In smokers the exposure to cigarette smoke alone led to an increased incidence of obstructive lung disease and the higher level occupational exposure to one or two times the threshold level of non-fibrogenic dust did not have an additive effect.

The results obtained from this extensive database on obstructive lung disease do not compare with those of specific epidemiological studies in precision of either diagnosis or evaluation of exposure levels. On the other hand, they demonstrate without doubt that prolonged exposure to various mixtures of dusts at shift-average levels above the threshold level for non-fibrogenic dusts (10 mg/m$^3$) results in a significant increase in the incidence of chronic obstructive lung disease relative to that in employees of similar age who are not exposed occupationally to dust. In the higher age groups of exposed persons, the incidence of chronic obstructive lung disease was practically twice that in persons not exposed to dust at work.

## 4.5 Other results

Oxman *et al.* (1993) reported an association between occupational dust exposure and the occurrence of chronic obstructive lung disease (COLD). In this review four studies are discussed, 3 studies of miners including the data of the DFG study "Chronische Bronchitis", and one study of workers from gold mines. The main conclusions are summarized below.

After exposure to an average fine dust concentration of 2 mg/m$^3$ for 35 years, more than 20 % reduction in the lung function parameter $FEV_1$, that is, obstructive disease was observed in 8 % of non-smokers (95 % CI: 3.4–13.7 %). In smokers after the same levels of exposure, this proportion was 6.6 % (95 % CI: 4.9–8.4 %). These values are above the value of 5 % assumed as "residual risk" (see Section 4.1). Analysis of the data for chronic bronchitis for the same conditions (2 mg/m$^3$ fine dust for 35 years) demonstrates that 4.5 % of the non-smokers (95 % CI: 2.1–7.4 %) and 7.4 % of smokers (95 % CI: 0–33.3 %) develop the occupational disease.

The evaluation of this study was carried out without assuming the existence of a threshold value. Under these conditions and for a "residual risk" below 5 %, the exposure limit would have to be below 2 mg/m$^3$.

The risk of developing lung disease is higher for the workers in gold mines than for the other miners. The authors suggest that this could be a result of the higher level of quartz in the dust from the gold mines.

# 5 Animal Studies

## 5.1 Long-term inhalation studies

Numerous animal studies of the effects of long-term exposure to dust have been carried out with rats as well as other animals and with various kinds of dust (reviewed by Morrow *et al.* 1991). In many of these studies, there was massive overloading of the lungs with dust. Unspecific effects such as reduced rates of alveolar clearance, chronic inflammatory reactions and interstitial fibrosis were observed; in some studies the incidence of lung tumours was even increased (Heinrich *et al.* 1994, 1995, Lee *et al.* 1985). Such effects were observed with the following poorly soluble dusts: coal mine dust, volcanic ash, flue ash from coal-fired power stations, petroleum coke, diesel engine emissions, PVC, titanium dioxide (Lee *et al.* 1985) and toner for photocopiers (Bellmann *et al.* 1991, Muhle *et al.* 1991). The results of these studies are summarized in Table 5.

## 5.2 Extrapolation from animal studies to man

The extrapolation of the results of animal studies to man depends on the following assumptions:

1. impairment of lung clearance—caused by overloading the lungs with dust—occurs in both rodents and in man
2. the overloading threshold for man is at the same dust concentration per gram lung tissue as for the rat
3. the reduction in the rate of particle clearance in man and rat depends decisively on the volume of dust taken up by the alveolar macrophages.

On the basis of these hypotheses and taking into account species-specific macrophage parameters and data for particle deposition and retention, extrapolation from the results of the studies with rats to the situation of persons exposed to dust can be carried out. However, it has still not been demonstrated unequivocally that the criteria for man are the same as those for the rat (see Section 3.4).

Assuming that the average volume of an alveolar macrophage is 1000 $\mu m^3$ in the rat and 2500 $\mu m^3$ in man and that the average number of macrophages in the lungs of the rat and man, respectively, is $2.6 \times 10^7$ and $7.0 \times 10^9$, 6 % of the macrophage volume is 1.04 $\mu l/g$ lung in the rat and 1.11 $\mu l/g$ lung in man (Crapo *et al.* 1983, Dethloff and Lehnert 1988).

**Table 5.** General effects of long-term exposure of rats to particles (cited from Morrow *et al.* 1991)

| Effect | Test material: | | | | | | |
|---|---|---|---|---|---|---|---|
| | titanium dioxide | volcanic ash | flue ash from coal-fired power stations | petroleum coke | diesel engine emissions | PVC | toner |
| increased lung weight | + | + | + | + | + | + | + |
| disproportionate increase in particle retention | ? | + | + | | + | + | + |
| reduced dust clearance | | + | + | | + | + | + |
| chronic inflammatory process | | + | | + | + | + | + |
| changes in pulmonary mechanics | | + | | | + | | + |
| increase in the number of macrophages loaded with particles | + | + | + | + | + | + | + |
| particles in the interstitium | + | + | + | + | + | + | + |
| thickening of the septa | + | + | | | + | + | + |
| alveolar proteinosis | + | + | | | + | | + |
| fibrosis | + | + | | + | + | ? | + |
| tumours | + | ? | | + | + | − | − |

+  positive results
?  questionable results
−  negative results

According to the results of studies with rats and according to the hypothesis of Morrow (1988), particle clearance in the lung is adversely affected when 6 % of the macrophage volume is filled with poorly soluble particles.

On the basis of specific deposition and clearance models for rat and man and assuming an alveolar retention half-time for poorly soluble particles in the rat of 75 days and in man of 400 days (Bailey *et al.* 1985), for particles with a density of 1 g/cm$^3$ and a diameter of 3 µm, the rat and man would have to be exposed continuously to particle concentrations of 3.3 mg/m$^3$ and 0.8 mg/m$^3$, respectively, to produce a lung load of about 1 mg/g lung and 1 µl/g lung (about 6 % of the macrophage volume). At this level the clearance coefficient in the rat is reduced by a factor of about 2 (see Figure 4); that is equivalent to a reduction in the half-time of alveolar clearance in the rat from about 60 to 120 days. For the rat in this calculation a tidal volume of 2 ml, a respiration rate of 73 per minute and a lung weight of 1.5 g were assumed. For man the corresponding data are 1250 ml, 16 per minute and 950 g; exposure times are taken as 8 hours daily and 240 days per year. For this calculation (Oberdörster 1994) obligatory nasal breathing was assumed for the rat and for man a mixture of nasal and oral breathing; mathematical particle deposition models (Schum and Yeh 1980, Yeh and Schum 1980) were used.

It is apparent that the calculated threshold concentrations are dependent not only on the density of the dust but also on the particle diameters (which affect the probability of deposition), on whether and to what extent nasal or oral breathing predominates, and on the tidal volume.


## 5.3 Threshold value for the respirable fraction

In this section, a threshold level for dust at which impairment of alveolar particle clearance begins is derived. The derivation is based on the method of Morrow *et al.* (1991).

A worker is exposed to a constant dust concentration c during 240 work shifts per year. The shift length is 8 h; outside the shift an exposure concentration of zero is assumed.

As detailed in Section 3.2, the retained mass of dust $m_{eq}$ in the lungs when dust deposition and clearance are in equilibrium is given by:

$$m_{eq} = \frac{D \cdot t_{1/2}}{\ln 2}$$

whereby D, the dust deposition per unit time, is assumed to be constant. On the other hand, the dust deposition per unit time during a work shift is given by:

$$D_{shift} = f \cdot c \cdot \dot{V}$$

(see Section 3.2), whereas outside the shift the deposition is zero. The deposition per unit time is thus not constant.

However, since it may be assumed that the clearance half-time is 400 days (see Section 3.4), it is permissible to use the weight of dust deposited per year as the constant D (deposition per unit time) and to base the calculation on a constant exposure on 365 days per year. This annual average value D is given by:

$$D = \frac{D_{shift} \cdot 8h \cdot 240}{1\ year}$$

The weight of dust deposited during an 8-hour shift is $D_{shift}\cdot 8$ h. The total weight of dust deposited during a year (given 240 shifts per year) is $D_{shift}\cdot 8$ h$\cdot 240$. Division by the total period of deposition (1 year) then yields the average deposition per time.
Thus

$$D = \frac{f \cdot c \cdot \dot{V} \cdot 8\ h \cdot 240}{1\ year}$$

where
c   is the (assumed constant) exposure concentration at the workplace and
$\dot{V}$   is the volume inhaled per unit time.

It is assumed that a worker inhales 15 litres per minute, i.e., $\dot{V} = 15$ l/min. In addition, it is assumed that the average fraction of the inhaled dust which is deposited in the lungs is $f = 0.3$ (Oberdörster 1994).
   Substituting for D in the above equation for $m_{eq}$ the aerosol concentration c in the air is given by:

$$c = \frac{m_{eq} \cdot \ln 2 \cdot 365\ days}{t_{1/2} \cdot f \cdot \dot{V} \cdot 8\ h \cdot 240}$$

   In experimental animal studies it was demonstrated that a retained dust volume of 1 µl/g lung (see Section 5.2) leads to marked "dust overloading" and to a reduced rate of particle clearance. Recalculation for the weight of dust requires knowledge of the density of the material. Thus, for a dust of density $\rho$ g/ml the rate of clearance is reduced by $\rho$ mg/g lung. Assuming that for man too the deposition of 1 µl/g lung can be seen as the threshold value, the threshold concentration c for a human lung weight of 1000 g is given by:

$$c = \frac{\rho\ mg/g\ lung \cdot 1000\ g \cdot \ln 2 \cdot 365\ days \cdot 1000\ l/m^3}{400\ days \cdot 0.3 \cdot 15\ l/min \cdot 8\ h \cdot 240 \cdot 60\ min/h} = \rho \cdot 1.2\ mg/m^3$$

   This is the aerosol concentration of particles of density $\rho$, which results in an equilibrium concentration of particles in the lung of 1 µl/g lung, equivalent to $\rho$ mg/g lung, assuming that alveolar clearance obeys first order kinetics.
   Since in animal studies a constant exposure concentration is used for periods up to two years, this concentration is equivalent to the annual average exposure concentration D.

### 5.3.1 Effect of particle density on the threshold for the respirable fraction

The density of dusts occurring at the workplace lies in the range between 1 g/cm$^3$ for plastic dusts and over 5 g/cm$^3$ for metal dusts (BIA 1989). Since, as described above, for a low density dust a low threshold may be derived, dusts are classified according to their density. For materials of density in the range between 2 and 5 g/cm$^3$ the value is derived on the basis of density 2 g/cm$^3$. This applies for most industrial dusts. For plastic dusts and other dusts of density between 1 and 2 g/cm$^3$ the calculated threshold is 1.2 mg/m$^3$.

The density of dusts found at real workplaces, apart from that of wood dust, lies between 2 and 5.

# 6 Manifesto (MAK value, classification)

The general threshold limit value for dust (fine dust, respirable dust) which was valid until 1996 was 6 mg/m$^3$ and was established on the basis of the data from the 1983 DFG study "Chronische Bronchitis".

There are no results of studies on exposed persons from which a dust concentration without effects in the sense of a NOEL could be deduced. The epidemiological studies provide data only about concentrations at which effects are observed.

For the present documentation, data from the DFG study "Chronische Bronchitis" (modified evaluation concept) and data from more recent epidemiological studies from the German Democratic Republic have been used. The target criterion was impairment of lung function.

It must also be pointed out that all of the data for fine and total dust concentrations on which the epidemiological studies are based were obtained with area sampling systems and so tend to be too low in comparison with the concentrations obtained by personal sampling.

In addition, data obtained in long-term experimental studies with rats exposed to insoluble dusts have also been used for the derivation of a threshold value for respirable dust; the target criterion in this case was a reduction in the rate of alveolar clearance. Assuming that similar criteria apply in man, the data for rats were extrapolated to the human situation.

## 6.1 Inhalable fraction (total dust)

As a basis for occupational medical examination of workers in the German Democratic Republic, the general threshold limit value for dust was set at 10 mg/m$^3$. For persons exposed to shift-average levels above 10 mg/m$^3$ the incidence of obstructive lung disease was markedly higher than for those exposed to lower levels. These studies make it clear that the threshold for the shift-average concentration must be below 10 mg/m$^3$. The analysis of the data from the DFG study with new statistical methods yielded a value of

3.8 mg/m$^3$, at which a slight increase (about 5 %) in the background incidence of the chronic bronchial reaction is to be expected. The longitudinal study of foundry workers observed for 15 years revealed values between 4 and 5.9 mg/m$^3$ for impairment of various lung function parameters. A study of persons employed in agriculture revealed that persons exposed to average levels of 5 mg/m$^3$ total dust or more had clearly reduced values for FEV$_1$ and FVC compared with the values found in persons exposed at lower levels.

On the basis of these results a general threshold limit value for inhalable dust (total dust) is established provisionally, until new data are available, at 4 mg/m$^3$. When such a threshold value is observed, the incidence of impairment of lung function as a result of the unspecific effects of dust—the so-called chronic bronchial reaction—is increased above the background level only little if at all.


## 6.2 Respirable fraction (fine dust)

From the data for the three factories included in the DFG study, a threshold value of 1.7 mg/m$^3$ for fine dust was derived from the data for the incidence of the chronic bronchial reaction. After exposures at this level, the incidence of the chronic bronchial reaction is expected to increase only slightly above the background level, by about 5 %. The longitudinal study of foundry workers yielded a threshold value of 1.6 mg/m$^3$ for its targeted changes in diagnostic lung function parameters or in a variable combining diagnostic lung function parameters.

The experimental clearance studies in animals suggest that marked impairment of clearance function is to be expected in man at concentrations in the region of 1.2 mg/m$^3$ respirable dust of density 1 g/cm$^3$. That means that exposure to insoluble dust of a single material at concentrations above $\rho \cdot 1.2$ mg/m$^3$ (Morrow 1991) should be avoided. The density of real dusts found at the workplace, apart from wood dust, is generally between 2 and 5 g/cm$^3$. For a density of 2, the experimental clearance studies with animals yield a threshold value of 2.4 mg/m$^3$ fine dust.

The MAK value for the respirable fraction (fine dust) is reduced provisionally to 1.5 mg/m$^3$. When such a threshold value is observed, the incidence of impairment of lung function as a result of the unspecific effects of dust—the so-called chronic bronchial reaction—is increased above the background level only little if at all.


## 6.3 Applicability

The general threshold limit value for dust is to apply for dusts for which no specific threshold value exists and for mixtures of dusts. If mixtures of dusts contain components for which specific MAK values exist, these are to be observed as well as the general threshold limit value for dust.

The cohorts discussed above (DFG study, foundries in the GDR, longitudinal study of agricultural workers) were exposed to mixtures of dusts which contained quartz and other toxic components, e.g. heavy metals, which may be assumed to have increased the severity of the effects. For safety's sake, the threshold values derived from these cohort studies have been adopted as the general threshold limit values for insoluble and poorly soluble dusts for which no other threshold value applies. Because the database is inadequate, genotoxic, carcinogenic, fibrogenic, allergenic and other toxic effects of these dusts cannot at present be excluded.

Epidemiological studies of persons exposed to dusts containing quartz have demonstrated that the incidence of quartz-induced lung tumours is increased relative to that in the general population in persons with silicosis. Since fibrogenic effects in the sense of pneumoconiosis are not to be expected if the general threshold limit value for mixtures of dusts is observed, it may be assumed that under these conditions any risk of developing lung cancer is small.

The MAK values for the inhalable and respirable dust fractions were derived from long-term exposure values. As described in the 1996 documentation "Derivation of MAK values for dusts from long-term threshold values" (in Volume 11 of this series), it is permissible for individual shift-average concentration values to exceed the shift-related MAK value by up to a value of K·MAK value, where K is the maximum permissible excursion factor. Since the general threshold limit value for dust applies for a variety of dusts, some of which have been only inadequately studied, it is recommended that the maximum size of this excursion factor be kept to K = 2.

For aluminium and its oxides, graphite, iron oxide, magnesium and titanium oxide—except when ultrafine aerosols are involved—the general threshold limit for dust will continue to be used until these substances have been studied individually in more detail.

# 7 References

Bailey MR, Fry RA, James AC (1985) Long-term retention of particles in the human respiratory tract. *J Aerosol Sci 16*: 295–305

Bellmann B, Muhle H, Creutzenberg O (1991) Lung clearance and retention of toner, utilizing a tracer technique during chronic inhalation exposure in rats. *Appl Toxicol 17*: 300–313

Bellmann BO, Creutzenberg HA, Mermelstein R (1994) Models of deposition, retention and clearance of particles after dust overloading of lungs in rats. *Ann Occup Hyg 38, Suppl*: 303–311

BIA (Berufsgenossenschaftliches Institut für Arbeitssicherheit) (1989) Arbeitsmappe, "*Messung von Gefahrstoffen*", Erich Schmidt Verlag, Bielefeld

BIA (Berufsgenossenschaftliches Institut für Arbeitssicherheit) (1995) Handbuch, *Schutzmaßnahmen in Putzereien der Gießereiindustrie*, Erich Schmidt Verlag, Bielefeld, 24th issue I/95

BIA (Berufsgenossenschaftliches Institut für Arbeitssicherheit) (1996) Report 5/96, *Stäube an Arbeitsplätzen in der DDR—Umrechnungsfaktoren der Meßverfahren, Meßergebnisse für mineralische (asbestfreie) Stäube, Bewertung*. Ed.: Hauptverband der gewerblichen Berufsgenossenschaften

Bolton RE, Vincent JH, Jones AD, Addison J, Beckett ST (1983) An overload hypothesis for pulmonary clearance of UICC amosite fibers inhaled by rats. *Br J Ind Med 40*: 264–272

Bräunlich A, Enderlein G, Heuchert G, Kersten N, Schneider WD (1993) Betriebsärztliche Möglichkeiten zur Prävention chronisch obstruktiver Lungenkrankheiten — Einfluß von Stäuben und chemischen Atemtraktirritantien. *Zentralbl Arbeitsmed Arbeitsschutz Ergonom 43*: 214–223

CEN (European Committee for Standardization) (1993) *EN 481 Workplace atmospheres — Size fraction definitions for measurement of airborne particles.* Brussels 1993, Beuth Verlag, Berlin 1993

Cohen DS, Arai, Brain JD (1979) Smoking impairs long-term dust clearance from the lung. *Science 204*: 514–517

Crapo JD, Young SL, Fram EK (1983) Morphometric characteristics of cells in the alveolar region of mammalian lungs. *Am Rev Respir Dis 128*: 42–46

Cuddihy RG, McClellan RO, Griffith WC (1979) Variability in target dose deposition among individuals exposed to toxic substances. *Toxicol Appl Pharmacol 49*: 179–187

Dethloff LA, Lehnert BE (1988) Pulmonary interstitial macrophages: Isolation and flow cytometric comparisons with alveolar macrophages and blood monocytes. *J Leukocyte Biol 43*: 80

Deutsche Forschungsgemeinschaft (1975) *DFG-Forschungsbericht "Chronische Bronchitis"*, Harald Boldt Verlag, Boppard

Deutsche Forschungsgemeinschaft (1981) *DFG-Forschungsbericht "Chronische Bronchitis", Teil 2*, Harald Boldt Verlag, Boppard

DFG (Deutsche Forschungsgemeinschaft) (1995) *List of MAK and BAT Values 1995*, Commission for the Investigation of Health Hazards of Chemical Compounds in the Work Area, Report No. 31, VCH Verlagsgesellschaft mbH; Weinheim

DFG (Deutsche Forschungsgemeinschaft) (1998) *List of MAK and BAT Values 1998*, Commission for the Investigation of Health Hazards of Chemical Compounds in the Work Area, Report No. 34, WILEY-VCH Verlag GmbH; Weinheim

Dietz E (1992) Estimation of heterogenicity — a GLM-approach. in: Fahrmeir L, Francis B, Gilchrist R, Hutz E (Eds) *Advances in GLIM and statistical Modelling*: Proceedings of the GLIM 92 Conference and the 7th International Workshop of Statistical Modelling, Munich, 13–17 July 1992, Springer Verlag, New York

Freedman AP, Robinson SE, Street MR (1988) Magnetopneumographic study of human alveolar clearance in health and disease. *Ann Occup Hyg 32, Suppl 1*: 809–820

Fruhmann G, Woitowitz HJ (1997) Chronisch-obstruktive Bronchitis und Lungenemphysem. *Dtsch Aerztebl 94*: A-235–236

Gore DJ, Patrick G (1982) A quantitative study of the penetration of insoluble particles into the tissue of the conducting airways. *Ann Occup Hyg 26*: 149–161

Greim H (Ed.) (1998) Toxikologisch-arbeitsmedizinische Begründungen von MAK-Werten, 27th issue, *Steinkohlengrubenstaub*, WILEY-VCH, Weinheim

Heinrich U, Dungworth DL, Pott F (1994) The carcinogenic effects of carbon black particles and tar/pitch condensation aerosol after inhalation exposure of rats. *Ann Occup Hyg, Suppl 38*: 351–356

Heinrich U, Fuhst R, Rittinghausen S, Creutzenberg O, Bellmann B, Koch W, Levsen K (1995) Chronic inhalation exposure of Wistar rats and two different strains of mice to diesel engine exhaust, carbon black, and titanium dioxide. *Inhal Toxicol 7*: 533–556

Hofmann S (1992) Arbeitsmedizinische Untersuchungen zum Atemtrakt bei Fahrern mobilen Landmaschinen in der Pflanzenproduktion. in: Schäcke G, Ruppe K, Vogel-Sührig C (Eds) *Arbeitsmedizin für eine gesunde Umwelt: Arbeitsmedizin als Schrittmacher für eine gesunde Umwelt*. Arbeitsmedizinisches Kolloquium der gewerblichen Berufsgenossenschaften (31st annual meeting of the Deutsche Gesellschaft für Arbeitsmedizin). Gentner Verlag, Stuttgart, 323–326

ICRP (International Commission on Radiological Protection) (1966) Task group on lung dynamics. *Health Phys 12*: 173–207

Karsten H, Schneider WD, Meyer B, Dietz E (1992) Dosis-Wirkungs-Beziehungen zwischen Staubexposition und Lungenfunktion bei Gießereiarbeitern. in: Schäcke G, Ruppe K, Vogel-Sührig Ch. (Eds) *Arbeitsmedizin für eine gesunde Umwelt: Arbeitsmedizin als Schrittmacher für eine gesunde Umwelt. Arbeitsmedizin in der Land- und Forstwirtschaft.* Arbeitsmedizinisches Kolloquium der gewerblichen Berufsgenossenschaften (31st annual meeting of the Deutsche Gesellschaft für Arbeitsmedizin). Gentner Verlag, Stuttgart, 107–109

Küchenhoff H, Ulm K (1997) Comparison of statistical methods for assessing threshold limiting values in occupational epidemiology. *Comput Stat 12*: 249–264

Kuthe C, Pernack EF (1994) Arbeitshygienische Untersuchungen zur Staubbelastung an mobilen Landmaschinen in landwirtschaftlichen Großbetrieben in Brandenburg. *Staub-Reinhaltung Luft 54*: 275–282

Lange HJ, Ulm K (1983) *Mathematische Modelle zur Frage eines Allgemeinen Staubgrenzwertes; epidemiologische Auswertungen mit Daten des DFG-Schwerpunktprogramms "Chronische Bronchitis" (CB).* Verlag Chemie, Weinheim

Lee KP, Trochimowicz HJ, Reinhardt CF (1985) Pulmonary response of rats exposed to titanium dioxide ($TiO_2$) by inhalation for two years. *Toxicol Appl Pharmacol 79*: 179–192

McClellan RO (1990) Particle overload in the lung: approaches to improve our knowledge. *J Aerosol Med 3*, Suppl 1: 197–207

Morrow PE (1988) Possible mechanisms to explain dust overloading of the lungs. *Fundam Appl Toxicol 10*: 369–384

Morrow PE (1992) Contemporary issues in toxicology. Dust overloading in the lungs. Update and appraisal. *Toxicol Appl Pharmacol 113*: 1–12

Morrow PE, Mermelstein R (1988) *Inhalation toxicology: the design and interpretation of inhalation studies and their use in risk assessment.* Springer Verlag, New York, 103–117

Morrow PE, Muhle H, Mermelstein R (1991) Chronic inhalation study findings as a basis for proposing a new occupational dust exposure limit. *J Am Coll Toxicol 10*: 279–290

Muhle H, Bellmann B, Heinrich U (1988) Overloading of lung clearance of experimental animals to particles. *Ann Occup Hyg 32, Suppl*: 141–148

Muhle H, Creutzenberg O, Bellmann B, Heinrich U, Mermelstein R (1990) Dust overload in lungs: investigations of various materials, species differences, and irreversibility of effects. *J Aerosol Med 3, Suppl 1*: 111–128

Muhle H, Bellman B, Creutzenberg O, Dasenbrock C, Ernst H, Kilpper R, MacKenzie J, Morrow P, Mohr U, Takenaka S, Mermelstein R (1991) Pulmonary response to toner upon chronic inhalation exposure in rats. *Fundam Appl Toxicol 17*: 280–299

Oberdörster G (1988) Lung clearance of inhaled insoluble and soluble particles. *J Aerosol Med 1*: 289–320

Oberdörster G (1994) Extrapolation of results from animal inhalation studies with particles to humans? in: Mohr U (Ed.) *Toxic and carcinogenic effects of solid particles in the respiratory tract,* ILSI Press, Washington, 335–353

Oxman AD, Muir DCF, Shannon HS, Stock SR, Hnizdo E, Lange HJ (1993) Occupational dust exposure and chronic obstructive pulmonary disease. A systematic overview of the evidence. *Am Rev Respir Dis 148*: 38–48

Patrick G, Sterling C (1977) The retention of particles in large airways of the respiratory tract. *Proc Roy Soc Lond B 198*: 455–462

Raabe OG (1967) Some important considerations in use of power functions to describe clearance data. *Health Phys 13*: 293–295

Rödelsperger K (1996) *Anorganische Fasern im menschlichen Lungengewebe. Lungenstaubfaseranalytik zur Epidemiologie der Risikofaktoren des diffusen malignen Mesothelioms (DMM).* Schriftenreihe der Bundesanstalt für Arbeitsmedizin, Fb 01 HK 076, Berlin

Scheuch G, Stalhofen W, Heyder J (1996) An approach to deposition and clearance measurements in human airways. *J Aerosol Med 9*: 35–41

Schlesinger RB (1989) Deposition and clearance of inhaled particles. in: McClellan RO, Henderson RF (Eds) *Conception Inhalation Toxicology.* Hemisphere, New York: 163–192

Schneider WD, Dietz E, Gierke E, Karsten H, Maintz G (1986) Dosis-Wirkungs-Beziehungen zwischen Staubexposition und Lungenfunktion bei Gießereiarbeitern. *Z Gesamte Hyg 32*: 688–692

Schneider WD, Karsten H, Gierke E (1994) Long-term dust exposure in foundry workers may lead to impaired lung function. in: Mohr U, Dungworth DL, Mauderly JL, Oberdörster G (Eds) *Toxic and carcinogenic effects of solid particles in the respiratory tract*, ILSI Press Washington: 491–494

Schneider WD, Karsten H, Dietz E, Gierke E, Lotz G (1995) Dosis-Wirkungs-Beziehungen zwischen Staubexposition und Lungenfunktion bei Gießereiarbeitern. in: Schneider WD, Bräunlich A, Lorenz A, Schöneich R, Thürmer H, Wallenstein G, *Dosis-Wirkungs-Beziehungen bei irritativer Atemtraktbelastung (Schlußbericht)*, Wirtschaftsverl. NW, Bremerhaven: 137–162

Schum M, Yeh HC (1980) Theoretical evaluation of aerosol deposition in anatomical models of mammalian lung airways. *Bull Math Biol 42*: 1–15

Snipes MB, Clem MF (1981) Retention of microspheres in the rat lung after intratracheal instillation. *Environ Res 24*: 33–41

Snipes MB, Chavez GT, Muggenburg BA (1984) Deposition of 3-, 7- and 13-μm microspheres instilled into lungs of dogs. *Environ Res 33*: 333–342

Stahlhofen W, Gebhart J, Rudolf G, Scheuch G (1986) Measurement of lung clearance with pulses of radioactively-labeled aerosols. *J Aerosol Sci 17*: 333–336

Stahlhofen W, Scheuch G, Bailey MR (1995) Investigations of retention of inhaled particles in the human bronchial tree. *Radiat Prot Dosimetry 60*: 311–319

Stöber W, Einbrodt HJ, Klosterkötter (1967) Quantitative studies of dust retention in animal and human lungs after chronic inhalation. in: Davies CN (Ed.) *Inhaled Particles and Vapours II*, Pergamon Press, Oxford, 409–417

Strom KA, Johnson JT, Chan TC (1989) Retention and clearance of inhaled submicron carbon black particles. *J Toxicol Environ Health 26*: 183–202

Thürmer H (1995) personal communication

Ulm K, Dannegger F, Spanier M (1996) *Bericht über die Auswertung der Daten der Studie "Chronische Bronchitis". Discussion paper No. 39*, Sonderforschungsbereich 386 der Ludwigs-Maximilians-Universität München

Ulmer WT (1979) *Bronchitis Asthma Emphysem*, Springer, Berlin Heidelberg New York

Valentin H, Woitowitz HJ (1967) Chronische Bronchitis und Lungenemphysem als arbeitsmedizinisches Problem. *Internist 8*: 165–172

WHO (World Health Organization) (1988) *Man-made mineral fibres*. Environmental Health Criteria 77, WHO, Genf

Woitowitz HJ (1972) Der chronisch Kranke am Arbeitsplatz. Diagnostik, Therapie, Betreuung. Das chronische unspezifische respiratorische Syndrom. *Arbeitsmed Sozialmed Arbeitshyg 7*: 22–26

Woitowitz HJ, Hillerdal G, Calavresoz A, Berghäuser KH, Rödelsperger K, Jöckel KH (1993) *Risiko- und Einflußfaktoren des diffusen malignen Mesothelioms (DMM)*, Schriftenreihe der Bundesanstalt für Arbeitsschutz, Fb 698, Bonn

Yeh HC, Schum M (1980) Theoretical evaluation of aerosol deposition in anatomical models of mammalian lung airways. *Bull Math Biol 42*: 461–480

Yu CP, Chen YK, Morrow PE (1989) An analysis of alveolar macrophage mobility kinetics at dust overloading of the lungs. *Fundam Appl Toxicol 13*: 452–459

completed 7.4.97

# Ethanol

| | |
|---|---|
| **MAK value (1998)** | **500 ml/m$^3$ (ppm) $\hat{=}$ 950 mg/m$^3$** |
| **Peak limitation (1998)** | **Category II,1** |
| **Absorption through the skin** | **–** |
| **Sensitization** | **–** |
| **Carcinogenicity (1998)** | **Category 5** |
| **Prenatal toxicity (1994)** | **Pregnancy risk group C** |
| **Germ cell mutagenicity (1998)** | **Germ cell mutagen group 2** |
| **BAT value** | **–** |
| Synonyms | absolute ethanol<br>alcohol<br>ethyl alcohol<br>methyl carbinol<br>spirits of wine |
| Chemical name (CAS) | ethanol |
| CAS number | 64-17-5 |
| Structural formula | $H_3C$–$CH_2OH$ |
| Molecular formula | $C_2H_6O$ |
| Molecular weight | 46.07 |
| Melting point | –114.1°C |
| Boiling point | 78.5°C |
| Density at 20°C | 0.79 g/cm$^3$ |
| Vapour pressure at 20°C | 59 hPa |
| logP$_{ow}$[1] | –0.32 |
| **1 ml/m$^3$ (ppm) $\hat{=}$ 1.9 mg/m$^3$** | **1 mg/m$^3$ $\hat{=}$ 0.53 ml/m$^3$ (ppm)** |

---

[1] $n$-Octanol/water distribution coefficient

*Essential MAK Value Documentations.* DFG, Deutsche Forschungsgemeinschaft
Copyright © 2006 WILEY-VCH Verlag GmbH & Co. KGaA, Weinheim
ISBN: 3-527-31394-X

# 1 Toxic Effects and Mode of Action

Ethanol is unusual among the substances to which people can be exposed occupationally because relevant levels of exposure can also be caused by alcoholic drinks consumed during and outside working hours. Ethanol is a very common industrial chemical. It is used, for example, as a solvent, it occurs as an intermediate in organic syntheses and is used as antifreeze, fuel or starting material for the synthesis of pharmaceuticals, plastics, resins and many other substances. Occupational exposure involves mainly inhalation and skin contact. About 60% of inhaled ethanol is retained; the amounts absorbed through the skin are small, at least under non-occlusive conditions. Absorbed ethanol is distributed mainly in the aqueous compartments of the body but can penetrate the blood-brain barrier and cross the placenta. Ethanol is oxidized almost completely in the liver by means of alcohol dehydrogenase, at higher blood concentrations also by cytochrome P4502E1, to yield acetaldehyde which is in turn metabolized by aldehyde dehydrogenase to acetic acid. The acetic acid enters intermediary metabolism or is broken down to $CO_2$ and water.

The acute toxicity of inhaled ethanol is low in man and animals. Single exposures to concentrations up to 5000 ml/m$^3$ have neither local nor systemic effects in man; the resulting blood ethanol concentrations are far below those required to produce first effects on the central nervous system, about 200 mg ethanol per litre blood. The toxicity of ethanol after repeated inhalation has not been investigated in appropriate studies. Long-term ingestion of ethanol in alcoholic drinks can—depending on the dose—have adverse effects on practically all organ systems. The primary target organ is the liver.

On the skin and mucous membranes of the eyes, liquid ethanol is moderately irritating. A significant sensitizing potential cannot be deduced from the data available for man or animals.

Both *in vitro* after metabolic activation and *in vivo*, ethanol has a weak genotoxic potential. Evidence of DNA damage (adducts, chromosomal aberrations) is more frequently found in persons who consume large amounts of alcohol than in control persons. Evidence of mutagenicity in germ cells of animals has so far only been obtained after administration of very high doses in the markedly toxic range.

In man, alcohol consumption can result in an increased incidence of tumours at various localizations. In animal studies, a carcinogenic effect of ethanol has not been demonstrated unambiguously. Data for the effects of inhalation of ethanol as it occurs during occupational exposure are not available.

Reduction in fertility, and effects on the levels of sexual hormones have been described in man and animals after ingestion of large amounts of ethanol.

The reproductive toxicity of ingested ethanol and the clinical symptoms of alcohol embryopathy are well established and have been verified in animal studies. When pregnant rats were exposed to ethanol concentrations up to 20000 ml/m$^3$, maternal toxic effects but no toxic effects on reproduction were seen.

# 2 Mechanism of Action

## 2.1 Genotoxicity

For the weak genotoxic effects of ethanol which have been detected in only few *in vitro* and *in vivo* tests, metabolic activation seems to be required. The activation probably involves metabolic conversion of ethanol to the reactive electrophile, acetaldehyde, for which chromosome-damaging effects have been demonstrated in various eukaryotic cell systems (IARC 1988, Obe and Anderson 1987, see also 1986 MAK documentation for acetaldehyde in *Occupational Toxicants Volume 3*). In human lymphocytes, acetaldehyde induces DNA single strand breaks and, at very high concentrations, also double strand breaks (Singh and Khan 1995) and mutations at the HPRT locus. The mutagenic effect is weak and caused mainly by deletions at the 3'-end of the gene (He and Lambert 1990). The mechanism suggested for the genotoxicity of acetaldehyde is the formation of DNA-protein crosslinks which have been detected *in vitro* and in the nasal mucosa of rats which had inhaled the substance. However, the capacity of acetaldehyde to form DNA-protein crosslinks when incubated with histone and a DNA plasmid *in vitro* is about 100000 times less than that of formaldehyde (Kuykendall and Bogdanffy 1992a) and the crosslinks are not very stable at 37°C (Kuykendall and Bogdanffy 1992b).

In addition, direct reaction of acetaldehyde with nucleophilic centres of DNA bases could also contribute to the genotoxicity of this substance. *In vitro*, for example, acetaldehyde reacts with the exocyclic amino group of guanine to form the corresponding relatively unstable imine. After reduction, for example with $NaBH_4$ at neutral pH, the stable adduct, $N^2$-ethyl-deoxyguanosine, is formed (Fang and Vaca 1995). Recently this adduct has also been detected in the blood cells of alcohol-dependent volunteers (Fang and Vaca 1997) and in the hepatocytes of mice exposed to ethanol, and the detection was possible without previous reduction of the sample (Fang and Vaca 1995). This indicates that the unstable imine can be reduced in the organism, at least to some extent, to yield the stable $N^2$-ethyl-deoxyguanosine. The ability of the organism to repair the $N^2$-ethyl-deoxyguanosine adduct and the mutagenic potential of this adduct have not yet been studied.

As well as the genotoxic effects of the main metabolite, acetaldehyde, it is conceivable that radicals formed during the biotransformation of ethanol also contribute indirectly to its genotoxicity (Bondy 1992). The radicals are mainly hydroxyl radicals which are formed during the oxidation of ethanol by cytochrome P4502E1. In addition, hydroxyethyl radicals and $H_2O_2$, which can be reduced to hydroxyl radicals in the presence of $Fe^{2+}$, for example, have been detected *in vivo* after repeated exposure to ethanol (Brooks 1997). The results of *in vitro* studies suggest that reactive oxygen species can also result from the activity of the enzyme acetaldehyde dehydrogenase (ALDH). The enzyme oxidizes not only its normal substrate, acetaldehyde, but also NADH, and this results in the production of superoxide anions (Mira *et al.* 1995). Reactive oxygen species can react directly with DNA, but can also attack membrane lipids and increase the amount of lipid peroxidation, which has been repeatedly demonstrated in experimental animals exposed for long periods to ethanol. The degradation products of lipid peroxidation, malondialdehyde and 4-hydroxynonenal, can react with DNA bases to yield the corresponding etheno-base adducts. These adducts are also formed endogenously, presumably as a result of the lipid peroxidation which takes place at low levels all the time. A mutagenic potential of the etheno-bases has been demonstrated at least *in vitro* (Brooks 1997).

## 2.2 Carcinogenicity

Ethanol, or its metabolites, is considered to be a decisive factor in the carcinogenic effect of alcoholic drinks on man. This was concluded from the observation that all kinds of alcoholic drinks, consumed in sufficiently large quantities, can increase the incidence of certain kinds of tumour (Blot 1992, IARC 1988). The mechanisms of ethanol-induced carcinogenesis are not well understood. Tumorigenesis caused by local effects of ethanol in the oral cavity, pharyngeal cavity, larynx and possibly also the oesophagus, on the one hand, must be distinguished from the carcinogenic effects in the liver and mammary gland which result from the systemic availability of ethanol or its metabolites. The existence of a local mechanism was postulated also because the incidence of cancer in the oral and pharyngeal cavities was increased in persons who used alcoholic mouth washes which contained at least 25% ethanol although they were generally not swallowed (Blot 1992). The physicochemical properties of ethanol suggest a possible mechanism: it could modify the barrier function of the cell membranes so that carcinogenic substances more readily penetrate the cells. This hypothesis is supported by the fact that ethanol and tobacco smoke, with its many carcinogenic components, have synergistic effects in the area of the oral and pharyngeal cavities. In addition, the effect of ethanol on the expression of certain xenobiotic-metabolizing enzymes, which could result in increased activation of carcinogens, could also be responsible (Blot 1992, IARC 1988). That ethanol causes an increase in the incidence of tumours induced by various carcinogens has been demonstrated in several animal studies. For example, ethanol increases the carcinogenicity of certain nitrosamines in the upper gastrointestinal and respiratory tracts and increases the hepatocarcinogenesis of vinyl chloride (Anderson *et al.* 1995, Blot 1992, IARC 1988, Mufti *et al.* 1997, Seitz *et al.* 1992).

To date it has not been demonstrated convincingly that ethanol can act as a complete carcinogen in the liver. It is conceivable that a causal role is played by the development of liver cirrhosis which, independent of the factors responsible, is considered to be a precarcinogenic lesion. The induction of cytochrome P4502E1 and the resulting increase in the production of reactive oxygen species and in lipid peroxidation is considered to play an essential part in the induction of liver damage by alcohol (Brooks 1997). It has also been suggested that ethanol could have a modulating effect on hepatocellular carcinogenesis resulting from other factors such as infection with hepatitis B or hepatitis C viruses. In addition, the induction of various xenobiotic-metabolizing enzymes by ethanol and its alteration of the properties of the membranes in the endoplasmic reticulum and the cell could modify the toxicokinetics and bioavailability of certain carcinogens or precarcinogens and thus favour the development of hepatocellular or extrahepatic carcinomas (Anderson *et al.* 1995, Farber 1996, Seitz *et al.* 1992).

The high incidence of mammary cancer which has been associated with the intake of alcoholic drinks could result from the effects of ethanol on the hormone system. However, the results of the animal and *in vitro* studies which have been carried out to date do not offer any clear mechanistic explanations for the epidemiological findings (Blot 1992, Longnecker 1995, Singletary 1997).

It is not known to what extent the weak genotoxic effects of ethanol, or rather of acetaldehyde, and the reactive oxygen species which are formed during the metabolism of ethanol are involved in producing the carcinogenic effect. At least in alcoholics and in persons with a genetically reduced activity of acetaldehyde dehydrogenase (ALDH), increased acetaldehyde levels were found in peripheral venous blood (Eriksson and Fukunaga 1993). Evidence for a critical role of acetaldehyde in the etiology of tumours caused by ethanol in the upper digestive tract was obtained in Japanese persons with the inactive ALDH2 genotype which results in increased concentrations of acetaldehyde after consumption of ethanol: 40 alcohol addicts and 29 patients not addicted to alcohol with oesophagal squamous epithelial carcinoma and, respectively, 55 and 28 control persons were tested for their ALDH2 genotype. Among both the alcoholics and the non-alcoholics with an ALDH2*2 allele, which codes for an inactive ALDH2, a markedly increased risk of oesophagus cancer was found. The odds ratio (OR) was 7.6 (95 % confidence interval (CI): 2.8–20.7) for alcoholics and 12.1 (95 % CI: 3.4–42.8) for non-alcoholics. The alcoholics consumed about 120 g ethanol daily, the non-alcoholics about 55 g. Because of the small size of the groups, confounders like smoking and diet could not be taken into account (Yokoyama *et al.* 1996a). In support of these findings, an association between incidence of primary oesophagus tumours and the ALDH2 genotype was observed in 33 male Japanese alcoholics with squamous epithelial carcinomas in the oesophagus. Of 17 patients with the genotype for inactive ALDH2, 13 had multiple primary tumours, but of 16 with active ALDH2 only 5 had multiple tumours. The difference was statistically significant (p < 0.01). Age, drinking and smoking habits of the patients with single carcinomas did not differ from those of the patients with multiple carcinomas. The prevalence of other tumours in the region of the upper airways and digestive tract was increased in the patients with inactive ALDH2 (29.4 %) relative to that in the patients with active ALDH2 (6.3 %). As the cells of the oesophagus epithelium do not have ALDH2 activity, the authors suggested that the observed differences between the patients with active and inactive ALDH2 is the result of an increased systemic availability of acetaldehyde in the latter collective (Yokoyama *et al.* 1996b).

# 3 Toxicokinetics and Metabolism

## 3.1 Uptake and distribution

The forensic aspects of the toxicokinetics of ethanol are described in the standard textbook by Schütz (1983).

Pulmonary retention was on average 62 % in persons exposed to ethanol concentrations of 5800 to 10000 ml/m$^3$ at respiration rates of 7 to 25 l/min. The blood ethanol level of 3 volunteers was determined at hourly intervals during a 3-hour or 6-hour exposure to 14000–16000 mg/m$^3$. The equilibrium concentrations of ethanol in blood were attained after about 2 hours and were in the range between 20 and 60 mg/l at a respiration rate of 7 l/min and about 80 mg/l blood at a respiration rate of 15 l/min. At higher respiration rates (22–25 l/min) the equilibrium concentration had still not been reached after 3 hours, and the blood ethanol levels were in the range between 250 and 450 mg/l. The levels determined before the exposure were 9–27 mg/l (Lester and Greenberg 1951). These high values suggest that the method used cannot have been reliable because in modern studies much lower values are obtained for endogenous ethanol concentrations (see below).

In a more recent study with 12 female and 12 male volunteers, the ethanol concentrations in the exhaled air resulting from a 4-hour exposure to 150, 750 and 1500 mg/m$^3$ (80, 390 and 790 ml/m$^3$) were determined. During the exposure (no other details) the average ethanol concentrations in the alveolar air were 34, 170 and 336 mg/m$^3$, with an average interindividual scatter of ± 38 %, and thus were related linearly to the exposure concentration (r = 0.99). The blood ethanol concentrations were approximately the same after 2-hour and 4-hour exposures and averaged 0.23, 0.85 and 2.18 mg/l, respectively, (1 ‰ = 1000 mg/l) for the three air concentrations with an interindividual scatter of ± 53 % for the lowest concentration, ± 20 % and ± 26 % for the higher concentrations. This indicates that the blood ethanol concentrations have already reached equilibrium at the end of a 2-hour exposure. The blood ethanol concentrations were also related linearly to the concentration in the inhaled air (r = 0.99); 20 minutes after the end of the exposure the ethanol concentrations in the alveolar air were 3, 20 and 57 mg/m$^3$ and in blood 0.1, 0.26 and 0.69 mg/l. The ethanol concentrations in the alveolar air and in blood correlated significantly (r = 0.89). For these parameters no clear sex-specific differences could be found (Blaszkewicz 1998, Golka *et al.* 1994, Seeber *et al.* 1994). These results indicate that in persons exposed to ethanol concentrations up to 800 ml/m$^3$, the equilibrium ethanol concentrations in blood would be below 5 mg/l.

The physiological blood alcohol concentration determined by various authors with the GC-MS method, which is considered to be reliable, lies in the range from 0.1 to 0.3 mg/l (Ostrovsky 1986). In a group of 130 fasting adults, most values were between 0.1 and 0.2 mg/l. All values were under 0.75 mg/l. From the individual values given in the publication, an average value ± standard deviation of 0.27 ± 0.17 mg/l was calculated. Likewise, in a collective of 30 patients who were being treated for metabolic disorders or to cure alcoholism, all the values were in this range (Sprung *et al.* 1981).

The source of endogenous ethanol has not been identified unambiguously. The suggested role of intestinal bacteria in endogenous ethanol production (Geertinger *et al.* 1982) seems, in the light of more recent findings, to be questionable (Ostrovsky 1986). It is conceivable that endogenous ethanol is formed from endogenous acetaldehyde which can be produced, for example, in the oxidative decarboxylation of pyruvate by pyruvate dehydrogenase, during breakdown of phosphoethanolamine by *O*-phosphorylethanolamine phosphorylase, during breakdown of the amino acid threonine by threonine aldolase or during the breakdown of $\beta$-alanine. The enzyme, alcohol dehydrogenase (ADH), keeps the endogenous equilibrium between ethanol and acetaldehyde displaced towards the ethanol and so converts the acetaldehyde to ethanol (Ostrovsky 1986).

With a physiologically based pharmacokinetic (PBPK) model, the results given above from Lester and Greenberg (1951) could be well simulated. However, the high endogenous blood ethanol concentrations described in this study were not used in the model. For men assumed to be exposed for 6 hours to ethanol concentrations of 50 or 600 ml/m$^3$, the model predicted blood ethanol concentrations between 0.3 and 1 mg/l and between 4 and 13 mg/l, respectively (Pastino *et al.* 1997).

In mice and rats exposed to 50, 200 or 600 ml/m$^3$, the changes in the blood ethanol levels with time were determined. The equilibrium concentrations in blood (50 and 200 ml/m$^3$) were reached after 5 minutes, that at 600 ml/m$^3$ after 30 minutes. The blood ethanol concentrations during the exposures to 200 and 600 ml/m$^3$ were at most 1.1 and 4.4 mg/l in the mouse and 0.7 and 3.7 mg/l in the rat. After the end of the exposures the blood ethanol concentrations decreased rapidly. A PBPK model calculated on the basis of this data yielded the best predictions for both species if the pulmonary retention was taken to be about 60 % (Pastino *et al.* 1997).

After oral uptake, ethanol is practically completely absorbed by simple diffusion, up to 80 % of the dose in the small intestine, the rest in the stomach. Within one hour most of the ethanol has been absorbed from the gastrointestinal tract; the process can, however, be delayed by various factors such as the fullness of the stomach and the fat content of the food (ECB 1995, IARC 1988).

The dermal absorption of ethanol was investigated in the 1940s in a study with volunteers from which it was deduced that ethanol is absorbed only poorly through the intact skin because occlusive application of the substance to large areas of skin did not result in an increase in the blood ethanol level (ECB 1995). Because of the rapid metabolism of ethanol in the low concentration range, in which first order kinetics apply, and because of the inadequate sensitivity of the analytical methods available at that time, this conclusion is not justified. From the known physicochemical properties of ethanol, a relatively high dermal penetration rate of 2.84 mg/cm$^2$ and hour was estimated for a saturated aqueous ethanol solution applied to human skin (Fiserova-Bergerova *et al.* 1990). Studies with the isolated dorsal skin of guinea pigs revealed that during a period of 19 hours, about 1 % of applied ethanol penetrated the skin after non-occlusive application and up to 30 % after occlusive application. The ethanol was applied to a skin area of 5 cm$^2$ and the applied volumes were 25 to 500 µl at an incubation temperature of 37°C (Gummer and Maibach 1986). With isolated human skin a permeability constant of 3.2 µm/h was determined for ethanol. For the dorsal skin of the rat the corresponding value was 4.15; for 8 skin samples taken from various parts of the marmoset monkey, values between 3.2 and 8.0 µm/h were found (Scott *et al.* 1991). Repeated skin contact can cause defatting and thus increased dermal absorption (no other details) (ECB 1995).

Absorbed ethanol is distributed mainly in the aqueous compartments of the body and therefore does not accumulate. The blood/air distribution coefficient of ethanol at 37°C was determined with human blood *in vitro* to be 1265 ± 83. With human tissue homogenates the tissue/air distribution coefficients at 37°C were determined to be 850 ± 114 for muscle tissue, 940 ± 89 for kidney tissue, 1172 ± 210 for lung tissue, 790 ± 97 for the white brain matter, 1044 ± 61 for the grey brain matter and 215 ± 60 for adipose tissue

(Fiserova-Bergerova and Diaz 1986). Because it is soluble in both water and lipids, ethanol can also cross the blood-brain barrier and the placenta (ECB 1995).

## 3.2 Metabolism and elimination

The ethanol-metabolizing capacity of the respiratory tract mucosa has not been determined quantitatively. In human lung tissue, the mRNA for several of the 5 known classes of alcohol dehydrogenases (ADH) has been detected. In comparison with the levels in liver, however, the lung mRNA levels were low (Estonius *et al.* 1996). One kind of ALDH activity (oxidation of benzaldehyde and propionaldehyde) has been detected in the nasal mucosa of man (Gervasi *et al.* 1991) and cynomolgus monkey; for the monkey enzyme the substrate conversion per unit time and weight of protein was higher in the nasal mucosa than in the liver (Longo *et al.* 1992).

When ethanol is ingested, only a small proportion of the dose is oxidized by the ADH of the gastrointestinal mucosa and this metabolic capacity is lower in women than in men. This fact, together with the lower distribution volume for alcohol in women, was suggested to contribute to the higher bioavailability of ethanol in women which, in the opinion of the authors, could in turn lead to a greater "sensitivity" of the female liver to alcohol (Frezza *et al.* 1990). More than 90 % of absorbed ethanol is metabolized in the liver, the rest is excreted unchanged via the kidneys or exhaled.

Three different enzyme systems with different cellular locations and substrate affinities are involved in the hepatic oxidation of ethanol to acetaldehyde: the cytosolic ADH (EC 1.1.1.1.), the microsomal cytochrome P4502E1 and the peroxisomal catalase, the last of which, however, seems to play a less important role. The rate of ethanol oxidation is independent of concentration, except at very low and very high blood ethanol levels, and is carried out mainly by ADH because of the low $K_M$ of this enzyme. The pharmacokinetic parameters of ethanol degradation were determined in male and female volunteers after intravenous injection of an ethanol dose of 0.78 mmol/kg body weight (36 mg/kg body weight). Independent of the sex of the persons, the Michaelis-Menten constant $K_M$ was about 1.6 mmol/l (about 75 mg/l) and the maximum elimination rate $V_{max}$ about 3.8 mmol/l and hour (about 175 mg/l and hour). From the course of the ethanol and acetate concentrations in blood, the authors estimated the elimination half-time for acetaldehyde to be 1.7 minutes, that for acetic acid to be at most 6.4 minutes (Kohlenberg-Müller and Bitsch 1990). Other authors have given the average $K_M$ and $V_{max}$ values for ethanol elimination as 95 mg/l and 228 mg/l and hour (Lewis 1985b). At blood ethanol concentrations of 3000 mg/l or more, cytochrome P4502E1 ($K_M$ for ethanol 8–10 mM, 370–460 mg/l) makes a noticeable contribution to ethanol degradation. Long-term high level alcohol consumption can result in a 4-fold to 10-fold induction of this P450-isoform; ADH, in contrast, is not inducible (IARC 1988, Lieber 1997).

Acetaldehyde is mainly oxidized to acetic acid by the constitutively expressed mitochondrial ALDH (ALDH2, EC 1.2.1.3.). The acetic acid enters intermediary metabolism or is degraded to water and $CO_2$ (Lieber 1997). In persons with "normal" alcohol consumption (no other details) practically 99 % of the acetaldehyde produced from ethanol is oxidized in the liver, and so no free or reversibly bound acetaldehyde is found in peripheral venous blood (detection limit 0.5 µM, about 0.022 mg/l). In the *vena hepatica*, acetaldehyde concentrations of 64 µM were detected, but had sunk to 4 µM by the time the blood reached the right atrium. In alcoholics the blood acetaldehyde concentration in the *vena hepatica* was 160 µM, in peripheral blood 100 µM. Inhibition of ALDH with calcium carbimide, for example, increased the blood acetaldehyde concentrations after alcohol consumption up to 210 µM (Eriksson and Fukunaga 1993).

Thus, for the maximum blood concentration of 5 mg/l which can be expected after inhalation of an ethanol concentration of 1000 ml/m$^3$, the oxidation of the ethanol to acetaldehyde will take place according to first order kinetics; also the oxidation of the acetaldehyde to acetic acid will not be saturated so that accumulation of acetaldehyde is not to be expected.

ADH and ALDH are subject to genetic polymorphism which has been repeatedly studied during recent years (reviews: Agarwal and Goedde 1992, Lindahl 1992, Yoshida 1992). At least 5 gene loci control the synthesis of the various subunits of ADH. "Atypical" ADH is found in Caucasians with a maximum incidence of 20%, whereas in Chinese and Japanese populations it has been detected in 85–89% of the individuals who have been studied. Three classes of ALDH have been identified to date in mammalian organisms. Enzymes of class 1 (ALDH1) are found in the cytoplasm and both constitutive and inducible forms have been characterized. Class 2 enzymes (ALDH2), which mainly catalyze the oxidation of acetaldehyde, have only been detected to date in liver mitochondria and are not inducible. ALDH3 forms have not yet been described in man. The affinity of purified ALDH2 from human liver for acetaldehyde was about 900-times higher than that of purified ALDH1 (Klyosov 1996, Rashkovetsky *et al.* 1994). Genetic polymorphism of ALDH is observed in members of Mongolian populations. The by far most frequent ALDH variant is characterized by a lack of activity of the mitochondrial ALDH2. After intake of alcohol even in quantities which have no effects or only slight effects on most Europeans, persons with ALDH2 deficiency react with marked intolerance symptoms such as flushing, drop in blood pressure, tachycardia, palpitations, muscle weakness, headaches, nausea and vomiting, symptoms which are presumably caused by the delayed degradation of acetaldehyde. The inheritance of ALDH2 deficiency has not yet been clarified in detail.

## 3.3 Estimation of the ethanol body burden

An appropriate parameter for the ethanol body burden is considered to be the area under the curve (AUC) obtained by plotting the blood ethanol concentration against time, that is, the product of the blood ethanol concentration and the corresponding exposure time.

From the endogenous ethanol concentration in the blood of unexposed control persons, 0.27 ± 0.17 mg/l, (Sprung *et al.* 1981) and an assumed life-time of 80 years, an AUC of 21.6 ± 13.6 (mg/l) × years may be calculated. It provides a measure of the life-time body burden of ethanol, on the assumption that the endogenous blood ethanol concentration remains constant for the whole life-time.

In the blood of volunteers (12 women, 12 men) exposed to ethanol concentrations of 80, 390 and 790 ml/m$^3$, the average blood ethanol concentrations at equilibrium were

0.23, 0.85 and 2.18 mg/l with coefficients of variation of $\pm 53\%$, $\pm 20\%$ and $\pm 26\%$, respectively (Golka *et al.* 1994). Thus, up to an exposure concentration of at least 790 ml/m$^3$, the relationship between the ethanol concentration in the inhaled air (x) and the average equilibrium concentration in blood (y) is a straight line (y = a + b × x), where a is the average ethanol concentration in the blood of unexposed control persons (0.27 mg/l) and b the slope of the regression line, 0.0022 (mg/l)/(ml/m$^3$). It may be concluded from this linear relationship that the elimination obeys first order kinetics in the concentration range studied. Thus the additional ethanol body burden resulting from inhalation of ethanol during occupational exposure may be calculated as the product of the increase in the blood ethanol concentration (b × x) and the exposure duration (8 hours per day, 5 days per week, 40 years). Depending on the exposure concentration, x,

$$AUC_x = 0.0022 \ (mg/l)/(ml/m^3) \times x \ (ml/m^3) \times 8/24 \times 5/7 \times 40 \ years$$

From this equation, for exposure concentrations of 500 and 1000 ml/m$^3$, additional body burdens of 10.5 and 21 (mg/l) × years may be calculated which, together with the average background body burden of 21.6 (mg/l) × years, yield life-time body burdens of 32.1 and 42.6 (mg/l) × years. The significance of these levels will be discussed in Section 6.


# 4 Effects in Man


## 4.1 Single exposures


### 4.1.1 Inhalation

The odour threshold for ethanol in the inhaled air has been given as about 80 ml/m$^3$, that for eye irritation as about 1000-times this level (Cometto-Munic and Cain 1995, Seeber *et al.* 1994). An early publication (Loewy and von der Heide 1918) describes studies of ethanol inhalation in which ethanol levels in the air of as little as 1000 to 2500 ml/m$^3$ (1900 to 4750 mg/m$^3$) led to symptoms of intoxication which became more severe at higher alcohol concentrations; after exposure for 45 to 90 minutes to an ethanol concentration as low as 7500 ml/m$^3$ (14250 mg/m$^3$) the persons felt dazed, weary and wanted to sleep. The findings described in two later reports of single inhalation studies (Campbell and Wilson 1986, Mason and Blackmore 1972) and a case report of inhalation of excess ethanol at the workplace (Lewis 1985a) are not in agreement with the earlier findings and suggest that inhalation of ethanol concentrations in the range of the then valid MAK value of 1000 ml/m$^3$ do not cause a biologically significant increase in the blood alcohol concentration and thus cannot cause systemic effects. This view was supported by model calculations (Lewis 1985b). Various other published statements as to the effects of inhaled ethanol on man (Browning 1953, Lehmann and Flury 1938, Lester and Greenberg 1951) are based on the above-mentioned study by Loewy and von der Heide

(1918). Because of the very few reports and the many years of experience with ethanol at the workplace it has been assumed to date that on exposure to ethanol concentrations in the air up to 5000 ml/m³ no signs of local irritation occur and that at the workplace threshold level of 1000 ml/m³, which applies in numerous countries, systemic effects are not to be expected either (ACGIH 1986).

To confirm this assumption, in the study of the toxicokinetics of inhaled ethanol which has been described above (Golka *et al.* 1994), local irritation and central nervous symptoms were also recorded. Volunteers were exposed for a period of 4 hours to constant concentrations of 80, 400 or 800 ml/m³ in one series of experiments and to 1000 ml/m³ or to hourly changing concentrations of 100 and 1900 ml/m³ in a second series. Thus both constant concentrations and concentrations which changed at hourly intervals were studied. In the first series 12 male and 12 female volunteers were exposed, in the second series 8 males and 8 females. In none of the experiments were exposure-related changes in performance parameters detected. They included simple reaction time, choice reaction time and short-term memory performance (2 to 5 items which had been studied earlier had to be recognized). The current symptoms were recorded according to the "Swedish Performance Evaluation System" (17 acute symptoms, 10 of which were on the scale "irritation"), the sense of well-being in parameters of tension, tiredness, symptoms and annoyance (Seeber *et al.* 1997). At constant exposure concentrations up to 1000 ml/m³, no significant effects on any of the above-mentioned variables were detected. When the exposure concentration was changed at hourly intervals from 100 to 1900 ml/m³, a corresponding statistically significant increase in the incidence of annoyance and irritation was recorded. The incidence of annoyance correlated significantly with the ethanol concentration in the inhaled air. The course of these changes in the sense of well-being is indicative of rapid reversibility of the effects. The results indicate that exposure to average ethanol concentrations of 1000 ml/m³ for several hours does not cause persistent effects on the sense of well-being or behaviour. This applies equally for women and for men. Local irritation was not considered significant (Seeber *et al.* 1994, 1997).

### 4.1.2 Ingestion

The acute toxic effects of ingested ethanol on the central nervous system are generally known. The reader is referred to the widely known reviews (Gerchow and Heberle 1980, Schütz 1983). According to the forensic medical literature (see Schütz 1983), the performance of the central nervous system can be impaired at a blood alcohol concentration as low as 0.2‰ to 0.3‰ (200–300 mg/l). From 0.6‰ to 0.7‰ performance is impaired enormously in most persons and above 1.0‰ to 1.1‰ there is nobody under any imaginable conditions who would not display significant disorders, for example, in driving ability. A fixed value cannot be given for a "lethal blood alcohol concentration" since numerous factors (e.g. general constitution, disease, temperature) can have an enormous effect. In the literature, values from about 4‰ are given as lethal. However, in individual cases much higher blood alcohol concentrations were not lethal.

## 4.2 Repeated exposures

There are no data for the effects of repeated inhalation of ethanol on man.

### Ingestion

Long-term consumption of large doses of alcohol causes toxic effects in almost all organ systems. The most affected target organ is the liver. Beginning with fatty degeneration, the damage can progress via necrotic and fibrotic stages to liver cirrhosis. The threshold dose for the induction of toxic liver damage is given for women as the regular daily consumption of 20–40 g ethanol and for men as 60–80 g. The critical dose and time for the development of liver cirrhosis has been estimated to be 120–180 g ethanol/day during at least 15 years (IARC 1988). Because this exposure route is not relevant for the workplace and because of the high doses required to produce these effects, they are not discussed in more detail here.

## 4.3 Effects on skin and mucous membranes

Because of its defatting effect on the skin, ethanol can cause irritative contact dermatitis (no other details) (ECB 1995). Direct contact with the mucosa of the eyes produces a sharp, burning pain and eyelid closure reflex. When a 40–50% ethanol solution was splashed onto the ocular mucosa it produced only surface lesions and hyperaemia. The effects were completely reversible (ECB 1995).

## 4.4 Allergenic effects

There are various case reports of patients with either occupational or non-occupational contact dermatitis which was shown in epicutaneous tests to be caused by an allergic reaction to alcohol. In some persons, skin reactions also developed after consumption of alcoholic drinks, also erythema, aphthous lesions and a burning sensation on the oral mucosa. Cross-reactions with other primary and secondary alcohols and with acetone were also observed in some cases (Fisher 1983, Fregert *et al.* 1969, Van Ketel and Tan-Lim 1975, Melli *et al.* 1986, Patruno *et al.* 1994). In a predictive epicutaneous test carried out with 93 volunteers, patches soaked in a 50% aqueous ethanol solution were applied three times weekly for three weeks occlusively to the forearm and left there for 24 hours. This induction treatment caused mild skin irritation in all persons. Provocation with 50% ethanol on the other arm was carried out 17 days after the last application and produced skin reactions in 15 of the persons. In 6 of these 15 persons, skin reactions also developed when the provocation treatment was repeated later and for these 6 persons the reactions were considered to be allergic (Stotts and Ely 1977). As well as allergic contact dermatitis, various cases of allergic contact urticaria induced by ethanol have been described (Ophaswongse and Maibach 1994).

The phenomenon of alcohol-induced bronchial asthma, that is, the intensification or induction of bronchial reactions in persons with asthma after consumption of alcoholic drinks, has been described to date only for Asians and not for members of Caucasian populations. The mechanism of this reaction is considered to involve the increased acetaldehyde levels in blood such as develop in about 50 % of Asians but only in a minority of Caucasians after alcohol consumption (Myou *et al.* 1996, Shimoda *et al.* 1996). In 2 patients with mild stable asthma, an inhalation provocation test with aerosolized ethanol resulted in severe bronchoconstriction with a 20 % to 40 % decrease in $FEV_1$. This result does not, however, indicate that the genesis of the reaction is allergic (Hooper *et al.* 1995).

## 4.5 Reproductive and developmental toxicity

### 4.5.1 Fertility

In women, the long-term consumption of large amounts of alcohol was shown to produce disorders of the hypothalamus-hypophysis axis and thus functional disorders of the sexual organs. The observed effects included early begin of the menopause, reduced gonadotropin levels and increased oestrogen levels in the plasma of women in the postmenopausal phase. Various studies demonstrate that excessive alcohol consumption reduces female fertility, leading in some cases to infertility. In men, direct effects of ethanol on the testes and on hormone secretion in the hypothalamus and hypophysis have been described (Emanuelle and Emanuelle 1997, Gavaler and Van Thiel 1987, Grodstein *et al.* 1994).

### 4.5.2 Developmental toxicity

The first report of damage in children of women who had drunk alcohol during pregnancy appeared in 1968 (Lemoine *et al.* 1968). Later, alcohol-induced prenatal damage in 8 other children was described (Jones *et al.* 1973).

Nowadays the terms "alcohol embryopathy" and "foetal alcohol syndrome" are generally used for the prenatally induced morphological and functional changes in the children of alcoholic women.

Experience in various western countries indicates that the incidence of alcohol embryopathy lies between 1:212 (Dehaene *et al.* 1981) and 1:1000 (Dehaene *et al.* 1977). Such studies have not been carried out for Germany. If an incidence like that in France is assumed, about 1800 newborn babies with alcohol embryopathy, including all grades of severity, would be expected annually in Germany (Majewski 1986). Such numbers reflect the alcohol consumption in the countries concerned. The annual per head consumption of alcohol in the population in Germany is at present about 13 litres pure ethanol. Among the 1.5 million alcoholics in Germany, there are increasingly also women under the age of 25 years: it has been estimated that this group contains about 30000 women (Neubert 1986).

Thus it can be said that ethanol is currently by far the most common exogenous cause of prenatal toxicity (Neubert 1986). However, the alcohol consumption must be excessive during pregnancy and the alcoholism of the mother must have reached an advanced stage before the risk of alcohol embryopathy becomes high.


## Symptoms of alcohol embryopathy

The following symptoms are seen in more than half of children with alcohol embryopathy: reduced intrauterine growth, microcephalus, statomotoric and mental retardation, hyperactivity, hypotonia of the musculature, epicanthus, nasolabial folds, narrow red part of the lips, hypoplasia of the mandibles, anomalous palmar furrows and clinodactyly V (Majewski 1986).

Other signs of alcohol embryopathy are less frequent: ptosis, antimongoloid eyelid axes, shortened bridge of the nose, high palate, cleft palate, funnel chest, heart defects, anomalies of the genital and coccygeal grooves (see, e.g., Jones 1986).


## Etiology and pathogenesis of alcohol embryopathy

The mechanisms which result in teratogenic effects in cases of chronic abuse of alcohol appear to be complex, and numerous other factors affecting the induction of malformations must be taken into account. The most important of these factors are changes in the nutritional status of the women, especially episodes of hypoglycaemia and zinc and folic acid deficiency; these occur in persons who suffer from chronic alcoholism and must be considered as well when assessing the actual prenatal toxic potential of ethanol. In addition, medicine consumption, smoking habits, ethnic affiliation, social situation and age also play a role.


## Threshold dose for ethanol-induced damage

Without doubt, the risk is highest for women who are chronic alcoholics and consume more than 90 g ethanol daily. Alcohol embryopathy is seen in about 30% of the children of such persons (Jones *et al.* 1974). If less severe cases without full expression of the typical symptoms are also included, anomalies can be detected in up to 70% of the offspring of women with chronic alcoholism (Ouellette *et al.* 1977).

Some studies have been carried out to determine what effects moderate alcohol consumption during pregnancy has on the development of the children. In the assessment of the effects of low levels of alcohol consumption during pregnancy, it is particularly difficult to distinguish direct effects of the substance from those of other factors.

An effect of moderate alcohol consumption during pregnancy on the children was found in a study of 353 pregnant women (Mau 1980). However, since no details were obtained of the amounts of alcohol consumed, the results of this study cannot be used to establish a limit for the amount of alcohol which is associated with an increased risk of alcohol embryopathy.

In a prospective study it was found that even moderate alcohol consumption during pregnancy (1–2 alcoholic drinks per day in the first trimester) was associated with a spontaneous abortion rate in the second trimester which was twice that found in women who were abstinent. On the other hand, no significant association with abortions in the first trimester was found (Harlap and Shiono 1980).

Similar results were reported in a retrospective study of 616 women with abortions and 632 controls, in which even the consumption of 2 alcoholic drinks per week increased the spontaneous abortion rate (Kline *et al.* 1980).

Both studies have, however, been criticized (Lancet 1980). Because of the high incidence of spontaneous abortions in man, it is difficult to demonstrate a causal relationship between exposure to an exogenous toxin and the abortion rate. It is also known that the social situation of the woman can influence the incidence of miscarriages.

In a prospective study, an effect of moderate alcohol consumption before and during pregnancy (n = 263) on the birth weights of the children was found. The results suggested that the intake of about 35 ml ethanol per day, even before pregnancy, was associated with an average reduction in birth weight of 91 g. The same amount of alcohol in later stages of pregnancy led to a weight reduction of 160 g. The authors themselves qualified their results and suggested a "cautious" interpretation of the data as, for example, the number of mother-child pairs studied was not large enough (Little 1977).

Dependence of birth weights on drinking habits of the mother was also demonstrated in a prospective study of 900 pregnant women. The pregnant women were divided into three groups according to the number of alcoholic drinks consumed per week. The results indicated that the probability of giving birth to a child with reduced birth weight was twice as high for women who consumed 10 alcoholic drinks per week (or more than 100 g alcohol, i.e. about 15 g alcohol/day) than for women who drank less than 5 drinks per week (Wright *et al.* 1983). On the basis of the results of this study it has been recommended that—until a clear threshold dose is known—all pregnant women should avoid alcohol during pregnancy or at least not drink more than 5 alcoholic drinks per week (Lancet 1983).

This threshold dose could not be confirmed in another study: the incidence of miscarriages or reduced birth weights in women who consumed up to 4 alcoholic drinks per day was not significantly different from that in those who drank alcohol less than once a month. Anomalies and developmental disorders were first apparent when an average of 174 ml ethanol/day was consumed (Ouellette *et al.* 1977).

In month 5 of pregnancy, 1529 women were asked about their alcohol consumption during the month before they knew they were pregnant and during pregnancy. After the birth, the children of mothers who had consumed more than 30 g alcohol/day or occasionally more than 4 alcoholic drinks per day were examined for signs of alcohol embryopathy. In addition, children of abstinent women were examined as controls. In 11 of the total of 163 children who were examined, clinical symptoms were found which were considered to be symptoms of prenatal alcohol toxicity. Only for two children was the diagnosis "alcohol embryopathy" established. In both cases, the mothers were heavy drinkers who had consumed more than 30 g alcohol per day also during the first 5 months of pregnancy. Seven of the nine remaining children with morphological signs of alcohol embryopathy were born to mothers whose alcohol consumption was about 30 g or more per day in the month before they knew they were pregnant. In only one of these mothers did the level of alcohol consumption remain as high also during the first 5 months of pregnancy (Table 1). The results showed that the appearance of morphological symptoms depends on the daily alcohol intake especially before diagnosis of pregnancy (Table 2). The authors conclude that even moderate alcohol consumption during the first month of pregnancy can affect growth and development of the foetus (Hanson *et al.* 1978).

**Table 1.** Alcohol consumption before diagnosis of pregnancy and during the pregnancy of 11 mothers whose children showed morphological signs of alcohol embryopathy (Hanson *et al.* 1978)

| | Number of mothers with daily alcohol consumption (g) | | |
| --- | --- | --- | --- |
| | > 30 | 3.3–27 | < 3 |
| before diagnosis of pregnancy | 9 | 0 | 2 |
| during the first 5 months of pregnancy | 3 | 4 | 4 |

**Table 2.** Clinical symptoms of children as a function of the daily alcohol consumption of the mother (Hanson *et al.* 1978)

| Clinical category | n | Daily alcohol consumption (g) | |
| --- | --- | --- | --- |
| | | before diagnosis of pregnancy | during the first 5 months |
| anomalous with signs of alcohol embryopathy | 11 | 129 (0–775) | 36 (0–258) |
| slightly anomalous | 7 | 42 (0–207) | 27 (0–150) |
| normal | 145 | 24 (0–258) | 18 (0–78) |

Behavioural tests were carried out with 500 children from 1529 pregnancies. The children were classified into several groups according to the alcohol consumption of the mother and examined on several occasions between the first day of life and age 4. Children of mothers who had consumed alcohol during pregnancy showed behavioural disorders. The differences were dose-dependent and could already be detected during the first days of life. Evidence of disturbance of psychomotoric development was obtained in the examination carried out in the eighth month of life. The effects were clearest in the tests providing information about cerebral information processing which were carried out during the child's fourth year. A significant effect was seen only when the mothers had consumed more than about 30 g alcohol daily during pregnancy. The proportion of clinically abnormal children diagnosed in the 2.5-hour psychological and anthropometric study carried out in the fourth year of life was significantly increased only when the mothers had consumed more than 60 g alcohol/day during pregnancy (Table 3; Streissguth *et al.* 1984).

**Table 3.** Proportion of psychologically and anthropometrically abnormal children as a function of alcohol consumption by the mother (Streissguth *et al.* 1984)

| Daily alcohol consumption during pregnancy | Proportion of clinically abnormal children (%) | Number of children examined |
| --- | --- | --- |
| < 3 | 5 | 202 |
| 3–27 | 4 | 164 |
| 30–60 | 4 | 69 |
| > 60 | 19* | 21 |

* significantly different from all groups with alcohol consumption < 60 g/day. (p = 0.02, Fisher exact test)

As a whole, because the exposure levels can be estimated only roughly, the data for the threshold dose are inconsistent and inadequate for the establishment of a precise limit. However, cautious interpretation of the findings indicates that adverse effects on the progeny develop if the mother consumes more than 30 g ethanol daily. The blood ethanol concentration resulting from such an exposure would rise above 1000 mg/l for short periods.

## 4.6 Genotoxicity

In various studies, the incidence of chromosomal aberrations or sister chromatid exchange (SCE) in peripheral lymphocytes has been described as higher in "alcoholics" than in "non-alcoholics". However, it is frequently not possible to find precise data for the level of alcohol consumption in the individual publications, and smoking habits and other confounders were not always taken into account sufficiently. Even with supposedly non-smoking alcoholics, considerations of life-style suggest a high level of exposure to passive smoke. In addition, the effect of chronic alcohol consumption on xenobiotic-metabolizing enzymes could have adverse effects on the detoxifying capacity for other foreign substances. Various components of alcoholic drinks apart from ethanol could also contribute to the described effects (IARC 1988, Obe and Anderson 1987, Sram *et al.* 1990).

The level of unscheduled DNA synthesis in isolated peripheral lymphocytes of alcoholics was determined after incubation of the cells with 1-methyl-3-nitro-1-nitroso-guanidine or *N*-methyl-*N*-nitrosourea and shown to be significantly less than in the lymphocytes of control persons. At the same time, significantly increased malondialdehyde levels—indicative of increased levels of lipid peroxidation—were detected in both the lymphocytes and the plasma of the alcoholics. The authors suggested that oxidative stress could account for the impaired DNA repair capacity in the lymphocytes of the alcoholics (Sram *et al.* 1990, Topinka *et al.* 1991).

In 24 alcoholics who were admitted to a clinic with marked symptoms of alcohol intoxication and from whom blood was taken immediately after admittance, acetaldehyde adducts with DNA ($N^2$-ethyl-3'-deoxyguanosine monophosphate) were detected in the lymphocytes and granulocytes. The average adduct incidence was $3.4 \pm 3.8$ per $10^7$ nucleotides in the granulocytes and $2.1 \pm 0.8$ per $10^7$ nucleotides in the lymphocytes, whereby interindividual scatter was conspicuously large. This was accounted for in terms of differences in alcohol consumption but also differences in metabolic capacity and thus different acetaldehyde body burdens. An effect of smoking could be excluded. In the lymphocytes and granulocytes of 6 persons who consumed no alcohol and of 6 persons with moderate alcohol consumption (at most 50 g ethanol/week), the levels of adducts were below the detection limit of 5 adducts per $10^8$ nucleotides (Fang and Vaca 1997).

## 4.7 Carcinogenicity

Studies of potential carcinogenicity of ethanol inhaled at the workplace are not available to date.

Consumption of alcoholic drinks has been shown to increase the incidence of tumours at a variety of locations. The epidemiological studies on the subject have been reviewed in detail by the IARC. In summary, numerous retrospective studies and some prospective cohort studies of alcoholics or brewery workers allowed to consume large amounts of alcohol during working hours have revealed a clear association between the consumption of alcoholic drinks and the development of tumours in the oral cavity, pharynx, larynx, oesophagus and liver. Numerous case-control studies have confirmed the association and produced evidence of dose-response relationships which include, however, a large element of uncertainty. They suggest that the relative risk for tumours in the oral or pharyngeal cavities or the larynx are significantly increased in some cases after consumption of as little as about 10 g ethanol daily. The effects were generally not dependent on the kind of alcoholic drink which was most frequently consumed (IARC 1988). In the meantime, evidence has also accumulated for an association between the consumption of alcoholic drinks and an increase in the risk of developing breast cancer and there is evidence for an association between the consumption of alcoholic drinks and an increased incidence of colorectal tumours (Blot 1992, Longnecker 1995, Singletary 1997).

The most recent studies on the relationship between chronic alcohol consumption and cancer risk were discussed at a symposium in 1998. A summary of this discussion and the papers presented by various authors have been published (in German) in the journal "Forum der Deutschen Krebsgesellschaft" (13. Jahrgang 1998).

# 5 Animal Experiments and *in vitro* Studies

## 5.1 Acute toxicity

### 5.1.1 Inhalation

The acute toxicity of inhaled ethanol is low. Mice were exposed for 7 hours to 29000 ml/m$^3$ before the first deaths occurred. A 4-hour LC$_{50}$ for the mouse was given as about 21000 ml/m$^3$. Visible symptoms of intoxication included ataxia, coordination disorders and narcosis (ACGIH 1986, ECB 1995). Not lethal for CD-1 mice were a 10-minute exposure to 60000 ml/m$^3$, a 30-minute exposure to 50000 ml/m$^3$ and a 1-hour exposure to 40000 ml/m$^3$ but they impaired motor coordination for more than 4 hours (Moser and Balster 1985). Exposure to an ethanol concentration of up to 48000 ml/m$^3$ for up to 40 minutes did not significantly affect the reactions of CD-1 mice in a test of operant performance (Moser and Balster 1986).

Among rats exposed to ethanol, deaths occurred first after 22-hour exposure to about 13000 ml/m$^3$. A 4-hour LC$_{50}$ was given as about 63000 ml/m$^3$. The symptoms of intoxication were ataxia, coordination disorders and narcosis (ACGIH 1986, ECB 1995). Unconditioned reflexes and conditioned avoidance reactions of rats were significantly affected by exposure to 8000 ml/m$^3$ from an exposure duration of 2 hours, at 16000 ml/m$^3$ from 30 minutes; 18 hours after the beginning of the 4-hour exposure, substance-related effects were no longer detectable (Mullin and Krivanek 1982).

With F344 rats, the effect of 2-hour to 5-hour exposure to ethanol concentrations of 102–1290 ml/m$^3$ on two acquired behavioural patterns was studied. The behavioural patterns were probably maintained by the positive effect of rewards: on the one hand, the rats could obtain a sugar solution at certain intervals by operating a lever, on the other an electroneurophysiological self-stimulation in the posterior hypothalamus could be attained. The effects on reinforcement rate were determined. The frequency of lever operations in the studies with periodic reward in the form of sugar solution decreased significantly after as little as 1 hour exposure to an ethanol concentration of 202 ml/m$^3$ but even after 5 hours exposure an ethanol concentration of 104 ml/m$^3$ had no effect. When the animals were exposed to 202 ml/m$^3$, 2 hours daily for 5 days, the inactivation was detected only on the first 2 days of the study. In a variation of this study the sugar solution could be obtained by every lever operation. The tested concentrations of 104 and 202 ml/m$^3$ did not produce any significant change in lever operation. This was also true for the studies with electrical self-stimulation up to the highest concentration tested, 1290 ml/m$^3$. At the end of the 2-hour exposure to 140 ml/m$^3$ a blood ethanol concentration of about 90 mg/l was found, at 600 ml/m$^3$ it was about 400 mg/l and at 1290 ml/m$^3$ about 600 mg/l (Ghosh *et al.* 1991). According to the available data for blood ethanol concentrations in man and animals after ethanol inhalation (see Sections 3 and 5.5.2) the values given by the authors cannot be achieved under the exposure conditions described. Therefore this study cannot be considered relevant for the present assessment.

Guinea pigs were exposed to an ethanol concentration of 22000 ml/m$^3$ for 9 hours before the first deaths occurred (ACGIH 1986, ECB 1995).


### 5.1.2 Oral and dermal uptake

Various studies with the mouse yielded oral LD$_{50}$ values in the range from 3450 to 9500 mg/kg body weight; for the rat the values were between 6200 and 15000 mg/kg body weight. Likewise, for the guinea pig, rabbit, cat and dog the LD$_{50}$ values were at least 5000 mg/kg body weight. In rabbits the lowest dermal dose with lethal effects was 20000 mg/kg body weight (no other details) (ECB 1995).


## 5.2 Subacute, subchronic and chronic toxicity


### 5.2.1 Inhalation

During continuous exposure of 15 rats, 15 guinea pigs, 3 rabbits, 3 monkeys and 2 dogs to an ethanol concentration of 86 mg/m$^3$ (about 40 ml/m$^3$) for 90 days, no visible signs of toxic effects were seen in any species. Similarly, the haematological tests and the histological examination of the liver, kidneys, lungs, heart and spleen and for dogs and monkeys also the brain, spinal cord and adrenals did not reveal any exposure-related effects (Coon *et al.* 1970).

Continuous exposure of male Sprague-Dawley rats to an ethanol concentration of 25000 mg/m$^3$ (about 12000 ml/m$^3$) for 14 days had no effect on body weight gains but produced a marked reduction in the cell count in the spleen, thymus and bone marrow. The erythrocyte count, haemoglobin level and white blood cell count remained unchanged. The differential blood count revealed a reduced number of polymorphonuclear granulocytes and an increased lymphocyte count. In the bone marrow the number of erythrocyte precursor cells was reduced. From the fourth until the last day of exposure, the blood ethanol levels averaged 1690 ± 140 mg/l. Other parameters were not studied (Marietta *et al.* 1988). In male Sprague-Dawley rats exposed for 5 weeks continuously to high ethanol concentrations which produced ataxia and blood ethanol levels of 2000–3000 mg/l, no signs of oxidative stress could be detected in the lungs. GSH and vitamin E levels were reduced in the liver but not in lung tissue. On the other hand, the activities of the Cu/Zn-containing superoxide dismutase and catalase were increased only in the lungs. The exposure to ethanol caused a reduction in body weight gain. Other parameters were not studied (Rikans and Gonzalez 1990).

Since in the studies with repeated exposure to ethanol only single concentrations were tested and only selected end points studied, the results cannot be used for the establishment of a threshold value.

### 5.2.2 Ingestion

In numerous animal studies it has been demonstrated that repeated ingestion of ethanol can cause damage in practically all organ systems. The toxic effects are most prominently manifested in the liver. These studies are not suitable for the derivation of a threshold value and therefore they are not described here in more detail. They have been reviewed elsewhere (ECB 1995, IARC 1988).

## 5.3 Effects on skin and mucous membranes

In a study carried out according to the OECD guideline 404 and in two other studies carried out according to American guidelines, 95%–99% ethanol proved to be not irritating on rabbit skin. Mild irritation of rabbit skin was produced with ethanol after 24-hour occlusive application (ECB 1995).

The irritant effects of ethanol in the rabbit eye were found to be moderate in a study carried out according to OECD guideline 405, in another they were considered to be nonexistent. In the Draize test on the rabbit eye, 96% ethanol was moderately irritating. However, in another study carried out with the methods of the OECD guideline, ethanol produced severe irritation in the rabbit eye (ECB 1995).

## 5.4 Allergenic effects

When 75% ethanol was used as the vehicle both for induction and provocation in a Magnusson-Kligman maximization test with guinea pigs, no skin reactions which could be ascribed to ethanol developed. Likewise, in the mouse ear swelling test with ethanol, no evidence of a sensitizing effect was found (ECB 1995).

## 5.5 Reproductive and Developmental Toxicity

### 5.5.1 Fertility

Exposure of groups of 18 male Sprague-Dawley rats to ethanol concentrations of 10000 or 16000 ml/m$^3$, 7 hours daily for 6 weeks had no effects on fertility during subsequent mating with untreated animals. The body weight gain and food and water consumption of the male animals were not affected by the treatment. Repeated study of the neuromotor coordination, activity and learning ability of the progeny in the age span between 16 and 58 days revealed no effects of the treatment. In the brains of 21-day old animals some neurochemical parameters were changed but because of the normal behavioural patterns of the animals, these changes are not considered to be adverse or relevant for the current evaluation (Nelson *et al.* 1985a, 1988).

Studies of the adverse effects of ingested ethanol on fertility are mentioned here only in summary because they are of less relevance for the evaluation of effects of occupational exposure; they have been discussed in detail elsewhere (Gavaler and Van Thiel 1987, IARC 1988). In female rats and mice, high doses of ethanol (more than 2000 mg/kg body weight and day) administered before and during gestation had no effect on fertility. The available studies of dominant lethal mutations document adverse effects of high ethanol doses on the fertility of male rats and mice (see Section 5.6.2). In concurrence with these results, it has been demonstrated that after repeated oral administration of high doses of ethanol to rats and mice of both sexes, damage to various tissues in the sexual organs can be detected. In the male rats the plasma levels of testosterone and luteinizing hormone were reduced, in female mice, rats, rabbits and monkeys disorders of oestrus and of ovarian function were observed.

## 5.5.2 Developmental toxicity

Groups of 15 pregnant Sprague-Dawley rats were exposed to ethanol concentrations of 10000, 16000 and 20000 ml/m$^3$ from day 1 to day 19 of gestation for 7 hours daily. Satellite groups of 3 non-pregnant rats were exposed simultaneously to determine the blood ethanol levels resulting from the exposure. At 10000 ml/m$^3$ the blood ethanol values were between the detection limit of 10 mg/l and 44 mg/l, at 16000 ml/m$^3$ between 330 and 1100 mg/l and at 20000 ml/m$^3$ between 900 and 2540 mg/l. In the animals of the middle concentration group, body weight gain was reduced in the first week only slightly, but the reduction was significant. The animals of the high concentration group were in narcosis at the end of each exposure period and body weight gains and food consumption were reduced during the first week of exposure. In the examination for skeletal and visceral malformations in the progeny born *per sectio* on day 20 of gestation, no significant differences were seen between the groups exposed to ethanol and the controls. This was also the case for the number of implantations, the number of resorptions and the body weight of the foetuses (Nelson *et al.* 1985b). In the progeny of Sprague-Dawley rats which were exposed to ethanol concentrations of 10000 or 16000 ml/m$^3$, 7 hours daily during the whole gestation period, repeated examination of neuromotor coordination, activity and learning capacity during the age span from 16 to 58 days revealed no effects of the treatment. In the brains of 21-day old animals some neurochemical parameters were changed but, as mentioned above, in particular because of the normal behaviour of the animals, these changes cannot be considered to be adverse effects or cannot be assessed. The body weight development of the dams and the numbers and body weight development of the progeny were not different in the treated and control groups (Nelson *et al.* 1985a, 1988).

Groups of 10 pregnant ICR mice were exposed continuously from day 7 to day 9 or from day 7 to day 12 of gestation to ethanol concentrations of about 15000 mg/m$^3$ (about 8000 ml/m$^3$). In non-pregnant animals this treatment resulted in blood ethanol levels of 26 mg/l on day 1 and of 32 mg/l on days 3 and 6 of exposure. On day 18 of gestation the foetuses were removed *per sectio* and examined for external and skeletal but not for visceral malformations. In the group which was exposed for the longer period, the incidence of skeletal malformations was slightly increased but the increase was not statistically significant. The resorption rate in this group was 40% and was significantly increased relative to the rates in the control group and the group exposed for the shorter period (both about 10%). Weights of foetuses and placenta were not different in the three groups (Ukita *et al.* 1993). The publication gives no details of maternal toxic effects and can for this reason not be included in the present assessment.

Since the numerous animal studies in which reproductive toxicity was studied after oral administration of ethanol are not well-suited for estimation of the risk associated with inhalation of the substance, they are not discussed in the present document. The reader is referred to a number of reviews in which these studies are described (IARC 1988, Löser 1991, Persaud 1988, Schardein 1985).

## 5.6 Genotoxicity

### 5.6.1 *In vitro*

Ethanol proved not to be mutagenic in several *Salmonella* mutagenicity tests in the strains TA97, TA98, TA100, TA1535, TA1537 and TA1538 both in the presence and the absence of a metabolic activation system (IARC 1988). In strain TA102 with metabolic activation, ethanol increased the number of revertants to 1.5 to 2 times the control value, however, only with very high concentrations of at least 160 mg/plate. In a DNA repair test with repair-proficient and repair-deficient strains of *E. coli*, ethanol produced weak positive results at a concentration of 5 mg/test. The authors did not consider that their results demonstrated genotoxic effects of ethanol, especially as the possibility could not be excluded that the effects were caused by impurities (De Flora *et al.* 1984). In the *E. coli* strain CHY832 (with the temperature-sensitive phenotype RK$^+$), a 10-minute incubation with an ethanol concentration of 180 mg/ml without metabolic activation resulted in an increase in RK$^-$ mutants which, because of a forward mutation in an integrated plasmid, are still capable of colony formation after an increase in temperature from 30°C to 42°C. Survival of the ethanol treatment, however, was only 15% (Hayes *et al.* 1984). Three other tests for DNA damaging effects in prokaryotes yielded negative results (Hellmér and Bolcsfoldi 1992, IARC 1988).

In *Saccharomyces cerevisiae*, ethanol was mutagenic but only at an incubation temperature of 30°C. At 4°C the test yielded negative results, from which the authors concluded that metabolic activation was necessary for the mutagenicity. In *Aspergillus nidulans*, ethanol in concentrations of about 5% in the test medium (40 mg/ml) caused aneuploidy (nondisjunction, mitotic recombination). Forward mutations were not induced (IARC 1988).

In mouse lymphoma L5178Y cells, ethanol was not mutagenic up to the highest tested concentration of 4.5% (36 mg/ml); however, the test was not carried out with metabolic activation (Amacher *et al.* 1980). After incubation of primary rat hepatocytes for 3 hours with 1% ethanol (8 mg/ml) no DNA damage could be detected with alkaline elution. In V79 cells incubated for 1 hour with 5% ethanol (40 mg/ml), no micronuclei were found, in CHO cells (a cell line from Chinese hamster ovary) after a 30-minute incubation with 1 % ethanol (8 mg/ml) no chromosomal aberrations. Various authors have determined sister chromatid exchange in CHO cells treated with ethanol, but the studies were carried out without the addition of an appropriate metabolic activation system. With the following incubation protocols no SCE was found: 0.5 hours with 1 % ethanol (8 mg/ml), 3 hours with about 0.6 % (4.8 mg/ml), 28 hours with 1 % (8 mg/ml) or 8 days with 0.1 % (0.8 mg/ml) daily. Likewise the treatment of mouse kidney fibroblasts for 44 hours with 0.1 % ethanol (0.8 mg/ml) did not induce SCE (IARC 1988). In contrast, in another study the incidence of SCE was increased relative to the control value after a 1-hour incubation of CHO cells with a metabolic activation system and 0.5 %, 1 %, 2 % or 4 % ethanol (4, 8, 16 or 32 mg/ml), even at the lowest concentration. The maximum increase in the incidence of SCE to 3 times the control value was attained with 2 % ethanol. Cytotoxic effects were not seen. Without metabolic activation, the incidence of SCE increased to at most 1.5 times the control value. The results were the same independent of whether the test was carried out with 96 % technical ethanol or with 100 % ethanol (De Raat *et al.* 1983).

In a total of 13 studies with human lymphocytes incubated for 48–72 hours with ethanol concentrations of 0.1 % to 2 % (0.8–16 mg/ml) no SCE was detected, and after incubation for 24 or 48 hours with ethanol concentrations of 0.5 % to 1 % (4–8 mg/ml) no chromosomal aberrations (CA); the tests were carried out only without metabolic activation. The only study in which a metabolic activation system was also added did not find an increase in the incidence of SCE with 0.1 % ethanol (0.8 mg/ml) and 1 hour incubation (IARC 1988). However, treatment of cultured human lymphocytes for 72 hours with 0.05 % ethanol (0.4 mg/ml) without metabolic activation produced an increase in the incidence of SCE to 1.4 times the control value. With 0.15 % and 0.5 % (1.2 and 4 mg/ml) the increase was about 1.7 times. Inhibition of replication or mitosis by ethanol could be excluded (Alvarez *et al.* 1980a). After a 50-hour incubation of human blood with an ethanol concentration of 1.16, 2.32 or 3.48 mg/ml, chromosome and chromatid aberrations were found in the lymphocytes. The effects were concentration-dependent. Gaps and genuine breaks were recorded separately (Badr *et al.* 1977). A 1-hour incubation of isolated human lymphocytes with 1.56–100 mM ethanol (0.07–4.6 mg/ml) did not cause single or double strand breaks in DNA (Singh and Khan 1995).

### 5.6.2 *In vivo*

### Somatic cells

*Mouse*
Ethanol administered to Swiss mice for 26 days in the drinking water (40 % v/v; about 50 g/kg body weight and day) did not cause an increase in the incidence of micronuclei in the polychromatic erythrocytes in the bone marrow. The same result was also obtained after subcutaneous injection of 0.05 ml 95 % ethanol into male ddy mice (IARC 1988). The intraperitoneal injection of ethanol doses of 0.62, 1.24 or 1.88 g/kg body weight on each of two consecutive days caused an increase in the incidence of micronuclei in the

maturing erythrocytes of the bone marrow of Parkes mice. The maximum value was about twice the control value; the effect was not dose-dependent. The cells were obtained 6 hours after the last injection and about 1000 cells/animal (8 animals/group) were examined (Badr *et al.* 1977). Oral administration of a 10% ethanol solution, once daily for 4 days, did not increase the incidence of SCE in male Swiss-Webster mice (no other details) (IARC 1988). In male CBA mice which were given drinking water containing 10% or 20% ethanol for a period of 3–16 weeks (ethanol doses about 12 or 25 g/kg body weight and day) the incidence of SCE in the bone marrow cells was increased to 1.5 times the control value from the first determination (3 weeks after the start of exposure) (Obe *et al.* 1979). The significance of this result is very uncertain because of the small numbers of animals (n = 2–4, 6 only in one group) and because so few metaphases were evaluated (maximum 30/animal). After intraperitoneal injection of 50% ethanol the incidence of SCE increased slightly in a dose-dependent manner in the bone marrow cells of male NIH mice. At the highest dose of 2.4 g/kg body weight, the incidence was increased to about 1.4 times the control value. The average generation time of the cells was unchanged. Details of the number of animals used were not given (Calva and Madrigal-Bujaidar 1993). In the DNA from hepatocytes of 7 mice which were given drinking water containing 10% ethanol for a period of 5 weeks (ethanol dose about 12 g/kg body weight and day), $N^2$-ethyl-deoxyguanosine adducts were detected at a frequency of 1.5 ± 0.8 per $10^8$ nucleotides. These adducts are formed in the reaction of acetaldehyde with the exocyclic amino group of guanine. In the untreated control animals such adducts could not be detected; the detection limit was 5 adducts/$10^9$ nucleotides (Fang and Vaca 1995).

In several studies with pregnant mice, evidence for a transplacental genotoxic effect of ethanol was found. In ICR mice given oral doses of ethanol in various administration patterns from day 16 to day 18 of gestation, a dose-dependent increase in the incidence of SCE up to more than twice the control value was found in the foetal hepatocytes 21 hours after the last dose. In each group 50 hepatocytes arrested in mitosis were evaluated per litter. Two to three pregnant animals per dose group were treated. The highest cumulative ethanol dose was 8 g/kg body weight. The maximum maternal blood ethanol level was found to be 450 mg/l (Alvarez *et al.* 1980b). Likewise, after administration of a single intraperitoneal injection of a 10% ethanol solution in a dose of 4 g/kg body weight to 4 pregnant ICR mice on day 10 of gestation, the incidence of SCE in the pooled embryonal cells from each litter was significantly increased to 1.5 times the control value. An ethanol dose of 2 g/kg body weight had no effect (Czajka *et al.* 1980).

*Rat*

In the erythrocytes from the bone marrow and in the liver cells of male Wistar rats which were given drinking water containing 10% or 20% ethanol for 3 or 6 weeks (ethanol doses about 4 and 8 g/kg body weight and day) the incidence of micronuclei was not increased. Chromosomal aberrations were not detectable either in the bone marrow cells or in lymphocytes. The incidence of SCE in the peripheral lymphocytes was increased in both concentration groups after 3 and 6 weeks to 1.5 times the control value; however, cells from only one or two animals from each group were investigated (Tates *et al.* 1980). In 10 male CD rats given liquid feed in which ethanol accounted for 36% of the calories (ethanol dose 12–16 g/kg body weight and day) for 6 weeks, the incidence of micronuclei in the erythrocytes of the bone marrow was slightly and significantly increased. At the same time, the proportion of nucleated cells in the bone marrow was decreased, the proportion of erythrocytes and mitotic cells increased. The diet produced blood alcohol levels of about 1500 mg/l and body weight loss of about 15%. The mean corpuscular volume of the erythrocytes in the peripheral blood was increased (Baraona *et al.* 1981).

*Hamster*
The incidence of chromosomal aberrations in bone marrow of Chinese hamsters given drinking water containing 10% ethanol for 9 weeks was unchanged. Likewise, after 46 weeks the incidence of SCE in bone marrow cells and of CA in lymphocytes was not increased. When the ethanol level in the drinking water was increased from 10% in the first week to 15% in the second and third weeks and then to 20% in the fourth to twelfth weeks, there were still no increases in CA or SCE in the bone marrow cells (IARC 1988).

## Germ cells

The results which are available from studies of the induction of germ cell mutations with ethanol, like those of the *in vivo* genotoxicity studies described above, are indicative of species-specific and strain-specific activity (Adler and Ashby 1989). The mutagenic potential seems to be weak, is limited to the induction of aneuploidy and could be demonstrated to date only with very high doses of at least 5 g/kg body weight, which produced systemic toxicity, and only in mice.

The hourly analysis of spermatocytes in the second metaphase of meiosis in CBA/CA mice, 2–6 hours after a single oral dose of 0.8 ml of a 12.5% or 15% ethanol solution (about 5 g/kg body weight) revealed a significantly increased incidence of aneuploidy. The dose groups comprised 7 and 10 animals so that for each analysis time only one or two animals could be investigated. The number of cells evaluated was also too small. The blood ethanol levels were in a range which is attained after administration of the $LD_{50}$ (Hunt 1987). When (C57BL×CBA)$F_1$ mice were given an oral dose of 1 ml of a 10% to 15% ethanol solution (about 4–6 g/kg body weight) shortly after fertilization, up to 17.9% of the female-derived chromosome sets in the zygotes were aneuploid. In each dose group, 22 to 93 ova were analyzed. (Kaufman 1983). These results could also be confirmed in mated CFLP mice which were given oral doses of 1.5 ml of a 12.5% ethanol solution (ethanol dose about 7.5 g/kg body weight) 4, 13.5 or 17 hours after induction of superovulation. Both in the embryonal single cell stage and also in the morula stage, the incidence of aneuploidy was increased to 19% and 13.5%, respectively, independent of the time at which the ethanol was administered. On day 10 or day 11 of gestation, an increase in the incidence of morphologically altered triploid and trisomal embryos was observed. In each dose group, at least 34 embryonal single cell stages and at least 66 morula stages were analyzed. Maximum serum ethanol levels of 2600–2800 mg/l were found 30 minutes after administration (Kaufman and Bain 1984).

Ten-week to twelve-week old (C3H×C57BL)F$_1$ mice were treated to induce superovulation and then mated with untreated male animals. The 19 fertilized females were treated about 1.5 to 2.5 hours after ovulation with an oral dose of 1 ml of a 12.5% ethanol solution (ethanol dose about 5 g/kg body weight) and a colchicine dose of 2.5 mg/kg body weight; 15 hours later the zygotes were isolated and subjected to cytogenetic examination. A total of 105 preparations from treated animals could be examined, from the control group 59. The proportion of zygotes with a normal chromosome set (n = 40) was not significantly different in the control (76.2%) and ethanol (88.1%) groups. Zygotes with 41 chromosomes occurred in the control group not at all and in the ethanol group once, zygotes with 39 chromosomes twice and eleven times, respectively. Thus the ethanol treatment resulted in an significant increase in aneuploid zygotes (Washington *et al.* 1985),

In the spermatogonia of Wistar rats given drinking water containing at least 10% ethanol (no other details) the incidence of chromosomal aberrations was not increased (IARC 1988). This finding was confirmed in Sprague-Dawley rats, which were given ethanol in the drinking water for 36 weeks. The concentration was increased from initially 7% to 20%; control and treated groups each comprised about 30 animals. The number of mitoses which could be analyzed varied markedly, however, and was mostly small (Halkka and Eriksson 1977). In Chinese hamsters given doses of 1.5 ml of a 12.5% ethanol solution, no aneuploidy was detectable in the spermatogonia or spermatocytes type I and II (IARC 1988).

Dominant lethal mutations were observed in mice given bolus doses of highly concentrated ethanol solutions from which very high blood ethanol levels must have resulted. For rats such findings have only been reported after very high doses which produced marked symptoms of systemic toxicity.

After oral administration of 0.1 ml of a 40% or 60% ethanol solution, daily for 3 days (ethanol dose about 1.24 or 1.86 g/kg body weight and day) to male CBA mice (n = 6–13) followed by repeated mating with untreated female animals during the subsequent 6 weeks, the litter sizes were only reduced significantly if mating took place 14–17 days after treatment. In two repeat studies with the same treatment scheme, the induction of dominant lethal mutations was also confirmed in the effects on the numbers of living and dead implants. However, these effects only developed when mating took place 4–13 days after treatment. The time span during which the effects occurred indicated that ethanol causes specific damage to early or late spermatids (Badr and Badr 1975, Badr *et al.* 1977). During a collaborative study in which 3 laboratories took part, 3 independent dominant lethal tests were carried out with ethanol administered to CFLP mice. Groups of 15 male animals were given oral doses of 10% or 40% ethanol (2 ml/kg body weight and day) during a period of 5 days. The high dose of about 0.6 g/kg body weight and day was found in a preliminary study to be the maximum tolerated dose (MTD). After the treatment the males were mated at weekly intervals, each with 2 untreated females, during a period of 8 weeks. Ethanol had no effect on fertility. The incidence of dominant lethal mutations was not increased in any of the 3 tests for most mating times. The few significantly increased individual results did not produce a consistent picture and were not considered to be evidence for the induction of dominant lethal mutations (James and Smith 1982). The induction of dominant lethal mutations in the ova of 10-week to

12-week old (C3H×C57BL)F$_1$ mice was studied after oral administration of 1 ml doses of a 12.5% ethanol solution (ethanol dose about 5 g/kg body weight) to the female animals 1, 1.5 or 2 hours after mating. The contents of the uterine horns were inspected after 12 or 17 days. At the time of the first inspection the number of dead implants was not different from the control value. At the examination on day 17 of gestation, the number of dead implants was significantly increased, but only in the group treated with ethanol 2 hours after mating. A significance level under 0.05 could only be attained with a group size of more than 40 animals (Washington *et al.* 1985).

After mating untreated Long-Evans rats with male animals (n = 10) which had been given drinking water containing 20% ethanol (ethanol dose about 10 g/kg body weight and day) for 60 days previously, preimplantation losses and resorptions were increased for up to three weeks after the last dose of ethanol. The ethanol consumption caused reduction in testis weights and, in one case, also testicular atrophy. In the surviving progeny, the incidence of malformations was high and body weights were reduced (Mankes *et al.* 1982). When 6 male Sprague-Dawley rats which had been given a liquid diet containing 6% ethanol for one week and 10% ethanol during the next 4 weeks were mated, each with 2 untreated animals, the incidence of resorptions was significantly increased. From the measured daily consumption of liquid food (about 50 ml), an ethanol dose of 10–15 g/kg body weight and day can be estimated. The treated animals showed marked signs of alcohol intoxication and dependence and had reduced body weights. The authors stated that the serum ethanol levels were only slightly increased (Klassen and Persaud 1976), but this seems unlikely in view of the described symptoms.

Administration of 2 ml doses of a 40% ethanol solution (about 2.5 g/kg body weight) to male rats and mating with untreated females 2 weeks after treatment caused a significant increase in preimplantation and postimplantation losses. The control group comprised 34 pregnant animals, the ethanol group 11. If the animals were mated 6 weeks after treatment, the incidence of dead embryos was not increased. This suggests that ethanol caused specific damage to spermatids (Barilyak and Kozachuk 1981). The cytogenetic studies which were also described in this publication cannot be evaluated because of various inadequacies in the methods. When Wistar rats were given 30% ethanol in the drinking water (ethanol dose about 15 g/kg body weight and day) for 4 days and then mated once weekly for 8 weeks with untreated females, determination of the number of *corpora lutea* and of living and dead implants yielded no evidence of dominant lethal mutations. Likewise after administration of ethanol for 35 days in the drinking water during which time the ethanol concentration was increased from 15% to 20% or 30%, no dominant lethal mutations developed. In each dose group and for each mating time 26–30 female animals were used. The number of treated male animals is not given (Chauhan *et al.* 1980).

## 5.7 Carcinogenicity

Studies of the carcinogenic effects of ethanol under conditions of inhalation exposure such as occur at the workplace are not available.

The available studies in which animals were exposed long-term to oral doses of ethanol or alcoholic drinks have been reviewed in detail elsewhere (IARC 1988). The majority of studies with the mouse, rat and hamster did not yield evidence of treatment-related increases in tumour incidences but, on the other hand, practically none of the studies met the requirements of a valid carcinogenicity study, for example, because of the absence of a control group, because the number of animals was too small, or because the histological examination was inadequate (IARC 1988).

Studies which meet the necessary criteria are described below in detail. In an investigation of the effect of ethanol on vinyl chloride-induced hepatocarcinogenesis, a (control) group of 80 male Sprague-Dawley rats was given drinking water containing 5% (v/v) ethanol for a period of 30 months. Assuming an average drinking water consumption of 15 ml/day and a body weight of 300 g, the average ethanol dose can be calculated to be 2 g/kg body weight and day. Another control group of the same size was given pure water; that is, there was no compensation for the probable higher calorie intake of the ethanol group. Survival after 18 months was similar in the two groups: 73% (ethanol) and 70% (water). By the end of the study, 8 hepatocellular carcinomas and 29 hyperplastic nodules had been found in the animals of the ethanol group, in the untreated group, however, only 1 such carcinoma and 10 nodules. Tumours of the endocrine system were found in 57 of the animals treated with ethanol, in the control group in only 8. The incidences were respectively 26/79 and 8/80 for pituitary tumours, 14/79 and 0/80 for adrenal tumours, 14/79 and 0/80 for pancreas tumours (not better specified), and 3/79 and 0/80 for testis tumours. In all, in the animals treated with ethanol, 91 tumours were diagnosed of which 44% were classified as malignant, and in the control group 16 tumours, of which 5 were malignant (Radike *et al.* 1981).

Sprague-Dawley rats (50 animals/dose and sex) were given a semi-synthetic liquid diet containing 1% or 3% ethanol for a period of 2 years. From the food consumption of the animals the ethanol doses can be estimated as about 1 and 3 g/kg body weight and day. Instead of ethanol, the diet of the control groups contained equicaloric amounts of glucose. From week 104 until the end of the study (120 weeks) all animals were given a normal liquid diet. Survival in the group treated with ethanol was not reduced relative to that in the control group. The body weights in the high dose group were significantly reduced relative to the control values, for males from week 13 and for females from week 69; the body weight reduction was at most about 15%. Liver and kidney weights of the animals treated with ethanol were not significantly different from the control values. The histological examination of the group treated with ethanol revealed a significantly higher incidence of various non-neoplastic lesions than in the control group. In the male animals of both dose groups there was cystic and focal degeneration of hepatocytes and chronic pancreas inflammation. In addition, in the high dose group males the incidence of bile duct fibrosis was increased; in the low dose group males, pancreas hyperplasia, C-cell hyperplasia in the thyroid, and demyelination of peripheral nerves were found. In the female animals of both dose groups there was conspicuous focal hyperplasia in the adrenal cortex and, only in the high dose group, C-cell hyperplasia in the thyroid, inflammation of the clitoris, pigmentation of the mandibular lymph nodes and demyelinization of peripheral nerves. Analysis of neoplasms produced no evidence of significant ethanol-related changes in the tumour spectrum or tumour incidence in males. In the female animals of the low dose group the incidence of fibromas, fibroadenomas and adenomas of the mammary gland was significantly increased, that of tumours of the islet cells of the pancreas decreased. In the high dose group the number of not better specified neoplasms of the pituitary gland was increased, that of adenomas of the adrenal cortex decreased. The total number of tumours in the high dose group was reduced. The authors concluded from their results that ethanol does not have carcinogenic effects (Holmberg and Ekström 1995).

The effect of life-long ethanol consumption on life-span was studied in male C57BL mice. Initially, from week 14 to week 19 of life, the animals were given 3.5 % (v/v) ethanol in the drinking water. Then they were divided into three groups which were given 3.5 %, 7.5 % or 12 % ethanol in the drinking water for the rest of their lives. Each group comprised 100 animals which were kept in individual cages. Two control groups also with 100 animals in each were included in the study; in one control group the animals were kept 1 per cage, in the other 5 animals per cage. From the fluid consumption, the authors calculated average ethanol doses for the three exposed groups: 2.8, 7.3 and 11.1 g/kg body weight and day. The blood ethanol concentrations were determined for the first time in week 13 and also on three later occasions at quarter year intervals and at different times of day. The resulting values were averaged and revealed concentrations which increased with increasing ethanol dose: 66, 142 and 268 mg/kg blood (approximately equal to mg/l blood). The body weight gains in the five groups did not differ. Average life-span in the middle dose group was increased significantly relative to that in the control group with single animals in the cages; the values for the other groups did not differ from the controls. When the animals died naturally or when they had been killed *in extremis* a detailed histological examination was carried out. Of the animals kept in individual cages, between 72 and 89 animals per group could be examined, of those kept in groups, only 55 because of cannibalism. The incidence, kind and severity of non-neoplastic liver lesions did not differ significantly in the 5 groups. Likewise in the other organs subjected to histological examination (brain, lungs, heart, pancreas, spleen, kidneys, small intestine and testes) no damage was observed which would be ascribed to the ethanol treatment. The kind and incidence of neoplasms in the various groups provided no evidence of an ethanol-related significant alteration in the tumour spectrum or tumour incidences (Schmidt *et al.* 1987).

The results of the studies carried out with appropriate methods are thus inconsistent. The tumour incidences were increased in male Sprague-Dawley rats after long-term ingestion of ethanol in the drinking water, but not in male or female animals of the same strain given ethanol in liquid feed. A drinking water study with mice also yielded negative results. Since the MTD was apparently not reached in any of the three studies, their significance for the evaluation of the carcinogenic potential of ethanol is anyway limited.

In September 1996, a 2-year carcinogenicity study with oral administration of ethanol to mice was begun by the NTP. The publication of the results cannot be expected for a number of years.

# 6 Manifesto (MAK value, classification)

Long term consumption of large amounts of ethanol in the form of alcoholic drinks results in the formation of tumours in the mouth and pharynx, larynx, oesophagus, liver, and probably also the mammary gland and large intestine. Since in the biotransformation of ethanol the genotoxic metabolite acetaldehyde and genotoxic radicals are produced, it is assumed that this is the mechanism of the carcinogenic action of the substance. Assuming linear kinetics, the body burden of genotoxic metabolites increases in proportion to the body burden of absorbed ethanol. Therefore a threshold value for the carcinogenic effects cannot be determined and any occupational exposure to ethanol should be kept as small as possible so that the life-time body burden of these substances is not increased significantly. Reliable parameters for body burden, such as the area under the curve obtained by plotting the blood concentrations against time, cannot be derived from the available data for the genotoxic metabolites of ethanol but can be obtained for the metabolic precursor, ethanol, itself (see Section 3).

It may be shown that the average life-time body burden of ethanol resulting from occupational exposure to a concentration of 500 ml/m$^3$ lies within the range of the standard deviation of the endogenous body burden, and that resulting from exposure to 1000 ml/m$^3$ lies above this range. Therefore workplace exposures to concentrations up to 500 ml/m$^3$ do not result in a significant contribution to cancer risk.

Therefore the MAK value for ethanol was reduced in 1998 to 500 ml/m$^3$ and ethanol was classified in Section III of the *List of MAK and BAT Values* (Carcinogenic substances) in Category 5.

According to two studies with volunteers, at concentrations corresponding to this MAK value, no local irritation is to be expected. Similarly, systemic toxic effects may be excluded because the blood ethanol concentration resulting from such exposures, about 2 mg/ml, is far below the threshold for the first effects on the central nervous system, about 200 mg/l blood.

Because ethanol is absorbed rapidly and has a short half-life, the substance is classified in Peak limitation category II,1.

Ethanol induces aneuploidy in the germ cells of the mouse and dominant lethal mutations in both rat and mouse. Therefore ethanol is classified in Germ cell mutagens group 2. However, as the described effects developed in animal studies only after very high, markedly toxic doses of ethanol and as no significant increase in the life-time body burden of ethanol is to be expected in persons observing the MAK value, under these conditions the potential for effects on the germ cells is expected to be negligible.

The prenatal toxicity of ingested alcohol has been demonstrated unambiguously in man and animals. However, the maternal blood ethanol concentrations responsible for these effects are of an order of magnitude which can never be achieved by inhalation of concentrations in the range of the MAK value. In animal studies, ethanol concentrations up to 20000 ml/m$^3$ had no effects on the progeny in spite of causing maternal toxicity. Therefore, provided the MAK value is observed, prenatal toxicity is not to be expected and ethanol remains in Pregnancy risk group C.

From the permeability coefficient determined for ethanol on human skin, 3.2 µm/hour, a flux of 0.25 mg/cm$^2$ and hour may be estimated for undiluted ethanol. Thus, if the skin of both hands and forearms were in contact with liquid ethanol (area about 2000 cm$^2$), about 500 mg ethanol could be absorbed dermally in one hour. During exposure to ethanol concentrations in the range of the MAK value of 500 ml/m$^3$ (950 mg/m$^3$), about 5700 mg ethanol could be absorbed by inhalation during one shift (volume of air inhaled in one shift is about 10 m$^3$, pulmonary retention of ethanol about 60 %). As the dermal uptake of ethanol under the described conditions is less than 10 % of the uptake via inhalation, the substance is not designated with an "H".

There are only very few reports of cases of ethanol-induced contact dermatitis. Therefore, because of the multitude of possibilities for contact with the substance and its ready availability, it may be concluded that any sensitizing potential of ethanol must be very small. There is no evidence of sensitizing effects on the airways. Designation with an "S" is not considered necessary.

# 7 References


ACGIH (American Conference of Governmental Industrial Hygienists) (1986) Ethyl alcohol. in: *Documentation of threshold limit values and biological exposure indices, 5th edition*, American Conference of Governmental Industrial Hygienists, Cincinnati, OH, USA

Adler ID, Ashby J (1989) The present lack of evidence for unique rodent germ-cell mutagens. *Mutat Res 212*: 55–66

Agarwal DP, Goedde HW (1992) Pharmacogenetics of alcohol metabolism and alcoholism. *Pharmacogenetics 2*: 48–62

Alvarez MR, Cimino LE, Cory MJ, Gordon RE (1980a) Ethanol induction of sister chromatid exchanges in human cells *in vitro. Cytogenet Cell Genet 27*: 66–69

Alvarez MR, Cimino LE, Pusateri TJ (1980b) Induction of sister chromatid exchanges in mouse foetuses resulting from maternal alcohol consumption during pregnancy. *Cytogenet Cell Genet 28*: 173–180

Amacher DE, Paillet SC, Turner GN, Ray VA, Salsburg DS (1980) Point mutations at the thymidine kinase locus in L5178Y mouse lymphoma cells. *Mutat Res 72*: 447–474

Anderson LM, Chhabra SK, Nerurkar PV, Souliotis VL, Kyrtopoulos SA (1995) Alcohol-related cancer risk: a toxicokinetic hypothesis. *Alcohol 12*: 97–104

Badr FM, Badr RS (1975) Induction of dominant lethal mutations in male mice by ethyl alcohol. *Nature 253*: 134–135

Badr FM, Badr RS, Asker RL, Hussain FH (1977) Evaluation of the mutagenic effects of ethyl alcohol by different techniques. *Adv Exp Med Biol 85A*: 25–46

Baraona E, Guerra M, Lieber CS (1981) Cytogenetic damage of bone marrow cells produced by chronic alcohol consumption. *Life Sci 29*: 1797–1802

Barilyak IR, Kozachuk SY (1981) Effects of ethanol on the genetic apparatus of mammalian germ cells. *Tsitologiya Genetika 15*: 29–32

Blot WJ (1992) Alcohol and cancer. *Cancer Res (Suppl) 52*: 2119s–2123s

Bondy SC (1992) Ethanol toxicity and oxidative stress. *Toxicol Lett 63*: 231–241

Brooks PJ (1997) DNA damage, DNA repair, and alcohol toxicity—a review. *Alcohol Clin Exp Res 21*: 1073–1082

Browning E (1953) *Toxicity of industrial organic solvents.* Medical Research Council, Industrial Health Research Board, Report No. 80, Her Majesty's Stationary Office, London, 217–222

Calva AP, Madrigal-Bujaidar E (1993) SCE frequencies induced by ethanol, tequila and brandy in mouse bone marrow cells *in vivo*. *Toxicol Lett 66*: 1–5

Campbell L, Wilson HK (1986) Blood alcohol concentrations following the inhalation of ethanol vapour under controlled conditions. *J Forensic Sci Soc 26*: 129–135

Chauhan PS, Aravindakshan M, Kumar NS, Sundaram K (1980) Failure of ethanol to induce dominant lethal mutations in Wistar male rats. *Mutat Res 79*: 263–275

Cometto-Munic JE, Cain WS (1995) Relative sensitivity of the ocular trigeminal, nasal trigeminal and olfactory systems to airborne chemicals. *Chem Senses 20*: 191–198

Coon RA, Jones RA, Jenkins Jr LJ, Siegel J (1970) Animal inhalation studies on ammonia, ethylene glycol, formaldehyde, dimethylamine, and ethanol. *Toxicol Appl Pharmacol 16*: 646–655

Czajka MR, Tucci SM, Kaye GI (1980) Sister chromatid exchange frequency in mouse embryo chromosomes after *in utero* ethanol exposure. *Toxicol Lett 6*: 257–261

De Flora S, Camoirano A, Zanacchi P, Bennicelli C (1984) Mutagenicity testing with TA97 and TA102 of 30 DNA-damaging compounds, negative with other strains. *Mutat Res 134*: 159–165

De Raat WK, Davis PB, Bakker GL (1983) Induction of sister-chromatid exchanges by alcohol and alcoholic beverages after metabolic activation by rat-liver homogenate. *Mutat Res 124*: 85–90

Dehaene P, Crepin G, Delahousse G, Querleu D, Walbaum R, Titran M, Samaille-Villette C (1981) Aspects épidémiologiques du syndrome d'alcoolisme foetal. *Nouv Presse Med 10*: 2639–2643

Dehaene P, Samaille-Villette C, Samaille P, Crépin G, Walbaum R, Deroubaix P, Blanc-Garin AP (1977) Le syndrome d'alcoolisme foetal dans le nord de la France. *Rev Alcool 23*: 145–158

ECB (European Chemicals Bureau) (1995) *IUCLID data sheet, Ethanol*, 23.10.95

Emanuelle N, Emanuelle MA (1997) The endocrine system. Alcohol alters critical hormonal balance. *Alcohol Health Res World 21*: 53–64

Eriksson CJP, Fukunaga T (1993) Human blood acetaldehyde (update 1992). *Alcohol Alcoholism, Suppl 2*: 9–25

Estonius M, Svensson S, Höög J-O (1996) Alcohol dehydrogenase in human tissues: localisation of transcripts coding for five classes of the enzyme. *FEBS Lett 397*: 338–342

Fang JL, Vaca CE (1995) Development of a [32]P-postlabelling method for the analysis of adducts arising through the reaction of acetaldehyde with 2′-deoxyguanosine-3′-monophosphate and DNA. *Carcinogenesis 16*: 2177–2185

Fang JL, Vaca CE (1997) Detection of DNA adducts of acetaldehyde in peripheral white blood cells of alcohol abusers. *Carcinogenesis 18*: 627–632

Farber E (1996) Alcohol and other chemicals in the development of hepatocellular carcinoma. *Clin Lab Med 16*: 377–394

Fiserova-Bergerova V, Diaz ML (1986) Determination and prediction of tissue-gas partition coefficients. *Int Arch Occup Environ Health 58*: 75–87

Fiserova-Bergerova V, Pierce JT, Droz PO (1990) Dermal absorption potential of industrial chemicals: criteria for skin notation. *Am J Ind Med 17*: 617–635

Fisher AA (1983) Topically applied alcohol as a cause of contact dermatitis. *Cutis 31*: 588–600

Fregert S, Groth O, Hjorth N, Magnusson B, Rorsman H, Övrum P (1969) Alcohol dermatitis. *Acta Derm Venereol 49*: 493–497

Frezza M, Di Padova C, Pozzato G, Terpon M, Baraona E, Lieber CS (1990) High blood alcohol levels in women. The role of decreased gastric alcohol dehydrogenase activity and first-pass metabolism. *N Engl J Med 322*: 95–99

Gavaler JS, Van Thiel DH (1987) Reproductive consequences of alcohol abuse: males and females compared and contrasted. *Mutat Res 186*: 269—277

Geertinger P, Bodenhoff J, Helweg-Larsen K, Lund A (1982) Endogenous alcohol production by intestinal fermentation in sudden infant death. *Z Rechtsmed 89*: 167–172

Gerchow J, Heberle B (1980) *Alkohol, Alkoholismus*. Lexikon. Neuland-Verlagsgesellschaft, Hamburg

Gervasi PG, Longo V, Naldi F, Panattoni G, Ursino F (1991) Xenobiotic-metabolizing enzymes in human respiratory nasal mucosa. *Biochem Pharmacol 41*: 177–184

Ghosh TK, Copeland Jr RL, Alex PK, Pradhan SN (1991) Behavioral effects of ethanol inhalation in rats. *Pharmacol Biochem Behav 38*: 699–704

Golka K, Blaskewicz M, Bandel T, Kiesswetter E, Vangala RR, Seeber A, Bolt HM (1994) Bio-monitoring of men and women exposed to ethanol vapours. *Naunyn-Schmiedeberg's Arch Pharmacol 349, Suppl*: R 120

Grodstein F, Goldman MB, Cramer DW (1994) Infertility in woman and moderate alcohol use. *Am J Public Health 84*: 1429–1432

Gummer CL, Maibach HI (1986) The penetration of [$^{14}$C]ethanol and [$^{14}$C]methanol through excised guinea-pig skin *in vitro*. *Food Chem Toxicol 24*: 305–309

Halkka O, Eriksson K (1977) The effects of chronic ethanol consumption on goniomitosis in the rat. in: Gross MM (Ed.) *Alcohol intoxication and withdrawal, IIIa, Biological aspects of ethanol*, Plenum Press, New York, 1–6

Hanson JW, Streissguth AP, Smith DW (1978) The effects of moderate alcohol consumption during pregnancy on fetal growth and morphogenesis. *J Pediatr 92*: 457–460

Harlap S, Shiono PH (1980) Alcohol, smoking and incidence of spontaneous abortions in the first and second trimester. *Lancet 2*: 173–176

Hayes S, Gordon A, Sadowski I, Hayes C (1984) RK bacterial test for independently measuring chemical toxicity and mutagenicity: Short-term forward selection assay. *Mutat Res 130*: 97–106

He A, Lambert B (1990) Acetaldehyde-induced mutation at the hprt locus in human lymphocytes *in vitro*. *Env Mol Mut 16*: 57–63

Hellmér K, Bolcsfoldi G (1992) An evaluation of the *E. coli* K-12 uvrB/recA DNA repair host-mediated assay I. *In vitro* sensitivity of the bacteria to 61 compounds. *Mutat Res 272*: 145–160

Holmberg B, Ekström T (1995) The effects of long-term oral administration of ethanol on Sprague-Dawley rats—a condensed report. *Toxicology 96*: 133–145

Hooper G, Steed KP, Gittins DP, Newman SP, Richards A, Rubin I (1995) Bronchoconstriction following inhaled ethanol solutions. *Respir Med 89*: 457–458

Hunt PA (1987) Ethanol-induced aneuploidy in male germ cells of the mouse. *Cytogenet Cell Genet 44*: 7–10

IARC (International Agency for Research on Cancer) (1988) *Alcohol drinking. IARC monographs on the evaluation of carcinogenic risks to humans, Bd 44*, International Agency for Research on Cancer, Lyon

James DA, Smith DM (1982) Analysis of results from a collaborative study of the dominant lethal assay. *Mutat Res 97*: 303–314

Jones KL (1986) Fetal alcohol syndrome. *Pediatrics Rev 8*: 122–126

Jones KL, Smith DW, Streissguth AP, Myrianthopoulos NC (1974) Outcome in offspring of chronic alcoholic women. *Lancet 1*: 1076–1078

Jones KL, Smith DW, Ulleland CN, Streissguth AP (1973) Pattern of malformation in offspring of chronic alcoholic mothers. *Lancet 1*: 1267–1271

Kaufman MH (1983) Ethanol induced chromosomal abnormalities at conception. *Nature 302*: 258–260

Kaufman MH, Bain IM (1984) The development potential of ethanol-induced monosomic and trisomic conceptuses in the mouse. *J Exp Zool 231*: 149–155

Klassen RW, Persaud TVN (1976) Experimental studies on the influence of male alcoholism on pregnancy and progeny. *Exp Pathol 12*: 38–45

Kline J, Shrout P, Stein Z, Susser M, Warburton D (1980) Drinking during pregnancy and sponta-neous abortion. *Lancet 2*: 176–180

Klyosov AA (1996) Kinetics and specificity of human liver aldehyde dehydrogenases toward aliphatic, aromatic, and fused polycyclic aldehydes. *Biochemistry 35*: 4457–4467

Kohlenberg-Müller K, Bitsch I (1990) Neue Methoden zur pharmakokinetischen Beschreibung des Alkohols und seiner Metaboliten bei weiblichen und männlichen Versuchspersonen. *Blutalkohol 27*: 40–48

Kuykendall JR, Bogdanffy MS (1992a) Efficiency of DNA-histone crosslinking induced by saturated and unsaturated aldehydes *in vitro*. *Mutat Res 283*: 131–136

Kuykendall JR, Bogdanffy MS (1992b) Reaction kinetics of DNA—histone crosslinking by vinyl acetate and acetaldehyde. *Carcinogenesis 13*: 2095–2100

Lancet (1980) Alcohol and spontaneous abortion. (Editorial) *Lancet 2*: 188

Lancet (1983) Alcohol and the fetus—is zero the only option? (Editorial) *Lancet 1*: 682–683

Lehmann KB, Flury F (Eds) (1938) *Toxikologie und Hygiene der technischen Lösungsmittel.* Verlag Julius Springer, Berlin, 152–155

Lemoine P, Harousseau H, Borteyru J-P, Menuet J-C (1968) Les enfants de parents alcooliques. Anomalies observées. A propos de 127 cas. *Ouest Méd 21*: 476–482

Lester D, Greenberg LA (1951) The inhalation of ethyl alcohol by man. *Q J Stud Alcohol 12*: 167–178

Lewis MJ (1985a) Inhalation of ethanol vapour: a case report and experimental test involving the spraying of shellac lacquer. *J Forensic Sci Soc 25*: 5–9

Lewis MJ (1985b) A theoretical treatment for the estimation of blood alcohol concentration arising from inhalation of ethanol vapour. *J Forensic Sci Soc 25*: 11–22

Lieber CS (1997) Ethanol metabolism, cirrhosis and alcoholism. *Clin Chim Acta 257*: 59–84

Lindahl R (1992) Aldehyde dehydrogenases and their role in carcinogenesis. *Crit Rev Biochem Mol Biol 27*: 283–335

Little RE (1977) Moderate alcohol use during pregnancy and decreased infant birth weight. *Am J Public Health 67*: 1154–1156

Loewy A, von der Heide R (1918) Über die Aufnahme des Äthylalkohols durch die Atmung. *Biochem Z 86*: 125–175

Longnecker MP (1995) Alcohol consumption and risk of cancer in humans: an overview. *Alcohol 12*: 87–96

Longo V, Mazzaccaro A, Ventura P, Gervasi PG (1992) Drug-metabolising enzymes in respiratory nasal mucosa and liver of cynomolgus monkey. *Xenobiotica 22*: 427–431

Löser H (1991) Alkoholeffekte und Schwachformen der Alkoholembryopathie. *Dtsch Aerztebl 88*: C-1921–C-1927

Majewski F (1986) Die Alkoholembryopathie. Eine häufige und vermeidbare teratogene Schädigung von Kindern. *Kinderarzt 8*: 1127–1138

Mankes RF, LeFevre R, Benitz KF, Rosenblum I, Bates H, Walker AIT, Abraham R (1982) Paternal effects of ethanol in the Long-Evans rat. *J Toxicol Environ Health 10*: 871–878

Marietta CA, Jerrells TR, Meagher RC, Karanian JW, Weight FF, Eckardt MJ (1988) Effects of long-term ethanol inhalation on the immune and hematopoietic systems of the rat. *Alcohol Clin Exp Res 12*: 211–214

Mason JK, Blackmore DJ (1972) Experimental inhalation of ethanol vapour. *Med Sci Law 12*: 205–208

Mau G (1980) Moderate alcohol consumption during pregnancy and child development. *Eur J Pediatr 133*: 233–237

Melli MC, Giorgini S, Sertoli A (1986) Sensitization from contact with ethyl alcohol. *Contact Dermatitis 14*: 315

Mira L, Maia L, Barreira L, Manso CF (1995) Evidence for free radical generation due to NADH oxidation by aldehyde oxidase during ethanol metabolism. *Arch Biochem Biophys 318*: 53–58

Moser VC, Balster RL (1985) Acute motor and lethal effects of inhaled toluene, 1,1,1-trichloroethane, halothane, and ethanol in mice: effects of exposure duration. *Toxicol Appl Pharmacol 77*: 285–291

Moser VC, Balster RL (1986) The effects of inhaled toluene, halothane, 1,1,1-trichloroethane, and ethanol on fixed-interval responding in mice. *Neurobehav Toxicol Teratol 8*: 525–531

Mufti SI, Nachiapan V, Eskelson CD (1997) Ethanol-mediated promotion of oesophageal carcinogenesis: association with lipid peroxidation and changes in phospholipid fatty acid profile of the target tissue. *Alcohol Alcoholism 32*: 221–231

Mullin LS, Krivanek ND (1982) Comparison of unconditioned reflex and conditioned avoidance tests in rats exposed by inhalation to carbon monoxide, 1,1,1-trichloroethane, toluene or ethanol. *Neurotoxicology 3*: 126–137

Myou S, Fujimura M, Nishi K, Watanabe K, Matsuda M, Ohka T, Matsuda T (1996) Effect of ethanol on airway caliber and nonspecific bronchial responsiveness in patients with alcohol-induced asthma. *Allergy 51*: 52–55

Nelson BK, Brightwell WS, Burg JR (1985a) Comparison of behavioral teratogenic effects of ethanol and *n*-propanol administered by inhalation to rats. *Neurobehav Toxicol Teratol 7*: 779–783

Nelson BK, Brightwell WS, MacKenzie DR, Khan A, Burg JR, Weigel WW, Goad PT (1985b) Teratological assessment of methanol and ethanol at high inhalation levels in rats. *Fundam Appl Toxicol 5*: 727–736

Nelson BK, Brightwell WS, MacKenzie-Taylor DR, Burg JR, Massari VJ (1988) Neurochemical, but not behavioral, deviations in the offspring of rats following prenatal or paternal inhalation exposure to ethanol. *Neurotoxicol Teratol 10*: 15–22

Neubert D (1986) Arzneimittel, Umweltchemikalien, ionisierende Strahlen und Schwangerschaft. in: Künzel W (Ed.) *Die normale Schwangerschaft. Klinik der Frauenheilkunde und Geburtshilfe, Band 4*, Urban & Schwarzenberg, München, Wien, Baltimore, 97–137

Obe G, Anderson D (1987) Genetic effects of ethanol. *Mutat Res 186*: 177–200

Obe G, Natarajan AT, Meyers M, Hertog D (1979) Induction of chromosomal aberrations in peripheral lymphocytes of human blood *in vitro*, and of SCEs in bone-marrow cells of mice *in vivo* by ethanol and its metabolite acetaldehyde. *Mutat Res 68*: 291–294

Okazawa H, Aihara M, Nagatani T, Nakajima H (1998) Allergic contact dermatitis due to ethyl alcohol. *Contact Dermatitis 38*: 233

Ophaswongse S, Maibach HI (1994) Alcohol dermatitis: allergic contact dermatitis and contact urticaria syndrome. A review. *Contact Dermatitis 30*: 1–6

Ostrovsky YM (1986) Endogenous ethanol—its metabolic, behavioral and biomedical significance. *Alcohol 3*: 239–247

Ouellette EM, Rosett HL, Rosman NP, Weiner L (1977) Adverse effects on offspring of maternal alcohol abuse during pregnancy. *N Engl J Med 297*: 528–530

Pastino GM, Asgharian B, Roberts K, Medinsky MA, Bond JA (1997) A comparison of physiologically based pharmacokinetic model predictions and experimental data for inhaled ethanol in male and female B6C3F$_1$ mice, F344 rats, and humans. *Toxicol Appl Pharmacol 145*: 147–157

Patruno C, Suppa F, Sarracco G, Balato N (1994) Allergic contact dermatitis due to ethyl alcohol. *Contact Dermatitis 31*: 124

Persaud TVN (1988) Fetal alcohol syndrome. *CRC Crit Rev Anat Cell Biol 1*: 277–325

Radike MJ, Stemmer KL, Bingham E (1981) Effect of ethanol on vinyl chloride carcinogenesis. *Environ Health Perspect 41*: 59–62

Rashkovetsky LG, Maret W, Klyosov AA (1994) Human liver aldehyde dehydrogenases: new method of purification of the major mitochondrial and cytosolic enzymes and re-evaluation of their kinetic properties. *Biochim Biophys Acta*: 1205: 301–307

Rikans LE, Gonzalez LP (1990) Antioxidant protection systems of rat lung after chronic ethanol inhalation. *Alcohol Clin Exp Res 14*: 872–877

Schardein JL (1985) *Chemically induced birth defects*. Marcel Dekker, New York, Basel, 779–783

Schmidt W, Popham RE, Israel Y (1987) Dose-specific effects of alcohol on the lifespan of mice and the possible relevance to man. *Br J Addict 82*: 775–788

Schütz H (1983) *Alkohol im Blut*. VCH Verlagsgesellschaft, Weinheim

Scott RC, Corrigan MA, Smith F, Mason H (1991) The influence of skin structure on permeability: an intersite and interspecies comparison with hydrophilic penetrants. *J Invest Dermatol 96*: 921–925

Seeber A, Blaszkewicz M, Golka K, Kiesswetter E (1997) Solvent exposure and ratings of well-being: dose-effect relationships and consistency of data. *Environ Res 73*: 81–91

Seeber A, Kiesswetter E, Bandel T, Blaszkewicz M, Golka K, Heitmann P, Vangala RR, Bolt HM (1994) Biomonitoring, Leistung und Befinden bei inhalativer Ethanolexposition. in: *Verhandlungen der Deutsche Gesellschaft für Arbeitsmedizin und Umweltmedizin*. 34th Annual Meeting, 16.05.–19.05., Wiesbaden, Gentner Verlag, Stuttgart, 205–209.

Seitz HK, Simanowski UA, Osswald BR (1992) Epidemiology and pathophysiology of ethanol-associated gastrointestinal cancer. *Pharmacogenetics 2*: 278–287

Shimoda T, Kohno S, Takao A, Fujiwara C, Matsuse H, Sakai H, Watanabe T, Hara K, Asai S (1996) Investigation of the mechanism of alcohol-induced bronchial asthma. *J Allergy Clin Immunol 97*: 74–84

Singh NP, Khan A (1995) Acetaldehyde: genotoxicity and cytotoxicity in human lymphocytes. *Mutat Res 337*: 9–17

Singletary K (1997) Ethanol and experimental breast cancer: a review. *Alcohol Clin Exp Res 21*: 334–339

Sram RJ, Topinka J, Binkova B, Kocisova J, Kubicek V, Gebhart JA (1990) Genetic damage in peripheral lymphocytes of chronic alcoholics. in: Garner RC, Hradec J (Eds) *Biochemistry of chemical carcinogenesis*, Plenum Press, New York, 219–226

Sprung R, Bonte W, Rüdell E, Domke M, Frauenrath C (1981) Zum Problem des endogenen Alkohols. *Blutalkohol 18*: 65–70

Stotts J, Ely WJ (1977) Induction of human skin sensitization to ethanol. *J Invest Dermatol 69*: 219–222

Streissguth AP, Barr HM, Martin DC (1984) Alcohol exposure *in utero* and functional deficits in children during the first four years of life. in: *Mechanisms of alcohol damage in utero. Ciba Foundation symposium 105*. Pittman, London, 176–196

Tates AD, de Vogel N, Neuteboom I (1980) Cytogenetic effects in hepatocytes, bone-marrow cells and blood lymphocytes of rats exposed to ethanol in the drinking water. *Mutat Res 79*: 285–288

Topinka J, Binkova B, Sram RJ, Fojtikova I (1991) DNA-repair capacity and lipid peroxidation in chronic alcoholics. *Mutat Res 263*: 133–136

Ukita K, Fukui Y, Shiota K (1993) Effects of prenatal alcohol exposure in mice: influence of an ADH inhibitor and a chronic inhalation study. *Reprod Toxicol 7*: 273–281

Van Ketel WG, Tan-Lim KN (1975) Contact dermatitis from ethanol. *Contact Dermatitis 1*: 7–10

Washington WJ, Cain KT, Cacheiro NLA, Generoso WM (1985) Ethanol-induced late fetal death in mice exposed around the time of fertilization. *Mutat Res 147*: 205–210

Wright JT, Waterson EJ, Barrison IG, Toplis PJ, Lewis IG, Gordon MG, MacRae KD, Morris NF, Murray-Lyon IM (1983) Alcohol consumption, pregnancy, and low birthweight. *Lancet 1*: 663–665

Yokoyama A, Muramatsu T, Ohmori T, Higuchi S, Hayashida M, Ishii H (1996a) Esophageal cancer and aldehyde dehydrogenase-2 genotypes in Japanese males. *Cancer Epidemiol Biomarkers Prev 5*: 99–102

Yokoyama A, Muramatsu T, Ohmori T, Makuuchi H, Higuchi S, Matsushita S, Yoshino K, Maruyama K, Nakano M, Ishii H (1996b) Multiple primary esophageal and concurrent upper aerodigestive tract cancer and the aldehyde dehydrogenase-2 genotype of Japanese alcoholics. *Cancer 15*: 1986–1990

Yoshida A (1992) Molecular genetics of human aldehyde dehydrogenase. *Pharmacogenetics 2*: 139–147

completed 25.05.1998

# Ethylene oxide

| | |
|---|---|
| **MAK value** | – |
| **Peak limitation** | – |
| **Absorption through the skin (1984)** | H |
| **Sensitization** | – |
| **Carcinogenicity (1984)** | **Category 2** |
| **Prenatal toxicity** | – |
| **Germ cell mutagenicity** | – |
| **BAT value** | . **see Section IX of the** *List of MAK and BAT Values* |
| Chemical name (CAS) | oxirane |
| CAS number | 75-21-8 |

The toxicology of ethylene oxide was reviewed in Volume 5 of the present series; this chapter is a supplement to the earlier review.

# 1 Allergenic Effects

## 1.1 Effects in man

Ethylene oxide is a potent alkylating agent and reacts with hydroxyl, sulfhydryl, amino and carboxyl groups in human macromolecules. As a hapten, it becomes an active allergen after binding to human proteins. For ethylene oxide, especially allergies of immediate type are well documented. In addition, there are case reports describing contact dermatitis caused by reactions to ethylene oxide.

### 1.1.1 IgE-mediated reactions to ethylene oxide

Anaphylactic reactions in dialysis patients with attacks of sneezing, retrosternal burning pains, larynx oedema, bronchial obstruction and hypersecretion, flushing and pruritus

*Essential MAK Value Documentations.* DFG, Deutsche Forschungsgemeinschaft
Copyright © 2006 WILEY-VCH Verlag GmbH & Co. KGaA, Weinheim
ISBN: 3-527-31394-X

and sometimes even anaphylactic shock have been described by several authors (Bommer *et al.* 1985, Röckel *et al.* 1988, Rumpf *et al.* 1985). Because of the large number of dialysis treatments carried out, these reactions are not unusual (4.3 per 100000 dialyses) (Bommer *et al.* 1985). As potential causes of these complications, IgE-mediated reactions to the disinfectants ethylene oxide and formaldehyde, to various isocyanates and various phthalates have been suggested. Direct complement activation by the cellulose filters used has also been discussed (Röckel *et al.* 1989). However, various authors came independently to the conclusion that by far the main aetiopathic factor in the provocation of such reactions is allergy of immediate type to ethylene oxide. In such cases the presence of IgE specific for conjugates of ethylene oxide with human serum albumin (HSA) have been demonstrated with the radioallergosorbent test (RAST) (Bommer *et al.* 1985, Grammer *et al.* 1984, Röckel *et al.* 1989, Rumpf *et al.* 1985).

In a study of 83 dialysis patients, 16 dialysis unit personnel and 44 healthy control persons, IgE specific for conjugates of ethylene oxide with HSA was detected in 35 of the dialysis patients but in only 2 persons from the control group and 2 dialysis unit personnel. Dialysis patients with IgE antibodies had allergic complications during dialysis more frequently than patients without antibodies. After use of materials which had not been sterilized with ethylene oxide for eight weeks, the antibodies in sensitized dialysis patients were present at much lower levels or were no longer detectable and the clinical symptoms had suddenly improved. Re-exposure to materials sterilized with ethylene oxide resulted in reappearance of the clinical symptoms (Bommer *et al.* 1985).

Ethylene oxide is considered to be a potential occupational sensitizer for nurses and other persons employed in the health services (Alberts 1993, Trotti and McCarthy 1989). However, the literature contains only very few reports of IgE-mediated reactions to ethylene oxide resulting from workplace exposure.

Occupational allergic rhinoconjunctivitis in a midwife caused by ethylene oxide has been described. The source of the allergen was gloves sterilized with ethylene oxide. The diagnosis was based on positive results in a cutaneous test and a provocation test. IgE specific for ethylene oxide could, however, not be detected. There was no evidence for a simultaneous latex allergy (Wendling *et al.* 1994). In contrast, in other reported cases the possibility cannot be excluded that the symptoms were induced by latex proteins, which must be seen as the much more potent allergens (Dugue *et al.* 1991, Goeters *et al.* 1993, Jacson *et al.* 1991, Meurice *et al.* 1990). Obstructive airway diseases caused by occupational exposure to ethylene oxide have been described as non-immunological, chemical-irritative effects (Deschamps *et al.* 1992).

## 1.1.2 Cell-mediated reactions to ethylene oxide

Aqueous solutions of ethylene oxide are highly irritating for human skin (maximum irritation with 50 % w/w solutions). There are several reports of contact dermatitis induced by ethylene oxide (WHO 1985). In some cases the observed reactions were interpreted as cell-mediated allergy to ethylene oxide. In one female patient with bronchial asthma, two days after use of a respiratory mask which had been sterilized with ethylene oxide, dermatitis developed at the site of contact. Thorough differential diagnostic investigation led the authors to the conclusion that an allergic mechanism must be responsible for the reaction and that ethylene oxide was the allergen (Alomar *et al.* 1981).

## 1.2 Results of animal studies

In mice and rats treated by parenteral application of ethylene oxide-protein conjugates, the formation of specific IgE antibodies was demonstrated. By means of transfer tests, the specificity of the IgE antibodies could be demonstrated *in vivo* (Chapman *et al.* 1986).

# 2 Manifesto (sensitization)

There are numerous reports of the allergenic effects of ethylene oxide in dialysis patients. However, the substance is not a relevant allergen in occupational medicine. Therefore, ethylene oxide is currently not designated with an "S".

# 3 References

Alberts WM (1993) Occupational asthma in the respiratory care worker. *Respir Care 38*: 997–1004

Alomar A, Camarasa JMG, Noguera J, Aspinolea F (1981) Ethylene oxide dermatitis. *Contact Dermatitis 7*: 205–207

Bommer J, Barth HP, Wilhelms OH, Schindele H, Ritz E (1985) Anaphylactoid reactions in dialysis patients: role of ethylene-oxide. *Lancet 28*: 1382–1384

Chapman J, Lee W, Youkilis E, Martis L (1986) Animal model for ethylene oxide (EtO) associated hypersensitivity reactions. *Trans Am Soc Artif Intern Organs 32*: 482–485

Deschamps D, Rosenberg N, Soler P, Maillard G, Fournier E, Salson D, Gervais P (1992) Persistent asthma after accidental exposure to ethylene oxide. *Br J Ind Med 49*: 523–525

Dugue P, Faraut C, Figueredo M, Bettendorf A, Salvadori JM (1991) Asthme professionnel à l'oxyde d'éthylène chez une infirmière. *Presse Med 30*: 1455

Goeters C, Theissen JL, Kästner H, Brunner W (1993) Anaphylaktische Reaktion unter Narkose aufgrund einer kombinierten Latex- und Äthylenoxidallergie. *Anästhesiol Intensivmed Notfallmed Schmerzther 28*: 326–329

Grammer LC, Roberts M, Nicholls AJ, Platts MM, Patterson R (1984) IgE against ethylene oxide-altered human serum albumin in patients who have had acute dialysis reactions. *J Allergy Clin Immunol 74*: 544–546

Jacson F, Beaudouin E, Hotton J, Moneret-Vautrin DA (1991) Allergie au formol, latex et oxyde d'éthylène: triple allergie professionelle chez une infirmière. *Rev Fr Allergol 31*: 41–43

Meurice JC, Breuil K, Perault MC, Doré P, Underner M, Patte F (1990) Allergènes professionnels en milieu hospitalier (latex – trypsine – oxyde d'éthylène) et allergies alimentaires associées. *Rev Fr Allergol 30*: 247–249

Röckel A, Hertel J, Wahn U, Perschel WT, Thiel C, Abdelhamid S, Fiegel P, Walb D (1988) Ethylene oxide and hypersensitivity reactions in patients on hemodialysis. *Kidney Int 33, Suppl 24*: 62–67

Röckel A, Klinke B, Hertel J, Baur X, Thiel C, Abdelhamid S, Fiegel P, Walb D (1989) Allergy to dialysis materials. *Nephrol Dial Transplant 4*: 646–652

Rumpf KW, Seubert A, Valentin R, Ippen H, Seubert S, Lowitz HD, Rippe H, Scheler F (1985) Association of ethylene-oxide-induced IgE antibodies with symptoms in dialysis patients. *Lancet 28*: 1385–1387

Trotti JC, McCarthy RT (1989) Occupational hazards in the OR. *AORN J 49*: 276–283

Wendling JM, Dietemann A, Oster JP, Pauli G (1994) Allergie professionnelle à l'oxyde d'éthylène. *Arch Mal Prof 55*: 287–289

WHO (World Health Organization) (1985) *Ethylene oxide. IPCS – Environmental Health Criteria 55*, WHO, Genf

completed 24.11.1995

# Formaldehyde

| | |
|---|---|
| **MAK value (2000)** | **0.3 ml/m$^3$ (ppm) $\hat{=}$ 0.37 mg/m$^3$** |
| **Peak limitation (2000)** | **Category I, excursion factor 2** |
| **Momentary value (2000)** | **1 ml/m$^3$ (ppm) $\hat{=}$ 1.24 mg/m$^3$** |
| **Absorption through the skin** | **–** |
| **Sensitization (1971)** | **Sh** |
| **Carcinogenicity (2000)** | **Category 4** |
| **Prenatal toxicity (1991)** | **Pregnancy risk group C** |
| **Germ cell mutagenicity (2000)** | **Category 5** |
| **BAT value** | **–** |
| Synonyms | methanal<br>oxomethane<br>oxymethylene<br>methylene oxide<br>formic aldehyde<br>methyl aldehyde |
| Chemical name (CAS) | formaldehyde |
| CAS number | 50-00-0 |

The toxicity of formaldehyde was reviewed by the Commission in 1971 and 1987. The MAK documentation from 1987 was published in Volume 3 of the present series. The present document is a supplement to the earlier reviews.

In the 1987 supplement to the MAK documentation from 1971, formaldehyde was classified in Carcinogen category 3 (formerly Section IIIB) and a MAK value of 0.5 ml/m$^3$ was established. The main reason for the substance being classified in Category 3 (suspected carcinogen) and not in Category 2 was the fact that at the time it could not be decided whether the tumours observed are the result of genotoxic damage or are formed above all as a result of chronic local irritation. In particular the significance of the mutagenic properties of formaldehyde was unclear. In the meantime, new studies of these two aspects have been published, and with the introduction of the new Carcinogenicity categories 4 and 5, the question of whether formaldehyde can be classified in one of these categories was raised.

Data for the toxicology and carcinogenicity of formaldehyde can be found in the MAK documentation from 1987 and in several other publications (ATSDR 1999, ECETOC 1995, IARC 1995, WHO 1989).

*Essential MAK Value Documentations.* DFG, Deutsche Forschungsgemeinschaft
Copyright © 2006 WILEY-VCH Verlag GmbH & Co. KGaA, Weinheim
ISBN: 3-527-31394-X

# 1 Toxic Effects and Mode of Action

Formaldehyde is formed endogenously during metabolism, is released into the environment from various sources and is found at the workplace. As a result of its high reactivity, in target tissues with direct contact with the substance formaldehyde causes local irritation, acute and chronic toxicity and has genotoxic properties. In long-term experiments with rats exposed by inhalation, formaldehyde caused tumours in the epithelium of the nasal mucosa. The reasons for this are probably toxic and genotoxic effects. Changes in the forestomach and glandular stomach were observed in rats given formaldehyde by long-term oral administration.

Studies with volunteers yielded threshold concentrations for odour perception of less than 0.5 ml/m$^3$, for eye irritation of 0.5 to 1 ml/m$^3$ and for nose and throat irritation of 1 ml/m$^3$; eye irritation was observed in some cases also at lower concentrations. In workers exposed long-term to formaldehyde at the workplace, lesions were observed in the nasal mucosa even at average exposure concentrations below 1 ml/m$^3$.

Formaldehyde causes sensitization of the skin.

# 2 Mechanism of Action

## 2.1 Histopathological changes

In animal studies, inhalation exposure to formaldehyde led to histopathological changes such as hyperplasia, metaplasia, dysplasia and tumours in the nose. While most authors regard simple squamous cell metaplasia as of subordinate importance for the pathogenesis of nasal tumours, dysplastic metaplasia was thought to be a preliminary stage in tumour development (Boysen *et al.* 1990). This is supported by the findings in experimental animals. These results were obtained with concentrations which led to an increase in the incidence of squamous cell metaplasia (rat 2–6 ml/m$^3$, mouse 15 ml/m$^3$) yet did not produce nasal tumours.

### Rats

It is evident from the numerous available studies (see Table 4) that from concentrations of above 1 ml/m$^3$ to 2 ml/m$^3$, independent of the duration of exposure, purulent rhinitis and histopathological lesions of the nasal mucosa develop. With increasing concentration the quality and distribution of the changes alter. With concentrations up to about 6 ml/m$^3$, regenerative hyperplasia and simple squamous cell metaplasia of the transitional and respiratory epithelia were found in the anterior third of the nose. At higher concentrations, changes occurred in the whole nasal cavity; their histomorphological appearance was increasingly like that of dysplastic metaplasia. At concentrations of about 20 ml/m$^3$, changes in the olfactory and larynx epithelia were also described (see Table 4). The nasal cavity of the rat is lined to a large extent with a cell population which resembles that of the bronchial tree in man (Schüller 1989).

## Monkeys

In rhesus monkeys, effects on histology and changes in the proliferation rate in the respiratory tract were investigated (exposure to 6 ml/m$^3$, for up to 6 weeks); such changes were found mainly in the nasal region and only to a very small extent in the lower airways. The histological changes were more pronounced than in rats under comparable conditions (Monticello 1990, Monticello *et al.* 1989).

## Man

Studies of morphological changes in the nose show that mucosal lesions similar to those described in experiments with animals exposed to concentrations of 2 ml/m$^3$ and more can develop in exposed persons. The reported changes were characterized by loss of cilia and transformation of the mucous membrane into squamous epithelium (Edling *et al.* 1985, 1987, 1988), hyperplasia of the nasal mucosa, keratinizing squamous cell metaplasia and dysplastic lesions (Boysen *et al.* 1990) and atypical squamous cell metaplasia (Berke 1987) (see also Section 4.3 Effects in man. Studies of workplace exposure).

## 2.2 Cell proliferation

The development of histopathological changes is accompanied by increased cell proliferation. Several studies have been carried out, mainly with rats, to determine levels of cell proliferation in the nasal cavity (Table 1). The increase in cell proliferation is different for the various regions of the nasal cavity. The areas with high cell proliferation correlate with those with increased levels of DNA-protein crosslinks (DPX). The highest proliferation rates were found in regions with metaplasia and dysplasia. A concentration-dependent increase in cell proliferation was described in animals exposed to 3 ml/m$^3$ or more.

The proliferation rate in the intact respiratory epithelium is in general low (Harkema 1991) and tends to be reduced even further by exposure to formaldehyde concentrations below 2 ml/m$^3$ (Monticello *et al.* 1989, 1991, 1996). This indicates a slowing down of the cell cycle, which possibly facilitates the repair of DNA damage (Kligerman *et al.* 1999) and, in analogy to effects seen after exposure to radiation, could be interpreted as an "adaptive response" (hormetic behaviour of cell proliferation). From concentrations of 6 ml/m$^3$ there was a marked and more than proportional increase in cell proliferation in the nasal epithelium and in the incidence of nasal tumours ("hockey-stick" function).

The proliferation rate in the metaplastic epithelium depends on the histological severity of the metaplasia and can be much higher than in the hyperplastic epithelium (Monticello and Morgan 1989). The incidence of tumours in certain sections of the nasal cavity correlates well with the proliferation rate at these locations expressed in terms of the size of the cell population (number of cells in the appropriate section of the nasal epithelium) (Monticello *et al.* 1996).

**Table 1.** Studies of cell proliferation in the nasal epithelium of rats, mice and monkeys after inhalation of formaldehyde

| Species, strain, number of animals per sex and group | Exposure (of the whole animal unless otherwise stated) | Parameters studied | Findings, comments | References |
|---|---|---|---|---|
| **rat,** Wistar, groups of 5–6 ♂ | **1 and 3 days,** 6 h/day 0, 1, 3.2, 6.4 ml/m$^3$ (nose-only exposure) | determination of cell proliferation (ULLI) in the nasal epithelium; after 20 h BrdU-labelling (mini pump) and staining for PCNA | **1 ml/m$^3$**: NOAEL; **3.2 ml/m$^3$**: exposure for 1 day: no significant increase, exposure for 3 days: statistically significant increase detected with PCNA method but not with BrdU-labelling. **comments**: data presented only in graphic form. The study aimed to investigate effects of exposure to mixtures of formaldehyde, acetaldehyde and acrolein. No synergism in the range of the NOAELs of the individual substances. Increased effects of mixtures in the concentration ranges causing clear effects with the individual substances. | Cassee 1995, Cassee *et al.* 1996b |
| **rat,** Wistar groups of 2 ♂ | **3 days,** 6 h/day 0, 1, 10, 20 ml/m$^3$ | determination of cell proliferation (% of labelled cells) in the nasal turbinates; cells isolated after exposure and incubated with $^3$H-thymidine | **1 ml/m$^3$**: no increase in the rate of cell proliferation; **10 and 20 ml/m$^3$**: twice the number of labelled cells in areas of the respiratory epithelium which appeared unchanged under the light microscope and a 20-fold increase in the areas with squamous cell metaplasia. | Woutersen *et al.* 1987 |
| **rat,** Sprague-Dawley, groups of 5 ♂ | **1 and 3 days,** 6 h/day 0, 0.2, 1, 6, 20 ml/m$^3$ (nose-only exposure) | determination of cell proliferation in nasal, tracheal and lung lavage cells (macrophages); flow cytometry after BrdU pulse-labelling | increase (max. 3-fold) in the proportion of labelled cells in all tissues not related to the concentration or duration of exposure. **comments**: the authors regard the effects as substance-related. Quantitative differences from the results obtained with other methods are explained by the dilution of the proliferating cells by cells from unaffected regions. | Roemer *et al.* 1993 |
| **rat,** Wistar, groups of 10 ♂ | **3 days,** 22 h/day 0, 0.1, 1, 3 ml/m$^3$ | determination of cell proliferation (% of labelled cells) in the nasal passages; single intraperitoneal injections of $^3$H-thymidine | **up to 1 ml/m$^3$**: tendency towards reduced proliferation rate; **3 ml/m$^3$**: statistically significant increase in cell proliferation in nasal plane level of section II, but not in plane level III. **comments**: modulation of cell proliferation as a result of co-exposure to ozone. Data presented only in graphic form. Lower labelling index in the controls. | Reuzel *et al.* 1990 |

**Table 1.** continued

| Species, strain, number of animals per sex and group | Exposure (of the whole animal unless otherwise stated) | Parameters studied | Findings, comments | References |
|---|---|---|---|---|
| **rat,** F344, no other details | **1, 3, 5 days,** 5 h/day 0, 0.5, 2, 6, 15 ml/m$^3$ or **c × t study:** 3 and 10 days with 36 ml/m$^3$ × h/day: 3 ml/m$^3$ × 12 h 6 ml/m$^3$ × 6 h 12 ml/m$^3$ × 3 h | determination of cell proliferation (% of labelled cells) in nasal passage, planes A (front, similar to plane I) and B (middle front, similar to plane II); single intraperitoneal injections of $^3$H-thymidine 2 h or 18 h after the end of exposure | **increase in LI in plane B:** **up 2 ml/m$^3$**: no increase, **6 ml/m$^3$** (× 1 day): increased about 5-fold, **6 ml/m$^3$** (× 3 days): increased about 25-fold, **6 ml/m$^3$** (× 3 days): increased about 6-fold (from c × t study), **6 ml/m$^3$** (× 10 days): increased about 2-fold (from c × t study), **15 ml/m$^3$** (× 1 day): increased about 13-fold, **15 ml/m$^3$** (× 3 days): increased about 13-fold, **15 ml/m$^3$** (× 5 days): increased about 23-fold. **comments**: labelling 18 h after the end of exposure produced a greater number of labelled cells in the controls and exposed animals; explanation suggested by the authors: circadian variation; no clear concentration–time–effect relationship. **c × t study:** plane A: proliferation increase about 5-fold, independent of exposure level; plane B: concentration-dependent increase in proliferation: about 3-fold, 6-fold and 17-fold after 3 and about 2-fold, 2-fold and 7-fold after 10 days. | Swenberg *et al.* 1983 |
| **rat,** Wistar, groups of 10 ♂ (3 ♂ examined after 3 days, 7 ♂ after 4 weeks) | **3 days, 4 weeks,** 8 h/day, 5 days/week 0, 5, 10 ml/m$^3$, continuous or 10 and 20 ml/m$^3$, intermittent: 30 minutes exposure, 30 minutes without | determination of cell proliferation (% of labelled cells) in the nasal passages; single intraperitoneal injections of $^3$H-thymidine after 3 exposures and at the end of the study | **5 ml/m$^3$** (**continuous** = 40 ml/m$^3$ × h/day): increased about 3-fold after 3 exposures, about 2-fold at the end of exposure; **10 ml/m$^3$** (**intermittent** = 40 ml/m$^3$ × h/day): increased about 10-fold after 3 exposures, about 5-fold at the end of exposure; **10 ml/m$^3$** (**continuous** = 80 ml/m$^3$ × h/day): increased about 10-fold after both exposures; **20 ml/m$^3$** (**intermittent** = 80 ml/m$^3$ × h/day): increased about 20-fold after both exposures. **comments**: histopathologically changed and unchanged areas not distinguished; conclusion of the authors: cell proliferation depends more on concentration than on dose; tendency towards a decrease in the proliferation rate with the duration of exposure. | Wilmer *et al.* 1987 |

**Table 1.** continued

| Species, strain, number of animals per sex and group | Exposure (of the whole animal unless otherwise stated) | Parameters studied | Findings, comments | References |
|---|---|---|---|---|
| **rat,** F344, 36 ♂ (4–6 ♂ examined per exposure period and concentration) | **6 weeks,** 6 h/day, 5 days/week 0, 0.69, 2.0, 6.2, 9.9, 14.8 ml/m$^3$ animals examined after 1, 4, 9 days and 6 weeks | determination of cell proliferation (ULLI) at 5 different sites in the nasal passages; single intraperitoneal injections of $^3$H-thymidine after the various exposure periods | **up to 2.0 ml/m$^3$**: no effects; **6.2 ml/m$^3$**: increase in cell proliferation less pronounced from the front towards the rear section of the nose, but not at higher concentrations; **> 6.2 ml/m$^3$**: increase in the ULLI after the first exposure at most sites investigated. **comments**: no clear concentration–effect relationship in groups with the same exposure duration except between 6.2 and 9.9 ml/m$^3$ in the rear sections of the nose; no clear effects of exposure duration on the cell proliferation rate. | Monticello 1990, Monticello and Morgan 1990, Monticello *et al.* 1991 |
| **rat,** F 344, groups of 10 ♂ | **12 weeks,** 6 h/day 5 days/week 0, 0.7, 2.1, 5.9, 14.5 ml/m$^3$, followed by nose-only exposure to $^{14}$C formaldehyde for 3 h | determination of cell proliferation (determination of the incorporation of $^{14}$C from $^{14}$C formaldehyde into the DNA of the nasal epithelium) | **up to 2.1 ml/m$^3$**: no effects; **5.9 ml/m$^3$**: increase in $^{14}$C-DNA in the lateral, but not in the medial or posterior nasal passages; **15 ml/m$^3$**: increase in $^{14}$C-DNA in the lateral, medial and posterior nasal passages. **comments**: the aim of the study was to determine DNA-protein crosslinks in non-exposed animals and animals previously exposed medium-term. | Casanova *et al.* 1994 |
| **rat,** Wistar, groups of 5 ♂ | **13 weeks,** 8 h/day, 5 days/week 0, 1, 2 ml/m$^3$ continuous 2 and 4 ml/m$^3$ intermittent: 30 minutes exposure, 30 minutes without | determination of cell proliferation (% of labelled cells) in the nasal passages; single intraperitoneal injections of $^3$H-thymidine after 3 exposures and at the end of the study | **1 ml/m$^3$ (continuous** = 8 ml/m$^3$ × h/day): no increase; **2 ml/m$^3$ (intermittent** = 8 ml/m$^3$ × h/day): no increase; **2 ml/m$^3$ (continuous** = 16 ml/m$^3$ × h/day): no increase; **4 ml/m$^3$ (intermittent** = 16 ml/m$^3$ × h/day): about 3-fold increase after 13 weeks (not statistically significant). **comments**: histopathologically changed and unchanged areas not distinguished; lower labelling index in the controls; increase in cell proliferation after 13 weeks, but not after 3 days (unusual results). | Wilmer *et al.* 1989 |

**Table 1.** continued

| Species, strain, number of animals per sex and group | Exposure (of the whole animal unless otherwise stated) | Parameters studied | Findings, comments | References |
|---|---|---|---|---|
| **rat**, Wistar, groups of 5 ♂ and 5 ♀ | **3 days, 13 weeks,** 6 h/day, 5 days/week 0, 0.3, 1, 3 ml/m$^3$ | determination of cell proliferation (% of labelled cells) in the nasal passages; single intraperitoneal injections of $^3$H-thymidine after 3 exposures and at the end of the study | **0.3 and 1 ml/m$^3$**: statistically significant trend towards logarithmic concentration–effect relationship in plane III after 3 days exposure; **3 ml/m$^3$**: histology: changes in planes II and III after 3 days, but not after 13 weeks; proliferation: increased about 10-fold in regions with histological changes, no increase after 13 weeks in nasal plane of section III. **comments**: histopathologically changed and unchanged areas not distinguished; lower labelling index in controls; data presented only in graphic form; high variability of the individual results is interpreted by the authors as evidence of differences in the individual sensitivity of the animals. | Zwart *et al.* 1988 |
| **rat**, F 344, groups of 6 ♂ | **18 months,** 6 h/day, 5 days/week 0, 0.7, 2.0, 6.0, 9.9, 14.9 ml/m$^3$ animals examined after 3, 6, 12 and 18 months | determination of cell proliferation (ULLI), $^3$H-thymidine infusion for 5 days before determination with osmotic mini pumps; inflammatory reaction evaluated by counting the intraepithelial neutrophilic granulocytes | **up to 6 ml/m$^3$**: no increase in cell proliferation, minimal histological changes; **10 and 15 ml/m$^3$**: significant increase in cell proliferation (max. 11-fold and 16-fold) dependent on time investigated and site; increase in the inflammation index; cell proliferation particularly pronounced in metaplastic or preneoplastic lesions. | Monticello 1990, Monticello and Morgan 1989, 1990; Monticello *et al.* 1992, 1996 |

**Table 1.** continued

| Species, strain, number of animals per sex and group | Exposure (of the whole animal unless otherwise stated) | Parameters studied | Findings, comments | References |
|---|---|---|---|---|
| **mouse,** B6C3F$_1$, no other details | **1, 3, 5 days,** 6 h/day 0, 0.5, 2, 6, 15 ml/m$^3$ or **c × t study:** 3 and 10 days with 36 ml/m$^3$ × h/day: 3 ml/m$^3$ × 12 h 6 ml/m$^3$ × 6 h 12 ml/m$^3$ × 3 h | determination of cell proliferation (% of labelled cells) in the nasal passages; planes A (front, similar to plane I) and B (middle front, similar to plane II), single intraperitoneal injections of $^3$H-thymidine 2 or 18 h after the end of exposure | **increase in the LI in plane B:** **up to 6 ml/m$^3$**: no increase; **15 ml/m$^3$** (× 1 day): increase about 8-fold; **15 ml/m$^3$** (× 3 days): increase about 8-fold; **15 ml/m$^3$** (× 5 days): increase about 13-fold. **comments:** labelling 18 h after the end of exposure produced a greater proportion of labelled cells in the controls and exposed animals; explanation of the authors: circadian variation. **c × t study:** plane A: proliferation increased after 10 days about 8-fold, 4-fold and 1.4-fold (decreasing with increasing concentration); plane B: no increase in the proliferation rate. **comments:** the authors explain the inverse concentration–effect relationship in the c × t study by reduction in respiration rate, a typical reaction of the mouse to sensory irritation. | Swenberg *et al.* 1983 |
| **monkey,** rhesus, groups of 3 ♂ | **1 or 6 weeks** 0 ml/m$^3$: 6 h/day, 5 days/week, 6 weeks 6 ml/m$^3$: 5 h/day, 5 days/week, 1 or 6 weeks | determination of cell proliferation in the nasal passages, larynx, trachea, *carina tracheae* (ULLI) and in the terminal bronchioles (LI); single intraperitoneal injections of $^3$H-thymidine after the various exposure periods | **6 ml/m$^3$** (1 week): nose: increased proliferation in the transitional and respiratory epithelia depending on the site (max. 14-fold), clear decrease in the labelling from the front to the rear sections; larynx, trachea, *carina tracheae*: increase about 2–3-fold; **6 ml/m$^3$** (6 weeks): nose: increased proliferation in the transitional and respiratory epithelia depending on the site (max. 16-fold); larynx, trachea, *carina*: 7–9-fold increase (not statistically significant as a result of high variation); no increase in proliferation in maxillary sinus and terminal bronchioles. **comments:** analytical results for the concentration of formaldehyde in the air were not reported; LIs usually reported only as graph. | Monticello 1990, Monticello *et al.* 1989 |

LI: labelling index, PCNA: proliferating cell nuclear antigen, ULLI: unit length labelling index, BrdU: bromodeoxyuridine

## 2.3 Mechanisms of formaldehyde inactivation

Several mechanisms are involved in the inactivation of formaldehyde. The inhaled hydrophilic gas dissolves first of all in the layer of mucous covering the nasal epithelium; reactions with components of the mucous (Bogdanffy *et al.* 1987) and mechanical clearance of the mucous represent the first barrier. From a certain exposure concentration mucociliary clearance is impaired. In inhalation studies with rats exposed to 15 ml/m$^3$, the mucociliary function in the frontal nasal region was inhibited and marked mucostasis was observed. After 6 ml/m$^3$ only certain areas were affected. After 2 ml/m$^3$ minimal changes in the mucous flow rate were observed. 0.5 ml/m$^3$ had no effect (Morgan *et al.* 1986). With sufficiently high exposure concentrations, a concentration gradient of free formaldehyde was established within the layers of the nasal epithelium. Under these circumstances, in the fully differentiated cells near the surface, the actual concentration is higher than in the lower-lying proliferating stem cells. In the rostral third of the respiratory epithelium, however, the epithelium consists of only two cell layers with few basal cells (Hermann 1997). In the epithelial cells there are several ways inactivation can take place. Direct reactions with protein and RNA in the cytosol probably remove a large amount of free formaldehyde (Casanova-Schmitz *et al.* 1984a). The molecule can enter the C1 pool of cell metabolism after GSH-dependent oxidation by formaldehyde dehydrogenase (Heck and Casanova-Schmitz 1984). If GSH is decreased by pretreatment with phorone, exposure to formaldehyde yields more DNA-protein crosslinks than without pretreatment. Exposure to 15 ml/m$^3$ for 9 days was, however, not sufficient to completely eliminate all the GSH (Casanova-Schmitz *et al.* 1984b).

## 2.4 Genotoxicity

Formaldehyde is a genotoxic substance. DNA-protein crosslinks are used in many studies as a parameter for determining the biologically effective concentration. They represent an early and comparatively sensitive parameter for genotoxic effects (Heck *et al.* 1990, Merk and Speit 1998, Olin *et al.* 1996) and are formed in mammalian cells mainly with the involvement of histones (Costa 1991, Miller and Costa 1989, 1990). DNA-protein crosslinks can, however, also contribute towards cytotoxicity (Conaway *et al.* 1996). Unlike for other parameters of the effects of formaldehyde (cell degeneration, necrosis, erosion, cell proliferation, metaplasia, hyperplasia, dysplasia and tumours), for DNA-protein crosslinks a concentration without effects has not been found.

The relationship between DNA-protein crosslinks and mutagenicity has not yet been satisfactorily explained. In parallel with the formation of DNA-protein crosslinks, also the *in vitro* induction of DNA single strand breaks (Bedford and Fox 1981, Cosma and Marchok 1988, Fornace *et al.* 1982, Grafström *et al.* 1983, 1984, Saladino *et al.* 1985) and mutations (Craft *et al.* 1987, Merk and Speit 1998) has been described. DNA-protein crosslinks can also interfere with DNA synthesis. DNA synthesis is reduced in the sections of DNA with DNA-protein crosslinks (Heck and Casanova 1999a), but not completely inhibited (Heck and Casanova 1999b). Replication continues to a considerable extent, but can lead to the observed deletions via a slipped-strand mispairing mechanism.

Formaldehyde may disturb replication by forming links within or between the DNA strands or by linking proteins to the DNA (Benyajati *et al.* 1983). The number of mutations depends on the rate of DNA synthesis (rate of cell proliferation) and the life span of the DNA-protein crosslinks.

Decisive for the evaluation of the genotoxic effects of formaldehyde is the fact that the level of these effects, like irritation and the increase in cell proliferation, is not a linear function of the dose; in other words, the dose–response relationship at low exposure concentrations is very flat and its gradient increases very steeply only at exposure levels which also produce tumours (Casanova *et al.* 1989, 1991, 1994).

## 2.5 Conclusions

The available information supports the idea that the formaldehyde concentrations in the epithelial cells, and the resulting toxic effects, increase in a more than proportional manner only when the physiological defence mechanisms in the target organ, the nose, and in the target epithelia can no longer cope. This is the case in rats exposed long-term to formaldehyde concentrations of 2 ml/m$^3$ and above.

In the meantime, numerous examples have been seen of tumour formation which was independent of the extent of DNA damage, but dependent on promoting effects. Above all, regenerative cell proliferation and histopathological changes induced by toxicity, in particular hyperplasia, metaplasia and dysplasia, play a role. Bladder tumours, for example, are found in the mouse only after exposure to 2-acetylaminofluorene in toxic doses which also increase cell proliferation. Thus, the toxic effects can become the main criterion for evaluating the tumour risk. Substances with comparatively low genotoxicity but marked cytotoxic effects produce tumours only under exposure conditions which cause tissue damage. The question of whether the increased proliferation is a prerequisite for fixation of the DNA damage or whether initiated cells grow better under these conditions, i.e. are promoted, remains unanswered. It is also to be expected that with continuous formaldehyde exposure, as a result of the increased cell proliferation, the probability of additional DNA damage increases and this promotes progression of the lesion.

# 3 Toxicokinetics and Metabolism

The toxicokinetics and metabolism of formaldehyde were reviewed in the documentation from 1987 (in Volume 3 of the present series). The most important of these early results and also new findings are described below.

In the blood of F344 rats, rhesus monkeys and adult volunteers similar formaldehyde concentrations were found (rat: 74.7 ± 0.2, monkey: 80.7 ± 0.3, man: 87 ± 5 µM). These concentrations were not significantly increased immediately after exposure to formaldehyde (rat: 14.4 ml/m$^3$ for 2 hours; monkey: 6 ml/m$^3$ for 4 weeks, 6 hours/day, 5 days weekly; man 1.9 ml/m$^3$ for 40 minutes) (Heck *et al.* 1985, Casanova *et al.* 1988). This is evidence for the rapid transformation and low bioavailability of free formaldehyde.

Exogenous formaldehyde reacts in the cell first with glutathione and other peptides and proteins containing SH groups to form hemithioacetals. Formaldehyde can, however, also react with other functional groups of proteins. The hemithioacetal *S*-hydroxymethyl glutathione formed from glutathione is a substrate of the $NAD^+$-dependent formaldehyde dehydrogenase, an alcohol dehydrogenase of type III (Holmquist and Vallee 1991). The oxidation product is *S*-formyl glutathione from which the enzyme *S*-formyl glutathione hydrolase cleaves free formate to regenerate glutathione (Uotila and Koivusalo 1974).

Other dehydrogenases which influence the metabolism of formaldehyde have been identified: three mitochondrial aldehyde dehydrogenases, five cytosolic dehydrogenases and at least one microsomal dehydrogenase (Tank *et al.* 1981). The enzymes have different $K_m$ values, and therefore a differentiated reaction to different levels of formaldehyde in the cell is to be expected.

Formaldehyde can also react directly with nucleotides in RNA and DNA. The first step of the reaction yields hydroxymethyl derivatives (Shabarova and Bogdanov 1994). In the case of adenine the hydroxymethyl group is bound to the exocyclic amino group. The product gives up water, forming reactive Schiff's bases; these can react with an amino group, e.g. of a protein (or of a second purine base), covalently linking the DNA and protein via a methylene bridge (Casanova *et al.* 1989, Shabarova and Bogdanov 1994, HEI 1994).

When considering the reaction mechanisms, it is important to bear in mind that in the respiratory mucosa of rats immediately after the inhalation of formaldehyde (6 ml/m$^3$, for 6 hours, on 10 days) there was no increase in free or bound acid-labile formaldehyde (e.g. *S*-hydroxymethyl glutathione) (Heck *et al.* 1982). This means that the initially formed *S*-hydroxymethyl glutathione, which could have been detected, had already been oxidized to *S*-formyl glutathione and other products. Compared with the rapid oxidation, the formation of DNA-protein crosslinks is, however, a slow reaction. It can therefore be assumed that DNA-protein crosslinks are formed by substances with the oxidation state of *S*-formyl glutathione. The oxidation of *S*-hydroxymethyl glutathione influences the formation of DNA-protein crosslinks. Thus, after inhalation exposure of rats to a mixture of [$^3$H]-formaldehyde and [$^{14}$C]-formaldehyde, more $^3$H than $^{14}$C accumulated in the DNA-protein crosslinks of the nasal mucosa (Casanova and Heck 1987, Casanova-Schmitz *et al.* 1984a, Heck and Casanova 1987). After exclusion of possible alternative explanations in control experiments, this accumulation was explained by an isotope effect on the oxidation of glutathione-bound formaldehyde (Casanova and Heck 1987). It was proposed that the [$^{14}$C]-labelled formaldehyde is oxidized more rapidly (and hydrolysed to formic acid or linked with tetrahydrofolate) than the [$^3$H]-labelled form. As a result, more of the [$^3$H]-labelled form is available for the formation of DNA-protein crosslinks. Whether this simple explanation is correct, is, however, unclear.

In the formation of DNA-protein crosslinks via *S*-formyl glutathione, it may be assumed that the first crosslink partner (DNA or protein) attaches with its primary amino groups to the aldehyde group of *S*-formyl glutathione. The second crosslink partner could then displace glutathione as leaving group and form the final crosslink. This DNA-protein crosslink product is not linked via a methylene bridge (see above), but via an H-C-OH bridge.

To understand the metabolic pathways of formaldehyde it should be noted that the *S*-hydroxymethyl glutathione initially formed in the cell is rapidly oxidized, but the resulting *S*-formyl glutathione does not react immediately to form formate and glutathione (catalysing enzyme: *S*-formyl glutathione hydrolase) and is thus not immediately detoxified (Uotila and Koivusalo 1974). The life span of *S*-formyl glutathione is evidently long enough to allow the formation of DNA-protein crosslinks; but it also binds to tetrahydrofolate and can in this way enter the C1 pool.

Activated formaldehyde, that is formaldehyde covalently bound to tetrahydrofolate, contributes to the biosynthesis of amino acids, purines and thymine. As a result of enzymatic redox reactions this bound formaldehyde can be transformed to substances with the oxidation states of methanol ($N^5$-methyl-tetrahydrofolate), formaldehyde ($N^5,N^{10}$-methylene-tetrahydrofolate) or formic acid ($N^5$-formyl-tetrahydrofolate, $N^{10}$-formyl-tetrahydrofolate and $N^5,N^{10}$-methenyl-tetrahydrofolate) (Voet and Voet 1992). By binding to tetrahydrofolate, exogenous formaldehyde is able to enter all the metabolic pathways available to C1 fragments from endogenous reactions: for example, the methylation of homocysteine to methionine, the formation of deoxythymidine monophosphate (dTMP) from deoxyuridine monophosphate (dUMP), the fixation of the C atoms 2 and 8 during purine synthesis and the synthesis of histidine.

Inhalation experiments with rats revealed that exogenous formaldehyde is used in the synthesis of the building blocks that make up RNA and DNA and is therefore found in macromolecular RNA and DNA (Casanova and Heck 1987, Casanova-Schmitz *et al.* 1984a). It is incorporated not only at the site of local effects, e.g. in the nasal respiratory mucosa, but also in distant organs, e.g. bone marrow. In rats exposed to [$^{14}$C]formaldehyde concentrations of 6 ml/m$^3$ for 6 hours, up to 47 % of the total radioactivity detectable in the bone marrow was in the form of DNA labelled via synthesis (Casanova and Heck 1987, Casanova-Schmitz *et al.* 1984a). At the same time, only a negligible amount of "interfacial material" (a measure of DNA-protein crosslinks) was observed in the bone marrow cells investigated (Casanova-Schmitz *et al.* 1984a). A definitive answer to the question of whether DNA-protein crosslinks are also formed in the bone marrow, was not, however, obtained as no investigation was carried out to detect or exclude DNA-protein crosslinks in "interfacial material", or in the aqueous phase containing DNA (Casanova and Heck 1987, Casanova-Schmitz *et al.* 1984a).

It is unclear in which form the inhaled formaldehyde is transported from the lungs into bone marrow and other organs (Chang *et al.* 1983), to be linked there with tetrahydrofolate (Casanova and Heck 1987). Free formate can be excluded, as—after its distribution (dilution) in the bloodstream—this could hardly be so enriched by the bone marrow cells that the observed level of [$^{14}$C]-labelling of bone marrow DNA could be attained. A plausible form of transport is *S*-formyl glutathione; a formaldehyde-related increase in *S*-hydroxymethyl glutathione (and free formaldehyde) in the blood of both man and experimental animals was excluded (Casanova *et al.* 1988, Heck *et al.* 1982).

That inhaled formaldehyde enters the above-mentioned metabolic pathways is also suggested by the observation that the incorporated radioactivity profile in plasma and erythrocyte components after inhalation of H$^{14}$CHO is identical with that after intravenous injection of $^{14}$C-formic acid (Heck *et al.* 1983).

# 4 Effects in man

## 4.1 Single and repeated exposures

Studies with the controlled exposure of volunteers must be distinguished from epidemiological studies of persons exposed at the workplace or under certain environmental conditions. The most reliable data are obtained in controlled studies with volunteers. Studies of persons exposed at the workplace are less suitable for making quantitative statements, mainly because of uncertain levels of exposure.

The available literature has been evaluated in a recent review written by a panel of experts (Paustenbach *et al.* 1997). This review was used as the basis for re-examination of the MAK value.

## 4.2 Studies with volunteers

All the relevant studies with volunteers in which the irritative effects of formaldehyde were investigated are shown in Table 2. The parameters studied were usually odour perception, eye irritation and irritation of the nose and throat. It must be borne in mind when evaluating the effects that the findings for irritation are usually the subjective symptoms felt by the volunteers, and there was usually no objective differentiation between odour perception and irritation. The methodological shortcomings of such studies were discussed in detail in the documentation from 1987 (in Volume 3 of the present series). The eye irritation reflects sensory irritation. In the documentation from 1987 it was suggested that sensory irritation from aldehydes could be an indicator for the formation of covalent bonds with proteins. The significance of sensory irritation is, however, still unclear. As the database for irritation of the nose and throat is still inadequate, the data for eye irritation are used in the establishment of the MAK value.

There are only 2 useful studies with volunteers in which concentrations below 1 ml/m$^3$ were tested. In a study of volunteers exposed to formaldehyde for 5 hours, discomfort (eye irritation, dryness in the nose and throat) was described by 3/16 persons (19 %) at concentrations of 0.3 ml/m$^3$, by 5/16 (31 %) at 0.5 ml/m$^3$, and by 15/16 (94 %) at 1 and 2 ml/m$^3$. At 0.3 ml/m$^3$ and 0.5 ml/m$^3$, eye irritation occurred only after exposure for more than 2 hours. The discomfort increased with the duration of exposure. The symptoms were, however, only slight. The symptoms at 1 and 2 ml/m$^3$ were described on average by the volunteers as "slight discomfort" and were no longer present the next morning (Andersen and Mølhave 1983). In another study in which persons were exposed for 3 hours, eye irritation was not reported at concentrations of 0.5 ml/m$^3$ although the odour was detected. Eye irritation increased in a dose-dependent manner from concentrations of 1 ml/m$^3$ and occurred in 5/19 volunteers (26 %) (Kulle 1993, Kulle *et al.* 1987).

**Table 2.** Irritative effects of formaldehyde in volunteers

| Concentration (ml/m$^3$) | Duration | Physical activity | Number of volunteers | Results | | | | References |
|---|---|---|---|---|---|---|---|---|
| 0.3<br>0.5<br>1.0<br>2.0 | 5 h | none | 16 (11 ♂, 5 ♀)<br>(5 smokers) | eye irritation<br>0.3 ml/m$^3$   3/16 (19 %)<br>0.5 ml/m$^3$   5/16 (31 %)<br>1.0 ml/m$^3$   15/16 (94 %)<br>2.0 ml/m$^3$   15/16 (94 %) | | | | Andersen and Mølhave 1983 |
| 0<br>0.35–1.0 | 6 min | none | 28 (0 ml/m$^3$)<br>12 (0.35 ml/m$^3$)<br>26 (0.56 ml/m$^3$)<br>7 (0.70 ml/m$^3$)<br>5 (0.90 ml/m$^3$)<br>27 (1.00 ml/m$^3$) | eye irritation<br>0   ml/m$^3$   0 %<br>0.35 ml/m$^3$   42 %<br>0.56 ml/m$^3$   54 %<br>0.70 ml/m$^3$   57 %<br>0.90 ml/m$^3$   60 %<br>1.00 ml/m$^3$   74 % (significant)<br>**comments**: as a result of the short exposure duration, this study is only of limited usefulness | | | | Bender *et al.* 1983 |
| 0<br>0.5–3.0 | 3 h | during exposure to 2 ml/m$^3$ intermittent moderate physical activity for 8 min every half hour | 19 (1 ♂, 9 ♀)<br>(non-smokers;<br>with 0.5 ml/m$^3$ only 10;<br>with 3.0 ml/m$^3$ only 9) |     eye irritation<br>0   ml/m$^3$   5 %<br>0.5   ml/m$^3$   0 %<br>1.0   ml/m$^3$   26 %<br>2.0   ml/m$^3$   53 %<br>3.0   ml/m$^3$   100 % | odour perception<br>5 %<br>40 %<br>26 %<br>58 %<br>78 % | nose/throat irritation<br>16 %<br>10 %<br>5 %<br>37 %<br>22 % | | Kulle 1993, Kulle *et al.* 1987 |
| | | | | **comments**: the authors' estimated threshold values:<br>odour perception:   < 0.5 ml/m$^3$<br>eye irritation:   0.5–1.0 ml/m$^3$<br>nose/throat irritation:   1.0 ml/m$^3$ | | | | |
| 1.0 | 90 min | none | 18 (9 had previous complaints after using formaldehyde at home) | 1.0 ml/m$^3$: eye irritation in 83 %, throat irritation in 28 % | | | | Day *et al.* 1984 |

**Table 2.** continued

| Concentration (ml/m$^3$) | Duration | Physical activity | Number of volunteers | Results | References |
|---|---|---|---|---|---|
| (A) 0–3.2 increasing (B) 0, 1, 2, 3, 4 discontinuous | (A) 37 min (B) 5 × 1.5 min | | (A) 33 (24 ♂, 9 ♀) (B) 48 (35 ♂, 13 ♀) | significant effects from the following concentrations: 1.2 ml/m$^3$: irritation of the eyes and nose, annoyance ("the wish to leave the room") 1.7 ml/m$^3$: increased eye blinking rate 2.1 ml/m$^3$: irritation of the throat **comments**: at the same concentration effects of discontinuous exposure more pronounced than effects of continuous exposure | Weber-Tschopp *et al.* 1977 |
| 0 2.0 | 40 min | (R) none, (W) 10 min moderate physical activity (on different days) | 15 (non-smokers) | slight to severe:<br><br>| | eye irritation | odour perception | nose irritation | throat irritation |<br>0 ml/m$^3$ (R): 0 %, 47 %, 27 %, 13 %<br>0 ml/m$^3$ (W): 7 %, 13 %, 13 %, 0 %<br>2.0 ml/m$^3$ (R): 53 %, 80 %, 40 %, 27 %<br>2.0 ml/m$^3$ (W): 53 %, 87 %, 33 %, 33 %<br>**comments**: interpretation of the results by Paustenbach *et al.* (1997): eye irritation a more sensitive parameter than nose and throat irritation; irritation incidence up to 20 % not relevant | Schachter *et al.* 1986 |
| 0 2.0 | 40 min | (R) none, (W) 10 min moderate physical activity (on different days) | 15 (laboratory workers, exposed long-term to formaldehyde) | slight to severe:<br><br>| | eye irritation | odour perception | nose irritation | throat irritation |<br>0 ml/m$^3$ (R): 0 %, 47 %, 7 %, 7 %<br>0 ml/m$^3$ (W): 0 %, 33 %, 0 %, 0 %<br>2.0 ml/m$^3$ (R): 47 %, 80 %, 0 %, 0 %<br>2.0 ml/m$^3$ (W): 40 %, 87 %, 7 %, 0 %<br>**comments**: persons exposed long-term to formaldehyde react in the same way as persons not previously exposed (see Schachter *et al.* 1986) | Schachter *et al.* 1987 |

**Table 2.** continued

| Concentration (ml/m$^3$) | Duration | Physical activity | Number of volunteers | Results | | | | | References |
|---|---|---|---|---|---|---|---|---|---|
| 0<br>2.0 | 40 min | (R) none,<br>(W) 10 min<br>moderate<br>physical activity | 15 (asthmatics) | slight to severe: | | | | | Witek *et al.* 1987 |
| | | | | | eye<br>irritation | odour<br>perception | nose<br>irritation | throat<br>irritation | |
| | | | | 0    ml/m$^3$ (R) | 7 % | 33 % | 20 % | 27 % | |
| | | | | 0    ml/m$^3$ (W) | 14 % | 57 % | 14 % | 21 % | |
| | | | | 2.0  ml/m$^3$ (R) | 73 % | 100 % | 47 % | 33 % | |
| | | | | 2.0  ml/m$^3$ (W) | 36 % | 100 % | 36 % | 43 % | |
| 0<br>3.0 | 1 hour | (H): intermittent strenuous physical activity,<br>(A): intermittent moderate physical activity | 22 (healthy persons, H),<br>16 (asthmatics, A) | moderate to severe: | | | | | Green *et al.* 1987 |
| | | | | | eye<br>irritation | odour<br>perception | nose/throat<br>irritation | | |
| | | | | 0    ml/m$^3$ | 0 % | 0 % | 0 % | | |
| | | | | 3.0  ml/m$^3$ (H) | 27 % | 23 % | 32 % | | |
| | | | | 3.0  ml/m$^3$ (A) | 19 % | 31 % | 31 % | | |
| | | | | **comments**: no differences between healthy and asthmatic volunteers; no asthma attacks provoked | | | | | |
| 0<br>3.0 | 3 h | intermittent physical activity | 9 (healthy) (non-smokers) | relative severity (0–3): | | | | | Sauder *et al.* 1986 |
| | | | | | eye<br>irritation | odour<br>perception | nose/throat<br>irritation | | |
| | | | | 0    ml/m$^3$ | 0.00 | 0.22 | 0.22 | | |
| | | | | 3.0  ml/m$^3$ | 0.78[*] | 1.22 | 1.33[*] | | |
| | | | | [*]significant | | | | | |
| 0<br>3.0 | 3 h | intermittent physical activity | 9 (asthmatics) (non-smokers) | relative severity (0–5): | | | | | Sauder *et al.* 1987 |
| | | | | | eye<br>irritation | odour<br>perception | nose/throat<br>irritation | | |
| | | | | 0    ml/m$^3$ | 0.00 | not determined | 0.55 | | |
| | | | | 3.0  ml/m$^3$ | 1.33[*] | not determined | 1.00[*] | | |
| | | | | [*]significant | | | | | |
| | | | | **comments**: no differences between healthy and asthmatic volunteers (see Sauder *et al.* 1986). | | | | | |

**Table 3.** Irritation and changes in airway parameters caused by exposure to formaldehyde at the workplace

| Exposure ($ml/m^3$) | Exposure duration | Volunteers | Results (irritation, airway parameters) | References |
|---|---|---|---|---|
| average 1.0 (0.036–2.27) | not specified | persons in medical buildings | **0.036–2.27 $ml/m^3$**: productive cough, breathlessness, tightness of chest, significantly reduced lung function parameters (maximum mid-expiratory flow rate, $FEV_1$) | Khamgaonkar and Fulare 1991 |
| 0.49–0.93 | 10 weeks 3 hours/day | 24 students | **0.49–0.93 $ml/m^3$**: irritation of the eyes, nose and throat, lung function parameters (peak expiratory flow rate) significantly reduced in the course of the term (normalisation 2 weeks after the end of exposure) | Kriebel *et al.* 1993 |
| average 1.24 (0.07–2.94) | 2–3 hours | 34 persons from anatomical laboratories; 12 students as controls | **0.07–2.94 $ml/m^3$**: irritation of the eyes (88 %), nose (74 %), throat (29 %) and airways (21 %), lung function parameters changed (FVC decreased, $FEV_3$ decreased, $FEV_1$/FVC increased) | Akbar-Khanzadeh *et al.* 1994 |
| 1.88 (0.3–4.45) | 3 hours | 50 anatomy students; 36 physiotherapy students | **1.88 $ml/m^3$**: irritation of the eyes (76 %), nose (82 %), throat (36 %), airways (14 %) and the skin (12 %), lung function parameters decreased (FVC, $FEV_1$, $FEV_3$, $FEF_{25–75\%}$) | Akbar-Khanzadeh and Mlynek 1997 |
| 8-hour mean value on the day of the investigation | 8 hours | 109 exposed workers and 254 control persons | ambient level:  0.05 $ml/m^3$<br>low level:    > 0.05 to < 0.4 $ml/m^3$<br>medium level:  0.4 to < 1.0 $ml/m^3$<br>high level:    1.0 to < 3.0 $ml/m^3$<br>significant dose-dependent excess of irritant symptoms (cough, chest pains, aching, tightness or burning of the chest, expectoration, burning and itching of the nose, stuffy nose, burning or watering and itching of the eyes, sore or burning of the throat) and decreased lung function parameters ($FEV_1$/$FVC_\%$, $FEF_{25\%–75\%}$, $FEF_{50\%,\,75\%}$) between the control group with ambient exposure level and the exposed groups | Horvath *et al.* 1988 |

FVC: forced vital capacity
$FEV_x$: forced expiratory volume in x seconds
$FEF_{25–75\,\%}$: forced expiratory flow at 25–75 % of the FVC that can still be expired

In addition it was found that persons who had been exposed long-term to formaldehyde did not react differently from not previously exposed persons (Schachter *et al.* 1987) and there were no differences in the sensitivity of healthy and asthmatic volunteers (Green *et al.* 1987).

The data available for eye irritation have been presented by the panel of experts as a concentration–effect curve (Paustenbach *et al.* 1997). This shows that at concentrations between 0.5 and 1 ml/m$^3$, exposure for up to 6 hours can produce eye irritation in 5 % to 25 % of the exposed persons. It should also be noted that in the presence of other chemicals or dust, sensitivity to the irritative effects of formaldehyde may be increased. It was concluded from the available data that with maximum concentrations at the workplace of 0.3 ml/m$^3$ for 8 hours almost all the workers are protected against eye irritation. Significant increases in eye irritation are reported, however, only at concentrations of at least 1 ml/m$^3$, which is the reason that this concentration, which corresponds with the momentary value, is regarded as the ceiling value.

The publications of one research group (Gorski *et al.* 1992, Krakowiak *et al.* 1998, Pazdrak *et al.* 1993) are not included in the present evaluation because of methodological shortcomings (e.g. incorrect data for exposure, unusual methods). Also the publication of Schuck *et al.* (1966) is not included, as, due to the method for generating formaldehyde, co-exposure to other irritative substances cannot be excluded.

## 4.3 Studies of workplace exposure

### 4.3.1 Studies of irritative effects and changes in respiratory parameters

The relevant studies are shown in Table 3. Persons who worked in laboratories for anatomical preparation with human cadavers preserved in a solution containing formaldehyde reported irritation of the eyes, nose, throat and airways. Respiratory parameters were also changed [e.g. reduced forced expiratory volume in 1 second (FEV$_1$) or forced vital capacity (FVC)]. The average concentrations of formaldehyde which caused such effects were given as 1 to 2 ml/m$^3$, the maximum concentrations as over 2 ml/m$^3$.

Workers exposed to moderate (0.4–1.0 ml/m$^3$) and high (1.0–3.0 ml/m$^3$) concentrations of formaldehyde reported irritation of the eyes and nose, coughing and a dry throat more often than did workers exposed to ambient formaldehyde concentrations (0.05 ml/m$^3$) (Horvath *et al.* 1988).

### 4.3.2 Histological studies of the nasal mucosa

Biopsies of the nasal mucosa from the middle nasal turbinate of the following groups of persons were subjected to histological examination: 70 employees (average age 36.9 years) from 1 factory, who were exposed to formaldehyde as the only irritative substance on average for about 10 years, 100 employees from 5 furniture factories (average age 40.5 years), who were exposed to wood dust in addition to formaldehyde for 9 years, and

36 office workers (average age 39.8 years, employed for an average of 11 years) as the control group. The level of exposure was determined by means of personal air sampling during the study. The histological changes in the biopsy material were rated on a scale of 0 to 8. In the group of workers exposed to formaldehyde alone (0.3 mg/m$^3$, range 0.05–0.5 mg/m$^3$, frequent short peak exposures > 1 mg/m$^3$) the histological score of 2.16 (0–4) was significantly increased above that of 1.56 found for the control group (0–4, exposure concentration 0.09 mg/m$^3$, range 0.09–0.17 mg/m$^3$). The histological score of 2.07 (0–6) for the persons exposed to formaldehyde (0.25 mg/m$^3$, rarely > 0.5 mg/m$^3$) and wood dust (1.65 mg/m$^3$) was not significantly increased. There was no correlation between the histological changes and the estimated individual levels of exposure to formaldehyde or wood dust, exposure duration, the calculated dose (mg/m$^3$ × year) or the prevalence of nasal discharge. Nor did smoking habits influence the results (Holmström *et al.* 1989).

In other studies by the research group of Edling (1985, 1987, 1988) altogether 75 workers from 2 pressboard-processing factories and from a laminating workshop were investigated. The average age of the persons was 38 years and the average duration of exposure 10.5 years. Sporadic determination of the exposure levels at the workplace over the years 1975 to 1983 yielded concentrations between 0.1 and 1 ml/m$^3$ (shift average values) with peak values up to 5 ml/m$^3$; there is no information about the methods used. In the wood-processing factories the dust concentrations were found to be between 0.6 and 1.1 mg/m$^3$. The nose and nasopharynx were examined clinically and a biopsy of the mucous membrane of the lower nasal turbinates (diameter 2 mm) was subjected to histological examination. The control collective was made up of 25 non-exposed persons, with an average age of 35 years. The proportion of non-smokers was 56 % in the exposed collective and 36 % in the control collective. The majority of exposed persons reported irritation of the eyes and nose. Clinical examination revealed swollen or dry nasal mucosa in 25 % of the exposed persons. Biopsy of the nasal mucosa revealed changes in 96 % of the exposed persons, characterized by the loss of cilia and transformation of the mucous membrane into squamous epithelium. Rating of the changes on a scale of 0 to 8 yielded a significantly increased average score of 2.9 compared to 1.8 for the control group. The score was independent of the exposure duration and dose. Smokers had a slightly higher score than non-smokers. Co-exposure to wood dust had no effect (Edling *et al.* 1985, 1987, 1988).

37 employees from a factory which produced formaldehyde and formaldehyde resins were compared with 37 persons who were not exposed. The average age of the exposed collective was 51 years, with an average duration of employment of 20 years. The average age of the control group was 49 years. According to the authors, the smoking habits of the groups were comparable. Rhinoscopic examination was carried out and a biopsy sample taken of the mucous membrane of the middle nasal turbinate. The exposed persons were divided into groups according to their probable level of exposure. 29 persons were exposed with differing frequency to concentrations of 0.5 to 2 ml/m$^3$, 8 persons regularly to over 2 ml/m$^3$. There are no details of the frequency with which determinations were carried out or of the methods used. Significantly more persons from the exposed group than from the control group complained of nasal irritation. Rhinoscopic examination revealed hyperplasia of the nasal mucosa in 38 % of the exposed persons and in 12 % of the control persons. Rating of the histological changes in the mucous

membrane biopsies on a scale of 0 to 5 yielded an average score of 1.9, compared with 1.4 for the control group; this did not represent a significant increase. The score was slightly higher for the smokers than for the non-smokers. While the loss of cilia and metaplasia of the mucous membranes were similar in the exposed persons and controls, one exposed person was found to have keratinizing squamous cell metaplasia, and three other exposed persons dysplastic lesions. These four persons belonged to the low exposure group; a concentration-time-response relationship for the lesions is therefore not apparent. The authors believe that simple squamous cell metaplasia tends to be unspecific and its incidence increases with age. It can, however, be the starting point for dysplastic metaplasia after exposure to toxic substances (Boysen *et al.* 1990).

In a cross-sectional study of 4 collectives from a paper processing factory, clinical examination of the upper respiratory tract was carried out and smears from the nasal mucosa were examined. Collectives 1 (18 workers, average age 44.5 years, 60 % smokers) and 2 (24 workers, average age 49 years, 60 % smokers) were exposed to formaldehyde on average for 16.2 and 17.5 years. Collective 2 had no longer been exposed during the previous 4 years. Sporadic determination of the formaldehyde concentrations at the workplace as a rule yielded values between 0.02 and 2 ml/m$^3$, with documented outliers of up to 9 ml/m$^3$. The incidence of rhinoscopically visible irritation of the mucosa was significantly higher in the non-smokers of groups 1 and 2 than in the control collectives 3 (10 workers, average age 43 years, 60 % smokers) and 4 (28 workers, average age 58 years, 20 % smokers). The cytological examination of mucosal smears revealed no differences between the exposed groups and the controls. An age-dependent increase in atypical squamous cell metaplasia was found for all groups. The authors question the value of atypical squamous cell metaplasia in the nose for predicting the development of nasal carcinoma (Berke 1987).

Another study investigated 18 workers (average age 38 years) who were exposed to formaldehyde on average for about 11 years in a factory which produced industrial felts. During the study in 1989 the formaldehyde concentrations in workplace air were determined using the chromotropic acid method which yielded values of 1.5 to 5.3 ml/m$^3$ (average value of all determinations 2.54 ± 0.4 ml/m$^3$). The concentrations were similar to those found in earlier sporadic determinations and were thus markedly higher than the MAK value valid at that time. Other hazardous substances, such as dust or chemicals, were, according to the authors, not present. The subjective symptoms of the persons were headaches, a dry nose, difficulties in breathing through the nose and shortage of breath. While rhinoscopic examination did not reveal changes in 4 persons, the mucous membrane was found to be hyperplastic in 8 persons and atrophic in 6 persons. The exposed persons were compared with a control collective of 120 non-exposed persons who on average were about 5 years younger. In 50 % of the exposed persons an increase in the number of squamous cells in the nasal mucosal smears (and in 25 % of the controls) was found, and in 72 % of the exposed persons delayed mucociliary clearance of colourant particles. When 6 persons were examined again 1 year after the end of exposure, the symptoms had improved or regressed completely (Reiche *et al.* 1992).

It is not possible to make a definitive statement about which exposure concentrations and peak exposures produce the histopathological lesions of the nose reported in workers exposed to formaldehyde. Further studies are needed to clarify this.

## 4.4 Genotoxicity

Studies of the genotoxic effects of formaldehyde in man are discussed in Section 5.5 Animal Experiments and *in vitro* Studies. Genotoxicity.

## 4.5 Carcinogenicity

The evaluation of 30 publications on the epidemiology of the carcinogenic effects of formaldehyde in exposed populations yielded no clear evidence of carcinogenic effects in man (McLaughlin 1994). A recent meta-analysis of the results of 47 epidemiological studies reached the same conclusion (Collins *et al.* 1997). The International Agency for Research on Cancer (IARC) described the available epidemiological data in man as "limited evidence in humans for carcinogenicity", as there are some, although not significant findings of tumours of the nose and lung in man (IARC 1995).

# 5 Animal Experiments and *in vitro* Studies

## 5.1 Acute toxicity

Studies of the sensory irritation caused by formaldehyde in mice and rats showed the mouse to be markedly more sensitive (Barrow *et al.* 1983, 1986, Chang *et al.* 1981, 1984). The concentration which after short-term exposure leads to a reduction in the respiration rate to 50 % ($RD_{50}$) in mice, was found to be between 3 and 5 ml/m$^3$ (Chang *et al.* 1981, Schaper 1993). In rats, $RD_{50}$ values between 10 and 30 ml/m$^3$ have been reported (Cassee 1995, Cassee *et al.* 1996a, Chang *et al.* 1981, 1984, Schaper 1993).

## 5.2 Subacute, subchronic and chronic toxicity

Studies of the subchronic and chronic toxicity of inhaled formaldehyde are shown in Table 4.

## 5.3 Local effects

In all animal experiments, the most noticeable toxic effects of formaldehyde were observed in the upper respiratory tract; these effects have been investigated in numerous studies (see Table 4).

**Table 4.** Morphological changes induced by medium-term and long-term inhalation of formaldehyde

| Species, strain, number of animals per sex and group | Exposure | Effects in the nasal region (unless stated otherwise) | References |
|---|---|---|---|
| **rat**, F344, groups of 36 ♂ | **6 weeks,** 6 h/day, 5 days/week, 0, 0.7, 2, 6, 10, 15 ml/m$^3$ | **2 ml/m$^3$**: NOAEL<br>**from 6 ml/m$^3$**: epithelial hyperplasia, squamous cell metaplasia<br>**from 10 ml/m$^3$**: also rhinitis, erosion and ulceration | Monticello *et al.* 1991 |
| **rat**, Wistar, groups of 10 ♂ and 10 ♀ | **13 weeks,** 6 h/day, 5 days/week, 0, 1, 10, 20 ml/m$^3$ | **1 ml/m$^3$**: NOAEL<br>**from 10 ml/m$^3$**: reduced body weight gains (♂, ♀), rhinitis, squamous cell metaplasia, epithelial keratinization | Woutersen *et al.* 1987 |
| **rat**, Wistar, groups of 50 ♂ and 50 ♀ | **13 weeks,** 6 h/day, 5 days/week, 0, 0.3, 1, 3 ml/m$^3$ | **1 ml/m$^3$**: NOAEL<br>**3 ml/m$^3$**: epithelial hyperplasia, squamous cell metaplasia | Zwart *et al.* 1988 |
| **rat**, Wistar, groups of 45 ♂, groups of 10 ♂ for follow-up period | **13 weeks,** 6 h/day, 5 days/week, 0, 10, 20 ml/m$^3$, **follow-up period up to 126 weeks** | **10 ml/m$^3$**: rhinitis, focal hyperplasia, squamous cell metaplasia<br>**20 ml/m$^3$**: marked changes as described above, also olfactory epithelium metaplastic.<br>At the end of the follow-up period after 10 ml/m$^3$ still slight changes, after 20 ml/m$^3$ still marked changes as described above<br>tumours of the nasal cavity<br>**0 ml/m$^3$**  2/134 (1.5 %)<br>**10 ml/m$^3$**  2/132 (1.5 %)<br>**20 ml/m$^3$**  10/132 (7.6 %) | Feron *et al.* 1988 |
| **rat**, Wistar, groups of 25 ♂ | **13 weeks,** 8 h/day, 5 days/week ("continuous"), 1, 2 ml/m$^3$ or 8 × 30 min/day, 5 days/week ("intermittent"), 2, 4 ml/m$^3$ | **2 ml/m$^3$** (continuous): NOAEL<br>**4 ml/m$^3$** (intermittent): hyperplasia, metaplasia | Wilmer *et al.* 1989 |
| **rat**, F344, groups of 20 ♂ and 20 ♀ | **26 weeks,** 22 h/day, 7 days/week, 0, 0.2, 1.0, 3.0 ml/m$^3$ | **1.0 ml/m$^3$**: NOAEL<br>**3.0 ml/m$^3$**: reduced body weight gains, reduced absolute and relative liver weights, basal cell hyperplasia, squamous cell metaplasia | Rusch *et al.* 1983a, 1983b |

**Table 4.** continued

| Species, strain, number of animals per sex and group | Exposure | Effects in the nasal region (unless stated otherwise) | References |
|---|---|---|---|
| **rat**, Wistar, groups of 40 ♂ | **52 weeks,** 6 h/day, 5 days/week, 0, 0.1, 1, 10 ml/m$^3$ | **0.1 ml/m$^3$**: in animals with damaged mucosa (by electrocoagulation): metaplasia (without a concentration-dependent increase in the incidence) **1 ml/m$^3$**: no effects on the undamaged nasal mucosa **10 ml/m$^3$**: reduced body weight gains, rhinitis, hyperplasia, metaplasia | Appelman *et al.* 1988 |
| **rat**, SD, groups of 100 ♂ | **84 weeks,** 6 h/day, 5 days/week, 0, 14.2 ml/m$^3$ | **14.2 ml/m$^3$**: squamous cell carcinomas (10 %), mortality (38 %) | Albert *et al.* 1982 |
| **rat**, F344, ♂ | **24 months** 6 h/day, 5 days/week, 0, 0.7, 2, 6, 10, 15 ml/m$^3$ | **2 ml/m$^3$**: NOAEL **from 6 ml/m$^3$**: epithelial hyperplasia, squamous cell metaplasia **from 10 ml/m$^3$**: also rhinitis and lesions regarded as pre-neoplastic, squamous cell carcinomas **15 ml/m$^3$**: reduced survival<br><br>squamous cell carcinomas<br>**0** ml/m$^3$ 0/90 (0 %)<br>**0.7** ml/m$^3$ 0/90 (0 %)<br>**2** ml/m$^3$ 0/90 (0 %)<br>**6** ml/m$^3$ 1/90 (1.1 %)<br>**10** ml/m$^3$ 29/90 (32 %)<br>**15** ml/m$^3$ 69/147 (47 %) | Monticello *et al.* 1996 |
| **rat**, F344, groups of 120 ♂ and 120 ♀ | **24 months,** 6 h/day, 5 days/week, 0, 2.0, 5.6, 14.3 ml/m$^3$, some animals killed after 6, 12, 18, 24, 27, 30 months | **2.0 ml/m$^3$**: rhinitis, hyperplasia and even papillary adenomas, squamous cell metaplasia **from 5.6 ml/m$^3$**: slightly reduced body weight gains, squamous cell carcinomas<br><br>      rhinitis      adenomas      squamous cell carcinomas<br>**0** ml/m$^3$ 0/40 (0 %) 1/232 (0.4 %) 0/232 (0 %)<br>**2.0** ml/m$^3$ 2/40 (5 %) 8/236 (3.4 %) 0/236 (0 %)<br>**5.6** ml/m$^3$ 7/40 (17.5 %) 6/235 (2.6 %) 2/235 (0.9 %)<br>**14.3** ml/m$^3$ 39/40 (97.5 %) 5/232 (2.2 %) 103/232 (44 %) | Kerns *et al.* 1983, Swenberg *et al.* 1980 |

**Table 4.** continued

| Species, strain, number of animals per sex and group | Exposure | Effects in the nasal region (unless stated otherwise) | References |
| --- | --- | --- | --- |
| **rat**, Wistar, groups of 30 ♂ | **28 months** 6 h/day, 5 days/week 0, 0.1, 1.0, 10 ml/m$^3$ | **1 ml/m$^3$**: no effects (NOAEL) with undamaged mucosa; in animals with damaged mucosa (as a result of electrocoagulation) rhinitis, hyperplasia and metaplasia<br>**10 ml/m$^3$**: reduced body weight gains, rhinitis, hyperplasia, squamous cell carcinomas | Woutersen *et al.* 1989 |
| **mouse**, B6C3F$_1$, groups of 10 ♂ and 10 ♀ | **13 weeks,** 6 h/day, 5 days/week, 0, 2, 4, 10, 20, 40 ml/m$^3$ | **4 ml/m$^3$**: NOAEL<br>**from 10 ml/m$^3$**: squamous cell metaplasia<br>**from 20 ml/m$^3$**: reduced body weight gains<br>**from 40 ml/m$^3$**: increased mortality, inflammation, keratinization | Maronpot *et al.* 1986 |
| **mouse,** C57BL/6×C3H, groups of 120 ♂ and 120 ♀ | **24 months,** 6 h/day, 5 days/week, 0, 2.0, 5.6, 14.3 ml/m$^3$, some animals killed after 6, 12, 18, 24, 27, 30 months | **from 2.0 ml/m$^3$**: rhinitis, epithelial hyperplasia<br>**from 5.6 ml/m$^3$**: dysplasia, metaplasia, atrophy of the olfactory epithelium<br>**14.3 ml/m$^3$**: squamous cell carcinomas 2/240 (0.8 %) | Kerns *et al.* 1983 |
| **hamster,** Syrian golden hamster, groups of 10 ♂ and 10 ♀ | **26 weeks,** 22 h/days, 7 days/week, 0, 0.2, 1.0, 3.0 ml/m$^3$ | **3 ml/m$^3$**: NOAEL | Rusch *et al.* 1983a, 1983b |
| **hamster,** Syrian golden hamster, 132 ♂ (0 ml/m$^3$), 88 ♂ (10 ml/m$^3$) | **116 weeks** (lifetime), 5 h/day, 5 days/week, 0, 10 ml/m$^3$ | **10 ml/m$^3$**: reduced survival, slightly increased incidences of hyperplasia and metaplasia (4/88, 5 %), no rhinitis, no tumours of the respiratory tract | Dalbey 1982 |
| **monkey**, rhesus, groups of 3 ♂ | **1 or 6 weeks,** 6 h/day, 5 days/week, 0, 6 ml/m$^3$ | **6 ml/m$^3$**: eye irritation, epithelial hyperplasia of the nose, squamous cell metaplasia, inflammation | Monticello *et al.* 1989 |
| **monkey,** cynomolgus, groups of 6 ♂ | **26 weeks,** 22 h/day, 7 days/week, 0, 0.2, 1.0, 3.0 ml/m$^3$ | **0.2 ml/m$^3$**: NOAEL<br>**1 ml/m$^3$**: squamous cell hyperplasia and metaplasia (1/6)<br>**3 ml/m$^3$**: squamous cell hyperplasia and metaplasia (6/6) | Rusch *et al.* 1983a, 1983b |

In rats exposed to formaldehyde concentrations of 10 ml/m$^3$, daily for 6 hours on 5 days a week, rhinitis, hyperplasia and squamous metaplasia of the respiratory epithelium of the nasal mucosa were described in all studies. In rats exposed to 1.0 ml/m$^3$ for 2 years no histopathological changes were observed (no observed adverse effect level (NOAEL)) (Woutersen *et al.* 1989). From concentrations of 2 ml/m$^3$, rhinitis, epithelial dysplasia and even papillomatous adenomas and squamous metaplasia of the respiratory epithelium of the nose were found, from 6 ml/m$^3$ squamous cell carcinomas (see also Section 5.6 Carcinogenicity; Kerns *et al.* 1983, Swenberg *et al.* 1980). At this concentration also the cell proliferation rate in the nasal mucosa was increased transiently, and from 10 ml/m$^3$ increased permanently (Monticello *et al.* 1996).

Uninterrupted exposure of rats for 8 hours/day ("continuous") was compared with 8 exposures for 30 minutes followed by a 30-minute phase without exposure ("intermittent") in two 13-week studies with the same total dose. Effects were seen only after intermittent exposure to formaldehyde concentrations of 4 ml/m$^3$, but not after continuous exposure to 2 ml/m$^3$. The authors concluded that the toxicity in the nose depends on the concentration and not on the total dose (Wilmer *et al.* 1989).

In mice exposed to formaldehyde concentrations of 2 ml/m$^3$ for 2 years (6 hours/day, 5 days/week), rhinitis and epithelial hyperplasia was observed, from 6 ml/m$^3$ dysplasia, metaplasia and atrophy. Squamous cell carcinomas were observed only after concentrations of 14.3 ml/m$^3$ and above (Kerns *et al.* 1983).

In hamsters exposed to formaldehyde concentrations of 10 ml/m$^3$ (5 hours/day, 5 days per week) for life, survival was reduced and the incidence of hyperplasia and metaplasia (4/88, 5 %) was slightly increased, but not that of tumours (Dalbey 1982).

In cynomolgus monkeys exposed almost continuously to formaldehyde concentrations of 1 or 3 ml/m$^3$ for 26 weeks, metaplasia and hyperplasia were observed in 1/6 and 6/6 animals. In the animals exposed to concentrations of 0.2 ml/m$^3$, no histopathological changes were found (Rusch *et al.* 1983a, 1983b).

## 5.4 Systemic effects

Reduced body weight gains were reported in rats exposed to formaldehyde concentrations from 10 ml/m$^3$ for 6 hours a day in a 13-week inhalation study (Woutersen *et al.* 1987) and in those exposed to concentrations from 5.6 ml/m$^3$ in a 2-year inhalation study (Kerns *et al.* 1983, Swenberg *et al.* 1980). In mice, reduced body weight gains were found in a 13-week inhalation study only at concentrations from 20 ml/m$^3$. Other systemic effects were not observed in these studies. Only in a 26-week inhalation study with continuous exposure (22 hours a day, 7 days a week) were—in addition to reduced body weight gains and lesions in the nasal region—reduced absolute and relative liver weights observed from concentrations as low as 3 ml/m$^3$ (Rusch *et al.* 1983a, 1983b).

# 5.5 Genotoxicity

### 5.5.1 *In vitro*

Genotoxic and mutagenic effects of formaldehyde were found in various *in vitro* test systems. A reactive compound, formaldehyde reacts with nucleic acids and proteins. DNA adducts, DNA-protein crosslinks, strand breaks and the induction of repair were detected *in vitro*. Formaldehyde also produced back mutation and forward mutation in bacteria. High concentrations of formaldehyde (4 mM) produced insertions, deletions and point mutations in GC base pairs in the gpt gene of *Escherichia coli* (Crosby *et al.* 1988). Gene mutations were detected also in lymphoblasts treated with formaldehyde (Liber *et al.* 1989). Most of the mutations were AT → CG transversions at specific sites. Tests with V79 cells from the Chinese hamster, on the other hand, showed that although cytotoxicity parallels sister chromatid exchange (SCE) and micronucleus formation results from the formation of DNA-protein crosslinks, no gene mutation occurred (Merk and Speit 1998). Chromosomal aberrations (Natarajan *et al.* 1983) and SCE (Schmidt *et al.* 1986) were reported. Thus the mutagenic effects of formaldehyde are well-documented from *in vitro* studies.

### 5.5.2 *In vivo*

The results of the *in vivo* studies are more difficult to evaluate. Of particular importance in this context is the question of whether cytogenetic effects can only occur as the result of local exposure or also as the result of the systemic availability of formaldehyde (see Section 3 Toxicokinetics and Metabolism). In the rat, chromatid breaks are described in cells from lung lavage after inhalation exposure (Dallas *et al.* 1992) and micronuclei in the gastrointestinal epithelium after gavage of formaldehyde (Migliore *et al.* 1989).

The incidence of micronuclei was increased in cells of the nasal mucosa in non-smoking workers of a plywood factory relative to that in controls. In this case, however, there was mixed exposure to formaldehyde and wood dust (Ballarin *et al.* 1992) and the effects did not correlate with the dose.

In cultures of lymphocytes isolated from blood samples from persons exposed to formaldehyde concentrations of up to 9.8–11.0 mg/m$^3$ in pathological laboratories, the incidences of chromosomal aberrations and SCE were not increased (Thomson *et al.* 1984). In workers from a paper factory, the incidence of chromosomal aberrations was increased, but not that of SCE (Bauchinger and Schmid 1985). A slight increase in SCE was observed in the lymphocytes of students of an anatomy course compared to the values before the beginning of the course in which they were exposed to average formaldehyde concentrations of 1.2 ml/m$^3$ (Yager *et al.* 1986). In all three studies the number of cases was small (6, 20, and 8) and the effects slight.

## 5.5.3 DNA-protein crosslinks

*In vivo*, DNA-protein crosslinks were detected in the epithelium of sections of the trachea (Cosma *et al.* 1988) and in the nasal epithelium of rats (Casanova and Heck 1987, Casanova *et al.* 1989, 1994, Casanova-Schmitz and Heck 1983, Casanova-Schmitz *et al.* 1984a, Heck and Casanova 1995, Lam *et al.* 1985). In monkeys, the levels of DNA-protein crosslinks were highest in the mucosa of the middle turbinates; lower concentrations were produced in the anterior lateral wall/septum and nasopharynx. Very low concentrations were found in the larynx, trachea and *carina tracheae* and in the proximal portions of the major bronchi  (Casanova *et al.* 1991). The incidences of DNA-protein crosslinks varied widely in the various regions of the nasal cavity, and in the monkey in the deeper sections of the respiratory passages (Casanova *et al.* 1991, 1994). The distribution of DNA-protein crosslinks correlated with the probability of deposition of formaldehyde dictated by the anatomy and physiology of the various sections of the nose (Hubal *et al.* 1997).

In the nasal epithelium of F344 rats, DNA-protein crosslinks were still detected at formaldehyde concentrations as low as 0.3 ml/m$^3$ (Casanova *et al.* 1994). In the experiments with rhesus monkeys they were also found at the lowest concentration of 0.7 ml/m$^3$ (Casanova *et al.* 1991).

In the DNA of white blood cells from persons occupationally exposed to formaldehyde (average concentrations determined by personal air sampling: 2.8–3.1 ml/m$^3$), the incidence of DNA-protein crosslinks was significantly higher than in control persons (p = 0.03). Assuming that formaldehyde reaches the blood cells via the lungs, it was suggested DNA-protein crosslinks be used as a biomarker for exposure to formaldehyde (Shaham *et al.* 1996). Because of methodological shortcomings, this study has, however, been heavily criticized (the blood samples were allowed to stand for 3 hours, the intraindividual and analytical variability were not determined, formaldehyde-induced DNA-protein crosslinks and DNA-protein crosslinks of other genesis were not differentiated; Casanova *et al.* 1996); the findings cannot, therefore, be evaluated definitively.

DNA-protein crosslinks induced by formaldehyde can be removed by repair. Half-lives of 2 to 4 hours have been reported. Accordingly, DNA-protein crosslinks can usually no longer be detected 24 hours after exposure (Cosma and Marchok 1988, Craft *et al.* 1987, Grafström and Harris 1983, Grafström *et al.* 1984, Magaña-Schwenke and Moustacchi 1980, Merk and Speit 1998). In sections of the tracheal epithelium of rats, the DNA-protein crosslinks had been almost completely removed within 48 to 72 hours after the treatment, depending on the concentration of the instilled aqueous formaldehyde solutions (1.7–66.7 mM) (Cosma *et al.* 1988). This corresponds to a half-life of about 7 hours. Histological examination revealed hyperproliferation in the tracheal epithelium. The accumulation of DNA-protein crosslinks was investigated; because of the methods used, however, the results cannot be evaluated conclusively (Casanova *et al.* 1994).

Using a physiologically based pharmacokinetic model, it was calculated that in man fewer DNA-protein crosslinks are formed in the nasal mucosa than in the rat or monkey (Casanova *et al.* 1991).

### 5.5.4 p53 mutations

In a long-term inhalation study with rats published by Monticello *et al.* (1996), point mutations were found in the p53 gene in 5 of 11 nasal tumours. The tumours expressed only the mutated gene. The role of formaldehyde in causing these mutations is, however, unclear (Recio *et al.* 1992). p53 mutations have been detected in man in tumours of various genesis. In rodents, however, they are rare (Wolf *et al.* 1995). Often the mutations are produced secondarily during the promotion or progression phase. The heterogeneous spectrum of mutations in the nasal tumours does not suggest direct induction by a genotoxic substance.

### 5.5.5 Toxic effects on germ cells

The sperm count, sperm morphology and the occurrence of fluorescent bodies were investigated in 11 employees who carried out autopsies and were exposed to average formaldehyde concentrations of 0.61 to 1.32 ml/m$^3$. No significant differences from the controls were found (Ward *et al.* 1983, 1984). The exposure levels were, however, low and the number of persons investigated small.

The toxic effects of formaldehyde on germ cells have been demonstrated in numerous tests with *Drosophila* (Alderson 1965, Herkowitz 1953, 1959), in particular after administration with the diet, and were limited to effects on early spermatocytes of the larvae (see IARC 1982). Gaseous formaldehyde had no effect. In tests for mosaic mutations in *Drosophila* and in the Müller-5 test for recessive lethal mutations, formaldehyde yielded positive results (Szabad *et al.* 1983). In a comparative test with the unstable zeste-white assay in *Drosophila melanogaster*, formaldehyde produced somatic mutations, but no germ cell mutations (Rasmuson and Larsson 1992). *In vitro*, during the reaction of formaldehyde with adenosine, an $N^6$-hydroxymethyl adduct was produced. This kind of nucleoside modification is thought to have marked germ-cell-stage-specific mutagenic effects in male *Drosophila* larvae (Alderson 1985).

Few studies have been carried out with mammals. In a review of the dominant lethal test, formaldehyde is listed with substances for which premature death of the foetuses and preimplantation losses were within the control range (Epstein *et al.* 1972). In mice (Q strain) given single intraperitoneal injections of a 35 % formaldehyde solution (dose: 50 mg/kg body weight) no chromosomal changes were found in the metaphase I spermatocytes (Fontignie-Houbrechts 1981). In the dominant lethal test, the number of preimplantation and postimplantation losses in the first week of mating was twice the control value (Fontignie-Houbrechts 1981). In albino rats, marked dose-dependent effects were observed in the dominant lethal test in mating weeks one to three after intraperitoneal administration of 0.125 to 0.5 mg/kg body weight (1/4 to 1/16 of the lethal dose) in the form of a 37 % solution stabilized with 10 % methanol. Also the fertility of the treated male animals decreased in a dose-dependent manner (Odeigah 1997). In another test the authors found an increase in the number of abnormal sperm.

Formaldehyde can therefore be regarded as a potential germ cell mutagen in rodents, with mutagenic effects when it reaches the target organ and the target structures in sufficient amounts, as was demonstrated in the dominant lethal test with intraperitoneal injection of high-percentage solutions. Exposure to exogenous formaldehyde at levels which do not significantly increase the endogenous bioavailability of the substance is not expected to produce mutagenic effects on the germ cells. This corresponds also with the estimation of the US Agency for Toxic Substances and Disease Registry (ATSDR 1999) that the results of studies with man and animals make it seem very unlikely that low level exposure to formaldehyde causes developmental or reproductive damage.

## 5.6 Carcinogenicity

### 5.6.1 Inhalation

In a 2-year inhalation study with F344 rats, squamous cell carcinomas of the nose were observed. Exposure was to formaldehyde concentrations of 0, 2.0, 5.6 and 14.3 ml/m$^3$, for 6 hours/day on 5 days/week. All the animals exposed to formaldehyde developed rhinitis, epithelial dysplasia and metaplasia in the nasal cavity. After 18 months, 15/40 animals of the high exposure group had developed hyperplasia. In all the groups exposed to formaldehyde, metaplasia preceded dysplasia. If the exposure was interrupted for longer than 3 months, the rhinitis and metaplasia began to regress. After 24 months, squamous cell carcinomas were found in the nasal cavities only in the middle dose group (0.9 %) and in the high dose group (44 %). In the high dose group, also undifferentiated carcinomas and sarcomas were found. Also the number of polypoid adenomas was slightly increased in the male animals. The total tumour incidence in the high dose group was 48.7 % (Kerns *et al.* 1983, Swenberg *et al.* 1980; see Table 4). The formation of nasal tumours in the rat after high level exposure to formaldehyde ($\geq$ 6 ml/m$^3$) has been confirmed in other studies (Feron *et al.* 1988, Monticello *et al.* 1996, Woutersen *et al.* 1989; see Table 4).

In another long-term study over 28 months, F344 rats were exposed to formaldehyde concentrations of 0, 0.3, 2.0 and 15 ml/m$^3$ for 6 hours/day, 5 days/week. Although keratinizing squamous cell carcinomas were found only in the high dose group (in 13 of 32 animals), the incidence of epithelial hyperplasia and metaplasia of the nasal respiratory mucosa was significantly increased in all exposed groups. As inflammatory infiltration of the nasal mucosa, erosion and oedema were described in both the controls and the exposed animals, the possibility cannot be excluded that the hyperplasia and metaplasia were caused by the interaction of formaldehyde and inflammatory damage to the nasal mucosa (Kamata *et al.* 1997). Therefore this study cannot be included in the present assessment.

In a 2-year inhalation study with B6C3F$_1$ mice exposed to formaldehyde concentrations of 0, 2.0, 5.6 or 14.3 ml/m$^3$ for 6 hours/day on 5 days/week, squamous cell carcinomas of the nasal cavity were found in only 2/240 animals (0.8 %) of the high dose group. Epithelial metaplasia and dysplasia of the respiratory epithelium were, however, also observed (Kerns *et al.* 1983).

In the hamster exposed to concentrations of 10 or 30 ml/m³, no tumours were found (Dalbey 1982, IARC 1995, WHO 1989) and the incidence of non-neoplastic changes of the nasal epithelium was low.


## 5.6.2 Oral administration

Formaldehyde was administered with the drinking water for 2 years to Wistar rats in doses of 0, 10, 50 or 300 mg/kg body weight and day (Tobe *et al.* 1989) and 0, 1.2, 15 or 82 mg/kg body weight and day for male animals and 0, 1.8, 21 or 109 mg/kg body weight and day for female animals (Til *et al.* 1989). No changes were produced with doses up to 10 mg/kg body weight and day, and 15 and 21 mg/kg body weight and day, respectively. In almost all animals given doses from 50 mg/kg body weight, and 82 and 109 mg/kg body weight, histopathological changes in the forestomach (hyperplasia, keratinization) and inflammation and ulcers of the glandular stomach were found. In addition, at doses of 82 and 109 mg/kg body weight and day, food and liquid consumption, and body weight gains were reduced. There was no increase in the incidence of tumours (Tobe *et al.* 1989, Til *et al.* 1989). Til *et al.* note, however, that some of the histopathological changes they classified as hyperplasia could have been classified as papillomas by other pathologists. In the study of Til *et al.* (1989), also renal changes (increased relative kidney weights, necrosis), and changes in the composition of the urine were observed in the female animals of the high dose group; the authors attribute this to the reduced drinking-water consumption.

In another drinking-water study, formaldehyde was administered to 7-week-old male and female Sprague-Dawley rats for 104 weeks in concentrations of 0, 10, 50, 100, 500, 1000 or 1500 mg/l drinking-water. In addition, 25-week-old male and pregnant female animals, and later their offspring were given formaldehyde in concentrations of 0 or 2500 mg/l. Reduced body weights were found only in the animals (offspring) exposed from the embryonal phase. In the animals of the groups exposed to formaldehyde concentrations of 50 mg/l and above and the animals of the 2500 mg/l group, the incidence of leukaemia (lymphoblastic leukaemia, lymphosarcomas) was increased in a dose-dependent manner (controls 3.5 %, 10 mg/l: 3.0 %, 50 mg/l: 9 %, 500 mg/l: 12 %, 1000 mg/l: 13 %, 1500 mg/l: 18 %, 2500 mg/l: 11.1 %). Data for the statistical significance of the findings or for the historical controls were not given by the authors (Sofritti *et al.* 1989). Despite the criticism of this study (see below), the IARC (1995) regarded these data as being dose-dependent and significantly different from the data for the controls. Benign and malignant gastrointestinal tumours, which according to Sofritti *et al.* are very rare in this strain of rat (all incidences ≤ 0.1 %), were increased in particular in the animals of the following groups: 1000 mg/l (1 %: leiomyosarcomas), 1500 mg/l (2 %: adenomas) and 2500 mg/l (parent animals: 2.8 %: papillomas and acanthomas, 2.8 %: adenocarcinomas; offspring: 1.4 %: adenomas, 1.4 %: squamous cell carcinomas, 1.4 %: adenocarcinomas, 2.7 %: leiomyosarcomas) (Sofritti *et al.* 1989). The validity of this study has been questioned as a result of its content and the methods used (Feron *et al.* 1990).

# 6 Manifesto (MAK value/classification)

In the supplement to the MAK documentation for formaldehyde from 1987 it was assumed that the occurrence of tumours in the nasal mucosa of rats and mice may be the result of chronic proliferative processes caused by the cytotoxic effects of the substance. Evaluation of the data now available for the carcinogenic effects confirms this assumption. It was found that the dose–response relationships for all the parameters investigated, such as damage to the nasal epithelium, cell proliferation, tumour incidence and also the formation of DNA-protein crosslinks, is very flat for low level exposures and becomes much steeper at higher levels of exposure. For all the parameters mentioned, with exception of the formation of DNA-protein crosslinks, concentrations which do not produce effects can be calculated from the study results. The possibility of formation of DNA-protein crosslinks can therefore not be excluded even with low levels of exposure. The data suggest, however, that with concentrations that do not lead to cell proliferation, only few DNA-protein crosslinks are formed. In addition, the physiological proliferation rate in the respiratory epithelium is low, and as long as this is not increased (which requires exposure to concentrations of more than 2 ml/m$^3$), the probability that DNA-protein crosslinks are transformed into mutations is low. In the low dose range, which does not lead to an increase in cell proliferation, the Commission therefore considers that the genotoxicity of formaldehyde plays no or at most a minor part in its carcinogenic potential so that no significant contribution to human cancer risk is expected. This conclusion is supported by the results of a risk assessment which, for persons exposed to concentrations of 0.3 ml/m$^3$ at the workplace for 40 years, yielded a very low additional cancer risk for non-smokers of $1.3 \times 10^{-8}$ and for smokers of $3.8 \times 10^{-7}$ (CIIT 1999). Formaldehyde is therefore classified in Carcinogen category 4 of Section III of the *List of MAK and BAT Values.*

For the evaluation of the MAK value for formaldehyde, which also takes the carcinogenic risk into account, the avoidance of cell proliferation is therefore decisive. The cause of cell proliferation is the irritative effect of formaldehyde on the upper respiratory tract. For this parameter, however, the database is insufficient for the establishment of a MAK value. However, the database for the irritative effects of formaldehyde on the eye, a more sensitive parameter, is sufficient. On the basis of the available studies with volunteers it was concluded (Paustenbach *et al.* 1997) that with daily exposure for 8 hours to maximum formaldehyde concentrations of 0.3 ml/m$^3$ practically all workers are protected against eye irritation. The MAK value has therefore been reduced to 0.3 ml/m$^3$. The classification in Peak limitation category I has been retained, with an excursion factor of 2, which ensures that in the short-term no irritation or only slight irritation occurs. The momentary value of 1 ml/m$^3$ should not be exceeded, as from this concentration irritative effects are to be expected.

During exposure to mixtures of substances, care must be taken that no irritative effects occur.

As a result of the local effects of formaldehyde, designation with an "H" (for substances for which there is danger from cutaneous absorption) is not necessary.

Because of the sensitizing effects of formaldehyde on the skin, designation with an "Sh" has been retained.

Classification in Pregnancy risk group C is still valid.

For the classification of formaldehyde as a germ cell mutagen, the question of systemic availability of effective concentrations of formaldehyde after exposure at the workplace must be taken into consideration. In experimental exposures, even with higher levels of exposure, the equilibrium concentration maintained in the blood by endogenous formaldehyde was not found to be increased. As a result of the rapid metabolism it is also very unlikely that even moderate external exposure levels of formaldehyde can increase the systemically available concentration noticeably. Given observance of the MAK value of 0.3 ml/m$^3$, external exposure to formaldehyde is therefore not expected to make a noticeable contribution to the genetic risk for man. Formaldehyde is therefore classified in Germ cell mutagenicity category 5.

# 7 References


Akbar-Khanzadek F, Mlynek JS (1997) Changes in respiratory function after one and three hours of exposure to formaldehyde in non-smoking subjects. *Occup Environ Med 54*: 296–300

Akbar-Khanzadek F, Vaquerano MU, Akbar-Khanzadek M, Bisesi MS (1994) Formaldehyde exposure, acute pulmonary responses, and exposure control options in a gross anatomy laboratory. *Am J Ind Med 26*: 61–75

Albert RE, Sellakumar AR, Laskin S, Kuschner M, Nelson N, Snyder CA (1982) Gaseous formaldehyde and hydrogen chloride induction of nasal cancer in the rat. *J Nat Cancer Inst 68*: 597–603

Alderson T (1965) Chemically induced delayed germinal mutation in *Drosophila*. *Nature 207*: 164–167

Alderson T (1985) Formaldehyde-induced mutagenesis: a novel mechanism for its action. *Mutat Res 154*: 101–110

Andersen I, Mølhave L (1983) Controlled human studies with formaldehyde. in: Gibson E (Ed.) *Formaldehyde Toxicity*, Hemisphere, Washington DC, USA, 154–165

Appelman LM, Woutersen RA, Zwart RA, Falke HE, Feron VJ (1988) One-year inhalation toxicity study of formaldehyde in male rats with damaged or undamaged nasal mucosa. *J Appl Toxicol 8*: 85–90

ATSDR (Agency for Toxic Substances and Disease Registry) (1999) Public Health Service, US Department of Health & Human Services: *Toxicological profile for formaldehyde*, 251–262

Ballarin C, Sarto F, Giacomelli L, Bartolucci GB, Clonfero E (1992) Micronucleated cells in nasal mucosa of formaldehyde-exposed workers. *Mutat Res 280*: 1–7

Barrow CS, Buckley LA, James RA, Steinhagen WH, Chang JCF (1986) Sensory irritation: studies on correlation to pathology, structure-activity, tolerance development, and prediction of species differences to nasal injury. in: Barrow CS (Ed.) *Toxicology of the Nasal Passages*, Hemisphere Publishing Corp., Washington, New York, London, 101–120

Barrow CS, Steinhagen WH, Chang JCF (1983) Formaldehyde sensory irritation. in: Gibson JE (Ed.) *Formaldehyde toxicity*, Hemisphere Publishing Corp., Washington, New York, London, 16–25

Bauchinger M, Schmid E (1985) Cytogenetic effects in lymphocytes of formaldehyde workers of a paper factory. *Mutat Res 158*: 195–199

Bedford P, Fox BW (1981) The role of formaldehyde in methylene dimethanesulphonate-induced DNA cross-links and its relevance to cytotoxicity. *Chem Biol Interact 38*: 119–126

Bender JR, Mullin L, Graepel GJ, Wilson WE (1983) Eye irritation response of humans to formaldehyde. *Am Ind Hyg Assoc J 44*: 463–465

Benyajati C, Place AR, Sofer W (1983) Formaldehyde mutagenesis in *Drosophila*. Molecular analysis of ADH-negative mutants. *Mutat Res 111*: 1–7

Berke JH (1987) Cytologic examination of nasal mucosa in formaldehyde-exposed workers. *J Occup Med 29*: 681–684

Bogdanffy MS, Morgan PH, Starr TB, Morgan KT (1987) Binding of formaldehyde to human and rat nasal mucus and bovine serum albumin. *Toxicol Lett 38*: 145–154

Boysen M, Zadig E, Digernes V, Abeler V, Reith A (1990) Nasal mucosa in workers exposed to formaldehyde: a pilot study. *Br J Ind Med 47*: 116–121

Casanova-Schmitz M, Heck H d'A (1983) Effects of formaldehyde exposure on the extractability of DNA from proteins in the rat nasal mucosa. *Toxicol Appl Pharmacol 70*: 121–132

Casanova M, Heck H d'A (1987) Further studies of the metabolic incorporation and covalent binding of inhaled [$^3$H]- and [$^{14}$C]formaldehyde in Fischer 344 rats: effects of glutathione depletion. *Toxicol Appl Pharmacol 89*: 105–121

Casanova-Schmitz M, Starr TB, Heck H d'A (1984a) Differentiation between metabolic incorporation and covalent binding in the labeling of macromolecules in the rat nasal mucosa and bone marrow by inhaled [$^{14}$C]- and [$^3$H]formaldehyde. *Toxicol Appl Pharmacol 76*: 26–44

Casanova-Schmitz M, David RM, Heck H d'A (1984b) Oxidation of formaldehyde and acetaldehyde by NAD$^+$-dependent dehydrogenases in rat nasal mucosal homogenates. *Biochem Pharmacol 33*: 1137–1142

Casanova M, Heck H d'A, Everitt JI, Harrington WW, Popp Jr JA (1988) Formaldehyde concentrations in the blood of rhesus monkeys after inhalation exposure. *Food Chem Toxicol 26*: 715–716

Casanova M, Deyo DF, Heck H d'A (1989) Covalent binding of inhaled formaldehyde to DNA in the nasal mucosa of Fischer 344 rats: analysis of formaldehyde and DNA by high-performance liquid chromatography and provisional pharmacokinetic interpretation. *Fundam Appl Toxicol 12*: 397–417

Casanova M, Morgan KT, Steinhagen WH, Everitt JI, Popp JA, Heck H d'A (1991) Covalent binding of inhaled formaldehyde to DNA in the respiratory tract of rhesus monkeys: pharmacokinetics, rat-to-monkey interspecies scaling, and extrapolation to man. *Fundam Appl Toxicol 17*: 409–428

Casanova M, Morgan KT, Gross EA, Moss OR, Heck H d'A (1994) DNA-protein cross-links and cell replication at specific sites in the nose of F344 rats exposed subchronically to formaldehyde. *Fundam Appl Toxicol 23*: 525–536

Casanova M, Heck H d'A, Janszen D (1996) Comments on "DNA-protein crosslinks, a biomarker of exposure to formaldehyde – *in vitro* and *in vivo* studies" by Shaham *et al. Carcinogenesis 17*: 2097–2101

Cassee FR (1995) *Upper respiratory tract toxicity of mixtures of aldehydes*. Doctoral thesis, University of Utrecht, NL

Cassee FR, Arts JHE, Groten JP, Feron VJ (1996a) Sensory irritation to mixture of formaldehyde, acrolein, and acetaldehyde in rats. *Arch Toxicol 70*: 329–337

Cassee FR, Groten JP, Feron VJ (1996b) Changes in the nasal epithelium of rats exposed by inhalation to mixtures of formaldehyde, acetaldehyde and acrolein. *Fundam Appl Toxicol 29*: 208–218

Chang JCF, Steinhagen WH, Barrow CS (1981) Effects of single or repeated formaldehyde exposures on minute volume of B6C3F$_1$ mice and F344 rats. *Toxicol Appl Pharmacol 61*: 451–459

Chang JCF, Gross EA, Swenberg JA, Barrows CS (1983) Nasal cavity deposition, histopathology and cell proliferation after single or repeated formaldehyde exposures in B6C3F$_1$ mice and F344 rats. *Toxicol Appl Pharmacol 68*: 161–176

Chang JCF, Steinhagen WH, Barrow CS (1984) Sensory irritation tolerance in F344 rats exposed to chlorine or formaldehyde gas. *Toxicol Appl Pharmacol 76*: 319–327

Collins JJ, Acquavella JF, Esmen NA(1997) An updated meta-analysis of formaldehyde exposure and upper respiratory tract cancers. *J Occup Environ Med 39*: 639–651

Conaway CC, Whysner J, Verna LK, Williams GM (1996) Formaldehyde mechanistic data and risk assessment: endogenous protection from DNA adduct formation. *Pharmacol Ther 71*: 29–55

CIIT (Chemical Industry Institute of Toxicology) (1999) *Formaldehyde: hazard characterization and dose-response assessment for carcinogenicity by the route of inhalation* (revised edition). CIIT, Research Triangle Park, NC, USA

Cosma GN, Marchok AC (1988) Benzo[a]pyrene- and formaldehyde-induced DNA damage and repair in rat tracheal epithelial cells. *Toxicology 51*: 309–320

Cosma GN, Wilhite AS, Marchok AC (1988) The detection of DNA-protein cross-links in rat tracheal implants exposed *in vivo* to benzo(a)pyrene and formaldehyde. *Cancer Lett 42*: 13–21

Costa M (1991) DNA-protein complexes induced by chromate and other carcinogens. *Environ Health Perspect 92*: 45–52

Craft TR, Bermudez E, Skopek TR (1987) Formaldehyde mutagenesis and formation of DNA-protein crosslinks in human lymphoblasts *in vitro*. *Mutat Res 176*: 147–155

Crosby RM, Richardson KK, Craft TR, Benforado KB, Liber HL, Skopek TR (1988) Molecular analysis of formaldehyde-induced mutations in human lymphoblasts and *E. coli*. *Environ Mol Mutagen 12*: 155–166

Dalbey WE (1982) Formaldehyde and tumours in hamster respiratory tract. *Toxicology 24*: 9–14

Dallas CE, Scott MJ, Ward JB, Theiss JC (1992) Cytogenetic analysis of pulmonary lavage and bone marrow cells of rats after repeated formaldehyde exposure. *J Appl Toxicol 12*: 199–203

Day JJ, Less REM, Clark RH, Patter PL (1984) Respiratory response to formaldehyde and off-gas of urea formaldehyde foam insulation. *Can Med Assoc J 131*: 1061–1065

ECETOC (European Centre for Ecotoxicology and Toxicology of Chemicals) (1995) *Technical report no. 65: Formaldehyde and human cancer risk*. ISSN-0773-8072-65

Edling C, Ödkvist L, Hellquist H (1985) Formaldehyde and the nasal mucosa. *Br J Ind Med 42*: 570–571

Edling C, Hellquist H, Ödkvist L (1987) Occupational formaldehyde exposure and the nasal mucosa. *Rhinology 25*: 181–187

Edling C, Hellquist H, Ödkvist L (1988) Occupational exposure to formaldehyde and histopathological changes in the nasal mucosa. *Br J Ind Med 45*: 761–765

Epstein SS, Arnold E, Andrea J, Bass W, Bishop Y (1972) Detection of chemical mutagens by the dominant lethal assay in the mouse. *Toxicol Appl Pharmacol 23*: 288–325

Feron VJ, Bruyntjes JP, Woutersen RA, Immel HR, Appelman LM (1988) Nasal tumours in rats after short-term exposure to a cytotoxic concentration of formaldehyde. *Cancer Lett 39*: 101–111

Feron VJ, Til HP, Woutersen RA (1990) Formaldehyde must be considered a multipotential experimental carcinogen. *Toxicol Ind Health 6*: 637–639

Fontignie-Houbrechts N (1981) Genetic effects of formaldehyde in the mouse. *Mutat Res 88*: 109–114

Fornace AJ, Lechner JF, Grafström RC, Harris CC (1982) DNA repair in human bronchial epithelial cells. *Carcinogenesis 3*: 1373–1377

Gorski P, Tarkowski M, Krakowiak A, Kiec-Swierczynska M (1992) Neutrophil chemiluminescence following exposure to formaldehyde in healthy subjects and in patients with contact dermatitis. *Allergol Immunopathol (Madr) 20*: 20–23

Grafström RC, Fornace AJ, Autrup H, Lechner JF, Harris CC (1983) Formaldehyde damage to DNA and inhibition of DNA repair in human bronchial cells. *Science 220*: 216–218

Grafström RC, Fornace A, Harris CC (1984) Repair of DNA damage caused by formaldehyde in human cells. *Cancer Res 44*: 4323–4327

Green DJ, Sauder LR, Kulle TJ, Bascom R (1987) Acute response to 3.0 ppm formaldehyde in exercising healthy nonsmokers and asthmatics. *Am Rev Respir Dis 135*: 1261–1266

Harkema JR (1991) Comparative aspects of nasal airway anatomy: relevance to inhalation toxicology. *Toxicol Pathol 19*: 321–336

Heck H d'A, Casanova-Schmitz M (1984) Biochemical toxicology of formaldehyde. *Rev Biochem Toxicol 6*: 155–189

Heck H d'A, Casanova M (1987) Isotope effects and their implications for the covalent binding of inhaled [3H]- and [14C]formaldehyde in rat nasal mucosa. *Toxicol Appl Pharmacol 89*: 122–134

Heck H d'A, Casanova M (1995) Nasal dosimetry of formaldehyde: modeling site specificity and the effects of preexposure. in: Dugger EL, Schweiter C (Eds) *Nasal toxicity and dosimetry of inhaled xenobiotics: implications for human health*, Taylor & Francis, Washington DC, USA, 159–175

Heck H d'A, Casanova M (1999a) Pharmacodynamics of formaldehyde: arrest of DNA replication by DNA protein crosslinks is the proximal cause of mutation. *Toxicologist 48*: 123 (Abstract)

Heck H d'A, Casanova M (1999b) Pharmacodynamics of formaldehyde: applications of a model for the arrest of DNA replication by DNA-protein cross links. *Toxicol Appl Pharmacol 160*: 86–100

Heck H d'A, White EL, Casanova-Schmitz M (1982) Determination of formaldehyde in biological tissues by gas chromatography/mass spectrometry. *Biomed Mass Spectrom 9*: 347–353

Heck H d'A, Chin TY, Casanova-Schmitz M (1983) Distribution of [$^{14}$C] formaldehyde in rats after inhalation exposure. in: Gibson JE (Ed.) *Formaldehyde toxicity*, Hemisphere Publishing Corporation, Washington DC, USA, 26–37

Heck H d'A, Casanova-Schmitz M, Dodd PB, Schachter EN, Witek TJ, Tosun T (1985) Formaldehyde ($CH_2O$) concentrations in the blood of humans and Fischer-344 rats exposed to $CH_2O$ under controlled conditions. *Am Ind Hyg Assoc J 46*: 1–3

Heck H d'A, Casanova M, Starr TB (1990) Formaldehyde toxicity – new understanding. *CRC Crit Rev Toxicol 20*: 397–426

HEI (Health Effects Institute) (1994) *Development of methods for measuring biological markers of formaldehyde exposure.* Research Report No. 67, June 1994

Herkowitz H (1953) Formaldehyde-induced mutation in mature spermatozoa and early developmental stages of *Drosophila melanogaster. Genetics 38*: 668–669

Herkowitz H (1959) The differential induction of lethal mutations by formalin in the two sexes of *Drosophila. Science 112*: 302–303

Hermann R (1997) *Karzinogenese der Nasenschleimhaut durch N-Nitrosodimethylamin – Inhalation bei der Ratte.* (Carcinogenesis of the nasal mucosa caused by *N*-nitrosodimethylamine-inhalation by the rat.) Inaugural-Dissertation (doctoral thesis) aus den Abteilungen Cytopathologie und Toxikologie und Krebsrisikofaktoren und Krebsprävention am Deutschen Krebsforschungszentrum Heidelberg (German)

Holmquist B, Valle BL (1991) Human liver class III alcohol and glutathione dependent formaldehyde dehydrogenase are the same enzyme. *Biochem Biophys Res Com 178*: 1371–1377

Holmström M, Wilhelmsson B, Hellquist H, Rosen G (1989) Histological changes in the nasal mucosa in persons occupationally exposed to formaldehyde alone and in combination with wood dust. *Acta Otolaryngol 107*: 120–129

Horvath EP, Anderson H, Pierce WE, Hanrahan L, Wendlick JD (1988) Effects of formaldehyde on the mucus membranes and lungs. *J Am Med Assoc 259*: 701–707

Hubal EA, Schlosser PM, Conolly RB, Kimbell JS (1997) Comparison of inhaled formaldehyde dosimetry predictions with DNA-protein cross-link measurements in the rat nasal passages. *Toxicol Appl Pharmacol 143*: 47–55

IARC (International Agency for Research on Cancer) (1982) *Monographs on the Evaluation of the Carcinogenic Risk of Chemicals to Humans, Formaldehyde*, Vol 29, IARC, Lyon

IARC (International Agency for Research on Cancer) (1995) *Monographs on the Evaluation of Carcinogenic Risks to Humans, Wood Dust and Formaldehyde*, Vol 62, IARC, Lyon

Kamata E, Nakadate M, Uchida O, Ogawa Y, Suzuki S, Kaneko T, Saito M, Kurokawa Y (1997) Results of a 28-month chronic inhalation toxicity study of formaldehyde in male Fisher-344 rats. *J Toxicol Sci 22*: 239–254

Kerns WD, Pavkov KL, Donofrio DJ, Gralla EJ, Swenberg JA (1983) Carcinogenicity of formaldehyde in rats and mice after long-term inhalation exposure. *Cancer Res 43*: 4382–4392

Khamgaonkar MB, Fulare MB (1991) Pulmonary effects of formaldehyde exposure – an environmental epidemiology study. *Indian J Chest Dis Allied Sci 33*: 9–13

Kligerman AD, Doerr CL, Tennant AH (1999) Cell cycle specificity of cytogenetic damage induced by 3,4-epoxy-1-butene. *Mutat Res 444*: 151–158

Krakowiak A, Gorski P, Pazdrak K, Ruta U (1998) Airway response to formaldehyde inhalation in asthmatic subjects with suspected respiratory formaldehyde sensitization. *Am J Ind Med 33*: 274–281

Kriebel D, Sama SR, Cocanour B (1993) Reversible pulmonary responses to formaldehyde. *Am Rev Respir Dis 148*: 1509–1515

Kulle TJ (1993) Acute odor and irritation response in healthy nonsmokers with formaldehyde exposure. *Inhal Toxicol 5*: 323–332

Kulle TJ, Sauder LR, Hebel JR, Green DJ, Chatham MD (1987) Formaldehyde dose-response in healthy nonsmokers. *J Air Pollut Control Assoc 37*: 919–924

Lam C-W, Casanova M, Heck H d'A (1985) Depletion of nasal mucosal glutathione by acrolein and enhancement of formaldehyde-induced DNA-protein cross-linking by simultaneous exposure to acrolein. *Arch Toxicol 58*: 67–71

Liber HL, Benforado K, Crosby RM, Simpson D, Skopek TR (1989) Formaldehyde-induced and spontaneous alterations in human hprt DNA sequence and mRNA expression. *Mutat Res 226*: 31–37

Magaña-Schwenke N, Moustacchi E (1980) Biochemical analysis of damage induced in yeast by formaldehyde. III. Repair of induced cross-links between DNA and proteins in the wild-type and in excision-deficient strains. *Mutat Res 70*: 29–35

Maronpot RR, Miller RA, Clarke WJ, Westerberg RB, Decker JR, Moss OR (1986) Toxicity of formaldehyde vapor in B6C3F$_1$ mice exposed for 13 weeks. *Toxicology 41*: 253–266

McLaughlin JK (1994) Formaldehyde and cancer: a critical review. *Int Arch Occup Environ Health 66*: 295–301

Merk O, Speit G (1998) Significance of formaldehyde-induced DNA-protein crosslinks for mutagenesis. *Environ Mol Mutagen 32*: 260–268

Migliore L, Ventura L, Barale, Loprieno N, Castellino S, Pulci R (1989) Micronuclei and nuclear anomalies induced in the gastro-intestinal epithelium of rats treated with formaldehyde. *Mutagenesis 4*: 327–334

Miller CA, Costa M (1989) Analysis of proteins cross-linked to DNA after treatment of cells with formaldehyde, chromate, and cis-diamminedichloroplatinum (II). *Mol Toxicol 2*: 11–26

Miller CA, Costa M (1990) Immunodetection of DNA-protein crosslinks by slot blotting. *Mutat Res 234*: 97–106

Monticello TM (1990) *Formaldehyde-induced pathology and cell proliferation*. Doctoral thesis, Duke University, USA

Monticello TM, Morgan KT (1989) Cell kinetics and characterization of preneoplastic lesions in nasal epithelium of rats exposed to formaldehyde. *Proc Am Assoc Cancer Res 30*: Abstract 772

Monticello TM, Morgan KT (1990) Cell proliferation in rat nasal respiratory epithelium following three months exposure to formaldehyde gas. *Toxicologist 10*: Abstract 724

Monticello TM, Morgan KT, Everitt JI, Popp JA (1989) Effects of formaldehyde gas on the respiratory tract of rhesus monkeys. *Am J Pathol 134*: 515–527

Monticello TM, Miller FJ, Morgan KT (1991) Regional increases in rat nasal epithelial cell proliferation following acute and subchronic inhalation of formaldehyde. *Toxicol Appl Pharmacol 111*: 409–421

Monticello TM, Miller FJ, Swenberg JA, Starr TB, Gibson JE, Morgan KT (1992) Association of enhanced cell proliferation and nasal cancer in rats exposed to formaldehyde. *Toxicologist 12*: Abstract 1017

Monticello TM, Swenberg JA, Gross EA, Leininger JR, Kimbell JS, Seilkop S, Starr TB, Gibson JE, Morgan KT (1996) Correlation of regional and nonlinear formaldehyde-induced nasal cancer with proliferating populations of cells. *Cancer Res 56*: 1012–1022

Morgan KT, Gross EA, Patterson DL (1986) Responses of the nasal mucociliary apparatus of F-344 rats to formaldehyde gas. *Toxicol Appl Pharmacol 82*: 1–13

Natarajan AT, Darroudi F, Bussman CJM, van Kasteren-van Leeuwen AC (1983) Evaluation of the mutagenicity of formaldehyde in mammalian cytogenetic assays *in vivo* and *in vitro*. *Mutat Res 122*: 355–360

Odeigah PGC (1997) Sperm head abnormalities and dominant lethal effects of formaldehyde in albino rats. *Mutat Res 389*: 141–148

Olin KL, Cherr GN, Rifkin E, Keen CL (1996) The effects of some redox-active metals and reactive aldehydes on DNA-protein cross-links *in vitro*. *Toxicology 110*: 1–8

Paustenbach D, Alarie Y, Kulle T, Schachter N, Smith R, Swenberg J, Witschi H, Harowitz SB (1997) A recommended occupational exposure limit for formaldehyde based on irritation. *J Toxicol Environ Health 50*: 217–263

Pazdrak K, Gorski P, Krakowiak A, Ruta U (1993) Changes in nasal lavage fluid due to formaldehyde inhalation. *Int Arch Occup Environ Health 64*: 515–519

Rasmuson A, Larsson J (1992) Somatic and germline mutagenesis assayed by the unstable zeste-white test in *Drosophila melanogaster*. *Mutagenesis 7*: 219–223

Recio L, Sisk S, Pluta L, Bermudez E, Gross EA, Chen Z, Morgan K (1992) p53 Mutations in formaldehyde-induced nasal squamous cell carcinoma in rats. *Cancer Res 52*: 6113–6116

Reiche K, Müller C, Börngen K (1992) Partielle Reversibilität der morphologischen und funktionellen Veränderungen der Nasenschleimhaut nach Beendigung langjähriger Exposition gegenüber Formaldehyd (Partial reversibility of the morphological and functional changes in the nasal mucosa in the period after years of exposure to formaldehyde). *Zbl Arbeitsmed 42*: 182–186 (German)

Reuzel PGJ, Wilmer JWGM, Woutersen RA, Zwart A (1990) Interactive effects of ozone and formaldehyde on the nasal respiratory lining epithelium in rats. *J Toxicol Environ Health 29*: 279–292

Roemer E, Anton HJ, Kindt R (1993) Cell proliferation in the respiratory tract of the rat after acute inhalation of formaldehyde or acrolein. *J Appl Toxicol 13*: 103–107

Rusch GM, Bolte HF, Rinehart WE (1983a) A 26-week inhalation toxicity study with formaldehyde in the monkey, rat and hamster. in: Gibson JE (Ed.) *Formaldehyde toxicity*, Hemisphere, New York, 98–110

Rusch GM, Clary JJ, Bolte HF, Rinehart WE (1983b) A 26-week inhalation toxicity study with formaldehyde in the monkey, rat, and hamster. *Toxicol Appl Pharmacol 68*: 329–343

Saladino AJ, Willey JC, Lechner JF, Grafström RC, LaVeck M, Harris CC (1985) Effects of formaldehyde, acetaldehyde, benzoyl peroxide, and hydrogen peroxide on cultured normal human bronchial epithelial cells. *Cancer Res 45*: 2522–2526

Sauder LR, Chathan MD, Green DJ, Kulle TJ (1986) Acute pulmonary response to formaldehyde exposure in healthy subjects. *J Occup Med 28*: 420–424

Sauder LR, Green DJ, Chatham MD, Kulle TJ (1987) Acute pulmonary response of asthmatics to 3.0 ppm formaldehyde. *Toxicol Ind Health 3*: 569–578

Schachter EN, Witek TJ, Brody DJ, Tosun T, Beck GJ, Leaderer BP (1986) A study of respiratory effects from exposure to 2 ppm formaldehyde in healthy subjects. *Arch Environ Health 41*: 229–239

Schachter EN, Witek TJ, Tosun T, Beck GJ (1987) A study of respiratory effects from exposure to 2.0 ppm formaldehyde in occupationally exposed workers. *Environ Res 44*: 188–205

Schaper M (1993) Development of a database for sensory irritants and its use in establishing occupational exposure limits. *Am Ind Hyg Assoc J 54*: 488–544

Schmid E, Göggelmann W, Bauchinger M (1986) Formaldehyde-induced cytotoxic, genotoxic and mutagenic response in human lymphocytes and *Salmonella typhimurium*. *Mutagenesis 1*: 427–431

Schuck EA, Stephens ER, Middleton JT (1966) Eye irritation response at low concentrations of irritants. *Arch Environ Health 13*: 570–575

Schüller HM (1989) Nitrosamine-induced nasal cavity carcinogenesis in rodents. in: Feron VJ, Bosland MC (Eds) *Nasal carcinogenesis in rodents: relevance to human health risk*, Pudoc Wageningen, 63–69

Shabarova Z, Bogdanov A (1994) *Advanced organic chemistry of nucleic acids*, VCH, Weinheim, 419–423

Shaham J, Bomstein Y, Meltzer A, Kaufman Z, Palma E, Ribak J (1996) DNA-protein crosslinks, a biomarker of exposure to formaldehyde – *in vitro* and *in vivo* studies. *Carcinogenesis 17*: 121–125

Sofritti M, Maltoni C, Maffei F, Biagi R (1989) Formaldehyde: an experimental multipotential carcinogen. *Toxicol Ind Health 5*: 699–730

Swenberg J, Kerns WD, Mitchell RI, Gralla EJ, Pavkow KL (1980) Induction of squamous cell carcinomas of the rat nasal cavity by inhalation exposure to formaldehyde vapour. *Cancer Res 40*: 3398–3402

Swenberg JA, Gross EA, Randall HW, Barrow CS (1983) The effect of formaldehyde exposure on cytotoxicity and cell proliferation. in: Clary JJ, Gibson JE, Waritz RS (Eds) *Formaldehyde, toxicology epidemiology and mechanisms*, Marcel Dekker Inc., New York, 225–236

Szabad J, Soos I, Polgar G, Hejja G (1983) Testing the mutagenicity of malonaldehyde and formaldehyde by the *Drosophila* mosaic and the sex-linked recessive lethal test. *Mutat Res 113*: 117–133

Tank AW, Weiner H, Thurman JA (1981) Enzymology and subcellular localization of aldehyde oxidation in rat liver. *Biochem Pharmacol 30*: 3265–3275

Thomson EJ, Shackleton S, Harrington JM (1984) Chromosome aberration and sister-chromatid exchange frequencies in pathology staff exposed to formaldehyde. *Mutat Res 141*: 89–93

Til HP, Woutersen RA, Feron VJ, Hollanders VHM, Falke HE (1989) Two-year drinking-water study of formaldehyde in rats. *Food Chem Toxicol 27*: 77–87

Tobe M, Naito K, Kurokawa Y (1989) Chronic toxicity study on formaldehyde administered orally to rats. *Toxicology 56*: 79–86

Uotila L, Koivusalo M (1974) Formaldehyde dehydrogenase from human liver. Purification, properties, and evidence for the formation of glutathione thiol esters by the enzyme. *J Biol Chem 249*: 7653–7663

Voet D, Voet JG (1992) *Aminosäurestoffwechsel*. Biochemie, VCH, Weinheim, 714–716

Ward Jr BJ, Legator MS, Chang LW, Periera MA (1983) Evaluation of occupational exposure to formaldehyde using a battery of tests for genetic damage. *Environ Mutagen 5*: 433–434

Ward Jr JB, Hokanson JA, Smith ER, Chang LW, Pereira MA, Whorton Jr EB, Legator MS (1984) Sperm-count, morphology and fluorescent body frequency in autopsy service workers exposed to formaldehyde. *Mutat Res 130*: 417–424

Weber-Tschopp A, Fischer T, Grandjean E (1977) Reizwirkungen des Formaldehyds (HCHO) auf den Menschen. (Irritating effects of formaldehyde in humans). *Int Arch Occup Environ Health 39*: 207–218

WHO (World Health Organisation) (1989) International Programme on Chemical Safety, *Environmental Health Criteria 89, Formaldehyde*, WHO, Geneva

Wilmer JWGM, Woutersen RA, Appelman LM, Leeman WR, Feron VJ (1987) Subacute (4-week) inhalation toxicity study of formaldehyde in male rats: 8-hour intermittent versus 8-hour continuous exposures. *J Appl Toxicol 7*: 15–16

Wilmer JWGM, Woutersen RA, Appelman LM, Leeman WR, Feron VJ (1989) Subchronic (13-week) inhalation toxicity study of formaldehyde in male rats: 8-h hour intermittent versus 8-hour continuous exposures. *Toxicol Lett 47*: 287–293

Witek TJ, Schachter EN, Tosun T, Beck GJ, Leaderer BP (1987) An evaluation of respiratory effects following exposure to 2.0 ppm formaldehyde in asthmatics: lung function, symptoms, and airway reactivity. *Arch Environ Health 42*: 230–237

Wolf DC, Gross EA, Lyght O, Bermudez E, Recio L, Morgan KT (1995) Immunohistochemical localization of p53, PCNA, and TGF-α proteins in formaldehyde-induced rat nasal squamous cell carcinomas. *Toxicol Appl Pharmacol 132*: 27–35

Woutersen RA, Appelman LM, Wilmer JWGM, Falke HE, Feron VJ (1987) Subchronic (13-week) inhalation toxicity study of formaldehyde in rats. *J Appl Toxicol 7*: 43–49

Woutersen RA, van Gardener-Hoetmer A, Bruijntjes JP, Feron VJ (1989) Nasal tumours in rats after severe injury to the nasal mucosa and prolonged exposure to 10 ppm formaldehyde. *J Appl Toxicol 9*: 39–46

Yager JW, Cohn KL, Spear RC, Fisher JM, Morse L (1986) Sister-chromatid exchanges in lymphocytes of anatomy students exposed to formaldehyde-embalming solution. *Mutat Res 174*: 135–139

Zwart A, Woutersen RA, Wilmer JWGM, Spit BJ, Feron VJ (1988) Cytotoxic and adaptive effects in rat nasal epithelium after 3-day and 13-week exposure to low concentrations of formaldehyde vapour. *Toxicology 51*: 87–99


completed 24.02.2000

# Germ Cell Mutagens

## I Introduction

Germ cell mutagens produce heritable gene mutations, and heritable structural and numerical chromosome aberrations in germ cells. The consequences of germ cell mutations in subsequent generations include genetically determined phenotypic alterations without signs of illness, or reduction in fertility, or embryonic or perinatal death, more or less severe congenital malformations, or genetic diseases with various degrees of health impairment. The term 'germ cell mutagenicity' refers specifically to mutagenicity in male and female germ cells and is distinguished from mutagenicity in somatic cells which can initiate cancer.

The existence of biological processes resulting in the elimination of genetically damaged germ cells, for example, during meiosis or as programmed cell death (apoptosis), has been demonstrated in animal studies. After fertilization has taken place, genetic damage is eliminated if it causes death of the embryo or foetus so that the child with an inherited disease is never born. The human genome is thus protected to some extent from the sequelae of germ cell mutations. During the long process of evolution of the species, the interaction of mutation and selection has resulted in the present species populations with their well-adapted, balanced gene pools. This also applies for man.

The balance of the gene pool can be disturbed in two ways. Firstly, it is assumed that exposure to mutagenic chemicals and ionizing radiation can increase the mutation frequency. Secondly, selection pressures have been reduced, for example, by the therapeutic capabilities of modern medicine and by the increased mobility of man. An increase in the incidence of the genetic component of diseases with multifactorial genesis is favoured especially by decreased selection pressure on the human population. As a consequence, such diseases with partially genetic origins are becoming more frequent. It is probable that mutagenic substances contribute to this increase although to date this could not be demonstrated in population studies. Even the demonstration that an increase in the frequency of monogenic diseases has been caused by exposure to ionizing radiation or mutagenic chemicals is very difficult.

## II Epidemiological Methods and their Limitations

A causal relationship between exposure of a population to chemicals or ionizing radiation and the occurrence of inherited diseases has not yet been demonstrated.

*Essential MAK Value Documentations.* DFG, Deutsche Forschungsgemeinschaft
Copyright © 2006 WILEY-VCH Verlag GmbH & Co. KGaA, Weinheim
ISBN: 3-527-31394-X

One of the reasons for this is that exposure to a certain mutagen does not induce a specific and therefore conspicuous disorder like the non-genetic malformations seen after thalidomide or the forms of cancer associated with certain workplace exposures. Because of the nature of the genetic code, DNA damage and the resulting mutations are distributed more or less randomly in the genome and therefore can cause the one or other heritable disease at random. Even among the progeny of the Japanese parents exposed to radiation in the towns of Hiroshima and Nagasaki, a significant increase in the incidence of genetic damage could not be demonstrated. These results appear to be in contradiction to the finding that an increase in the incidence of structural chromosome aberrations was found in the germ cells of men exposed voluntarily to radiation in the USA (Brewen *et al.* 1975, Martin *et al.* 1986). In Japan, however, it was not inherited diseases in the narrow sense that were studied in the progeny but rather indicators with a genetic component, namely embryonal and neonatal mortality including congenital malformations, survival of the children to age 17, and alterations in the sex ratio (Neel *et al.* 1974, Schull *et al.* 1966). Only later were more specifically genetic end points added to the studied parameters, such as structural and numerical chromosomal anomalies (balanced translocations, inversions, trisomy 21 and aneuploidy of the sex chromosomes), mutations affecting protein charge and enzyme activity and certain childhood cancers resulting from germ cell mutations (Awa *et al.* 1987, Neel *et al.* 1982, 1988, Schull *et al.* 1981a, 1981b, 1982, Yoshimoto *et al.* 1991). But even that did not change the conclusion that a statistically significant increase in inherited diseases after exposure to radiation could not be demonstrated. The reason, however, does not lie in the lack of effectiveness of ionizing radiation but in the difficulty of detecting germ cell mutations with epidemiological methods. The especially critical factors are the determination of exposure levels, the size of the sample population and an adequate control population, as well as appropriate indicators of genetic damage (Sankaranarayanan 1988, 1993).

Of the multitude of genetic diseases, few are suitable for investigation in epidemiological studies, namely dominantly inherited diseases which are manifested during childhood. In addition, familial occurrence of an inherited disease must be differentiated from newly occurring mutations. Although a doubling of the mutation frequency is to be regarded as a marked effect, it cannot be expected that doubling of the incidence of such rare events would be recognized in population samples of a few hundred to a few thousand children. In Hungary, a programme for assessing the mutation frequencies for 25 exemplary dominantly inherited diseases (sentinel phenotypes) has been in operation for some years (Mulvihill and Czeizel 1983). This study was possible because Hungary has a registry of inherited diseases and malformations. On the basis of the Hungarian data it has been calculated that, for example, 47500 live-born children would have to be examined for the presence of the 25 inherited diseases in order to detect a doubling of the mutation rate with 95 % confidence (Czeizel and Kis-Varga 1987). This study can provide an appropriate basis for future epidemiological investigations.

It is still too early for an estimation of the genetic consequences of the reactor accident in Chernobyl, but it is to be expected that the application of appropriate genetic epidemiological methods will provide new insights.

In summary it is concluded that the probability of demonstrating mutagenic effects of a substance in human germ cells after workplace exposures is very small, even if the appropriate population studies are carried out optimally. The classification of substances as germ cell mutagens must therefore be based on experimental data from *in vivo* studies with appropriate animal models and relevant exposures.

# III Categories for Germ Cell Mutagens

The categories for germ cell mutagens have been established in analogy to the categories for carcinogenic chemicals at the workplace.

1. Germ cell mutagens which have been shown to increase the mutant frequency in the progeny of exposed humans
2. Germ cell mutagens which have been shown to increase the mutant frequency in the progeny of exposed mammals
3A. Substances which have been shown to induce genetic damage in germ cells of humans or animals, or which produce mutagenic effects in somatic cells of mammals *in vivo* and have been shown to reach the germ cells in an active form
3B. Substances which are suspected of being germ cell mutagens because of their genotoxic effects in mammalian somatic cells *in vivo*, in exceptional cases, substances for which there are no *in vivo* data but which are clearly mutagenic *in vitro* and structurally related to *in vivo* mutagens
4. not applicable[1]
5. Germ cell mutagens, the potency of which is considered to be so low that, provided the MAK value is observed, their contribution to genetic risk for man is expected not to be significant

# IV Classification of Germ Cell Mutagens

**1. Germ cell mutagens which have been shown to increase the mutant frequency in the progeny of exposed humans**

In Section II it was explained why epidemiological studies have to date not been able to prove that the exposure of a particular human population to a particular substance has resulted in an increase in the incidence of inherited mutations. This is true both for

---

[1] Category 4 carcinogenic substances are those with non-genotoxic mechanisms of action. By definition, germ cell mutagens are genotoxic. Therefore, a Category 4 for germ cell mutagens cannot exist. At some time in the future it is conceivable that a Category 4 could be established for genotoxic substances with primary targets other than the DNA (e.g. purely aneugenic substances) if research results make this seem sensible.

ionizing radiation and chemical mutagens. And even if epidemiological methods are improved further, it is unlikely that such proof will be available in the foreseeable future. Category 1 will therefore probably remain without any entries.


**2. Germ cell mutagens which have been shown to increase the mutant frequency in the progeny of exposed mammals**

In this situation, the results of animal studies must be given particular attention. It is necessary to assume that human germ cells react to the mutagenic effects of chemicals in the same way as do those of animals. There are several reasons for this assumption and they have been discussed in detail (Favor 1989). The most important is the universality of the genetic code for the information carried in the DNA. The effects of changes in the genetic information in the germ cells are largely similar in experimental animals and man. In addition, the course of the biological and physiological processes in the various phases of germ cell development is the same for all mammals including man. Finally, the spectrum of mutants known for the mouse includes several which can serve as models with which the development and effects of human inherited diseases can be studied (Bedell *et al.* 1997). Many homologies in the genomes of mouse and man have been identified (Seldin 1997, Serikawa *et al.* 1998).

To be classified in Category 2 are substances which increase the incidence of genetically modified live progeny in animal studies, for example, in the specific locus test (Ehling *et al.* 1985) or in the test for heritable translocations (Adler 1984a). Likewise, in Category 2 should be classified substances which increase the incidence of embryos which die *in utero*, for example, in the dominant lethal test (Ehling 1977). The dominant lethal test detects progeny which are not viable because of chromosomal damage. The viability of foetuses with chromosomal damage is very different in experimental animals and man: human foetuses with unbalanced chromosome complements are more likely to survive until birth and beyond than are those of the mouse (Schinzel 1984). Therefore an increase in dominant lethal mutations in an animal study is considered relevant for the classification of a substance in Category 2.


**3A. Substances which have been shown to induce genetic damage in germ cells of humans or animals, or which produce mutagenic effects in somatic cells of mammals *in vivo* and have been shown to reach the germ cells in an active form**

For many substances data appropriate for classification in Category 2 are not available but the substances have been tested with methods which detect genetic changes in germ cells. These methods include tests for genotoxicity in the germ cells of exposed animals, such as tests for induction of structural chromosomal changes in spermatogonia or spermatocytes (Adler 1984b), for sister chromatid exchange in spermatogonia (Allen and Latt 1976), for micronuclei in round spermatids (Tates *et al.* 1983), for numerical chromosome changes in secondary spermatocytes (Pacchierotti *et al.* 1983) or in spermatozoa (Lowe *et al.* 1995), for DNA single strand breaks (Anderson *et al.* 1997, Sega and

Generoso 1988) and for repair synthesis (Sega *et al.* 1976) or for covalent binding to the DNA (Sega and Owens 1978). Also relevant are the observations obtained from exposed human populations which provide evidence for structural or numerical chromosome changes in spermatozoa of exposed persons (Robbins *et al.* 1997). The development of new methods, especially molecular genetic methods for the detection of gene mutations in germ cells is to be expected. Substances which yield positive results in tests with germ cells are classified in Category 3A. Also taken into account are clearly positive results from *in vivo* tests for mutagenicity in somatic cells, for example, chromosomal aberrations or micronuclei in bone marrow cells (Adler *et al.* 1971, Schmid 1975), somatic mutations in the mammalian spot test (Fahrig 1977) or in transgenic animals (Gossen and Vijg 1993), provided that it has been demonstrated that under relevant exposure conditions the active substance or a reactive metabolite reaches the germ cells. Such substances are suspected of being mutagenic in germ cells too and therefore they also are classified in Category 3A.

**3B. Substances which are suspected of being germ cell mutagens because of their genotoxic effects in mammalian somatic cells *in vivo*, in exceptional cases, substances for which there are no *in vivo* data but which are clearly mutagenic *in vitro* and structurally related to *in vivo* mutagens**

If the available data are not sufficient for a classification in Category 3A but the substance is clearly genotoxic in somatic cells of exposed animals or man, it is suspected that the substance is also mutagenic in germ cells. Substances which have yielded positive results only in one or several *in vitro* mutagenicity tests are generally not classified in Category 3B. An exception is made for substances for which there are no relevant *in vivo* data but which are clearly genotoxic *in vitro* and also structurally related to substances known to be genotoxic *in vivo*. Such substances raise concern and are classified in Category 3B.

**5. Germ cell mutagens, the potency of which is considered to be so low that, provided the MAK value is observed, their contribution to genetic risk for man is expected not to be significant**

In analogy to the categorization of carcinogenic substances, also for germ cell mutagens a category has been conceived which takes into account the potency of the mutagenic effect. For substances classified in Category 5 it is considered that, provided the MAK value is observed, their contribution to genetic risk for man is expected not to be significant. The classification is supported by information on the spectrum of effects, their dose-dependence and toxicokinetic data pertinent to species comparison. Biochemical and biological markers can be used to characterize the contribution to genetic risk.

The contribution to genetic risk is considered not to be significant when, after exposure at the workplace, the body burden of the substance or its biomarker lies in the range of the background burden in a not specifically exposed reference population:

- under workplace conditions the levels of biochemical markers of systemic exposure and effects such as DNA and protein adducts are not significantly increased above the background levels, and
- physiological-toxicokinetic model calculations based on experimental animal data reveal no significant genetic risk for man.


# V References



Adler ID (1984a) The heritable translocation test. in: Baß R, Glocklin V, Grosdanoff P, Henschler D, Kilbey B, Müller D, Neubert D (Eds) *Critical evaluation of mutagenicity tests*, BGA-Schriften 3/84, MMV Verlag, München, 291–298

Adler ID (1984b) Cytogenetic tests in mammals. in: Venitt S, Parry JM (Eds) *Mutagenicity testing. A practical approach*, IRL Press, Oxford, 275–306

Adler ID, Ramarao G, Epstein SS (1971) *In vivo* cytogenetic effects of trimethylphosphate and of TEPA on bone marrow cells of male rats. *Mutat Res 13*: 263–273

Allen JW, Latt SA (1976) Analysis of sister chromatid exchange formation *in vivo* in mouse spermatogonia as a new test system for environmental mutagens. *Nature (London) 260*: 449–451

Anderson D, Dobrzynka MM, Jackson LI, Yu TW, Brinkworth MH (1997) Somatic and germ cell effects in rats and mice after treatment with 1,3-butadiene and its metabolites, 1,2-epoxybutene and 1,2,3,4-diepoxybutane. *Mutat Res 391*: 233–242

Awa AA, Honda T, Neriishi S, Sufuni T, Shimba H, Ohtaki K, Nakano M, Kodama Y, Itoh M, Hamilton HB (1987) Cytogenetic study of the offspring of atomic bomb survivors, Hiroshima and Nagasaki. in: Obe G, Basler A (Eds) *Cytogenetics: basis and applied aspects*, Springer, Berlin, 166–183

Bedell MA, Largaespada DA, Jenkins NA, Copeland NG (1997) Mouse models of human disease. Part II: recent progress and future directions. *Genes Dev 11*: 11–43

Brewen JG, Preston RJ (1975) Analysis of X-ray-induced chromosomal translocations in human and marmoset spermatogonial stem cells. *Nature (London) 253*: 468–470

Czeizel A, Kis-Varga A (1987) Mutation surveillance of sentinel anomalies in Hungary, 1980–1984. *Mutat Res 186*: 73–79

Ehling UH (1977) Dominant lethal mutations in male mice. *Arch Toxicol 38*: 1–11

Ehling UH, Charles DJ, Favor J, Graw J, Kratochvilova J, Neuhäuser-Klaus A, Pretsch W (1985) Induction of gene mutations in mice: the multiple endpoint approach. *Mutat Res 150*: 393–401

Fahrig R (1977) The mammalian spot test (Fellfleckentest) with mice. *Arch Toxicol 38*: 87–98

Favor J (1989) Risk estimation based on germ-cell mutations in animals. *Genome 31*: 844–852

Gossen J, Vijg J (1993) Transgenic mice as model systems for studying gene mutations *in vivo*. *Trends Genet 9*: 27–31

Lowe X, Collins B, Allen J, Titenko-Holland N, Breneman J, van Beek M, Bishop J, Wyrobek AJ (1995) Aneuploidies and micronuclei in the germ cells of male mice of advanced age. *Mutat Res 338*: 59–76

Martin RH, Hildebrand K, Yamamoto J, Rademaker A, Barnes M, Douglas G, Arthur K, Ringrose T, Brown IS (1986) An increased frequency of human sperm chromosomal abnormalities after radiotherapy. *Mutat Res 174*: 219–225

Mulvihill JJ, Czeizel A (1983) Perspectives in mutation epidemiology, 6: a 1983 view of sentinel phenotypes. *Mutat Res 123*: 345–361

Neel JV, Kato H, Schull WJ (1974) Mortality in the children of atomic bomb survivors and controls. *Genetics 76*: 311–326

Neel JV, Schull WJ, Otake M (1982) Current status of genetic follow-up studies in Hiroshima and Nagasaki. in: Bora KC, Douglas GR, Nestmann ER (Eds) *Progress in Mutation Research, Vol 3*, Elsevier Biomedical, Amsterdam, 39–51

Neel JV, Satoh C, Goriki K, Asakawa J, Fujita M, Takahashi N, Kageoka T, Hazama R (1988) Search for mutations altering protein charge and/or function in children of atomic bomb survivors: final report. *Am J Hum Genet 42*: 663–676

Pacchierotti F, Bellincampi D, Civitareale D (1983) Cytogenetic observations, in mouse secondary spermatocytes, on numerical and structural chromosome aberrations induced by cyclophosphamide in various stages of spermatogenesis. *Mutat Res 119*: 177–183

Robbins WA, Meistrich ML, Moore D, Hagemeister FB, Weier HU, Cassel MJ, Wilson G, Eskenazi B, Wyrobek AJ (1997) Chemotherapy induces transient sex chromosomal and autosomal aneuploidy in human sperm. *Nature Genet 16*: 74–78

Sankaranarayanan K (1988) Invited review: prevalence of genetic and partially genetic diseases in man and the estimation of genetic risks of exposure to ionizing radiation. *Am J Hum Genet 42*: 651–662

Sankaranarayanan K (1993) Ionizing radiation, genetic risk estimation and molecular biology: impact and inferences. *Trends Genet 9*: 79–84

Schinzel A (1984) *Catalogue of unbalanced chromosome aberrations in man*, Walter de Gruyter, Berlin, New York

Schmid W (1975) The micronucleus test. *Mutat Res 31*: 9–15

Schull WJ, Neel JV, Hashizume A (1966) Some further observations on the sex ratio among infants born to survivors of the atomic bombings of Hiroshima and Nagasaki. *Am J Hum Genet 18*: 328–338

Schull WJ, Otake M, Neel JV (1981a) Hiroshima and Nagasaki: a reassessment of the mutagenic effect of exposure to ionizing radiation. in: Hook EB, Porter IH (Eds) *Population and biological aspects of human mutation*, Academic Press, New York, London, Toronto, Sydney, San Francisco, 277–303

Schull WJ, Otage M, Neel JV (1981b) Genetic effects of the atomic bombs: a reappraisal. *Science 213*: 1220–1227

Schull WJ, Neel JV, Otake M, Awa A, Satoh C, Hamilton HB (1982) Hiroshima and Nagasaki: three and a half decades of genetic screening. in: Sugimura T, Kondo S, Takebe H (Eds) *Environmental mutagens and carcinogens*, Alan R. Liss, New York, 687–700

Sega GA, Generoso EE (1988) Measurement of DNA breakage in spermiogenic germ-cell stages of mice exposed to ethylene oxide, using an alkaline elution procedure. *Mutat Res 197*: 93–99

Sega GA, Owens JG (1978) Ethylation of DNA and protamine by ethyl methanesulfonate in the germ cells of male mice and the relevancy of these molecular targets to the induction of dominant lethals. *Mutat Res 52*: 87–106

Sega GA, Owens JG, Cumming RB (1976) Studies on DNA repair in early spermatid stages of male mice after *in vivo* treatment with methyl-, ethyl-, propyl-, and isopropyl methanesulfonate. *Mutat Res 36*: 193–212

Seldin MF (1997) Genome surfing: using internet-based information tools toward functional genetic studies in mouse and humans. *Methods 13*: 445–457

Serikawa T, Cui Z, Yokoi N, Kuramoto T, Kondo Y, Kitada K, Guenet JL (1998) A comparative genetic map of rat, mouse and human genomes. *Exp Anim 47*: 1–9

Tates AD, Dietrich AJJ, de Vogel N, Neuteboom I, Bos A (1983) A micronucleus method for detection of meiotic micronuclei in male germ cells of mammals. *Mutat Res 121*: 131–138

Yoshimoto Y, Schull WJ, Kato H, Neel JV (1991) Mortality among the offspring ($F_1$) of atomic bomb survivors, 1946–1985. *J Radiat Res 32*: 327–351

completed 24.02.2000

# Hexachlorobenzene

| | |
|---|---|
| **MAK value** | – |
| **Absorption through the skin (1998)** | **H** |
| **Sensitization** | – |
| **Carcinogenicity (1998)** | **Category 4** |
| **Prenatal toxicity** | – |
| **Germ cell mutagenicity** | – |
| **BAT value (1984)** | **150 µg/l serum**<br>**Sampling time: not fixed** |
| Synonyms | HCB<br>perchlorobenzene |
| Chemical name (CAS) | hexachlorobenzene |
| CAS number | 118-74-1 |
| Structural formula | |

$$C_6Cl_6$$

| | |
|---|---|
| Molecular formula | $C_6Cl_6$ |
| Molecular weight | 284.78 g/mol |
| Melting point | 227–231°C |
| Boiling point | 322–326°C |
| Density at 20°C | 2.075 g/cm$^3$ |
| Vapour pressure at 20°C | 1.1–1.68 x 10$^{-3}$ Pa |
| log $P_{ow}$* | 5.5 |

**1 ml/m$^3$ (ppm) $\hat{=}$ 11.82 mg/m$^3$**     **1 mg/m$^3$ $\hat{=}$ 0.085 ml/m$^3$ (ppm)**

The present document is based on a report on hexachlorobenzene by the "Beratergremium für umweltrelevante Altstoffe" (BUA 1994).

---

* *n*-octanol/water distribution coefficient

Hexachlorobenzene was introduced in the 1940s as a fungicide for cereal seed. It was also used for pyrotechnics. Its use has been forbidden since 1975 in western Germany and since 1984 in eastern Germany. It is, however, formed in large quantities during the production of chlorinated solvents, occurs as a contaminant in pesticides and is a metabolite of lindane (BUA 1994).

# 1 Toxic Effects and Mode of Action

Hexachlorobenzene is a lipophilic substance which accumulates in adipose tissues and is released again only very slowly. The biological half-life in man is about 2 years and is much longer than the half-life in the rat (up to about 5 months).

Ingested hexachlorobenzene is absorbed readily from oily solutions but poorly from aqueous suspensions or when administered in crystalline form. It has been shown to be absorbed through the skin.

Mortality was high among the persons affected during the 4 to 5 years of a mass poisoning with fungicide containing hexachlorobenzene. The symptoms described were general weakness, skin lesions, porphyria, hyposomia, osteoporosis, arthritis and neuritis. Enlargement of the liver, thyroid and lymph nodes was also reported.

In animal studies the acute toxicity is low. The $LD_{50}$ for the rat is of the order of 3500–10000 mg/kg body weight. The symptoms of intoxication include tremor, convulsions and ataxia.

In animals given the substance for longer periods by oral administration, dermal lesions such as hair loss, blister and scab formation and hyperpigmentation were seen. In experimental animals the main target organ is the liver. Porphyria is observed especially in female rats and pigs (in both cases after doses of 0.5 mg/kg body weight and day administered for periods of 15 and 13 weeks, respectively). After long-term administration of hexachlorobenzene to rats in the diet, degenerative effects are seen in the livers of animals treated with concentrations of 8 ppm (doses of about 0.8 mg/kg body weight and day), preneoplastic foci in those treated with 40 ppm (about 4 mg/kg body weight and day) and liver tumours at 75 ppm (about 7.5 mg/kg body weight and day). Liver tumours also develop in mice given 100 ppm hexachlorobenzene in the diet (about 12 mg/kg body weight and day) and in hamsters given 50 ppm (about 4 mg/kg body weight and day). In rats, bile duct tumours, renal adenomas and adrenal pheochromocytomas are also seen. The liver, bile duct and kidney tumours may be attributed to the induction of cytochrome P450 (CYP) isozymes and the consequent formation of oxygen radicals which results in cell damage and compensatory hyperplasia. Tumour-promoting effects of the substance in the liver have been demonstrated in rats.

The thyroid gland is also a target organ. Hexachlorobenzene causes increased excretion of thyroxine (T4) which can result in stimulation of the thyroid and thyroid hyperplasia. In hamsters, thyroid adenomas develop (200 ppm, about 20 mg/kg body weight and day).

Hexachlorobenzene causes immunomodulation. In rats, hexachlorobenzene causes an increase in humoral and cell-mediated immune responses and autoimmune reactions. In mice, on the other hand, a suppression of immune responses was generally observed.

Hexachlorobenzene doses of 0.01 mg/kg body weight and day, administered to female monkeys for 13 weeks, caused ultrastructural changes in the ovaries. However, effects on fertility were not seen even after administration of doses as high as 10 mg/kg body weight and day for 90 days.

Hexachlorobenzene crosses the placenta and accumulates in the foetus. In a study with rats an increased incidence of 14th ribs was observed, and renal anomalies were seen in the offspring of mice given maternally toxic doses. These findings are conceivably a result of the general toxicity of hexachlorobenzene. During lactation, hexachlorobenzene is mobilized with the fat reserves and excreted in the milk. In breast-fed babies and suckling animal offspring whose mothers are exposed to hexachlorobenzene, severe toxic effects and increased mortality is observed.

Mutagenic effects have not been detected but there is evidence of weak clastogenicity.


# 2 Mechanism of Action


Hexachlorobenzene induces CYP1A1 in the liver, lungs and kidneys of the rat. CYP1A2 is induced mainly in the liver and lungs and CYP2B1 in the liver (Gulyaeva *et al.* 1994). The induction of cytochrome P450 isozymes and the redox reactions involving hexachlorobenzene metabolites such as tetrachlorohydroquinone result in the formation of oxygen radicals and in lipid peroxidation (Almeida *et al.* 1997, Billi de Catabbi *et al.* 1997). The consequent cell damage and compensatory hyperplasia have been suggested as the mechanism of induction of the liver tumours seen in experimental animals (Carthew and Smith 1994). In monkeys, ultrastructural alterations of the ovarian follicle membrane could be considered to be a result of the induction of cytochrome P450 (Bourque *et al.* 1995).

The porphyria induced by hexachlorobenzene is a result of changes in activities of the enzymes of haem biosynthesis, especially the inhibition of uroporphyrinogen decarboxylase. This results in an accumulation of intermediates of haem biosynthesis, especially in the liver, and in increased porphyrin excretion (Billi de Catabbi *et al.* 1997, BUA 1994, Mylchreest and Charbonneau 1997). The oxidative biotransformation and thus the formation of the very reactive tetrachloro-1,4-benzoquinone is directly related to the porphyrinogenic action of hexachlorobenzene (van Ommen *et al.* 1989). Later studies demonstrated, however, that a reactive intermediate formed during the cytochrome P450-mediated oxidation of hexachlorobenzene to pentachlorophenol could be responsible for the porphyrinogenic effects (Rietjens *et al.* 1995). There is also evidence that thyroid hormones play an important role in hexachlorobenzene-induced porphyria (Soprena de Kracoff *et al.* 1994).

Reduced T4 concentrations in serum were observed in man and rats exposed to hexa-chlorobenzene. This may be due in part to the induction of various UDP glucuronosyl transferases in the rat liver and the consequent increase in glucuronidization and excretion of T4 in the bile (van Raaij *et al.* 1993a). In addition, the hexachlorobenzene metabolite pentachlorophenol was shown to compete with T4 for the binding site on its transport protein (transthyretin) and to have an affinity for the transport protein which was twice that of T4 (van Raaij *et al.* 1993b). The increased excretion of T4 resulted in TSH-mediated stimulation of the thyroid and thus in hypertrophy and hyperplasia.

There is evidence that immunomodulation reactions are involved in the aetiology of the hexachlorobenzene induced skin lesions (Michielsen *et al.* 1997).

The mechanisms of the neurological and behavioural effects of the substance have not yet been clarified. Prenatal and postnatal exposure to hexachlorobenzene impairs scheduled-controlled behaviour in rats (Lilienthal *et al.* 1996). It has been suggested that hexachlorobenzene interferes with myelination during development or that it modifies the activity of neurotransmitters. Effects of altered thyroid hormone concentrations on the developing nervous system are also conceivable (see Lilienthal *et al.* 1996).

Studies of the mechanism of action of hexachlorobenzene on the endocrine and immune systems have not been published. In analogy to 2,3,7,8-tetrachlorodibenzo-*p*-dioxin, hexachlorobenzene could act by Ah receptor-mediated mechanisms.

# 3 Toxicokinetics and Metabolism

Two days after oral administration of a hexachlorobenzene dose of 20 mg/kg body weight in olive oil to female Wistar rats, 83.2 % of the dose had been absorbed. When the substance was administered in water (16 mg/kg body weight) only 19.7 % of the dose had been absorbed after 3 days (BUA 1994).

Hexachlorobenzene was administered by occlusive application of a dermal dose of 10 mg/kg body weight in tetrachloroethylene to male Fischer rats. After 6 hours 1.05 % of the dose had been absorbed, after 24 hours 2.67 % and after 72 hours 9.71 % (BUA 1994).

Hexachlorobenzene is distributed into all organs. The time required to reach equilibrium depends markedly on the administration route; after oral administration of hexachlorobenzene in oil, it is 2 to 5 days. The highest concentrations are found in adipose tissue (BUA 1994). In the rat, high concentrations are also found in endocrine tissues such as the thyroid gland, adrenal glands and ovaries (Foster *et al.* 1993).

Hexachlorobenzene is oxidized by CYP3A to pentachlorophenol and then to tetra-chlorohydroquinone. In addition, the conjugation products *N*-acetyl-*S*-(pentachloro-phenyl)cysteine and mercaptotetrachlorothioanisole are found (den Besten *et al.* 1994). The rate of metabolism is, however, low (BUA 1994).

Hexachlorobenzene is excreted only very slowly. Fourteen days after intraperitoneal administration of a hexachlorobenzene dose of 4 mg/kg body weight in oil to Wistar rats,

34 % of the dose had been excreted with the faeces (80 % as unchanged hexachlorobenzene) and 5 % with the urine (96 % as metabolites) (Koss and Koransky 1975). The excretion kinetics are biphasic or multiphasic. The elimination half-time in the rat was up to 5 months, in the Rhesus monkey 2.5 to 3 years (BUA 1994). The biological half-life in workers exposed to hexachlorobenzene was found to be about 2 years (Henschler and Lehnert 1984).

# 4 Effects in Man

There are no data available for the effects of single exposures of persons to hexachlorobenzene nor for its effects on skin or mucous membranes or allergenic effects.

## 4.1 Repeated exposure

### 4.1.1 Workers

In a group of 50 workers in the production of chlorinated hydrocarbons who were monitored in the years 1974 to 1977, the average plasma hexachlorobenzene level varied around 325 µg/l, 65 times the average concentration of 5 µg/l found in the general population. Because the biochemical and liver function parameters in these workers were not statistically different from those of control collectives, the BAT value for hexachlorobenzene was established at 150 µg/l plasma (Henschler and Lehnert 1984).

Workers who were exposed until 1980 to hexachlorobenzene concentrations of 2.1 to 10.8 mg/m$^3$ and during the study period between 1983 and 1990 to 0.012 to 0.022 mg/m$^3$ still had serum hexachlorobenzene levels of more than 500 µg/l even 9 years after the exposure concentrations had been reduced (1980: 534 µg/L, 1989: 575 µg/L). Significant changes, but within the normal biological range, were found in the levels of immunoglobulin (IgG, IgA, IgM and IgE increased) and transport proteins (including transferrin) and there was evidence of effects on kidney function and of reduced immune responses (Richter *et al.* 1994).

### 4.1.2 General population

In the years between 1955 and 1959, the use of seed corn treated with hexachlorobenzene for flour and bread production resulted in poisoning of more than 3000 persons. The seed had been treated with 0.2 % of a fungicide containing 10 % hexachlorobenzene. The exact composition of the fungicide is, however, unknown. The symptoms included photosensitive lesions on skin areas exposed to light, bald skin areas, arthritic hands and fingers, hypertrichosis, enlargement of the liver and thyroid, weight loss, general weakness, anorexia and, mainly in men, porphyria (*porphyria cutanea tarda*). The daily dose of hexachlorobenzene was estimated to be 50 to 200 mg (about 0.7–4 mg/kg body weight). About 10 % of the affected persons died. Late sequelae such as scarring, hyperpigmentation, arthritis, osteoporosis, myotonia, paresthesia and enlargement of the thyroid gland were still recognizable in some cases 30 years later. Children under 4 years of age, especially breast-fed children of mothers exposed to hexachlorobenzene, were particularly sensitive (BUA 1994).

## 4.2 Reproductive toxicity

Breast-fed babies and children whose mothers had eaten seed corn treated with hexachlorobenzene in the years 1955 to 1959 (Section 4.1) had more severe symptoms of intoxication (including cutaneous erythema ("pink sores"), weakness, tremor and convulsions) than did adults. Mortality was as high as 95 % in some areas (BUA 1994).

## 4.3 Genotoxicity

In 41 workers who were exposed to tetrachloromethane, tetrachloroethylene and hexachlorobenzene, the incidence of micronuclei in the peripheral lymphocytes was increased significantly above that in 28 control persons but was not correlated with the serum hexachlorobenzene concentration (da Silva Augusto *et al.* 1997).

## 4.4 Carcinogenicity

A hepatocellular carcinoma was diagnosed in a worker who was employed from 1955 to 1981 in an aluminium smelter. Until 1973, hexachloroethane was added to degas the molten aluminium (700°C, open container); this resulted in the formation of large amounts of gas with a chlorine-like odour. Ventilation and protective clothing were not available. In another plant it was demonstrated that during this process not only chlorobenzenes, chlorophenols, dioxins and dibenzofurans are formed but also particularly high concentrations of hexachlorobenzene (Seldén *et al.* 1989). However, because it must be assumed that the man was also exposed to the other chlorinated compounds, it is not possible to ascribe the tumour formation to the hexachlorobenzene exposure alone.

In a town with a population of 5003 which was close to an old factory producing chlorinated organic compounds and chlorinated solvents, significantly increased hexachlorobenzene concentrations were found in the air (average 35 $\mu$g/m$^3$, reference value 0.3 $\mu$g/m$^3$); in earlier years the concentrations must have been much higher, at times over 100 $\mu$g/m$^3$. The concentrations of polychlorinated biphenyls (PCB), dichlorobischlorophenylethene (DDE), chloroform (trichloromethane), tetrachloromethane, trichloroethylene and tetrachloroethylene in the air were not increased above the reference concentration. The concentrations of chlorinated compounds were measured in the serum of 21 persons. The concentration of hexachlorobenzene was 26 µg/l (7.5–69 µg/l) and was significantly increased above the values in a reference population (4.8 µg/l (1.5–15 µg/l). The average concentration in the 9 men who were studied was 32 µg/l and was higher than the 22 µg/l found in the 12 women. The concentrations of DDE (12 µg/l) were increased relative to those in the control population (1.6 µg/l) but were not significantly higher than those in other populations (e.g. 1978–1982 in the Netherlands 3.9–22 µg/l). Concentrations of other substances were not increased. In the study period from 1980 to 1989, the incidences of "unknown tumours" (ICD 199), tumours of the thyroid (ICD 193), the brain (ICD 191) and of soft-tissue sarcomas (ICD 171) were increased. The increase was significant or of borderline significance for males (Table 1). All the men with tumours were employed in the factory (Grimalt *et al.* 1994). The authors suggest that there could be an association between the observed tumours and the exposure to chlorinated organic compounds, especially hexachlorobenzene, because the thyroid gland is a target organ of hexachlorobenzene in experimental animals and soft tissue sarcomas have been associated with the exposure to chlorinated compounds.

**Table 1.** Incidences of tumours in the years 1980–1989 in a population exposed to hexachlorobenzene (from Grimalt *et al.* 1994)

| Tumour localization (ICD-9) | Men number | SIR (95% CI) | Women number | SIR (95% CI) |
|---|---|---|---|---|
| thyroid gland (193) | 2 | 6.7 (1.6–28)* | 1 | 1.0 (0.14–7.4) |
| soft tissue (171) | 3 | 5.5 (1.7–17.5)* | 1 | 2.2 (0.3–16) |
| brain (191) | 4 | 2.7 (0.99–7.2) | 1 | 0.93 (0.13–6.7) |
| unknown (199) | 10 | 2.35 (1.25–4.4)* | 4 | 1.2 (0.45–3.2) |
| urinary bladder (188) | 12 | 1.3 (0.75–2.3) | 2 | 1.1 (0.28–4.6) |
| female breast (174) | 0 | – | 19 | 1.3 (0.84–2.1) |
| colo-rectum (153–154) | 10 | 1.0 (0.54–1.9) | 7 | 0.82 (0.39–1.7) |
| lungs (162) | 11 | 0.76 (0.42–1.4) | 2 | 1.6 (0.39–1.7) |
| lymphomas (196) | 1 | 0.56 (0.08–4.02) | 1 | 0.67 (0.09–4.8) |
| liver (155) | 0 | – | 1 | 0.82 (0.11–5.8) |
| Total (140–208) | 74 | 0.96 (0.76–1.2) | 55 | 0.93 (0.72–1.2) |

* significant; level of significance not specified
ICD: International Classification of Disease
SIR: standardized incidence ratio
CI: confidence interval

There are no studies which have investigated the development of tumours in the population which was poisoned with the hexachlorobenzene-containing seed corn (Section 4.1).

# 5 Animal Experiments and *in vitro* Studies

## 5.1 Acute toxicity

The acute toxicity of hexachlorobenzene is low.

In inhalation studies, the $LC_{50}$ in the rat was found to be 3600 mg/m$^3$, in the mouse 4000 mg/m$^3$, in the rabbit 1800 mg/m$^3$ and in the cat 1600 mg/m$^3$.

The oral $LD_{50}$ values reported for the rat after administration of hexachlorobenzene were 3500 and 10000 mg/kg body weight, for the mouse 4000 mg/kg body weight, for the rabbit 2600 mg/kg body weight and for the cat 1700 mg/kg body weight. The symptoms of intoxication include weakness, convulsions, tremor and ataxia (BUA 1994).

A single oral hexachlorobenzene dose of 50 mg/kg body weight caused a significant reduction in the T4 level in female rats (Foster *et al.* 1993).

## 5.2 Subacute, subchronic and chronic toxicity

There are no inhalation studies available with hexachlorobenzene.

### 5.2.1 Ingestion

A description of all the studies with repeated administration of hexachlorobenzene to experimental animals may be found in the BUA review (BUA 1994). Here only a few selected studies are described.

Oral administration of hexachlorobenzene doses of 485 or 750 mg/kg body weight to male WAG-RIJ rats, three times weekly for 2 weeks, resulted respectively in slight and significant reductions in the blood T4 levels. At 256 mg/kg body weight, no changes were observed. Administration of hexachlorobenzene doses of about 1000 mg/kg body weight for 4 weeks resulted also in increased TSH concentrations (van Raaij *et al.* 1993b) and in significantly reduced T3 concentrations (Kleimann de Pisarev *et al.* 1995).

Porphyrinogenic effects of hexachlorobenzene have been observed in the rat, mouse, rabbit and pig. The NOEL for the porphyrinogenic effects was found to be 0.5 mg/kg body weight and day for the rat (treated for 9 months) and 0.05 mg/kg body weight and day for the pig (treated for 13 weeks) (BUA 1994). In the hamster too, especially in males, after administration of hexachlorobenzene in the diet in a concentration of 100 ppm (about 10 mg/kg body weight and day) for a period of 5 or 10 months, the relative liver weight was increased and the relative spleen weight reduced. In both organs deposition of iron was observed. The porphyrin level in the liver was increased (Ríos de Molina *et al.* 1996).

After administration of hexachlorobenzene to Sprague-Dawley rats and golden Syrian hamsters in concentrations of 0, 200 or 400 ppm in the diet (about 0, 20 and 40 mg/kg body weight and day) and to Swiss mice in concentrations of 0, 100 or 200 ppm (about 0, 15 and 30 mg/kg body weight and day) for up to 13 weeks, the pathological examination revealed mainly effects on the lymphohaematopoietic system (hyperplasia with lympho-cytic infiltration in the liver, kidneys and in some other organs, lymphosarcomas), the liver (hepatitis, liver cirrhosis, metaplasia, preneoplastic foci, hepatomas) and the kidneys (severe hyperaemia, degeneration, nephritis, renal cell adenomas) (Ertürk *et al.* 1986). In the publication it is not clear which effects are associated with which concentrations; the study can therefore not be used in the present assessment.

In Sprague-Dawley rats given hexachlorobenzene in the diet in doses of 0, 0.5, 2, 8 and 32 mg/kg body weight and day for up to 15 weeks, the liver proved to be the most sensitive organ. Female rats were more affected than males. In the females of the group given 0.5 mg/kg body weight and day, the liver porphyrin levels were increased during the recovery period in week 19. The NOEL for porphyrin excretion was therefore below 0.5 mg/kg body weight and day. In animals given 2 mg/kg body weight and day or more, the induction of xenobiotic-metabolizing enzymes was observed. From doses of 8 mg/kg body weight and day, relative liver weights were increased and ultrastructural changes were detected in the liver (increased amount of smooth endoplastic reticulum). In addition, the porphyrin concentrations in the liver, kidneys and spleen were increased. In the animals given 32 mg/kg body weight and day, mortality was increased and clinical symptoms such as tremor, irritability, ataxia and hair loss were seen. Relative weights of kidneys, spleen and adrenal glands were increased (Kuiper-Goodman *et al.* 1977, Kuiper-Goodman and Grant 1986).

In a 2-generation study with Sprague-Dawley rats, hexachlorobenzene was administered in the diet in concentrations of 0, 0.32, 1.6, 8.0 and 40 ppm (about 0.03, 0.16, 0.8 and 4 mg/kg body weight and day) to the parent animals ($F_0$) for 3 months before mating and to the progeny ($F_1$) for 130 weeks. In the $F_0$ generation males given doses of 0.8 mg/kg body weight and day or more, statistically significant increases in relative liver weights and in absolute heart and brain weights were recorded. In the $F_1$ generation males given doses from 0.03 mg/kg body weight and day, the incidences of peribiliary lymphocytosis and fibrosis and biliary distention were increased. These effects were, however, not dose-related and can therefore not be evaluated. In animals given doses from 0.8 mg/kg body weight and day, the incidence of centrilobular basophilic chromogenesis in the liver was increased. With doses from 4 mg/kg body weight and day increased incidences of severe chronic nephrosis, parathyroid adenomas, neoplastic nodules in the liver (female animals) and phaeochromocytomas of the adrenal glands (female animals) were observed (Arnold *et al.* 1985, Arnold and Krewski 1988). The publication does not state whether the excretion of porphyrins or the induction of xenobiotic-metabolizing enzymes was determined. The NOAEL (no observed adverse effect level) can be given as 1.6 ppm hexachlorobenzene in the diet, that is about 0.16 mg/kg body weight and day.

Groups of 5 male pigs were given hexachlorobenzene in the diet for 13 weeks in doses of 0, 0.05, 0.5, 5 and 50 mg/kg body weight and day. At doses of 0.5 mg/kg body weight and day or more, the excretion of porphyrins was increased as was the induction of microsomal liver enzymes, hypertrophy and alterations in liver cells. From 5 mg/kg body weight and day, weights of liver, kidneys and thyroid glands were increased and the histopathological examination revealed changes in the liver. At 50 mg/kg body weight and day, anorexia, tremor, shortness of breath, and mortality were recorded and the histopathological examination revealed changes in the kidneys, testes and lymph nodes. The NOEL found in this study was 0.05 mg/kg body weight and day (den Tonkelaar *et al.* 1978).

In several studies by one research group, hexachlorobenzene was administered for 90 days in capsules to groups of 4 cynomolgus monkeys. At all doses from the lowest dose of 0.01 mg/kg body weight and day, the mitochondria in the developing egg cells were condensed and in the follicles deep indentations and an abnormal accumulation of lipid droplets in the cytoplasm were seen. The authors described these changes as reversible. In animals given 0.1 mg/kg body weight and day or more, degeneration of follicles was observed (Bourque *et al.* 1995). In the egg cell membranes, nuclei and cytoplasm, alterations were seen which became more severe with increasing dose (Jarrell *et al.* 1993). There were changes in the surface epithelium of the ovary such as stratification of the epithelial cells, columnar cells and irregular shapes, accumulation of lysosomes and lipid vesicles and signs of degeneration (Babineau *et al.* 1991, Sims *et al.* 1991). Histopathological examination of animals given doses of 1 mg/kg body weight and day or more revealed changes in the liver. At 10 mg/kg body weight and day the weights of the liver and adrenal glands were increased. Increased excretion of porphyrins was not detected. The number of primordial follicles was reduced significantly. There were no adverse effects on menstruation or fertility (Jarrell *et al.* 1993). A NOEL cannot be derived from the results of this study.

## 5.3 Local effects on skin and mucous membranes

Hexachlorobenzene has been described as having little irritating effect on the skin and none in the eyes, but this claim is inadequately documented (BUA 1994).

## 5.4 Allergenic effects

In a study by Landsteiner and Jacobs (1936), hexachlorobenzene (0.05 mg in 0.05 ml olive oil) was administered intracutaneously to guinea pigs (numbers not specified), once weekly for 10 weeks. After a 2-week pause, the provocation treatment was carried out with "1 drop of a 20 % solution of the substance in alcohol". Sensitization was not detected (BUA 1994).

## 5.5 Reproductive and developmental toxicity

### 5.5.1 Fertility

**Fertility of males**

In a dominant lethal test, male rats were treated with hexachlorobenzene doses of 70 or 221 mg/kg body weight and day for 5 days; the only effect was dose-dependent reduction in the number of matings (Simon *et al.* 1979), presumably a result of the fact that the doses were in the toxic range.

**Fertility of females**

Administration of oral doses of hexachlorobenzene of 0, 1, 10 or 100 mg/kg body weight and day for 30 days to adult female Sprague-Dawley rats resulted in reduced concentrations of corticosterone in the blood of ovarectomized animals given 1 mg/kg body weight. There were no effects on the levels of progesterone or aldosterone (Foster *et al.* 1995b). In superovulated animals (21 days exposure, ovulation stimulated with gonadotropin) the ovary weights were increased only in the animals given 1 mg/kg body weight and day. However, the serum progesterone level was increased in all dose groups (Foster *et al.* 1992). Treatment with 10 mg/kg body weight and day led to ultrastructural changes in the granulosa-lutein cells (MacPhee *et al.* 1993).

In cynomolgus monkeys given oral doses of hexachlorobenzene for 13 weeks and then treated with follicle stimulating hormone (FSH) and luteinizing hormone (LH) to stimulate follicle development and gonadotropin to stimulate ovulation, even the lowest hexachlorobenzene dose of 0.01 mg/kg body weight and day produced ultrastructural changes in the developing egg cells and follicles (Bourque *et al.* 1995). In the animals given 0.1 mg/kg body weight and day, ultrastructural changes in the epithelium of the ovary and in the oocytes were seen (Babineau *et al.* 1991, Jarrell *et al.* 1993, Sims *et al.* 1991). In addition, the length of the menstrual cycle increased with the dose. After doses of 10 mg/kg body weight and day, the concentration of oestradiol during ovulation was significantly reduced (Foster *et al.* 1995a) but fertility was not affected (Jarrell *et al.* 1993; see also Section 5.2.2).

### 5.5.2 Developmental toxicity

**Prenatal toxicity**

Hexachlorobenzene readily passes the placental barrier. Accumulation in foetal tissue has been demonstrated in the rat (Nakashima *et al.* 1997), rabbit, pig and monkey (BUA 1994).

Prenatal toxic effects such as an increased incidence of unilateral and bilateral 14th ribs and sternal defects were seen in the progeny of rats given hexachlorobenzene doses of 40 mg/kg body weight and day for at least 10 days but not in those given doses up to 80 mg/kg body weight and day for up to 8 days (in both cases from day 6 of gestation). Therefore these results require confirmation. Toxic effects in the dams in the form of reduced body weights were first seen after administration of 80 mg/kg body weight and day for at least 16 days (Khera 1974). In the mouse, only maternally toxic doses have been studied. Doses of 100 mg/kg body weight administered to pregnant mice on days 7 to 16 of gestation caused increased liver weights in the dams and an increased incidence of kidney anomalies in the progeny, in one litter also an increase in the incidence of cleft palate (Courtney *et al.* 1976). In dams given doses of 125 mg/kg body weight and day for 5 days before mating and also on days 8 to 12 of gestation, mortality was increased. Of

30 litters, 2 were totally resorbed. Only 6 dams gave birth to pups; the litter sizes were significantly reduced (Kavlock *et al.* 1987).

**Perinatal and postnatal toxicity; multigeneration studies**

Because of the mobilization of lipids during lactation, a large amount of the hexachlorobenzene in the dams is transferred via the milk to the suckling pups and can result in toxicity and increased postnatal mortality. The available studies have been reviewed by BUA (1994). Therefore only the most important and more recent studies are described below.

In lactating sheep it was shown that, 37 days after administration of hexachlorobenzene in a single intramuscular injection of 0.2 mg/kg body weight, the concentrations of hexachlorobenzene in the milk (84 ng/g lipid) and blood (76 ng/g lipid) were similar, but the concentration in the blood of the lambs (216 ng/g lipid) was much higher (Vrecl *et al.* 1996).

In a 4-generation study, Sprague-Dawley rats were given hexachlorobenzene in the diet in concentrations up to 640 ppm (about 1, 2, 4, 8, 16, 32, 64 mg/kg body weight and day). In the dams given 16 mg/kg body weight or more, mortality was increased and from 32 mg/kg body weight the fertility was decreased (not clear whether male or female fertility). In the $F_1$ generation, birth weights were reduced from doses of 8 mg/kg body weight and survival during the lactation period from 16 mg/kg body weight; 32 mg/kg body weight resulted in an increased incidence of resorptions. In later generations, these effects were also seen in the groups given 4 mg/kg body weight. No significant effects were seen at 2 mg/kg body weight (BUA 1994).

In a study to examine neurobehavioural functions in rats after prenatal and postnatal exposure, female Wistar rats were given hexachlorobenzene in the diet in concentrations of 0, 4, 8, or 16 mg/kg diet (about 0, 0.4, 0.8, 1.6 mg/kg body weight and day) from 90 days before mating until the end of the lactation period. The progeny was given hexachlorobenzene in the diet until they were 150 days old. The relative liver weights in the dams were increased during the lactation period. In the progeny exposed to hexachlorobenzene, neither open-field activity (postnatal day 21) or active avoidance learning (postnatal day 90) was different from that of the control animals. There were significant changes in trained operant behaviour (from postnatal day 150) for animals given 0.8 mg/kg body weight or more (Lilienthal *et al.* 1996).

In sows given 1 or 20 ppm hexachlorobenzene in the diet (about 0.025 and 0.5 mg/kg body weight and day) from 1 week before mating until the birth (maximum 170 and 190 days), body weight gains were increased and mild haematological effects (increased total leukocyte count, reduced lymphocyte count, increased incidence of neutrophilia in the high dose group), a slight dose-dependent increase in relative liver weight and evidence of stomach irritation were seen. Mortality of the piglets was increased during the first week of lactation in both dose groups; the incidence of stillbirths was increased in the high dose group (BUA 1994).

## 5.6 Genotoxicity

### 5.6.1 *In vitro*

The studies of *in vitro* genotoxicity of hexachlorobenzene are listed in Table 2.

**Table 2.** *In vitro* genotoxicity studies with hexachlorobenzene

| Test system | Test organism | Concentration | Results without S9 | with S9 | References |
|---|---|---|---|---|---|
| *Salmonella* mutagenicity test (Ames test) | *S. typhimurium* TA98, TA100, TA1535, TA1537, TA1538 | up to 10000 µg/plate | – | – | BUA 1994 |
| mutation, DNA repair | *Escherichia coli* WP2, WPuvrA | up to 1000 µg/plate | – | | BUA 1994 |
| SOS chromotest | *E. coli* K12 PQ37 | no data | – | – | Raabe *et al.* 1993 |
| HPRT (8-azaguanine) | V79 cells | about 2–8 µg/ml | (+) | | Kuroda 1986 |
| Na$^+$K$^+$-ATPase (ouabain) | V79 cells | about 2–8 µg/ml | – | | Kuroda 1986 |
| chromosomal aberrations | human lymphocytes | 0.001–0.1 mM | – | | Siekel *et al.* 1991 |
| micronucleus test | rat hepatocytes | 0.10–0.18 mM 0.32 mM | – + | | Canonero *et al.* 1997 |
| micronucleus test | human hepatocytes | 0.10 mM 0.18 mM 0.32 mM | –/–[1] +/(+) (+)/(+) | | Canonero *et al.* 1997 |
| DNA strand breaks | rat hepatocytes | 0.18–0.56 mM | – | | Canonero *et al.* 1997 |
| DNA strand breaks | human hepatocytes | 0.18-0.56 mM | (+)[1,2] | | Canonero *et al.* 1997 |

+ positive and significant, (+) weakly positive (not significant), – negative
[1] results obtained with hepatocytes from two donors
[2] designated by the authors as biologically significant


In the *Salmonella* mutagenicity test, hexachlorobenzene proved not to be mutagenic (BUA 1994). An SOS chromotest with *Escherichia coli* yielded negative results (Raabe *et al.* 1993). Selection of V79 cells with ouabain revealed no increase in mutation frequency, with 8-azaguanine a minimal increase (Kuroda 1986). Hexachlorobenzene did not induce chromosomal aberrations in human lymphocytes (Siekel *et al.* 1991). An increased incidence of micronuclei in rat hepatocytes was seen at the highest hexachloro-benzene concentration tested. In human hepatocytes the incidence was slightly but not significantly increased. The incidence of DNA single strand breaks was not increased in rat hepatocytes but in human hepatocytes it was slightly (from 24 % to 32 % of eluted DNA) but not significantly increased (Canonero *et al.* 1997).

### 5.6.2 *In vivo*

Two dominant lethal tests with rats given hexachlorobenzene doses of 0, 20, 40 or 60 mg/kg body weight and day for 10 days (Khera 1974) or 0, 70 or 221 mg/kg body weight and day for 5 days (Simon *et al.* 1979) revealed no significant effects on preg-nancy, fertility, number of implantations or resorptions (BUA 1994).

In one inadequately documented study, a test for sister chromatid exchange in the bone marrow cells of male mice yielded negative results. In the rat liver, a slight dose-dependent increase in DNA strand breaks was reported (Górski *et al.* 1986).

A test for replicative DNA synthesis in the livers of B6C3F$_1$ mice given single oral hexachlorobenzene doses of 500 or 1000 mg/kg body weight and examined after 24, 39 or 48 hours revealed a slight but not significant increase in the incidence of replicative DNA synthesis in the 500 mg/kg group animals from 0.18 % to 0.36 %, 0.35 % and 0.29 %, respectively. In the 1000 mg/kg group, the incidences were 0.16 %, 0.66 % and 0.01 %; the 0.66 % incidence proved statistically significant (Miyagawa *et al.* 1995).

In male mice given hexachlorobenzene in doses of 170 mg/kg body weight and day for 5 days, the liver weights were significantly increased. The incidence of DNA single-strand breaks and of 8-hydroxydeoxyguanosine formation were not increased (Umegaki *et al.* 1993).

# 5.7 Carcinogenicity

## 5.7.1 Short-term studies

In initiation-promotion studies, initiating potential of hexachlorobenzene in the rat liver could not be detected but a tumour-promoting effect was found (Table 3). This was evi-dent after initiation with a single diethylnitrosamine dose of 10 mg/kg body weight fol-lowed by hexachlorobenzene doses of 2 to 5 mg/kg body weight and day administered for 11 weeks (Oesterle and Deml 1994, 1998).

The development of the observed tumours can be explained in terms of known mechanisms (see Sections 2 and 6).

**Table 3.** Initiation-promotion studies with hexachlorobenzene

| Species | Initiation (I) | Promotion (P) | Results | | References |
|---|---|---|---|---|---|
| *Initiation studies* | | | | | |
| rat, F344 10 or 4 ♂ | PH, after 12 h oral HCB dose 5000 mg/kg body weight | week 2–12: 0.15 % cholic acid in the diet; week 4: one oral $CCl_4$ dose 1 ml/kg body weight | **number of GST-positive liver foci per $cm^2$:** P: I (HCB): I (HCB) + P: | 0.72 ± 0.41 0.79 ± 0.41 0.91 ± 0.65 | Tsuda *et al.* 1993 |
| *Promotion studies* | | | | | |
| rat, F344 at least 10 ♂ per group | DEN: 200 mg/kg body weight, i.p., once | 6 weeks 0.6, 3, 15, 75, 150 ppm HCB in the diet (about 0.06, 0.3, 1.5, 7.5, 15 mg/kg body weight and day); after 1 week PH | **number of GST-positive liver foci per $cm^2$:** I: I + P (HCB 0.06): I + P (HCB 0.3): I + P (HCB 1.5): I + P (HCB 7.5): I + P (HCB 1.5): | 8.8  ± 2.00 9.95 ± 2.60 10.84 ± 2.19* 11.86 ± 1.97** 16.96 ± 3.40** 21.18 ± 4.53** | Cabral *et al.* 1996 |
| rat, SD at least 5 ♀ per group | 1 × 10 mg DEN oral | 11 weeks (3 × weekly) HCB: 2, 5, 10 mg/kg body weight and day, oral | **number of GGT-positive liver foci per $cm^2$:** 0: I: I + P (HCB 2): I + P (HCB 5): I + P (HCB 10): | 0.2–0.3 1.0 1.3 2.6 2.5 | Oesterle and Deml 1994, 1998 |
| rat F344, 10 ♂, 10 ♀ | 3 weeks 0.015 % DEN in the drinking water | 30 weeks 0.02 % HCB in the diet (12 mg/kg body weight and day) | **% liver foci which were GGT-positive:** 0: I: P (HCB 12): I + P (HCB 12): | < 0.1 (♂), < 0.1 (♀) 3.4 (♂),   5.5 (♀) < 0.1 (♂), < 0.1 (♀) 18.8 (♂),   43.4 (♀) | Stewart *et al.* 1989 |
| | DEN: 20 mg/kg body weight, i.p., once | 30 weeks 0.02 % HCB in the diet (12 mg/kg body weight and day) | **% liver foci per $cm^2$ which were GGT-positive:** 0: I: P (HCB 12): I + P (HCB 12): | < 0.01 (♂), 0.32 (♀) 0.06 (♂), 0.75 (♀) 1.41 (♂), 1.71 (♀) 17.9 (♂), 12.36 (♀) | |

DEN: diethylnitrosamine, F344: Fischer-344, GGT: γ-glutamyl transpeptidase, GST: glutathione-*S*-transferase, HCB: hexachlorobenzene, I: initiation, P: promotion, PH: partial hepatectomy

* $p < 0.05$, ** $p < 0.01$

## 5.7.2 Long-term studies

After long-term administration of hexachlorobenzene in the diet, liver tumours were found in all the animal species studied (rat, mouse, golden Syrian hamster) (Table 4). In rats, especially in female rats, the incidence of neoplastic nodules was significantly increased from hexachlorobenzene concentrations of 40 ppm in the diet, equivalent to about 4 mg/kg body weight and day (Arnold *et al.* 1985, Arnold and Krewski 1988) and that of haemangiomas and hepatocarcinomas at 75 ppm, equivalent to about 7.5 mg/kg body weight and day (Ertürk *et al.* 1986). In the mouse, liver tumours were seen in animals given 100 ppm or more (about 6 mg/kg body weight and day; Cabral *et al.* 1979) and in golden Syrian hamsters from 50 ppm (about 6 mg/kg body weight and day; Cabral *et al.* 1977).

In addition, in the rats, especially in the females, phaeochromocytomas in the adrenal glands and bile duct tumours were found and, especially in the males, adenomas in the parathyroid glands and the kidneys (Arnold *et al.* 1985, Arnold and Krewski 1988, Ertürk *et al.* 1986). In golden Syrian hamsters, especially in the male hamsters, the incidence of thyroid tumours was increased (Cabral *et al.* 1977).

**Table 4.** Carcinogenicity studies with hexachlorobenzene

| Author: | Arnold *et al.* 1985, Arnold and Krewski 1988 |
|---|---|
| Substance: | hexachlorobenzene (analytical grade) |
| Species: | rat, Sprague-Dawley, at least 40 ♂, 40 ♀ per group |
| Administration route: | (*in utero*, via lactation) in the diet |
| Concentration: | 0, 0.32, 1.6, 8.0, 40 ppm<br>(about 0, 0.032, 0.16, 0.8, 4 mg/kg body weight and day) |
| Duration: | (*in utero*, via lactation) 130 weeks in the diet |
| Toxicity: | at concentrations of 8 ppm or more: slight effects on the liver (centrilobular basophilic chromogenesis)<br>40 ppm: peribiliary lymphocytosis and fibrosis (♂), chronic nephrosis (♂) |

| Tumours: | | Concentration (ppm in the diet) | | | | |
|---|---|---|---|---|---|---|
| | | 0 | 0.32 | 1.6 | 8 | 40 |
| number of animals | ♂ | 48 | 48 | 48 | 49 | 49 |
| | ♀ | 49 | 49 | 50 | 49 | 49 |
| liver, | ♂ | 2 (4 %) | 0 (–) | 0 (–) | 2 (4 %) | 1 (2 %) |
| neoplastic nodules | ♀ | 0 (–) | 0 (–) | 2 (4 %) | 2 (4 %) | 10 (20 %)** |
| adrenal glands, | ♂ | 10 (21 %) | 12 (25 %) | 7 (15 %) | 13 (27 %) | 17 (35 %) |
| phaeochromocytomas | ♀ | 2 (4 %) | 4 (8 %) | 4 (8 %) | 5 (10 %) | 17 (35 %)** |
| parathyroid glands, | ♂ | 2 (4 %) | 4 (8 %) | 2 (4 %) | 1 (2 %) | 12 (24 %)* |
| adenomas | ♀ | 0 (–) | 0 (–) | 0 (–) | 1 (2 %) | 2 (4 %) |

* p < 0.05, ** p < 0.01

**Table 4.** continued

| Author: | Ertürk *et al.* 1986 |
| --- | --- |
| Substance: | hexachlorobenzene (> 99.5 %) |
| Species: | rat, Sprague-Dawley, at least 50 ♂, 50 ♀ per group |
| Administration route: | diet |
| Concentration: | 0, 75, 150 ppm (about 0, 7.5, 15 mg/kg body weight and day) |
| Duration: | 104 weeks |
| Toxicity: | no data |

| Tumours: | | Concentration (ppm in the diet) | | |
| --- | --- | --- | --- | --- |
| | | 0 | 75 | 150 |
| number of animals | ♂ | 54 | 52 | 56 |
| | ♀ | 52 | 56 | 55 |
| liver, haemangiomas* | ♂ | 0 (–) | 10 (20 %) | 11 (20 %) |
| | ♀ | 0 (–) | 23 (41 %) | 35 (64 %) |
| liver, carcinomas* | ♂ | 0 (–) | 3 (6 %) | 4 (7 %) |
| | ♀ | 0 (–) | 36 (66 %) | 48 (87 %) |
| bile duct tumours* | ♂ | 0 (–) | 2 (4 %) | 2 (4 %) |
| | ♀ | 1 (2 %) | 19 (34 %) | 29 (53 %) |
| kidney, | ♂ | 7 (13 %) | 41 79 %) | 42 (75 %) |
| renal-cell adenomas* | ♀ | 1 (2 %) | 7 (13 %) | 15 (27 %) |

* the authors claimed that the incidence of these tumours was increased; level of significance not stated

| Author: | Smith *et al.* 1985 |
| --- | --- |
| Substance: | hexachlorobenzene (analytical grade) |
| Species: | rat, Fischer 344, 15 ♂, 15 ♀ |
| Administration route: | diet |
| Concentration: | 0.02 % (about 20 mg/kg body weight and day) |
| Duration: | 90 weeks |
| Toxicity: | 0.02 %: decreased body weight gains, increased liver and kidney weights, decreased spleen weights in ♀ |

| Tumours: | | Concentration (ppm in the diet) | |
| --- | --- | --- | --- |
| | | 0 | 0.02 % |
| number of animals | ♂ | 10 | 12 |
| | ♀ | 10 | 10 |
| liver tumours, neoplastic nodules | ♂ | 0 | 0 |
| | ♀ | 0 | 5 |
| hepatocellular carcinomas | ♂ | 0 | 0 |
| | ♀ | 0 | 5 |

**Table 4.** continued

| Author: | Smith and Cabral 1980 |
|---|---|
| Substance: | hexachlorobenzene (> 99.5 %) |
| Species: | rat, Agus, 14 ♀, Wistar, 6 ♀ |
| Administration route: | diet |
| Concentration: | 100 ppm: (about 6–8 mg/kg body weight and day) |
| Duration: | 90 weeks (Agus), 75 weeks (Wistar) |
| Toxicity: | 100 ppm: decreased body weight gains, porphyria |

| Tumours: | Concentration (ppm in the diet) | | | |
|---|---|---|---|---|
| | Agus | | Wistar | |
| | 0 | 100 | 0 | 100 |
| number of animals | 12 | 14 | 4 | 6 |
| liver cell tumours | 0 | 14 | 0 | 4 |

| Author: | Cabral *et al.* 1979 |
|---|---|
| Substance: | hexachlorobenzene (> 99.5 %) |
| Species: | mouse, Swiss,<br>50 ♂ and 50 ♀ (0, 200 ppm) and 30 ♂ and 30 ♀ (50, 100 ppm) |
| Administration route: | diet |
| Concentration: | 0, 50, 100, 200 ppm (0, 6, 12, 24 mg/kg body weight and day) |
| Duration: | 101–120 weeks |
| Toxicity: | at concentrations of 50 ppm or more: decreased body weight gains (♀)<br>at concentrations of 100 ppm or more: decreased body weight gains (♀)<br>200 ppm: increased mortality |

| Tumours: | | Concentration (ppm in the diet) | | | |
|---|---|---|---|---|---|
| | | 0 | 50 | 100 | 200 |
| number of animals | ♂ | 47 | 30 | 29 | 44 |
| | ♀ | 49 | 30 | 30 | 41 |
| survival (90 weeks) | ♂ | 50 % | 30 % | 27 % | 4 % |
| | ♀ | 48 % | 40 % | 30 % | 0 % |
| liver tumours | ♂ | 0 (–) | 0 (–) | 3 (10 %)* | 7 (16 %)* |
| | ♀ | 0 (–) | 0 (–) | 3 (10 %)* | 14 (34 %)* |

* increased incidence, level of significance not specified

| Author: | Cabral *et al.* 1977 |
|---|---|
| Substance: | hexachlorobenzene (> 99.5 %) |
| Species: | golden Syrian hamster, 40 ♂ and 40 ♀ (0 ppm),<br>30 ♂ and 30 ♀ (50, 100 ppm), 60 ♂ and 60 ♀ (200 ppm) |
| Administration route: | diet |
| Concentration: | 0, 50, 100, 200 ppm (0, 4, 8, 16 mg/kg body weight and day) |
| Duration: | for life (> 70 weeks) |
| Toxicity: | no data |

**Table 4.** continued

| Tumours: | | Concentration (ppm in the diet) | | | |
|---|---|---|---|---|---|
| | | 0 | 50 | 100 | 200 |
| number of animals | ♂ | 40 | 30 | 30 | 57 |
| | ♀ | 39 | 30 | 30 | 60 |
| tumour-bearing animals | ♂ | 3 (8 %) | 18 (60 %) | 27 (90 %) | 56 (98 %) |
| | ♀ | 5 (13 %) | 16 (53 %) | 18 (60 %) | 52 (87 %) |
| liver, hepatomas* | ♂ | 0 (–) | 14 (47 %) | 26 (87 %) | 49 (86 %) |
| | ♀ | 0 (–) | 14 (47 %) | 17 (57 %) | 51 (85 %) |
| haemangioendotheliomas* | ♂ | 0 (–) | 1 (3 %) | 6 (20 %) | 20 (35 %) |
| | ♀ | 0 (–) | 0 (–) | 2 (7 %) | 7 (12 %) |
| thyroid gland, tumours | ♂ | 0 (–) | 0 (–) | 1 (3 %) | 8 (14 %)** |
| | ♀ | 0 (–) | 2 (7 %) | 1 (3 %) | 3 (5 %) |
| spleen, haemangioendotheliomas | ♂ | 0 (–) | 1 (3 %) | 3 (10 %) | 4 (7 %) |
| | ♀ | 1 (3 %) | 0 (–) | 3 (10 %) | 4 (7 %) |
| other tumours | ♂ | 3 (8 %) | 11 (37 %) | 9 (30 %) | 6 (11 %) |
| | ♀ | 4 (10 %) | 5 (17 %) | 9 (30 %) | 8 (13 %) |

* increased incidence, level of significance not specified; ** $p < 0.05$

# 5.8. Other effects

## 5.8.1 Effects on the immune system

In persons who had eaten bread contaminated with hexachlorobenzene for a long period (see Section 4.1) enlarged thyroid glands and arthritis were observed, symptoms suggestive of an autoimmune reaction (Vos and Van Loveren 1995).

In rats, hexachlorobenzene caused amplification of the humoral and cell-mediated immune responses; the developing immune system was shown to be particularly sensitive (see also BUA 1994).

In Wistar rats exposed prenatally and postnatally to hexachlorobenzene, even the lowest concentration tested (4 ppm in the diet; about 0.4 mg/kg body weight and day) caused an increase in the immune response (IgM, IgG) to tetanus toxoid. The weight of the lymph nodes was slightly increased at 4 ppm and significantly increased from 20 ppm. In the lymph nodes the endothelial veins were enlarged. The spleen weights increased with the dose. B-lymphocyte hyperplasia was detected (BUA 1994), a sign of selective activation of B-1 cells (Schielen *et al.* 1995b). Macrophage accumulation was observed in the lungs. From 100 ppm the IgM level was increased (BUA 1994). There is also evidence that hexachlorobenzene causes effects like autoimmune reactions in rats (IgM directed, for example, against the autoantigens single strand DNA and native DNA), the significance of which is still unclear (Schielen *et al.* 1993, Vos and Van Loveren 1995).

The observed dermal lesions resulting from exposure to hexachlorobenzene are attributed at least in part to autoimmune reactions (Schielen *et al.* 1995a).

In mice, on the other hand, a suppression of immune responses was generally observed (BUA 1994).

### 5.8.2 Neurotoxicity

Weakness, tremor, seizures and high mortality were seen in breast-fed babies and small children whose mothers were exposed to hexachlorobenzene. In exposed adults, weakness but no tremor or seizures developed (see Section 4.1). Suckling rat pups whose dams were exposed to hexachlorobenzene died in convulsions. In adult rats seizures were rarely seen. Epileptiform seizures were, however, induced with ultrasound in 3 of 11 rats given oral hexachlorobenzene doses of 100 mg/kg body weight, daily for 13 days (Mylchreest and Charbonneau 1994).

# 6 Manifesto (MAK value/classification)

An epidemiological study of a population exposed to hexachlorobenzene has provided evidence of carcinogenic effects in man; clarification is still required.

Hexachlorobenzene is a substance with typical tumour-promoting properties. The liver, bile duct and kidney tumours observed in experimental animals may be attributed to the cytochrome P450-mediated formation of oxygen radicals resulting in cell damage and subsequent compensatory hyperplasia. Thyroid tumours result from induction of glucuronyl transferases which causes increased glucuronidation and excretion of thyroxine and subsequent enlargement of the thyroid gland. As for the majority of substances with tumour-promoting properties, if hexachlorobenzene has any clastogenic effects they are very weak. Because of its tumour-promoting properties, hexachlorobenzene is classified in Section III, Category 4 of the *List of MAK and BAT Values*.

Derivation of a MAK value for the substance on the basis of the data available for effects in man is not possible. Extrapolation from the results of animal studies to the human situation is associated with considerable uncertainty because of the different half-lives of the substance in man and animals (about 5 months in the rat, more than 2 years in man). Because of the highly cumulative effects of the substance in man, it is not to be expected that reliable data which would permit the establishment of a MAK value will be available in the future. There are, however, data available for the systemic levels of hexachlorobenzene in workers from which the BAT value of 150 µg hexachlorobenzene per litre plasma was established (Henschler and Lehnert 1984). Because no statistically significant changes in biochemical parameters such as markers of porphyrin metabolism or liver function were found in these workers, it may be assumed that observance of the BAT value also provides protection from the tumour-promoting effects of the substance.

Hexachlorobenzene can penetrate the skin and is therefore designated with an "H". The data available from animal studies yield no evidence of sensitizing effects of hexachlorobenzene. The substance has therefore not been designated with an "S". There is no evidence that the substance has potential to cause germ cell mutations.


# 7 References



Almeida MG, Fanini F, Davino SC, Aznar AE, Koch OR, Barros SBM (1997) Pro- and antioxidant parameters in rat liver after short term exposure to hexachlorobenzene. *Hum Exp Toxicol 16*: 257–261

Arnold DL, Krewski D (1988) Long-term toxicity of hexachlorobenzene. *Food Chem Toxicol 26*: 169–174

Arnold DL, Moodie CA, Charbonneau SM, Grice HC, McGuire PF, Bryce FR, Collins BT, Zawidzka ZZ, Krewski DR, Nera EA (1985) Long-term toxicity of hexachlorobenzene in the rat and the effect of dietary vitamin A. *Food Chem Toxicol 23*: 779–793

Babineau KA, Singh A, Jarrell JF, Villeneuve DC (1991) Surface epithelium of the ovary following oral administration of hexachlorobenzene to the monkey. *J Submicrosc Cytol Pathol 23*: 457–464

den Besten C, Bennik MM, van Iersel M, Peters MA, Teunis C, van Bladeren PJ (1994) Comparison of the urinary metabolite profiles of hexachlorobenzene and pentachlorobenzene in the rat. *Chem Biol Interact 90*: 121–137

Billi de Catabbi S, Sterin-Speziale N, Fernandez MC, Minutolo C, Aldonatti C, San Martin de Viale L (1997) Time course of hexachlorobenzene-induced alterations of lipid metabolism and their relation to porphyria. *Int J Biochem Cell Biol 29*: 335–344

Bourque AC, Singh A, Lakhanpal N, McMahon A, Foster WG (1995) Ultrastructural changes in ovarian follicles of monkeys administered hexachlorobenzene. *Am J Vet Res 56*: 1673–1677

BUA (Beratergremium für umweltrelevante Altstoffe der Gesellschaft deutscher Chemiker) (1994) *Hexachlorbenzol, Bericht Nr. 119*, Hirzel Verlag, Stuttgart

Cabral JRP, Shubik P, Mollner T, Raitano F (1977) Carcinogenic activity of hexachlorobenzene in hamsters. *Nature 269*: 510–511

Cabral JRP, Mollner T, Raitano F, Shubik P (1979) Carcinogenesis of hexachlorobenzene in mice. *Int J Cancer 23*: 47–51

Cabral R, Hoshiya T, Hakoi K, Hasegawa R, Ito N (1996) Medium-term bioassay for the hepatocarcinogenicity of hexachlorobenzene. *Cancer Lett 100*: 223–226

Canonero R, Campart GB, Mattioli F, Robbiano L, Martelli A (1997) Testing of *p*-dichlorobenzene and hexachlorobenzene for their ability to induce DNA damage and micronucleus formation in primary cultures of rat and human hepatocytes. *Mutagenesis 12*: 35–39

Carthew P, Smith AG (1994) Pathological mechanisms of hepatic tumour formation in rats exposed chronically to dietary hexachlorobenzene. *J Appl Toxicol 14*: 447–452

Courtney KD, Copeland MF, Robbins A (1976) The effects of pentachloronitrobenzene, hexachlorobenzene, and related compounds on fetal development. *Toxicol Appl Pharmacol 35*: 239–256

Ertürk E, Lambrecht RW, Peters HA, Cripps DJ, Gocmen A, Morris CR, Bryan GT (1986) Oncogenicity of hexachlorobenzene. *IARC Scientific Publ No 77*: 417–423

Foster WG, Pentick JA, McMahon A, Lecavalier PR (1992) Ovarian toxicity of hexachlorobenzene (HCB) in the superovulated female rat. *J Biochem Toxicol 7*: 1–4

Foster WG, Pentick JA, McMahon A, Lecavalier PR (1993) Body distribution and endocrine toxicity of hexachlorobenzene (HCB) in the female rat. *J Appl Toxicol 13*: 79–83

Foster WG, McMahon A, Younglai EV, Jarrell JF, Lecavalier P (1995a) Alterations in circulating ovarian steroids in hexachlorobenzene-exposed monkeys. *Reprod Toxicol 9*: 541–548

Foster WG, Mertineit C, Yagminas A, McMahon A, Lecavalier P (1995b) The effects of hexachlorobenzene on circulating levels of adrenal steroids in the ovariectomized rat. *J Biochem Toxicol 10*: 129–135

Górski T, Górska E, Górecka D, Sikora M, (1986) Hexachlorobenzene is non-genotoxic in short-term tests. *IARC Scientific Publ No 77*: 399–401

Grimalt JO, Sunyer J, Moreno V, Amaral OC, Sala M, Rosell A, Anto JM, Albaiges J (1994) Risk excess of soft-tissue sarcoma and thyroid cancer in a community exposed to airborne organochlorinated compound mixtures with a high hexachlorobenzene content. *Int J Cancer 56*: 200–203

Gulyaeva LF, Grishanova AY, Lyakhovich VV (1994) Induction of cytochromes P-4501A and P-4502B in various organs of rats treated with hexachlorobenzene and Arochlor-1254. *Biochemistry (Moskow) 59*: 383–387

Henschler D, Lehnert G (Eds) (1994) Hexachlorbenzol. in: *Biologische Arbeitsstoff-Toleranz-Werte (BAT-Werte) und Expositionsäquivalente für krebserzeugende Arbeitsstoffe (EKA)*, 7th issue (1994), and in English translation in: *Biological Exposure Values for Occupational Toxicants and Carcinogens Volume 2*, VCH-Verlagsgesellschaft, Weinheim

Jarrell JF, McMahon A, Villeneuve D, Franklin C, Singh A, Valli VE, Bartlett S (1993) Hexachlorobenzene toxicity in the monkey primordial germ cell without induced porphyria. *Reprod Toxicol 7*: 41–47

Kavlock RJ, Short Jr RD, Chernoff N (1987) Further evaluation of an *in vivo* teratology screen. *Teratogen Carcinogen Mutagen 7*: 7–16

Khera KS (1974) Teratogenicity and dominant lethal studies on hexachlorobenzene in rats. *Food Cosmet Toxicol 12*: 471–477

Kleiman de Pisarev DL, Ferramola de Sancovich AM, Sancovich HA (1995) Hepatic indices of thyroid status in rats treated with hexachlorobenzene. *J Endocrinol Invest 18*: 271–276

Koss G, Koransky W (1975) Studies on the toxicology of hexachlorobenzene. I. Pharmacokinetics. *Arch Toxicol 34*: 203–212

Kuiper-Goodman T, Grant DL (1986) Subchronic toxicity of hexachlorobenzene in the rat: clinical, biochemical, morphological and morphometric findings. *IARC Scientific Publ No. 77*: 343–346

Kuiper-Goodman T, Grant DL, Moodie CA, Korsrud GO, Munro IC (1977) Subacute toxicity of hexachlorobenzene in the rat. *Toxicol Appl Pharmacol 40*: 359–375

Kuroda Y (1986) Genetic and chemical factors affecting chemical mutagenesis in cultured mammalian cells. *Basic Life Sci 39*: 359–375

Landsteiner BK, Jacobs J (1936) Studies on the sensitization of animals with simple chemical compounds II. *J Exp Med 64*: 625–639

Lilienthal H, Benthe C, Heinzow B, Winneke G (1996) Impairment of schedule-controlled behavior by pre- and postnatal exposure to hexachlorobenzene in rats. *Arch Toxicol 70*: 174–181

MacPhee IJ, Singh A, Wright GM, Foster WG, LeBlanc NN (1993) Ultrastructure of granulosa lutein cells from rats fed hexachlorobenzene. *Histol Histopathol 8*: 35–40

Michielsen CP, Bloksma N, Ultee A, van Mil F, Vos JG (1997) Hexachlorobenzene-induced immunomodulation and skin and lung lesions: a comparison between brown Norway, Lewis, and Wistar rats. *Toxicol Appl Pharmacol 144*: 12–26

Miyagawa M, Takasawa H, Sugiyama A, Inoue Y, Murata T, Uno Y, Yoshikawa K (1995) The *in vivo–in vitro* replicative DNA synthesis (RDS) test with hepatocytes prepared from male B6C3F$_1$ mice as an early prediction assay for putative nongenotoxic (Ames-negative) mouse hepatocarcinogens. *Mutat Res 343*: 157–183

Mylchreest E, Charbonneau M (1994) Ultrasound-induced epileptiform activity in rats treated with hexachlorobenzene. *Neurotoxicology 15*: 273–278

Mylchreest E, Charbonneau M (1997) Studies on the mechanism of uroporphyrinogen decarboxylase inhibition in hexachlorobenzene-induced porphyria in the female rat. *Toxicol Appl Pharmacol 145*: 23–33

Nakashima Y, Ohsawa S, Umegaki K, Ikegami S (1997) Hexachlorobenzene accumulated by dams during pregnancy is transferred to suckling rats during early lactation. *J Nutr 127*: 648–654.

Oesterle D, Deml E (1994) Different potency of promoting agents in three rat liver foci bioassays. 35th Spring Meeting of the German Society for Experimental and Clinical Pharmacology and Toxicology, Mainz, Germany, March 15–17, *Naunyn-Schmiedeberg's Arch Pharmacol 349, Suppl*: R127

Oesterle D, Deml E (1998) Personal communication to the secretariat of the Commission, 19.05.1998

van Ommen B, Hendriks W, Bessems JGM, Geesink G, Müller F, van Bladeren PJ (1989) The relation between the oxidative biotransformation of hexachlorobenzene and its porphyrinogenic activity. *Toxicol Appl Pharmacol 100*: 517–528

Raabe F, Janz S, Wolff G, Merten H, Landrock A, Birkenfeld T, Herzschuh R (1993) Genotoxicity assessment of waste products of aluminum plasma etching with the SOS chromotest. *Mutat Res 300*: 99–109

van Raaij JAGM, Kaptein E, Visser TJ, van den Berg KJ (1993a) Increased glucuronidation of thyroid hormone in hexachlorobenzene-treated rats. *Biochem Pharmacol 45*: 627–631

van Raaij JAGM, Frijters CM, van den Berg KJ (1993b) Hexachlorobenzene-induced hypothyroidism. Involvement of different mechanisms by parent compound and metabolite. *Biochem Pharmacol 46*: 1385–1391

Richter J, Landa K, Reznicek J (1994) Immune response in persons occupationally exposed to hexachlorobenzene. *Prac Lek 46*: 151–154

Rietjens IM, Steensma A, den Besten C, van Tintelen G, Haas J, van Ommen B, van Bladeren PJ (1995) Comparative biotransformation of hexachlorobenzene and hexafluorobenzene in relation to the induction of porphyria. *Eur J Pharmacol 293*: 293–299

Ríos de Molina MC, Iglesias S, Lauria L, San Martin de Viale LC (1996) Precancerous pathology evoked by hexachlorobenzene treatment. *Acta Physiol Pharmacol Ther Lat Am 46*: 71–81

Schielen P, Schoo W, Tekstra J, Oostermeijer HH, Seinen W, Bloksma N (1993) Autoimmune effects of hexachlorobenzene in the rat. *Toxicol Appl Pharmacol 122*: 233–243

Schielen P, den Besten C, Vos JG, van Bladeren PJ, Seinen W, Bloksma N (1995a) Immune effects of hexachlorobenzene in the rat: role of metabolism in a 13-week feeding study. *Toxicol Appl Pharmacol 131*: 37–43

Schielen P, Van Rodijnen W, Pieters RHH, Seinen W (1995b) Hexachlorobenzene treatment increases the number of splenic B-1-like cells and serum autoantibody levels in the rat. *Immunology 86*: 568–574

Schielen P, van der Pijl A, Bleumink R, Pieters RH, Seinen W (1996) Local popliteal lymph node reactions to hexachlorobenzene and pentachlorobenzene: comparison with systemic effects. *Immunopharmacology 31*: 171–181

Seldén A, Jacobson G, Berg P, Axelson O (1989) Hepatocellular carcinoma and exposure to hexachlorobenzene: a case report. *Br J Ind Med 46*: 138–140

Sickel P, Chalupa I, Beno J, Blasko M, Novotny J, Burian J (1991) A genotoxicological study of hexachlorbenzene and pentachloroanisole. *Teratogen Carcinogen Mutagen 11*: 55–60

Simon GS, Tardiff RG, Borzelleca JF (1979) Failure of hexachlorobenzene to induce dominant lethal mutations in the rat. *Toxicol Appl Pharmacol 47*: 415–419

da Silva Augusto LG, Lieber SR, Ruiz MA, Souza CAD (1997) Micronucleus monitoring to assess human occupational exposure to organochlorides. *Environ Mol Mutagen 29*: 46–52

Sims DE, Singh A, Donald A, Jarrell J, Villeneuve DC (1991) Alteration of primate ovary surface epithelium by exposure to hexachlorobenzene: a quantitative study. *Histol Histopathol 6*: 525–529

Smith AG, Cabral JR (1980) Liver-cell tumors in rats fed hexachlorobenzene. *Cancer Lett 11*: 169–172

Smith AG, Francis JE, Dinsdale D, Manson MM, Cabral JRP (1985) Hepatocarcinogenicity of hexachlorobenzene in rats and the sex difference in hepatic iron status and development of porphyria. *Carcinogenesis 6*: 631–636

Soprena de Kracoff YE, Ferramola de Sancovich AM, Sancovich HA, Kleiman de Pisarev DL (1994) Effect of thyroidectomy and thyroxine on hexachlorobenzene induced porphyria. *J Endocrinol Invest 17*: 301–305

Stewart FP, Manson MM, Cabral JRP, Smith AG (1989) Hexachlorobenzene as a promoter of diethylnitrosamine-initiated hepatocarcinogenesis in rats and comparison with induction of porphyria. *Carcinogenesis 10*: 1225–1230

den Tonkelaar EM, Verschuuren HG, Bankovska J, de Vries T, Kroes R, van Esch GJ (1978) Hexachlorobenzene toxicity in pigs. *Toxicol Appl Pharmacol 43*: 137–145

Tsuda H, Matsumoto K, Ogino H, Ito M, Hirono I, Nagao M, Sato K, Cabral R, Bartsch H (1993) Demonstration of initiation potential of carcinogens by induction of preneoplastic glutathione *S*-transferase P-form-positive liver cell foci: possible *in vivo* assay system for environmental carcinogens. *Jpn J Cancer Res 84*: 230–236

Umegaki K, Ikegami S, Ichikawa T (1993) Hepatic DNA damage in mice given organochlorine chemicals. *J Food Hyg Soc Jpn 34*: 68–73

Vos JG, van Loveren H (1995) Markers for immunotoxic effects in rodents and man. *Toxicol Lett 82–83*: 385–394

Vrecl M, Jan J, Pogacnik A, Bavdek SV (1996) Transfer of planar and non-planar chloro-biphenyls, 4,4'-DDE and hexachlorobenzene from blood to milk and to suckling infants. *Chemosphere 33*: 2341–2346

completed 31.03.1998

# Ozone

| | |
|---|---|
| **Classification/MAK value:** | **see Section IIIB**<br>**of the List of MAK and BAT Values** |
| **Classification dates from:** | **1995** |
| Synonyms: | triatomic oxygen |
| Chemical name (CAS): | ozone |
| CAS number: | 10028-15-6 |
| Structural formula: | |
| Molecular formula: | $O_3$ |
| Molecular weight: | 48 |
| Melting point: | $-192.1°C$ |
| Boiling point: | $-112°C$ |
| **1 ml/m$^3$ (ppm) $\approx$ 2 mg/m$^3$** | **1 mg/m$^3$ $\approx$ 0.5 ml/m$^3$ (ppm)** |

## Note

The MAK value of 0.2 mg/m$^3$ which was established for ozone in 1958 was described in the MAK documentation of 1973 (Henschler 1973) as being without a satisfactory scientific basis. At that time the necessity was emphasized for field studies with sufficiently large collectives and precise analysis of the concentrations in the workplace air. Such field studies are still not available. The studies which have been carried out have been short-term experimental human exposures and numerous animal studies. Epidemiological studies have investigated effects of ozone where it occurs in the environment, especially in the so-called summer smog. Here ozone is only one of the toxicologically relevant substances, even though it is the one used as reference substance (Lippmann 1993, Tager 1993). For this reason such studies do not provide data which is appropriate for the establishment of a workplace threshold concentration for ozone.

The present documentation is based mainly on recent reviews; original studies and more recent relevant literature are cited where necessary.

*Essential MAK Value Documentations.* DFG, Deutsche Forschungsgemeinschaft
Copyright © 2006 WILEY-VCH Verlag GmbH & Co. KGaA, Weinheim
ISBN: 3-527-31394-X

# 1 Toxic Effects and Modes of Action

Ozone is a very reactive molecule which reacts with all biological systems. It is highly irritating for the eyes and mucous membranes of the respiratory tract. Directly or indirectly, it causes permeability changes in cell and tissue membranes which result in release of various plasma components or even of the contents of whole cells which can then be detected, for example, in nasal and lung lavage fluid. The occurrence and the activation of inflammatory cells are important factors determining the toxicity of ozone. The mechanism which is involved in the observed lung function changes has not yet been elucidated.

Long-term exposure results in alterations in the cells of the respiratory tract with degeneration, hyperplasia and fibrous changes in the lungs. Organs other than the airways are only affected after exposure to very high concentrations of ozone. This is mainly a result of the high reactivity of the substance which means that most of the inhaled ozone takes part in chemical reactions in the respiratory tract. The effects of ozone are attributed to oxidative stress, the formation of aldehydes and ketones and the production of free radicals. *In vitro* ozone has unequivocal genotoxic potential and cell-transforming activity. In mice exposed for long periods to high concentrations of ozone, the incidence of pulmonary adenomas and carcinomas was increased. It may be assumed that all effects of ozone induce adaptation reactions which involve the induction of protective mechanisms. High concentrations overload or inactivate these protective mechanisms. Low concentrations of ozone have immunotoxic effects. In concentrations relevant at the workplace, the substance has no effects on reproduction.

# 2 Mechanism of Action

Ozone is a highly reactive molecule which can oxidize many of the molecules found in cells. The reaction with olefinic structures takes place according to the following overall reaction:

$$R-CH=CH-R' + O_3 + H_2O \longrightarrow R-CH=O + R'-CH=O + H_2O_2$$

The mechanism of this so-called Criegee reaction is shown in Figure 1. The reaction of ozone with electron donors such as glutathione is shown below:

$$GSH + O_3 \longrightarrow GS^\bullet + O_3^{\bullet-} + H^+$$

$$O_3^{\bullet-} + H^+ \longrightarrow HO^\bullet + O_2$$

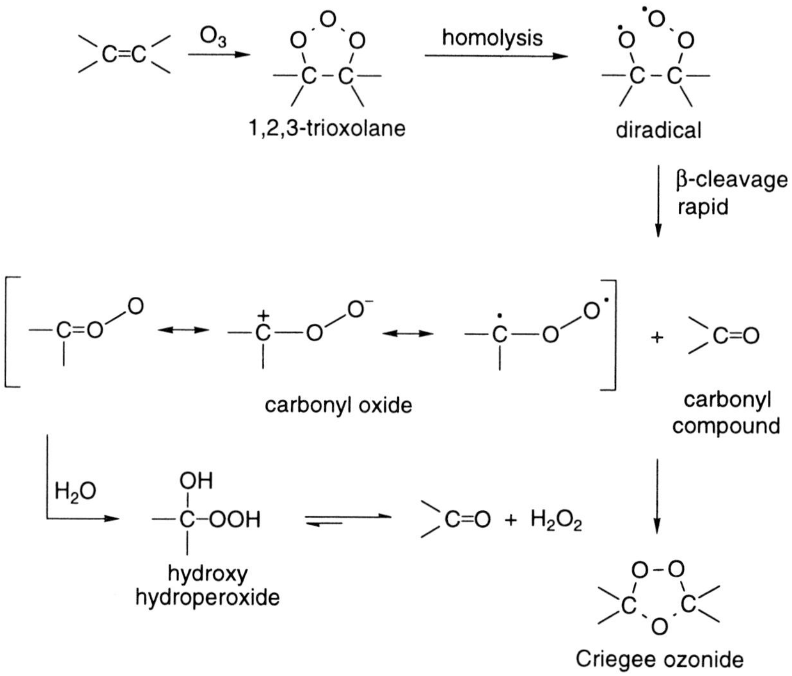

**Figure 1.** Criegee reaction in the presence and absence of water (from Pryor 1994)

Both reactions yield reactive oxygen species and aldehydes which could be responsible for the toxicity of ozone. The hydrogen peroxide is considered to be responsible for lipid peroxidation. The hydroxyl radical can react, for example, with DNA to produce strand breaks and oxidized bases (Steinberg *et al.* 1990). After incubation of DNA with ozone *in vitro*, thymine glycol, 8-hydroxydeoxyguanosine and hydroxymethyluracil were identified (Cajigas *et al.* 1994). After exposure of rats to ozone, spin trapping demonstrated that radicals can also be formed *in vivo*; the C-radical trapped *in vivo* has, however, not yet been identified (Kennedy *et al.* 1992). Which of the above-mentioned reactive species is responsible for the protein changes has not yet been fully clarified. In solution, especially the amino acids cysteine, methionine, tyrosine, tryptophan and histidine are seen to react with ozone (Pryor *et al.* 1991, Pryor 1994).

That ozone exerts its effects on cells by producing oxidative stress (Menzel 1994) can be deduced from the fact that many toxic effects of ozone can be prevented by antioxidants such as vitamin E and by deferoxamine, an iron chelating agent which also inhibits lipid peroxidation *in vivo* (Matsui *et al.* 1991, Louie *et al.* 1993, Pryor 1993). *In vitro* and *in vivo*, ozone induces numerous enzymes which are associated with mechanisms for protection of cells against oxidative stress. These include superoxide dismutase, glutathione peroxidase, glutathione reductase and catalase and also glutathione (Boehme *et al.*

1992, Rahman *et al.* 1991). In addition, phospholipases and factors which are induced during lipid peroxidation are activated (Wright *et al.* 1994). The concentration of ozone applied determines whether enzymes are induced or not because at higher concentrations toxic effects predominate.

Because so many intermediates are produced in the reactions of ozone with cellular components, it is practically impossible to allocate a particular reaction pathway as the mechanism of a particular toxic effect. Nonetheless, attempts in this direction have been made in a recent review article (Pryor 1994) (Figure 2).

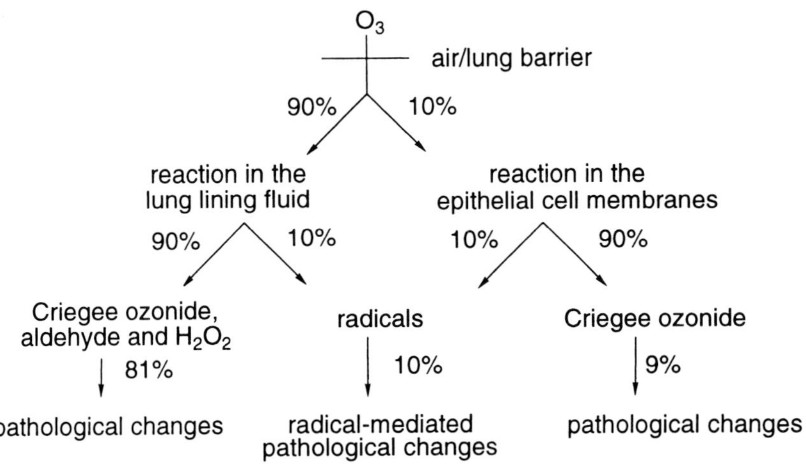


**Figure 2.** Potential reaction pathways for ozone in the lungs and their estimated proportions (from Pryor 1994)

The individual reaction pathways are highly dependent on the ozone concentration and on the kind of exposure (continuous or intermittent). Many parameters used to assay ozone toxicity are the results of secondary reactions; an example is the appearance of plasma constituents in the lungs after membrane destruction. It must be assumed that ozone cannot cross cell membranes because it reacts completely on the cell surface. Which of the products of such reactions are able to penetrate into the cells and perhaps to diffuse as far as the cell nucleus is not known. The stoichiometric production of hydrogen peroxide and the formation of reactive aldehydes and ketones can, however, account for a variety of intracellular effects of ozone exposure.

# 3 Toxicokinetics

The absorption of ozone in the lungs is dependent on the exposure concentration. During normal breathing about 40 % to 50 % is absorbed (Wiester *et al.* 1988). When ozone is applied directly to its site of action (in the region of the bronchio-alveolar junction) more than 90 % can be absorbed (Hu *et al.* 1992, 1994). The dose of ozone which reaches this part of the lung is highly dependent on the respiratory tidal volume (Postlethwait *et al.* 1994). Ozone is also very readily absorbed by the nasal mucosa (Hatch *et al.* 1994). However, oxygen derived from $^{18}O_3$ is found in blood only in extremely low concentrations. This is because the highly reactive ozone becomes bound in the lung tissue (Section 2 above). $^{18}O$ derived from $^{18}O_3$ accumulated in the lungs of mice where it had a half-life of about 6 hours (Santrock *et al.* 1989). The $^{18}O$ concentration in the lung lavage fluid obtained from rats exposed to ozone while at rest was only about one fifth of that obtained from physically active male test persons (Hatch *et al.* 1994). This suggests that the tidal volume determines the effective dose in the lungs. The differences are probably not only a result of differences in physical activity but also of differences in the respiratory physiology.

Until recently the toxicity of ozone was generally considered to be adequately described by Haber's rule (effect = concentration × time). Recently, however, the validity of this association has been questioned (Gelzleichter *et al.* 1992, Rajini *et al.* 1993a, 1993b). The effects of ozone depend not only on the tidal volume (which is constant in many studies) but also very definitely on the efficiency of absorption in the various sections of the respiratory tract (Gerrity *et al.* 1994, Kabel *et al.* 1994, Pinkerton *et al.* 1992). The latter is itself influenced by the toxic effects (Chang *et al.* 1991) which are also affected by adaptation; there can be marked differences between species (Oosting *et al.* 1991, Tepper *et al.* 1993). Recently developed mathematical models attempt to process the known data and to establish correlations between effective doses at the target tissue and the effects actually observed. These models have, however, not yet been validated (Kriebel and Smith 1990, McDonnell and Smith 1994).

The establishment of dose-response relationship is not only difficult because the effective dose cannot be calculated with any certainty but also because the various effects which are observed do not involve identical mechanisms. For example, there is evidence that the acute effects in the lungs result directly from reactions of ozone but subacute, subchronic and chronic effects from secondary reactions such as inflammatory processes. This is also the reason why no definite values for a NOEL (no observed effect level) for ozone can be given yet, neither for man nor for experimental animals.

# 4 Effects in Man

## 4.1 Single inhalation exposures

### 4.1.1 Effects on the nose

Healthy test persons were exposed to ozone concentrations in the range between 0.8 and 1.0 mg/m$^3$. Protein, inflammatory cells and inflammation mediators were detected in the nasal lavage fluid after as little as 2 hours exposure. Even 18 hours after the end of the exposure, the values of these parameters had not returned to their initial levels (Graham *et al.* 1988, Koren *et al.* 1990). The composition of the nasal lavage fluid did not differ from that of the lung lavage fluid except for the presence of some inflammation mediators (see below) (Bascom *et al.* 1990, Graham and Koren 1990).

In a lower concentration range between 0.24 and 0.48 mg/m$^3$ such effects were observed only in asthma patients (McBride *et al.* 1994). In the lung lavage fluid of these test persons, however, no increase in inflammatory cells or inflammation mediators was seen.

## 4.1.2 Effects on the lungs

Not until recently have short-term exposure studies been carried out with healthy test persons exposed for periods longer than two hours. In exposure chamber studies with and without physical activity, significant concentration-dependent changes in lung function were observed at ozone concentrations down to 0.16 mg/m$^3$. After exposure for 6.6 hours with moderate physical activity to an ozone concentration as low as 0.16 mg/m$^3$, forced expiratory volume in one second (FEV$_1$), forced vital capacity (FVC) and expiratory flow rate (FEF) were reduced (Table 1).

At ozone concentrations down to 0.24 mg/m$^3$ physical performance was also reduced. The reactivity towards substances with bronchoconstrictor activity increased after ozone exposure even at the lowest concentration studied, 0.16 mg/m$^3$. Airway permeability, on the other hand, was increased only after exposure to higher concentrations (Table 1). The release of inflammation mediators and inflammatory cells into the lung lavage fluid was also shown to be significantly increased after 6.6-hour exposures to ozone concentrations down to 0.16 mg/m$^3$ (Table 1). An increase in the tracheo-bronchial clearance of particles from the lungs was seen at concentrations down to 0.4 mg/m$^3$. Lower concentrations were not included in this study (Table 1).

Since ozone concentrations below 0.16 mg/m$^3$ were not tested in any of the studies, it is not possible to derive a NOEL for these effects.

The effects of ozone on the lungs of smokers, older adults, asthma patients and patients with lung diseases are not different from those on healthy adults. The only exception is patients with allergic rhinitis who are more sensitive to the effects of ozone on lung function parameters. Smokers who had given up smoking six months before the exposure to ozone were less sensitive to ozone-induced changes in lung function. Women seem to be a little more sensitive than men. Such comparative studies have, however, not been carried out in the low concentration range below 0.2 mg/m$^3$ (Lippmann 1992, McDonnell *et al.* 1993, Seal *et al.* 1993, Weymer *et al.* 1994).

An initial reduction in lung function parameters of 20 % after 70 minute exposure (including 30 minutes physical activity) to an ozone concentration of 0.6 mg/m$^3$ was only 4 % 18 hours later and after 42 hours 2 %. The lung function changes are generally considered to be reversible (Folinsbee *et al.* 1994, Lippmann 1992).

**Table 1.** Effects of single exposures to ozone in purified air on the lungs of test persons*

| Effects | Test persons | Exposure conditions | | | References |
|---|---|---|---|---|---|
| | | $O_3$ conc. (mg/m$^3$) | duration (h) | physical activity | |
| mean FEV$_1$ decreased by 5 %–10 % | healthy young men | 0.36 | 2 | high level, intermittent | McDonnell *et al.* 1983 |
| | | 0.16 | 6.6 | 5 h intermittent | McDonnell *et al.* 1991 |
| | | 0.2 | 6.6 | 5 h moderate, intermittent | Horstman *et al.* 1990 |
| | | 0.24 | 8 | 4 h moderate, intermittent | Hazucha *et al.* 1992 |
| FVC decreased, mean FEF decreased | healthy young men | 0.16 | 6.6 | 5 h intermittent | McDonnell *et al.* 1991 |
| coughing increased | healthy young men and women | 0.24–0.26 | *c* 0.3–0.5 | high level | Linder *et al.* 1988 |
| | | 0.24 | 2 | high level, intermittent | McDonnell *et al.* 1983 |
| | | 0.16 | 6.6 | 5 h moderate, intermittent | Horstman *et al.* 1990 |
| physical performance decreased | healthy young men | 0.36 | 0.5 plus 0.5 | at 54 l/min plus at 120 l/min respiratory volume | Schelegle and Adams 1986 |
| | healthy young men and women | 0.24–0.26 | *c* 0.3–0.5 | at 30–120 l/min respiratory volume | Linder *et al.* 1988 |

**Table 1.** continued

| Effects | Test persons | Exposure conditions | | | References |
|---|---|---|---|---|---|
| | | $O_3$ conc. (mg/m$^3$) | duration (h) | physical activity | |
| airway reactivity increased | young men with allergic rhinitis | 0.36 | 2 | high level | McDonnell *et al.* 1987 |
| | healthy young men | 0.16 | 6.6 | 5 h moderate, intermittent | Horstman *et al.* 1990; McDonnell *et al.* 1991 |
| airway permeability increased | healthy young men | 0.8 | 2 | high level, intermittent | Kehrl *et al.* 1987 |
| specific airway resistance increased | | 0.16 | 6.6 | 5 h intermittent | McDonnell *et al.* 1991 |
| airway inflammation increased | | 0.16 | 6.6 | 5 h, moderate intermittent | Devlin *et al.* 1991 |
| particle retention unchanged in spite of changes in lung function | | 0.8 | 1 | low level, continuous | Gerrity *et al.* 1993 |
| tracheo-bronchiolar particle clearance increased | healthy young men | 0.4, 0.66 | 2 | low level, continuous | Foster *et al.* 1987, 1993 |

FEV$_1$: forced expiratory volume in 1 second
FVC:  forced vital capacity
FEF:  expiratory flow rate
* modified from Lippmann 1992 (shortened and supplemented)

The cyclooxygenase products (thromboxanes, prostaglandins) detected in lung lavage fluid do not originate from inflammatory cells (neutrophils were investigated) but probably from pulmonary epithelial cells. This could be simulated *in vitro* with epithelial cells from bovine trachea and with human pulmonary epithelial cells (Lippmann 1992). It could be demonstrated that thromboxane is not responsible for the lung function changes induced by ozone although thromboxane $B_2$ was detected in the lung lavage fluid after ozone exposure. Significantly increased levels of certain proteins and other substances such as albumin, IgE, fibronectin, clotting factors, plasminogen activator, leukotrienes, interleukins, substance P and complement were also detected; some of these were still present at increased levels 18 hours after the end of exposure (Aris *et al.* 1993, Hazbun *et al.* 1993, Lippmann 1992).

The mechanism by which ozone causes changes in lung function has not been elucidated. It has been demonstrated that there is no association between bronchoconstriction and effects of ozone on the lung. The effects on certain parameters, e.g. the increase in airway resistance, seem to be evidence for a parasympathomimetic mechanism whereas the effects on others, such as forced vital capacity, are not in accordance with this thesis.

The changes in lung function do not appear to involve β-adrenergic mechanisms. An involvement of cyclooxygenase products is suggested by the fact that some of the lung function changes in man can be inhibited with indomethacin. After exposure of test persons to ozone, increases in prostaglandin $E_2$, $F_{2\alpha}$ and 6-keto-$F_{1\alpha}$ were found. Some of these cyclooxygenase metabolites are known to stimulate afferent neurons in the lung (Lippmann 1992, Hazbun *et al.* 1993). Unlike the situation in animals, no information is available on structural and morphological lung changes in man after short-term exposure to ozone.

## 4.2 Repeated inhalation exposures

It is well documented that functional changes caused by ozone are subject to adaptation (Lippmann 1992). During repeated daily exposure to low ozone concentrations the effects are most severe on day 2 and then become less pronounced every day until day 5 when practically no more effects on functional parameters can be detected. Whether the other lung changes in man are also subject to adaptation reactions is not known.

# 5 Animal Experiments and *in vitro* Studies

## 5.1 Acute and subacute inhalation toxicity

### 5.1.1 Effects on the nose

In rats exposed to an ozone concentration of 0.24 mg/m$^3$ for 6 hours, an increased number of inflammatory cells (neutrophils) was still found in the nasal lavage fluid 18 hours after the end of exposure. At higher concentrations and longer exposure times the effects were increased (Hotchkiss *et al.* 1989). As well as an increase in DNA synthesis, epithelial hyperplasia was seen (Henderson *et al.* 1993, Hotchkiss *et al.* 1991, Johnson *et al.* 1990).

### 5.1.2 Effects on the lungs

The effects of short-term exposures to ozone have been well documented in numerous animal studies (Lippmann 1992): the observed effects on lung function are like those known for man. If the product of ozone concentration and exposure time (mg/m$^3$ × hours) is plotted against the observed effects, a saturation curve is obtained. Mathematical model calculations revealed for the rat a saturation of the effects on lung function at 12 mg/m$^3$ × hours and a saturation of the release of inflammation mediators into the bronchio-alveolar fluid of about 8 mg/m$^3$ × hours.

Short-term effects of ozone are dependent on the respiratory tidal volume (Tepper *et al.* 1991, van Bree *et al.* 1992). Physical activity of the experimental animals and thus also tidal volume is reduced with increasing duration of exposure.

Functional changes can be detected after exposure to ozone concentrations down to about 0.2 mg/m$^3$. In animal studies the increase in airway reactivity could also be simulated but there is not enough data available to quote a lowest still effective concentration. As in man, the airway permeability in experimental animals is modified by ozone exposure. Albumin injected intravenously was found in the lung lavage fluid after exposure to an ozone concentration of 0.4 mg/m$^3$ continuously for 2 days. The increased permeability persisted in some animals for longer than 24 hours. Inflammation of the airways was observed in the rat and mouse after as little as four hours exposure to ozone concentrations down to 0.76 mg/m$^3$. More than 24 hours after the end of the exposure the protein levels in the lung lavage fluid were still increased. In contrast to the results obtained in man, the tracheobronchiolar clearance of particles from the lungs of experimental animals was reduced in all studies. The particle clearance dependent on the activity of the alveolar macrophages was accelerated after exposure to ozone concentrations as low as 0.2 mg/m$^3$ for as little as two hours (Chang *et al.* 1992, Kleeberger *et al.* 1993, Lippmann 1992).

# 5.2 Subchronic and chronic toxicity

## 5.2.1 Effects on the nose

Morphological changes were demonstrated in the nasal epithelial cells of monkeys exposed to an ozone concentration of 0.3 mg/m$^3$, 8 hours daily for 90 days (Dimitriadis 1992). The effects on the nasal epithelium have not been studied in as much detail as those on the pulmonary epithelium.

## 5.2.2 Effects on the lungs

Adaptation to ozone is well documented in animal studies. Daily exposure to an ozone concentration of 0.7 mg/m$^3$ (2.25 hours/day) resulted in lung function changes which were most severe on days one and two and then decreased continuously until day 5 when they were hardly detectable. With the same exposure pattern the adaptation was less at an exposure concentration of 1.0 mg/m$^3$ and was no longer present at 2.0 mg/m$^3$. In contrast, morphological changes and the release of protein did not show adaptive reactions at these ozone concentrations (0.7–2.0 mg/m$^3$). In fact, in this concentration range these effects were more pronounced after longer exposure periods (Lippmann 1992).

In animal studies long-term inhalation exposure produced adverse effects on lung function in all species investigated at ozone concentrations down to 0.2 mg/m$^3$ (Table 2). In this concentration range after a few weeks, cellular and morphological changes were also detectable in the lungs. Most severely affected were the ciliated cells and the Clara cells of the small airways (Table 2). Repair phenomena are manifested in hypertrophy and hyperplasia of various cells. Higher concentration exposures result in cell necrosis, collagen deposition, fibrosis and inflammation. In young animals higher concentrations inhibit growth and development of the lung (Table 2). Attention must be drawn to the fact that continuous exposure produced less pronounced alterations than did intermittent exposure, although the total dose administered intermittently was only half that administered during continuous exposure (Table 2).

To date it is not possible to state with any certainty where the threshold concentration for cumulative lung damage is to be found. It is also not completely clear which of the cellular and structural lung changes may be considered reversible (Plopper *et al.* 1994a). Exposure to high ozone concentrations (from 1.0 mg/m$^3$) causes irreversible lung changes with fibrosis (Table 2) (Lippmann 1992).

As well as functional and structural changes in the lungs, numerous biochemical changes have been detected, e.g. an increase in the activities of glutathione peroxidase and glutathione reductase and a decrease in $\alpha$-tocopherol levels (Chang *et al.* 1992, Lippmann 1992, Plopper *et al.* 1994b).

**Table 2.** Effects of medium-term and long-term exposure to ozone on the lungs of experimental animals*

| Effects | Exposure conditions | | References |
| --- | --- | --- | --- |
| | $O_3$ concentration (mg/m$^3$) | duration | |
| **juvenile rats** | | | |
| relative and absolute lung weights increased | 1.0, 2.0 | 6 h/day, 5 days/week, 4 weeks | NTP 1994 |
| **juvenile and adult rats** | | | |
| vital capacity increased, end-expiratory volume increased | 0.24–0.48 | 12 h/day, 6 weeks | Raub *et al.* 1983 |
| pneumocyte I hyperplasia | 0.24–0.48 | 12 h/day, 6 weeks | Barry *et al.* 1985 |
| alterations in ciliated and Clara cells in small airways | 0.24–0.48 | 12 h/day, 6 weeks | Barry *et al.* 1988 |
| **adult rats** | | | |
| pneumocyte I hyperplasia | 0.12 plus peak to 0.36 | 13 h/day plus peak: 9 h/day, 3 weeks or 12 weeks | Huang *et al.* 1988 |
| retention of inhaled asbestos fibres increased | 0.12 plus peak to 0.36 | 13 h/day plus peak: 9 h/day, 30 days; then 1 day asbestos | Pinkerton *et al.* 1989 |
| epithelial changes, endocrine cells increased, neuroepithelial bodies in the bronchioles increased | 0.5 | 5 h/day, 7 days/week, 20 months | Ito *et al.* 1994 |
| degeneration of ciliated cells in the bronchioles (also repair processes), hyperplasia of bronchiolar epithelial cells, increase of pneumocytes type II, fibrous changes in the alveoli after 3 months, in the bronchioles after 5 months, increase in collagen fibres | 1.0 | 6 h/day, 6 days/week, 12 months | Hiroshima *et al.* 1989 |
| epithelial hyperplasia, fibroblast proliferation, interstitial deposition of basal membranes and collagen fibres, damage to ciliated and Clara cells in the bronchio-alveolar region, changes more pronounced with increasing duration of exposure; changes only partially reversible | 0.12 plus peak to 0.5 | 13 h/day plus peak 9 h/day, 5 days/week, 78 weeks | Chang *et al.* 1992 |

**Table 2.** continued

| Effects | Exposure conditions | | References |
| --- | --- | --- | --- |
| | $O_3$ concentration (mg/m$^3$) | duration | |
| mild persistent fibrosis with collagen deposition in centri-azinar regions, no effect at 0.24 mg/m$^3$ | 1.0, 2.0 | 6 h/day, 5 days/week, 20 months | Last *et al.* 1993 |
| moderate changes in the functional residual capacity, the residual volume and the diffusion capacity | 1.0 | 20 h/day, 12 months | Gross and White 1987 |
| increased N$_2$-washout, residual volume decreased, lung capacity decreased, GSH peroxidase increased, GSH reductase increased, protein level in the lung lavage fluid increased | 0.12 plus peak to 0.36 | 13 h/day plus peak 9 h/day, 12 months | Grose *et al.* 1989 |
| expiratory resistance increased, respiration rate decreased especially during CO$_2$-induced hyperventilation, no detectable morphological changes | 0.12 plus peak to 0.36 | 13 h/day plus peak 9 h/day, 78 weeks | Tepper *et al.* 1991 |
| **juvenile mice** | | | |
| relative lung weight increased | 2.0 | 6 h/day, 5 days/week, 4 weeks | NTP 1994 |
| **juvenile and adult mice** | | | |
| increased areas, perimeters and linear intercepts of the alveolar wall, number and areas of LDH-staining pneumocytes type II increased | 0.6 | 7 h/day, 5 days/week, 6 weeks | Sherwin *et al.* 1983, Sherwin and Richters 1985 |
| **rabbits** | | | |
| secretory cells in the small airways increased after 4 months, then normalization, mucociliary clearance increased, after 6 months normalization | 0.2 | 2 h/day, 5 days/week, 12 months | Schlesinger *et al.* 1992 |
| **monkeys** | | | |
| hypertrophy and increase in the number of non-ciliated bronchiolar cells | 0.3 | 8 h/day, 90 days | Hyde *et al.* 1989 |

**Table 2.** continued

| Effects | Exposure conditions | | References |
|---|---|---|---|
| | $O_3$ concentration (mg/m$^3$) | duration | |
| hyperplasia of epithelial cells, accumulation of macrophages in the bronchioles, thickening of the cellular and non-cellular components of the epithelia, quantitative changes in the relative proportions of the various kinds of cells | 0.3, 0.6 | 8 h/day, 90 days | Harkema *et al.* 1993 |
| necrosis of ciliated cells, shortening of cilia, respiratory bronchiolitis | 0.3 or 0.6 | 8 h/day, 6 days or 90 days | Hyde *et al.* 1989 |
| chest wall compliance, inspiratory capacity and collagen content of the lungs increased, lung growth decreased, respiratory bronchiolitis | 0.5 | 8 h/day, 5 days/week, during alternate months, 18 months | Tyler *et al.* 1988 |
| thickening of the walls of the respiratory bronchioles, narrowing of the epithelial lumina and the interstitial compartments, increase in non-ciliated cells, interstitial fibres and amorphous ground substance, decrease in pneumocytes type I, increase in volume of macrophages, mast cells and neutrophils | 0.8, 1.28 | 8 h/day, 90 days | Moffatt *et al.* 1987 |
| diameter of the lumina of the terminal bronchioles decreased, volume of the respiratory bronchioles increased | 1.9 | 8 h/day, 90 days | Last *et al.* 1984 |
| lung growth decreased, respiratory bronchiolitis, effects similar to or less pronounced than after intermittent exposure | 0.5 | 8 h/day, 5 days/week, 18 months | Tyler *et al.* 1988 |
| lung collagen altered, effects still persistent 6 months after exposure | 1.22 then filtered air | 8 h/day, 12 months 6 months | Reiser *et al.* 1987 |
| volume of the respiratory bronchioles increased, but diameters of their lumina decreased, thickening of media and intima of the peribronchiolar arterioles and the epithelial and peribronchiolar connective tissue, cell volume of the cuboidal bronchiolar cells increased, ciliated cells, pneumocytes type II and amorphous extracellular matrix increased, infiltration of neutrophils and lymphocytes into peribronchiolar connective tissue | 1.28 | 8 h/day, 12 months | Fujinaka *et al.* 1985 |

* modified from Lippmann 1992 (shortened and supplemented)

### 5.2.3 Effects on other organs

In rats exposed for 5 days to an ozone concentration of 1.4 mg/m$^3$, extracellular and intracellular cardiac oedema was seen (Rahman *et al.* 1992). After shorter exposure periods and lower ozone concentrations (1.0 mg/m$^3$) reduced levels of protein synthesis and an increase in ANF (atrial natriuretic factor) were detected in the heart tissue of mice and rats (Kelly and Birch 1993, Vesely *et al.* 1994). The demonstration of lipid peroxidation products suggested that the effects could be considered to be indirect effects of ozone.

Effects on other organs have not been found at low ozone concentrations, that is at concentrations which are relevant for assessing the effects of human exposure.

Most results from animal studies which were obtained under conditions like those found at the workplace cannot be used to derive a NOEL, even when the sensitivity of the animals was like that of man, because the low concentration range (at and below 0.2 mg/m$^3$) and the toxicokinetics have not been studied systematically.

## 5.3 Allergenic effects

There is no evidence that ozone has sensitizing effects. The lung function changes, inflammatory reactions and other effects are generally considered not to be the result of allergic reactions (Lippmann 1992).

## 5.4 Reproductive and developmental toxicity

The early studies were reviewed in 1982 (Barlow and Sullivan 1982): there are no studies of effects on fertility. In rats which inhaled ozone in concentrations below 1.0 mg/m$^3$ during pregnancy, no embryotoxic effects were found. Exposure to 2.0 mg/m$^3$ for 8 hours per day led to slight increases in resorptions and signs of postnatal growth retardation. At ozone concentrations of 3.0 mg/m$^3$ increased embryo mortality and slightly reduced foetal weights were found; behavioural changes were also seen in the newborn pups. Ozone had no teratogenic effects in any of the animal studies.

In a more recent study of rats in which the dams were exposed during pregnancy (no other details) to an ozone concentration of 2.0 mg/m$^3$ for 12 hours daily, disorders of the sleeping-waking rhythm were seen in the offspring (Haro and Pas 1993). Another study in which mice were exposed continuously to ozone concentrations of 0.8, 1.6 or 2.4 mg/m$^3$ from day 7 to day 17 of gestation did not find such effects if the pups were separated from the dams immediately after birth (Bignami *et al.* 1994). In this study the only substance-related finding was a slight reduction in birth weights at the ozone concentration of 2.4 mg/m$^3$.

## 5.5 Genotoxicity

Ozone has unequivocal genotoxic potential. The reaction with isolated DNA and RNA is well known; the main points of attack are the bases thymine and guanine. Both single and double strand breaks are found. The reaction products have not yet been identified. Ozone is mutagenic in practically all known *in vitro* test systems (Table 3).

Negative results are obtained when insensitive test strains are used or when the ozone concentrations are too low or the high reactivity of ozone is not taken into account, e.g. bubbling through buffer solution before the addition of the cells. In plant cells too (not included in Table 3) ozone has genotoxic properties (Victorin 1992).

The genotoxic potential of ozone can also be demonstrated after exposure of animals *in vivo* (Table 4a). Here, however, negative results predominate even though very high concentrations were applied in some cases. It is noteworthy that negative results were obtained especially in bone marrow cells. The studies yielding positive results for chromosome and chromatid changes in peripheral lymphocytes and alveolar macrophages must be repeated to investigate the dependence of these effects on the ozone concentration.

The *in vivo* studies carried out to date with persons exposed to ozone cannot be interpreted because in human lymphocytes both negative and positive results have been obtained. It is conceivable that in the study which yielded positive results (Table 4b), the total dose of ozone (concentration × time) was decisive for the induction of the chromatid changes in the lymphocytes.

The *in vitro* genotoxicity of ozone may not be simply assumed to apply to the *in vivo* situation because ozone does not reach the appropriate targets *in vivo*. In the *in vivo* studies, clear relationships between exposure concentrations and effects have not been established. However, it is also conceivable that the genotoxic effects of ozone are mediated by products of the reactions of ozone with cell components, by aldehydes, for example (Victorin 1992). On the other hand, the *in vivo* results are not so clearly negative that an *in vivo* genotoxicity of ozone can be excluded.

## 5.6 Carcinogenicity

### 5.6.1 Short-term studies

In addition to its genotoxic effects, ozone also has clear cell transforming activity. This was demonstrated *in vitro* in primary hamster embryo cells, in mouse fibroblasts and in primary epithelial cells from rat trachea (Borek *et al.* 1986, 1989, Thomassen *et al.* 1991). However, the transforming effects of ozone could not be reproduced *in vivo* in tracheal epithelial cells of rats exposed to an ozone concentration of 2.4 mg/m$^3$, 6 hours daily, 5 days per week for up to 4 weeks (Thomassen *et al.* 1991).

**Table 3.** Genotoxic effects of ozone *in vitro*[*]

| Test system | End point | Result | References |
|---|---|---|---|
| *Escherichia coli* K12 | reversions and forward mutations | + | L'Hérault and Chung 1984a, 1984b, Hamelin *et al.* |
| various strains | DNA strand breaks | + | 1981, Hamelin and Chung 1989, Hamelin 1985, |
|  | toxicity in DNA repair deficient strains | + | Nover and Botzenhart 1985, Parduez Nagy-Gyorgy and Chung 1988 |
| *Bacillus subtilis* | toxicity in DNA repair deficient strains | + | Song and Chung 1983 |
| *Salmonella typhimurium* TA98, TA100, TA104, TA1535 | reversions, tested with and without metabolic activation | − | Dillon *et al.* 1992, Victorin and Stahlberg 1988 |
| *S. typhimurium* TA102 | reversions, tested with and without metabolic activation | + | Dillon *et al.* 1992 |
| *Saccharomyces cerevisiae* | reversions and forward mutations | + | Dubeau and Chung 1979, 1982 |
| various strains | gene conversion | + |  |
|  | mitotic "crossing-over" | + |  |
| Chinese hamster cells (V79) | inhibition of DNA synthesis | + | Rasmussen and Crocker 1982, Rasmussen 1986, |
|  | sister chromatid exchange (SCE) | weak + or − | Shiraishi and Bandow 1985 |
| mouse fibroblasts | DNA strand breaks | + | van der Zee *et al.* 1987 |
| (L 929) | DNA cross-links | + |  |
| cultured peripheral human | chromatid deletions | + | Gooch *et al.* 1976, Hsueh and Xiang 1984 |
| lymphocytes | sister chromatid exchange (SCE) | + |  |
|  | chromosomal aberrations | − |  |
| human foetal lung cells | chromatid deletions | + | Guerrero *et al.* 1979 |
| (WI-38) | sister chromatid exchange (SCE) | + |  |
|  | chromosomal aberrations | − |  |
| human epidermal cells (RHEK) | DNA strand breaks | − | Borek *et al.* 1988 |
| human lung fibroblasts (CCD-18Lu) | DNA strand breaks | − | Kozumbo and Agarwal 1990 |
| human lung fibroblasts (A549) | DNA strand breaks | − | Kozumbo and Agarwal 1990 |

[*] modified from Victorin 1992 (shortened and supplemented)

**Table 4a.** Genotoxic effects of ozone on experimental animals *in vivo* *

| Species | Cell type | End point | Result | References |
|---|---|---|---|---|
| Chinese hamster | peripheral lymphocytes | chromosomal aberrations | contradictory | Tice *et al.* 1978; Zelac *et al.* 1971a, 1971b |
|  |  | sister chromatid exchange (SCE) | – |  |
|  | bone marrow cells | structural chromosomal aberrations | – |  |
| mouse | peripheral lymphocytes | structural chromosomal aberrations | – | Gooch *et al.* 1976; Tice *et al.* 1978 |
|  |  | sister chromatid exchange (SCE) | – |  |
|  | bone marrow cells | structural chromosomal aberrations | – |  |
|  |  | sister chromatid exchange (SCE) | – |  |
| rat | lung macrophages | chromatid deletions | + | Rithidech *et al.* 1990; Zhurkov *et al.* 1979 |
|  |  | chromosomal aberrations | – |  |
|  | bone marrow cells | chromosomal aberrations | – |  |
| *Drosophila virilis* |  | dominant lethal mutations | + | Erdman and Hernandez 1982 |

* modified from Victorin 1992 (shortened and supplemented)

**Table 4b.** Genotoxic effects of ozone on test persons exposed *in vivo* (human peripheral lymphocytes)*

| Exposure | Test persons | End point | Result | References |
|---|---|---|---|---|
| 1.0 mg/m$^3$, 6 h or 10 h | 6 healthy persons (sex not specified) | achromatic lesions (gaps) | + | Merz *et al.* 1975 |
|  |  | structural chromosomal aberrations | – |  |
| 0.8 mg/m$^3$, 4 h | 30 healthy men | structural chromosomal aberrations | – | McKenzie *et al.* 1977 |
| 1.0 mg/m$^3$, 2 h | 31 healthy men and women | sister chromatid exchange (SCE) | – | Guerrero *et al.* 1979 |

* modified from Victorin 1992 (shortened and supplemented)

### 5.6.2 Long-term studies

A number of carcinogenicity studies have been carried out with ozone (Lippmann 1992, Victorin 1992, Witschi 1988, 1991). The early studies in which animals were exposed to ozone concentrations above 2.0 mg/m$^3$ do not, in the opinion of the experts, meet present day standards.

After exposure of female A/J mice to ozone concentrations of 0.62 or 1.0 mg/m$^3$, 103 hours weekly for 6 months, an increase in the incidence of lung adenomas was observed (no other details) (Hassett *et al.* 1985). In male A/J mice the incidence of lung adenomas was also increased after exposure to 1.6 mg/m$^3$ (8 hours daily for 18 weeks) but not after exposure to 0.8 mg/m$^3$ (Last *et al.* 1987). In male Swiss-Webster mice such an increase in the incidence of lung adenomas was not observed (Last *et al.* 1987).

In a more recent study with rats which were exposed to an ozone concentration of 0.4 mg/m$^3$, 6 hours daily, 5 days per week for 6 months, and then observed for the rest of their lives, the incidence of lung tumours increased from 0.9 % (historical control) to 6 % (Monchaux *et al.* 1994). Because there was no experimental control group, this study cannot be assessed conclusively.

Ozone has cocarcinogenic effects which are dependent on the dose scheme (Witschi 1991). That ozone can increase the tumour yield produced by known carcinogens but can also reduce it was ascribed to the fact that ozone has proliferation-promoting activity but is cytotoxic at higher concentrations.

In a recent NTP study (NTP 1994) the incidence of pulmonary adenomas and carcinomas was clearly increased in mice exposed to ozone concentrations of 1.0 mg/m$^3$ or more (Table 5). The tumour incidence was not increased in rats exposed to ozone but, as in the mice, metaplasia developed in the airways (Table 5). In mice exposed to an ozone concentration of 0.24 mg/m$^3$, the tumour incidence was not increased either (Table 5). These results were considered to provide equivocal evidence of carcinogenicity of ozone in male mice and some evidence in female mice (NTP 1994). In a parallel study of cocarcinogenic effects with the lung carcinogen NNK (4-[*N*-nitroso-*N*-methylamino]-1-[3-pyridyl]-1-butanone) there was no increase in the incidence of NNK-induced tumours in the rat lung (Boorman *et al.* 1994, NTP 1994).

## 5.7 Effects on the immune system

The effects of ozone on the immune system have been described frequently (Wright *et al.* 1990). The exposure of lymphocytes, monocytes, T-cells and macrophages *in vitro* caused suppression of the specific cell activity independent of whether the cells originated from man or from animals (Becker *et al.* 1989, 1991a, 1991b, Devlin *et al.* 1994, Harder *et al.* 1990, Soukup *et al.* 1993, Weideman and Schlesinger 1994). This could result in a reduced production of immunoglobulins and cell-specific mediators or in inhibition of cell activity such as phagocytosis of macrophages and killer activity of T-cells against tumour cells (Zelikoff *et al.* 1991). The activity of such immune cells is also reduced when they are isolated from persons or animals exposed to ozone (Gilmour and Jakab 1991, Orlando *et al.* 1988). It is not known whether the mechanism of the *in vitro* inhibition is identical with that of the *ex vivo* effects.

**Table 5.** Results of the NTP carcinogenicity study with rats and mice exposed to ozone (NTP 1994)

| | |
|---|---|
| Species: | rat (F344/N), 50 ♂ and 50 ♀ per group |
| Administration route: | inhalation |
| Concentration: | 0.24, 1.0, 2.0 mg/m$^3$ |
| Duration: | 104 weeks, 6 h/day, 5 days/week |
| Toxicity: | from 1 mg/m$^3$: metaplasia in the epiglottis, nose and lungs |
| Tumours: | no increase in the incidence of bronchioalveolar adenomas and carcinomas |

| | |
|---|---|
| Species: | rat (F344/N), 50 ♂ and 50 ♀ per group |
| Administration route: | inhalation |
| Concentration: | 1.0, 2.0 mg/m$^3$ |
| Duration: | 124 weeks (life-time exposure), 6 h/day, 5 days/week |
| Toxicity: | metaplasia in the nose, larynx and lungs (including fibrosis) |
| Tumours: | no increase in the incidence of bronchioalveolar adenomas and carcinomas |

| | |
|---|---|
| Species: | mouse (B6C3F$_1$), 50 ♂ and 50 ♀ per group |
| Administration route: | inhalation |
| Concentration: | 0.24, 1.0, 2.0 mg/m$^3$ |
| Duration: | 105 weeks, 6 h/day, 5 days/week |
| Toxicity: | from 1 mg/m$^3$: metaplasia in the nose and lungs, hyperplasia in the nose and epiglottis |
| Tumours: | increased incidence of bronchioalveolar adenomas and carcinomas |

| Ozone concentration (mg/m$^3$) | | 0 | 0.24 | 1.0 | 2.0 |
|---|---|---|---|---|---|
| survivors (104 weeks) | ♂ | 30/50 | 34/50 | 25/50 | 27/50 |
| | ♀ | 29/50 | 37/50 | 33/50 | 40/50 |
| bronchioalveolar adenomas and carcinomas | ♂ | 14/50 (28 %) | 13/50 (26 %) | 18/50 (36 %) | 19/50 (38 %) |
| | ♀ | 6/50 (12 %) | 7/50 (14 %) | 9/49 (18 %) | 16/50* (32 %) |

* significantly different from the control group (p < 0.05)

| | |
|---|---|
| Species: | mouse (B6C3F$_1$), 50 ♂ and 50 ♀ per group |
| Administration route: | inhalation |
| Concentration: | 1.0, 2.0 mg/m$^3$ |
| Duration: | 130 weeks (life-time exposure), 6 h/day, 5 days/week |
| Toxicity: | metaplasia in the nose, larynx and lungs, hyperplasia in the nose and epiglottis |
| Tumours: | increased incidence of bronchioalveolar adenomas and carcinomas |

| Ozone concentration (mg/m$^3$) | | 0 | 1.0 | 2.0 |
|---|---|---|---|---|
| survivors (130 weeks) | ♂ | 14/50 | 11/50 | 12/50 |
| | ♀ | 9/50 | 12/50 | 10/50 |
| bronchioalveolar adenomas and carcinomas | ♂ | 16/49 (33 %) | 22/49 (45 %) | 21/50 (42 %) |
| | ♀ | 6/50 (12 %) | 8/49 (16 %) | 12/50* (24 %) |

* significantly different from the control group (p < 0.01)

*In vivo* studies with rats and mice have also shown that ozone has immunotoxic effects. In these studies the animals were exposed to ozone concentrations of 0.8 and 1.6 mg/m$^3$ for 3 hours and then infected with bacteria. The ozone treatment increased the severity of the bacterial infection (Gilmour and Selgrade 1993). A threshold value for the immunotoxic effects of ozone has not yet been determined. After exposure of test persons to an ozone concentration of 0.16 mg/m$^3$ for 6.6 hours the *in vitro* phagocytosis activity of the alveolar macrophages in the lung lavage fluid was significantly reduced (Devlin *et al.* 1991).

# 6 Manifesto (MAK value, classification)

Recent studies demonstrated that exposure of healthy test persons to an ozone concentration of 0.16 mg/m$^3$ for 6.6 hours produced changes not only in lung function parameters but also in permeability of the lung with the release of inflammatory cells and inflammation mediators. These results made it clear that the MAK value of 0.2 mg/m$^3$ (0.1 ml/m$^3$), which had applied for ozone until 1995, required revision. Since these changes are reversible and also subject to adaptation and since the exposures took place under conditions of mild to moderate physical activity, that is, under conditions like those at the workplace, it was concluded that such effects would be avoided if the cumulative dose, (concentration × time: 0.16 mg/m$^3$ × 6.6 hours ≈ 1.0 mg/m$^3$ × hour), remained under 1.0 mg/m$^3$ × hour. Such requirements could be met by observance of a MAK value of 0.2 mg/m$^3$ for a period of 4 hours or 0.1 mg/m$^3$ for a period of 8 hours. However, it is not clear from which cumulative dose an effect must actually be expected. In addition, it has recently been suggested that the effects of ozone may not actually correlate with the cumulative dose.

On the other hand, such a MAK value could be established on the basis of results of animal studies. For example, on the basis of the effects which no longer developed in rats exposed to an ozone concentration of 0.48 mg/m$^3$ for 6 hours, a NOEL for the rat (nasal epithelium) of 2.88 mg/m$^3$ × hour has been calculated (Henderson *et al.* 1993). A MAK value of 0.2 mg/m$^3$ and 4-hour exposures or 0.1 mg/m$^3$ and 8-hour exposures (0.2 mg/m$^3$ × 4 hours = 0.8 mg/m$^3$ × hour) would then be lower than the NOEL for the rat by a factor of 3.6. In earlier studies, however, significant effects of ozone on the rat nose were found at a concentration as low as 0.24 mg/m$^3$ for a 6-hour exposure (Hotchkiss *et al.* 1989). In addition, it is unclear whether effects on the lung are also covered by these values because after a cumulative dose of 2.88 mg/m$^3$ × hour effects were still seen in the rat lung (Rajini and Witschi 1995). There are also indications that ozone has no effect threshold. This, however, is not in accordance with known adaptation phenomena which have not yet been successfully quantified. Therefore the available data are still not adequate for the derivation of a MAK value for ozone.

Since ozone caused a significant increase in the tumour incidence in female mice in the NTP study, the possibility that ozone has carcinogenic potential cannot be excluded (NTP 1994). This suspicion is supported by the results of other animal studies. There is evidence that ozone has tumour-promoting effects. At very high ozone concentrations. the tumours could be a result of the persistent inflammatory reactions in the lungs. This is in agreement with the fact that the increased incidence of lung tumours in the NTP study was seen only at the two highest concentrations. This mechanism of tumour induction, like that for formaldehyde, would in any case involve a dose-response relationship with an induction threshold concentration. That is not in accordance with the observation that ozone is unequivocally genotoxic *in vitro*. *In vivo* the genotoxicity is not as well documented and reliable dose-response relationships are not available. However, the suggested mechanism of genotoxicity via oxidative DNA damage or the formation of genotoxic reactants during oxidative stress could also involve a genotoxicity threshold because cells are equipped with protective mechanisms against oxidative stress. Overloading of these mechanisms in the mouse lung could account for the increased tumour incidence at high ozone concentrations. Differences in availability or inducibility of these protective mechanisms would also account for the species differences. All the studies which have been carried out with persons exposed to ozone and especially the observed adaptation phenomena suggest that such protective mechanisms also operate and can be induced in man. Therefore it is likely that a threshold concentration for genotoxicity and carcinogenicity of ozone could be derived for man. Until these matters have been clarified satisfactorily, ozone is to be classified in Category IIIB in the "List of MAK and BAT Values" and the MAK value withdrawn.


# 7 References


Aris RM, Christian D, Hearne PQ, Kerr K, Finkbeiner WE, Balmes JR (1993) Ozone-induced airway inflammation in human subjects as determined by airway lavage and biopsy. *Am Rev Respir Dis 148*: 1363–1372

Barlow SM, Sullivan FM (1982) *Reproductive hazards of industrial chemicals. An evaluation of animal and human data*, Academic Press, London, 422–430

Barry BE, Mercer RR, Miller FJ, Crapo JD (1988) Effects of inhalation of 0.25 ppm ozone on the terminal bronchioles of juvenile and adult rats. *Exp Lung Res 14*: 225–245

Barry BE, Miller FJ, Crapo JD (1985) Effects of inhalation of 0.12 and 0.5 ppm ozone on the proximal alveolar region of juvenile and adult rats. *Lab Invest 53*: 692–704

Bascom R, Naclerio RM, Fitzgerald TK, Kagey-Sobotka A, Proud D (1990) Effect of ozone inhalation on the response to nasal challenge with antigen of allergic subjects. *Am Rev Respir Dis 142*: 594–601

Becker S, Jordan RL, Orlando GS (1989) In vitro ozone exposure inhibits mitogen-induced lymphocyte proliferation and IL-2 production. *J Toxicol Environ Health 26*: 469–483

Becker S, Madden MC, Newman SL, Devlin RB, Koren HS (1991a) Modulation of human alveolar macrophage properties by ozone exposure in vitro. *Toxicol Appl Pharmacol 110*: 403–415

Becker S, Quay J, Koren HS (1991b) Effect of ozone on immunoglobulin production by human B cells in vitro. *J Toxicol Environ Health 34*: 353–366

Bignami G, Musi B, dell'Omo G, Laviola G, Alleva E (1994) Limited effects of ozone exposure during pregnancy on physical and neurobehavioral development of CD-1 mice. *Toxicol Appl Pharmacol 129*: 264–271

Boehme DS, Hotchkiss JA, Henderson RF (1992) Glutathione and GSH-dependent enzymes in bronchoalveolar lavage fluid cells in response to ozone. *Exp Mol Pathol 56*: 37–48

Boorman GA, Hailey R, Grumbein S, Chou BJ, Herbert RA, Goehl T, Mellick PW, Roycroft JH, Haseman JK, Sills R (1994) Toxicology and carcinogenesis studies of ozone and ozone 4-(N-nitrosoethylamino)-1-(3-pyridyl)-1-butanone in Fischer-344/N rats. *Toxicol Pathol 22*: 545–554

Borek C, Ong A, Cleaver JE (1988) DNA damage from ozone and radiation in human epithelial cells. *Toxicol Ind Health 4*: 547–553

Borek C, Ong A, Zaider M (1989) Ozone activates transforming genes in vitro and acts as a synergistic co-carcinogen with γ-rays only if delivered after radiation. *Carcinogenesis 10*: 1549–1551

Borek C, Zaider M, Augustinus O, Mason H, Witz G (1986) Ozone acts alone and synergistically with ionizing radiation to induce in vitro neoplastic transformation. *Carcinogenesis 7*: 1611–1613

van Bree L, Marra M, Rombout PJA (1992) Differences in pulmonary biochemical and inflammatory responses of rats and guinea pigs resulting from daytime or nighttime, single and repeated exposure to ozone. *Toxicol Appl Pharmacol 116*: 209–216

Cajigas A, Gayer M, Beam C, Steinberg JJ (1994) Ozonation of DNA forms adducts: A $^{32}$P-DNA labeling and thin-layer chromatography technique to measure DNA environmental biomarkers. *Arch Environ Health 49*: 25–36

Chang L-Y, Huang Y, Stockstill BL, Graham JA, Grose EC, Menache MG, Miller FJ, Costa DL, Crapo JD (1992) Epithelial injury and interstitial fibrosis in the proximal alveolar regions of rats chronically exposed to a simulated pattern of urban ambient ozone. *Toxicol Appl Pharmacol 115*: 241–252

Chang L, Miller FJ, Ultman J, Huang Y, Stockstill BL, Grose EC, Graham JA, Ospital JJ, Crapo JD (1991) Alveolar epithelial cell injuries by subchronic exposure to low concentrations of ozone correlate with cumulative exposure. *Toxicol Appl Pharmacol 109*: 219–234

Devlin RB, McDonnell WF, Mann R, Becker S, House DE, Schreinemachers D, Koren HS (1991) Exposure of human to ambient levels of ozone for 6.6 hours causes cellular and biochemical changes in the lung. *Am J Respir Cell Mol Biol 4*: 72–81

Devlin RB, McKinnon KR, Noah T, Becker S, Koren HS (1994) Ozone-induced release of cytokines and fibronectin by alveolar macrophages and airway epithelial cells. *Am J Physiol 266*: L612–L619

Dillon D, Combes R, McConville M, Zeiger E (1992) Ozone is mutagenic in Salmonella. *Environ Mol Mutagen 19*: 331–337

Dimitriadis VK (1992) Carbohydrate cytochemistry of bonnet monkey (Macaca radiaca) nasal epithelium. Response to ambient levels of ozone. *Histol Histopathol 7*: 479–488

Dubeau H, Chung YS (1979) Ozone response in wild type and radiation-sensitive mutants of Saccharomyces cerevisiae. *Mol Gen Genet 176*: 393–398

Dubeau H, Chung YS (1982) Genetic effects of ozone. Induction of point mutation and genetic recombination in Saccharomyces cerevisiae. *Mutat Res 102*: 249–259

Erdman HE, Hernandez T (1982) Adult toxicity and dominant lethals induced by ozone at specific stages in spermatogenesis in Drosophila virilis. *Environ Mutagen 4*: 657–666

Folinsbee LJ, Horstman DH, Kehrl HR, Harder S, Abul-Salaam S, Ives PJ (1994) Respiratory responses to repeated prolonged exposure to 0.12 ppm ozone. *Am J Respir Crit Care Med 149*: 98–105

Foster WM, Costa DL, Langenback EG (1987) Ozone exposure alters tracheobronchial mucociliary functions in humans. *J Appl Physiol 63*: 996–1002

Foster WM, Silver JA, Groth ML (1993) Exposure to ozone alters regional function and particle dosimetry in the human lung. *J Appl Physiol 75*: 1938–1945

Fujinaka LE, Hyde DM, Plopper CG, Tyler WS, Dungworth DL, Lollini LO (1985) Respiratory bronchiolitis following long-term ozone exposure in bonnet monkeys: A morphometric study. *Exp Lung Res 8*: 167–190

Gelzleichter TR, Witschi H, Last JA (1992) Concentration-response relationships of rat lungs to exposure to oxidant air pollutants: A critical test of Haber's Law for ozone and nitrogen dioxide. *Toxicol Appl Pharmacol 112*: 73–80

Gerrity TR, Bennett WD, Kehrl H, DeWitt PJ (1993) Mucociliary clearance of inhaled particles measured at 2 h after ozone exposure in humans. *J Appl Physiol 74*: 2984–2989

Gerrity TR, McDonnell WF, House DE (1994) The relationship between delivered ozone dose and functional responses in humans. *Toxicol Appl Pharmacol 124*: 275–283

Gilmour MI, Jakab GJ (1991) Modulation of immune function in mice exposed to 0.8 ppm ozone. *Inhalat Toxicol 3*: 293–308

Gilmour MI, Selgrade MJK (1993) A comparison of the pulmonary defenses against streptococcal infection in rats and mice following $O_3$ exposure: Differences in disease susceptibility and neutrophil recruitment. *Toxicol Appl Pharmacol 123*: 211–218

Gooch PC, Creasia DA, Brewen JG (1976) The cytogenetic effects of ozone: Inhalation and in vitro exposures. *Environ Res 12*: 188–195

Graham DE, Henderson F, House D (1988) Neutrophil influx measured in nasal lavages of humans exposed to ozone. *Arch Environ Health 43*: 228–233

Graham DE, Koren HS (1990) Biomarkers of inflammation in ozone-exposed human. Comparison of the nasal and bronchoalveolar lavage. *Am Rev Respir Dis 142*: 152–156

Grose EC, Stevens MA, Hatch CE, Jaskot RH, Selgrade MJK, Stead AG, Costa DL, Graham JA (1989) The impact of a 12-month exposure to a diurnal pattern of ozone pulmonary function, antioxidant biochemistry and immunology. in: Schneider T, Lee SD, Wolters GJR, Grant LD (Eds) *Atmospheric ozone research and its policy implications*, Elsevier, Nijmegen

Gross KB, White HJ (1987) Functional and pathologic consequences of a 52-week exposure to 0.5 ppm ozone followed by a clean air recovery period. *Lung 165*: 283–295

Guerrero RR, Rounds DE, Olson RS, Hackney JD (1979) Mutagenic effects of ozone on human cells exposed in vivo and in vitro based on sister chromatid exchange analysis. *Environ Res 18*: 336–346

Hamelin C (1985) Production of single- and double-strand breaks in plasmid DNA by ozone. *Int J Radiat Oncol Biol Phys 11*: 253–257

Hamelin C, Chung YS (1989) Repair of ozone-induced DNA lesions in Escherichia coli B cells. *Mutat Res 214*: 253–255

Hamelin C, Poliquin L, Chung YS (1981) Mutagenicity of ozone relative to other chemical and physical agents in Escherichia coli K12. *Rev Can Biol 40*: 305–307

Harder SD, Harris DT, House D, Koren HS (1990) Inhibition of human natural killer cell activity following in vitro exposure to ozone. *Inhalat Toxicol 2*: 161–173

Harkema JR, Plopper CG, Hyde DM, St. George JA, Wilson DW, Dungworth DL (1993) Response of macaque bronchiolar epithelium to ambient concentrations of ozone. *Am J Pathol 143*: 857–866

Haro R, Paz C (1993) Effects of ozone exposure during pregnancy on ontogeny of sleep in rats. *Neurosci Lett 164*: 67–70

Hassett C, Mustafa MG, Coulson WF, Elashoff RM (1985) Murine lung carcinogenesis following exposure to ambient ozone concentrations. *J Nat Cancer Inst 75*: 771–777

Hatch GE, Slade R, Harris LP, McDonnell WF, Devlin RB, Koren HS, Costa DL, McKee J (1994) Ozone dose and effect in humans and rats. A comparison using oxygen-18 labeling and bronchoalveolar lavage. *Am J Respir Crit Care Med 150*: 676–683

Hazbun ME, Hamilton R, Holian A, Eschenbacher WL (1993) Ozone-induced increases in substance P and 8-epi-prostaglandin $F_{2\alpha}$ in the airways of human subjects. *Am J Respir Cell Mol Biol 9*: 568–572

Hazucha MJ, Folinsbee J, Seal jr E (1992) Effects of steady-state and variable ozone concentration profiles on pulmonary function. *Am Rev Respir Dis 146*: 1487–1493

Henderson RF, Hotchkiss JA, Chang IY, Scott BR, Harkema JR (1993) Effect of cumulative exposure on nasal response to ozone. *Toxicol Appl Pharmacol 119*: 59–65

Henschler D. (Ed.) (1973) Ozon. in: *Gesundheitsschädliche Arbeitsstoffe. Toxikologisch-arbeitsmedizinische Begründungen von MAK-Werten*, 2nd issue VCH-Verlagsgesellschaft, Weinheim

Hiroshima K, Kohno T, Ohwada H, Hayashi Y (1989) Morphological study of the effects of ozone on rat lung. II. Long-term exposure. *Exp Mol Pathol 50*: 270–280

Horstman DH, Folinsbee LJ, Ives PJ, Abdul-Salaam S, McDonnell WF (1990) Ozone concentration and pulmonary response relationships for 6.6-hour exposures with five hours of moderate exercise to 0.08, 0.10, and 0.12 ppm. *Am Rev Respir Dis 142*: 1158–1163

Hotchkiss JA, Harkema JR, Sun JD, Henderson RF (1989) Comparison of acute ozone-induced nasal and pulmonary inflammatory responses in rats. *Toxicol Appl Pharmacol 98*: 289–302

Hotchkiss JA, Harkema JR, Henderson RF (1991) Effect of cumulative ozone exposure on ozone-induced nasal epithelial hyperplasia and secretory metaplasia in rats. *Exp Lung Res 15*: 589–600

Hsueh JL, Xiang W (1984) Environmental mutagenesis research at Fudan University. *Environ Sci Res 31*: 755–769

Hu S-C, Ben-Jebria A, Ultman JS (1992) Longitudinal distribution of ozone absorption in the lung: Quiet respiration in healthy subjects. *J Appl Physiol 73*: 1655–1661

Hu S-C, Ben-Jebria A, Ultman JS (1994) Longitudinal distribution of ozone absorption in the lung: Effects of respiratory flow. *J Appl Physiol 77*: 574–583

Huang Y, Chang LY, Miller FJ, Graham JA, Ospital JJ, Crapo JD (1988) Lung injury caused by ambient levels of oxidant air pollutants: Extrapolation from animal to man. *Am J Aerosol Med 1*: 180–183

Hyde DM, Plopper CG, Harkema JR, St. George JA, Tyler WS, Dungworth DL (1989) Ozone-induced structural changes in monkey respiratory system. in: Schneider T, Lee SD, Wolters GJR, Grant LD (Eds) *Atmospheric ozone research and its policy implications*, Elsevier, Nijmegen

Ito T, Ikemi Y, Ohmori K, Kitamura H, Kanisawa M (1994) Airway epithelial cell changes in rats exposed to 0.25 ppm ozone for 20 months. *Exp Toxicol Pathol 46*: 1–6

Johnson NF, Hotchkiss JA, Harkema JR, Henderson RF (1990) Proliferative responses of rat nasal epithelia to ozone. *Toxicol Appl Pharmacol 103*: 143–155

Kabel JR, Ben-Jebria A, Ultman JS (1994) Longitudinal distribution of ozone absorption in the lung: Comparison of nasal and oral quiet breathing. *J Appl Physiol 77*: 2584–2592

Kehrl HR, Vincent LW, Kowalsky RJ, Horstman DH, O'Neil JJ, McCartney WH, Bromberg PA (1987) Ozone exposure increases respiratory epithelial permeability in humans. *Am Rev Respir Dis 135*: 1124–1128

Kelly FJ, Birch S (1993) Ozone exposure inhibits cardiac protein synthesis in the mouse. *Free Rad Biol Med 14*: 443–446

Kennedy CH, Hatch GE, Slade R, Mason RP (1992) Application of the EPR spin-trapping technique to the detection of radicals produced in vivo during inhalation exposure of rats to ozone. *Toxicol Appl Pharmacol 114*: 41–46

Kleeberger SR, Levitt RC, Zhang L-Y (1993) Susceptibility to ozone-induced inflammation. I. Genetic control of the response to subacute exposure. *Am J Physiol 264*: L15–L20

Koren HS, Devlin RB, Graham DE, Mann R, McGee MP, Horstman DE, Kozumbo WJ, Becker S, House DE, McDonnell WF, Bromberg PA (1989) Ozone-induced inflammation in the lower airways of human subjects. *Am Rev Respir Dis 139*: 407–415

Koren HS, Hatch GE, Graham DE (1990) Nasal lavage as a tool in assessing acute inflammation in response to inhaled pollutants. *Toxicology 60*: 15–25

Kozumbo WJ, Agarwal S (1990) Induction of DNA damage in cultured human lung cells by tobacco smoke arylamines exposed to ambient level of ozone. *Am J Respir Cell Mol Biol 3*: 611–618

Kriebel D, Smith TJ (1990) A nonlinear pharmacologic model of the acute effects of ozone on the human lungs. *Environ Res 51*: 120–146

Last JA, Gelzleichter T, Harkema J, Parks WC, Mellick P (1993) Effects of 20 months of ozone exposure on lung collagen in Fischer 344 rats. *Toxicology 94*: 83–102

Last JA, Reiser KM, Tyler WS, Bucker RB (1984) Long-term consequences of exposure to ozone. I. Lung collagen content. *Toxicol Appl Pharmacol 72*: 111–118

Last JA, Warren DL, Pecquet-Goad E, Witschi H (1987) Modification by ozone of lung tumor development in mice. *J Nat Cancer Inst 78*: 149–154

L'Hérault P, Chung YS (1984a) Mutagenicity of ozone in different repair-deficient strains of Escherichia coli. *Mol Gen Genet 197*: 472–477

L'Hérault P, Chung YS (1984b) Effect of ozone on prophage induction in different strains of Escherichia coli K-12 lysogenic for lambda. *Can J Genet Cytol 26*: 706–709

Linder J, Herren D, Monn C, Wanner HU (1988) Die Wirkung von Ozon auf die körperliche Leistungsfähigkeit/The effect of ozone on physical activity. *Schweiz Z Sportmed 36*: 5–10

Lippmann M (1992) Ozone. in: Lippmann M (Ed.) *Environmental toxicants. Human exposures and their health effects*, Van Nostrand Reinhold, New York, 465–519

Lippmann, M (1993) Health effects of tropospheric ozone: Review of recent research findings and their implications to ambient air quality standards. *J Expos Anal Environ Epidemiol 3*: 103–129

Louie S, Arata MA, Offerdahl SD, Halliwell B (1993) Effect of tracheal insufflation of deferoxamine on acute ozone toxicity in rats. *J Lab Clin Med 121*: 502–509

Matsui S, Jones GL, Woolley MJ, Lane CG, Gontovnick LS, O'Byrne PM (1991) The effect of antioxidants on ozone-induced airway hyperresponsiveness in dogs. *Am Rev Respir Dis 144*: 1287–1290

McBride DE, Koenig JQ, Luchtel DL, Williams PV, Henderson WR (1994) Inflammatory effects of ozone in the upper airways of subjects with asthma. *Am J Respir Crit Care Med 149*: 1192–1197

McDonnell WF, Horstman DH, Hazucha MJ, Seal E, Haak ED, Abdul-Salaam S, House DE (1983) Pulmonary effects of ozone exposure during exercise: Dose-response characteristics. *J Appl Physiol 54*: 1345–1352

McDonnell WF, Horstman DH, Abdul-Salaam S, Raggio LJ, Green JA (1987) The respiratory responses of subjects with allergic rhinitis to ozone exposure and their relationship to non-specific airway reactivity. *Toxicol Ind Health 3*: 507–517

McDonnell WF, Kehrl HR, Abdul-Salaam S (1991) Respiratory response of humans exposed to low levels of ozone for 6.6 hours. *Arch Environ Health 46*: 145–150

McDonnell WF, Muller KE, Bromberg PA, Shy CM (1993) Predictors of individual differences in acute response to ozone exposure. *Am Rev Respir Dis 147*: 818–825

McDonnell WF, Smith MV (1994) Description of acute ozone response as a function of exposure rate and total inhaled dose. *J Appl Physiol 76*: 2776–2784

McKenzie WH, Knelson JH, Rummo NJ, House DE (1977) Cytogenetic effects of inhaled ozone in man. *Mutat Res 48*: 95–102

Menzel DB (1994) The toxicity of air pollution in experimental animals and humans: The role of oxidative stress. *Toxicol Lett 72*: 269–277

Merz T, Bender MA, Kerr HD, Kulle TJ (1975) Observation of aberrations in chromosomes of lymphocytes from human subjects exposed to ozone at a concentration of 0.5 ppm for 6 and 10 hours. *Mutat Res 31*: 299–302

Moffatt RK, Hyde DM, Plopper CG, Tyler WS, Putney LF (1987) Ozone-induced adaptive and reactive cellular changes in respiratory bronchioles of bonnet monkeys. *Exp Lung Res 12*: 57–74

Monchaux G, Morlier J-P, Morin M, Fritsch P, Tredaniel J, Masse R (1994) Carcinogenic and cocarcinogenic effects of ozone in rats: Preliminary results. *Pollut Atmos 142*: 84–88

Nover H, Botzenhart K (1985) Bactericidal effects of photochemical smog constituents produced by a flow reactor. III. Communication: Determination of mutagenic effects of photochemical smog on E. coli K 12 343/113. *Zentralbl Bakteriol Parasitenk Infektionskr Hyg, Abt 1, Orig, Reihe B 181*: 71–80

NTP (National Toxicology Program) (1994) *Toxicology and carcinogenesis. Studies of ozone (CAS No. 10028-15-6) and ozone/NNK (CAS No. 10028-15-6/64091-91-4) in F344/N rats and B6C3F₁ mice (inhalation studies)*. Technical Report Series 440, U.S. Department of Health and Human Services, Public Health Service, National Institutes of Health

Oosting RS, van Golde LMG, Verhef J, van Bree L (1991) Species differences in impairment and recovery of alveolar macrophage functions following single and repeated ozone exposures. *Toxicol Appl Pharmacol 110*: 170–178

Orlando GS, House D, Daniel EG, Koren HS, Becker S (1988) Effect of ozone on T-cell proliferation and serum levels of cortisol and beta-endorphin in exercising males. *Inhalat Toxicol 1*: 53–63

Parduez Nagy-Gyorgy S, Chung YS (1988) Sensitivity of polA mutants of Escherichia coli K-12 to ozone and radiations. *Mutagenesis 3*: 257–261

Pinkerton KE, Brody AR, Miller FJ, Crapo JD (1989) Exposure to low levels of ozone results in enhanced pulmonary retention of inhaled asbestos fibers. *Am Rev Respir Dis 140*: 1075–1081

Pinkerton KE, Mercer RR, Plopper CG, Crapo JD (1992) Distribution of injury and microdosimetry of ozone in the ventilatory unit of the rat. *J Appl Physiol 73*: 817–824

Plopper CG, Chu F-P, Haselton CJ, Peake J, Wu J, Pinkerton KE (1994a) Dose-dependent tolerance to ozone. I. Tracheobronchial epithelial reorganization in rats after 20 months' exposure. *Am J Pathol 144*: 404–421

Plopper CG, Duan X, Buckpitt AR, Pinkerton KE (1994b) Dose-dependent tolerance to ozone. IV. Site-specific elevation in antioxidant enzymes in the lungs of rats exposed for 90 days or 20 months. *Toxicol Appl Pharmacol 127*: 124–131

Postlethwait EM, Langford SD, Bidani A (1994) Determinants of inhaled ozone absorption in isolated rat lungs. *Toxicol Appl Pharmacol 125*: 77–89

Pryor WA (1993) The role of vitamin E in the protection of in vitro systems and animals against the effects of ozone. in: Packer L, Fuchs J (Eds) *Vitamin E in health and disease*, M. Dekker, New York, 715–736

Pryor WA (1994) Mechanisms of radical formation from reactions of ozone with target molecules in the lung. *Free Rad Biol Med 17*: 451–465

Pryor WA, Das B, Church DF (1991) The ozonation of unsaturated fatty acids: Aldehydes and hydrogen peroxide as products and possible mediators of ozone toxicity. *Chem Res Toxicol 4*: 341–348

Rahman I-U, Clerch LB, Massaro D (1991) Rat lung antioxidant enzyme induction by ozone. *Am J Physiol 260*: L412–L418

Rahman I-U, Massaro GD, Massaro D (1992) Exposure of rats to ozone: Evidence of damage to heart and brain. *Free Rad Biol Med 12*: 323–326

Rajini P, Gelzleichter TR, Last JA, Witschi H (1993a) Alveolar and airway cell kinetics in the lungs of rats exposed to nitrogen dioxide, ozone, and a combination of the two gases. *Toxicol Appl Pharmacol 121*: 186–192

Rajini P, Gelzleichter TR, Last JA, Witschi H (1993b) Airway epithelial labeling index as an indicator of ozone induced lung injury. *Toxicology 83*: 159–168

Rajini P, Witschi H (1995) Cumulative labeling indices in epithelial cell populations of the respiratory tract after exposure to ozone at low concentrations. *Toxicol Appl Pharmacol 130*: 32–40

Rasmussen RE (1986) Inhibition of DNA replication by ozone in Chinese hamster V79 cells. *J Toxicol Environ Health 17*: 119–128

Rasmussen RE, Crocker TT (1982) Lung cells grown on cellulose membrane filters as an in vitro model of the respiratory epithelium. *Environ Sci Res 25*: 105–120

Raub JA, Miller FJ, Graham JA (1983) Effects of low level ozone exposure on pulmonary function in adult and neonatal rats. *Adv Mod Environ Toxicol 5*: 363–367

Reiser KM, Tyler WS, Hennessy SM, Dominguez JJ, Last JA (1987) Long-term consequences of exposure to ozone. II. Structural alterations in lung collagen of monkeys. *Toxicol Appl Pharmacol 89*: 314–322

Rithidech K, Hotchkiss JA, Griffith WC, Henderson RF, Brooks AL (1990) Chromosome damage in rat pulmonary alveolar macrophages following ozone inhalation. *Mutat Res 241*: 67–73

Santrock J, Hatch GE, Slade R, Hayes JM (1989) Incorporation and disappearance of oxygen-18 in lung from mice exposed to 1 ppm $^{18}O_3$. *Toxicol Appl Pharmacol 98*: 75–80

Schelegle ES, Adams WC (1986) Reduced exercise time in competitive simulations consequent to low level ozone exposure. *Med Sci Sports Exercise 18*: 408–414

Schlesinger RB, Gorczynski JE, Dennison J, Richards L, Kinney PL, Bosland MC (1992) Long-term intermittent exposure to sulfuric acid aerosol, ozone, and their combination: Alterations in tracheobronchial mucociliary clearance and epithelial secretory cells. *Exp Lung Res 18*: 505–534

Seal Jr E, McDonnell WF, House DE, Salaam SA, Dewitt PJ, Butler SO, Green J, Raggio L (1993) The pulmonary response of white and black adults to six concentrations of ozone. *Am Rev Respir Dis 147*: 804–810

Sherwin RP, Richters V, Okimoto D (1983) Type 2 pneumocyte hyperplasia in the lungs of mice exposed to an ambient level (0.3 ppm) of ozone. *Adv Mod Environ Toxicol 5*: 289–297

Sherwin RP, Richters V (1985) Effect of 0.3 ppm ozone exposure on type II cells and alveolar walls of newborn mice: An image-analysis quantitation. *J Toxicol Environ Health 16*: 535–546

Shiraishi F, Bandow H (1985) The genetic effects of the photochemical reaction products of propylene plus $NO_2$ on cultured Chinese hamster cells exposed in vitro. *J Toxicol Environ Health 15*: 531–538

Song JM, Chung YS (1983) Effect of ozone on DNA-repair deficient mutants of Bacillus subtilis. *Rev Can Biol Exp 4*: 83–86

Soukoup J, Koren HS, Becker S (1993) Ozone effect on respiratory syncytial virus infectivity and cytokine production by human alveolar macrophages. *Environ Res 60*: 178–186

Steinberg JJ, Gleeson JL, Gil D (1990) The pathobiology of ozone-induced damage. *Arch Environ Health 45*: 80–87

Tager IB (1993) Summary of papers and research recommendations of working group on tropospheric ozone, Health Effects Institute Environmental Epidemiology Planning Project. *Environ Health Perspect 101, Suppl 4*: 237–239

Tepper JS, Costa DL, Lehmann JR (1993) Extrapolation of animal data to humans: Homology of pulmonary physiological responses with ozone exposure. in: Gardner DE (Ed.) *Toxicology of the Lung*, Raven Press, New York, 217–251

Tepper JS, Wiester MJ, Weber MF, Fitzgerald S, Costa DL (1991) Chronic exposure to a stimulated urban profile of ozone alters ventilatory responses to carbon dioxide challenge in rats. *Fundam Appl Toxicol 17*: 52–60

Thomassen DG, Harkema JR, Stephens ND, Griffith WC (1991) Preneoplastic transformation of rat tracheal epithelial cells by ozone. *Toxicol Appl Pharmacol 109*: 137–148

Tice RR, Bender MA, Ivett JL, Drew RT (1978) Cytogenetic effects of inhaled ozone. *Mutat Res 58*: 293–304

Tyler WS, Tyler NK, Last JA, Gillespie MJ, Barstow TJ (1988) Comparison of daily and seasonal exposures of young monkeys to ozone. *Toxicology 50*: 131–144

Vesely DL, Giordano AT, Raska-Emery P, Montgomery MR (1994) Ozone increases amino- and carboxy-terminal atrial natriuretic factor prohormone peptides in lung, heart, and circulation. *J Biochem Toxicol 9*: 107-112

Victorin K (1992) Review of the genotoxicity of ozone. *Mutat Res 277*: 221–238

Victorin K, Stahlberg M (1988) A method for studying the mutagenicity of some gaseous compounds in Salmonella typhimurium. *Environ Mol Mutagen 11*: 65–77

Weideman PA, Schlesinger RB (1994) Effect of in vitro exposure to ozone on eicosanoid metabolism and phagocytic activity of human and rabbit neutrophils. *Inhalat Toxicol 6*: 43–55

Weymer AR, Gong H jr, Lyness A, Linn WS (1994) Pre-exposure to ozone does not enhance or produce exercise-induced asthma. *Am J Respir Crit Care Med 149*: 1413–1419

Wiester MJ, Tepper JS, King ME, Ménache MG, Costa DL (1988) Comparative study of ozone ($O_3$) uptake in three strains of rats and in the Guinea pig. *Toxicol Appl Pharmacol 96*: 140–146

Witschi H (1988) Ozone, nitrogen dioxide and lung cancer: A review of some recent issues and problems. *Toxicology 48*: 1–20

Witschi H (1991) Effects of oxygen and ozone on mouse lung tumorigenesis. *Exp Lung Res 17*: 473–483

Wright DT, Adler KB, Akley NJ, Dailey LA, Friedman M (1994) Ozone stimulates release of platelet activating factor and activates phospholipases in guinea pig tracheal epithelial cells in primary culture. *Toxicol Appl Pharmacol 127*: 27–36

Wright ES, Dziedzic D, Wheeler, CS (1990) Cellular, biochemical and functional effects of ozone: New research and perspectives on ozone health effects. *Toxicol Lett 51*: 125–145

van der Zee J, van Beek E, Dubbelman TMAR, van Steveninck J (1987) Toxic effects of ozone on murine L929 fibroblasts. Damage to DNA. *Biochem J 247*: 69–72

Zelac RE, Cromroy HL, Bolch WE jr, Dunavant BG, Bevis HA (1971a) Inhaled ozone as a mutagen. I. Chromosome aberrations induced in Chinese hamster lymphocytes. *Environ Res 4*: 262–282

Zelac RE, Cromroy HL, Bolch WE, Dunavant BG, Bevis HA (1971b) Inhaled ozone as a mutagen. II. Effect on the frequency of chromosome aberrations observed in irradiated Chinese hamsters. *Environ Res 4*: 325–342

Zelikoff JT, Kraemer G-L, Vogel MC, Schlesinger RB (1991) Immunomodulating effects of ozone on macrophage functions important for tumor surveillance and host defence. *J Toxicol Environ Health 34*: 449–467

Zhurkov VS, Pechennikova EV, Feldt EG, Garibian LK, Tkhiem TK (1979) Analysis of chromosome aberrations in bone marrow cells of rats after inhalation exposure to ozone. *Gig Sanit 9*: 12–14

completed 23.03.1995

# Passive Smoking

The term "passive smoking" is understood to mean the inhalation of tobacco smoke in the ambient air (environmental tobacco smoke, ETS), "mainstream smoke" is the smoke actively inhaled by a smoker when drawing on a cigarette, and "sidestream smoke" is the smoke formed by a smouldering cigarette. Tobacco smoke in the air of rooms is a mixture of sidestream smoke and the air exhaled by smokers.

This chapter presents a re-evaluation of the damage caused by passive smoking at the workplace based on the results of recent epidemiological and toxicological studies and improvements in the quantitative data for systemic exposure and effects.

Some of the studies mentioned are cited from the reviews of passive smoking by the US EPA (1993) and the California EPA (1997).

# 1 Toxic Effects and Mode of Action

ETS irritates the eyes and the respiratory tract. The level of carboxyhaemoglobin in the blood is increased to an extent dependent on the exposure concentration. In particularly sensitive persons, sidestream smoke can have concentration-dependent adverse effects on lung function; there is evidence for a slight increase in the incidence of respiratory diseases and symptoms in healthy persons. Exposure to ETS can cause respiratory disease in children and can increase the frequency and severity of asthma attacks in persons with asthma. Passive smoking results in an increase in the mutagenic activity of the urine and in the urinary excretion of 4-(methylnitrosoamino)-1-(3-pyridyl)-1-butanone. There is evidence for an increase in the formation of haemoglobin adducts with carcinogenic components of ETS such as aminobiphenyls. Epidemiological studies and their meta-analyses reveal significant increases in the relative risk of developing lung cancer, increases which are dependent on the level of exposure to ETS. The relative risk of developing tumours in the nasal cavity or nasal sinuses and that for cardiovascular diseases were also shown in several studies to be increased significantly.

In studies of reproductive toxicity in the rat, the body weights of the pups were reduced after exposures which did not cause maternal toxicity. Prenatal and postnatal exposure increased mortality in rats above that seen after postnatal exposure alone. Exposure to sidestream smoke resulted in the formation of DNA-adducts in the lung tissue of mice and rats, micronuclei in mice and an increased mutagenic activity in rat urine. In strain A mice, the incidence of lung tumours was increased after exposure to ETS for 5 months. Condensate of sidestream smoke (containing benzo[*a*]pyrene in doses up to 2.5 µg/kg body weight) applied epicutaneously to mice for 3 months caused a dose-dependent increase in the incidence of precancerous lesions and an increased number of skin tumours and mammary carcinomas.

*Essential MAK Value Documentations.* DFG, Deutsche Forschungsgemeinschaft
Copyright © 2006 WILEY-VCH Verlag GmbH & Co. KGaA, Weinheim
ISBN: 3-527-31394-X

# 2 Components of Mainstream and Sidestream Smoke

The concentrations of the most important toxic substances in mainstream smoke and the ratios of their concentrations in sidestream and mainstream smoke are shown in Table 1.

**Table 1.** Components of mainstream and sidestream smoke (IARC 1986, US EPA 1993)

| | Mainstream smoke [μg/cigarette] | Sidestream smoke/ mainstream smoke | Classification in the *List of MAK and BAT Values* 1998 |
|---|---|---|---|
| acetaldehyde | 500–1200 | no data | Section III Category 3 |
| acetic acid | 330–810 | 1.9–3.6 | not classified |
| acetone | 100–250 | 2–5 | not classified |
| acrolein | 60–100 | 8–15 | Section III Category 3 |
| 4-aminobiphenyl | 0.003–0.005 | 31 | Section III Category 1 |
| ammonia | 50–130 | 3.5–5.1 | not classified |
| aniline | 0.36 | 29.7 | Section III Category 3 |
| benz[*a*]anthracene | 0.003–0.05 | 2.7 | Section III Category 2 |
| benzene | 12–48 | 5–10 | Section III Category 1 |
| benzo[*a*]pyrene | 0.038 | 2.1–3.5 | Section III Category 2 |
| 1,3-butadiene | 69 | 3–6 | Section III Category 1 |
| cadmium | 0.1–0.12 | 3.6–7.2 | Section III Category 2 |
| carbon monoxide | 13000–22000 | 2.5–4.7 | not classified |
| carbonyl sulfide | 12–42 | 0.03–0.13 | not classified |
| diethylnitrosamine | 0.025 | < 40 | Section III Category 2 |
| dimethylamine | 7.8–10 | 3.7–5.1 | not classified |
| dimethylnitrosamine | 0.01–0.04 | 20–100 | Section III Category 2 |
| ethylmethylnitrosamine | 0.001–0.002 | 10–20 | Section III Category 2 |
| formaldehyde | 70–100 | 0.1–50 | Section III Category 3 |
| formic acid | 210–490 | 1.4–1.6 | not classified |
| hydrazine | 0.032 | 3 | Section III Category 2 |
| hydrogen cyanide | 400–500 | 0.1–0.25 | not classified |
| methylamine | 11–29 | 4.2–6.4 | not classified |
| methyl chloride | 150–600 | 1.7–3.3 | Section III Category 3 |
| 2-naphthylamine | 0.001–0.022 | 30 | Section III Category 1 |
| nickel | 0.02–0.08 | 12–31 | Section III Category 1 |
| nicotine | 1330–1830 | 2.6–3.3 | not classified |
| nitrogen monoxide | 100–600 | 4–10 | not classified |
| nitrosopyrrolidine | 0.006–0.03 | 6–30 | Section III Category 2 |
| pyridine | 16–40 | 6.5–20 | not classified |
| toluene | 100–200 | 5.6–8.3 | not classified |
| 2-toluidine | 0.03–0.2 | 19 | Section III Category 2 |

# 3 Exposure

## 3.1 External exposure

### 3.1.1 Nicotine

In the USA, nicotine concentrations of 0.3–30 $\mu g/m^3$ have been found in rooms in a variety of buildings. In 50 % of the studies the concentrations were 30 $\mu g/m^3$ or less, in 90 % below 100 $\mu g/m^3$ (Table 2) (California EPA 1997). The average concentrations of nicotine in homes with one or more smokers were in the range between 2 and 14 $\mu g/m^3$. The nicotine concentrations at workplaces where smoking was permitted were as high or higher than those in homes of smokers. The highest concentrations were measured in bars and in smoking areas in aeroplanes (50–110 $\mu g/m^3$) (California EPA 1997). In private and public buildings in Germany, concentrations of 0.1–15 $\mu g/m^3$ and 0.3–100 $\mu g/m^3$, respectively, were found (Adlkofer 1992).

### 3.1.2 Particles that can enter the alveoli

The particles in sidestream smoke differ in size and density from those in mainstream smoke: diameters of particles in mainstream smoke were 0.2 to 0.4 $\mu m$, in sidestream smoke 0.1 to 0.2 $\mu m$. Even the highest concentrations which have been measured indoors, about 1 $mg/m^3$, do not cause overloading of the lungs (Adlkofer 1992). The highest average particle concentrations were found in bars. Indoor concentrations of particles with an aerodynamic diameter of 10 $\mu m$ or less were significantly higher in the homes of smokers (126 $\mu g/m^3$, n = 28) than in those of non-smokers (88 $\mu g/m^3$, n = 139) (California EPA 1997). In a sports club house the concentration of respirable particles (that is, particles that can enter the alveoli) in the air was about 57 $\mu g/m^3$. The authors stated that there was only a moderate level of smoking. After smoking had been banned the concentration dropped to 77 % of the previous value (California EPA 1997).

### 3.1.3 Polycyclic aromatic hydrocarbons

In the air in the rooms of 3 restaurants, the concentrations of 17 different polycyclic aromatic hydrocarbons (PAHs) were determined in the particle fractions. The total PAH concentration was 34 to 168 $ng/m^3$ (Husgafvel-Pursiainen *et al.* 1986).

In private homes in California, the concentrations of 13 different PAHs including benz[*a*]anthracene, benzo[*a*]pyrene, benzo[*k*]fluoranthene, indenol[1,2,3 cd]pyrene, and chrysene were determined. The concentrations of most of the PAHs were significantly higher (by a factor of 1.5 to 2) in the homes of smokers than in those of non-smokers. The 12-hour average benzo[*a*]pyrene concentration was 0.51 $ng/m^3$ in the homes of smokers and 0.2 $ng/m^3$ in those of non-smokers (California EPA 1997).

In another study in California, the concentrations of 14 PAHs in the room air of 280 homes were measured. Other sources of PAHs such as fireplaces, wood stoves and gas heaters were also recorded. The PAH concentrations in the air in 64 homes with heavy smokers were significantly higher by a factor of 1.5 to 4 than in 39 homes in which there were no smokers and no other sources of PAHs. The average 24-hour benzo[*a*]pyrene concentrations measured were: tobacco smoke 2.2 ng/m$^3$, heating with a wood stove 1.2 ng/m$^3$, open fire 1.0 ng/m$^3$, not otherwise specified 0.83 ng/m$^3$ (California EPA 1997).

**Table 2.** Concentrations of toxic and carcinogenic components in indoor air contaminated with tobacco smoke (Adlkofer 1992, Hoffmann and Wynder 1994, Siegel 1993)

| Substance | Rooms | Concentration [µg/m$^3$] |
|---|---|---|
| acetaldehyde | restaurants | 170–630 |
|  | bars | 180–200 |
| acrolein | restaurants | 30–100 |
| benz[*a*]anthracene | restaurants | 2–9 |
| benzo[*a*]pyrene | restaurants | 0.002–0.76 |
|  | workplaces | 0.003–0.025 |
| hydrogen cyanide | living rooms | 8–120 |
|  | offices | 3–49 |
|  | private homes | 7–17 |
|  | restaurants | 50–150 |
|  | bars | 30–40 |
| dimethylnitrosamine | restaurants | 0.01–0.05 |
|  | smoking compartments in trains | 0.11–0.13 |
|  | bars | 0.07–0.24 |
| formaldehyde | private homes | 8–280 |
|  | offices | 12–1300 |
| carbon monoxide | offices | 1160–3830 |
|  | restaurants | 580–11480 |
|  | bars | 3600–19720 |
| nicotine | offices | 0.8–37 |
|  | public buildings | 1–37 |
|  | restaurants | 1–80 |
|  | bars | 7.4–110 |
|  | private homes | 1.6–21 |
| particles | offices | 6–256 |
|  | private homes | 32–700 |
|  | restaurants | 27–690 |
|  | bars | 75–1370 |
| phenols (volatile) | cafeterias | 0.007–0.012 |
| nitrogen dioxide | workplaces | 68–410 |
|  | restaurants | 40–190 |
|  | bars | 2–116 |
|  | cafeterias | 67–200 |
| nitrogen monoxide | workplaces | 50–440 |
|  | restaurants | 17–270 |
|  | bars | 80–520 |
|  | cafeterias | 2.5–48 |

### 3.1.4 Nitrosamines

During the burning of a cigarette, volatile (e.g. *N*-nitrosodimethylamine) and non-volatile (e.g. *N*-nitrosodiethanolamine) nitrosamines are formed; tobacco-specific nitroso compounds such as 4-(methylnitrosoamino)-1-(3-pyridyl)-1-butanone and nitrosonornicotine are formed by nitrosation of nicotine and other pyridine alkaloids. Most of the nitrosamines which have been identified have been shown in animal studies to be carcinogenic; their concentrations in sidestream smoke are 10 to 200 times those in mainstream smoke (Table 1) (California EPA 1997). The concentrations determined in restaurants and other workplaces are shown in Table 2.

### 3.1.5 Other organic compounds

In private homes (n = 15) with heavy smokers (> 20 cigarettes per day), higher concentrations of benzene, dichlorobenzene, tetrachloroethene, trichloroethene and xylene were found than in homes without smokers (California EPA 1997).

## 3.2 Systemic dose and effects

### 3.2.1 Systemic dose

#### Nicotine and cotinine

Nicotine and cotinine are specific biomarkers for exposure to tobacco smoke and can be readily detected in saliva, blood, urine and hair. The concentration of cotinine in the plasma and saliva of non-smokers is in the range between 0.5 and 15 ng/ml. The average cotinine concentration in the urine of persons who described themselves as non-smokers was 8.84 ng/ml with single values up to about 85 ng/ml, the average concentration in the urine of smokers was about 1200 ng/ml (California EPA 1997, NIOSH 1991). The average increase in the cotinine concentration in urine after intake of food containing nicotine (tomatoes, potatoes, cauliflower, black tea) is 0.6 ng/ml, with a maximum value of 6.2 ng/ml (US EPA 1993).

10 non-smokers were exposed for 80 minutes in a 16 m$^3$ room to a constant stream of sidestream smoke produced by a smoking machine (2–4 cigarettes). The concentration of nicotine in the air of the room after 10 to 15 minutes was 280 µg/m$^3$ (California EPA 1997). The concentrations of nicotine and cotinine in the saliva, plasma and urine of the test persons are shown in Table 3.

**Table 3.** Concentrations of nicotine and cotinine in the saliva, plasma and urine of test persons exposed to the sidestream smoke from 4 cigarettes (California EPA 1997)

| Sampling time (minutes) | Saliva (ng/ml) | | Plasma (ng/ml) | | Urine (ng/mg creatinine) [ng/ml] | |
|---|---|---|---|---|---|---|
| | nicotine | cotinine | nicotine | cotinine | nicotine | cotinine |
| during exposure | | | | | | |
| 0 | 3 | 1.0 | 0.2 | 0.9 | 17 [25] | 14 [21] |
| 40 | 830 | 1.1 | 0.3 | 0.9 | – | – |
| 60 | 880 | 2.1 | 0.3 | 1.2 | – | – |
| 80 | 730 | 1.4 | 0.5 | 1.3 | 84 [126] | 28 [42] |
| after exposure | | | | | | |
| 30 | 148 | 1.7 | 0.4 | 1.8 | – | – |
| 150 | 17 | 3.1 | 0.7 | 2.9 | 100 [150] | 46 [69] |
| 240 | 3 | 2.0 | 1.1 | 3.3 | – | – |
| 300 | 7 | 3.5 | 0.6 | 3.4 | 48 [72] | 55 [83] |

– not determined

In 24 children aged 3 to 36 months who were frequently exposed to ETS, the nicotine concentrations in hair (15.4 ± 6.7 ng/mg hair) were significantly increased above those in unexposed children (1.3 ± 1.7 ng/mg hair). The cotinine concentrations were below the detection limit of 0.1 ng/mg hair (Pichini *et al.* 1997a, 1997b).

The concentrations of nicotine and cotinine in the hair of 94 mothers and their children were determined during weeks 31 to 42 of pregnancy in the mothers and days 1 to 3 after birth in the children. In passive smoking mothers, the concentrations were significantly increased above those in non-smoking mothers but were not increased in the children (Eliopoulos *et al.* 1994, 1996).

The concentrations of nicotine in mother's milk from 3 non-smokers who were exposed to ETS at work were in the range between 1 and 7 ng/ml (California EPA 1997). The cotinine concentrations in the mother's milk from 7 non-smokers who lived with partners who smoked reached values up to 277 ng/ml (California EPA 1997).

The concentration of cotinine in the amniotic fluid of 31 pregnant women was 8 times higher in smokers and 2.5 times higher in passive smokers than in non-smokers. In the urine of children of passive smoking women the cotinine concentrations were higher than those in the urine of children of non-smoking women (Jordanov 1990).

## Nitrosamines

Five non-smokers in a 16 m³ room were exposed to sidestream smoke twice on one day for 90 minutes each time (6 cigarettes in 90 minutes, 12 cigarettes in 90 minutes). After 6 months the exposure was repeated. After the exposure the concentrations of 4-(methylnitrosoamino)-1-(3-pyridyl)-1-butanol and its glucuronide conjugate were determined in 24-hour urine samples and shown to be increased. 4-(Methylnitrosoamino)-1-(3-pyridyl)-1-butanol has been shown to induce lung cancer in the rat and mouse (Hecht *et al.* 1993).

## Thiocyanates

In 27 non-smokers who were exposed to high concentrations of tobacco smoke for 40 hours per week (no other details) the thiocyanate concentration in plasma was significantly increased to 12 µmol/l above that in unexposed persons (Husgafvel-Pursiainen *et al.* 1987).

## 3.2.2 Systemic effects

## Protein and DNA adducts

The binding of components of tobacco smoke to macromolecules can be used to estimate the dose with significant biological effects. The formation of protein and DNA adducts has been demonstrated in smokers (Bryant *et al.* 1988, Cuzick *et al.* 1990, Falter *et al.* 1994). There is also evidence of the formation of protein adducts in passive smokers (see Tables 4 and 5) and DNA adducts too, but detection of the latter is difficult because of the low exposure levels. The isomers 3-aminobiphenyl and 4-aminobiphenyl form persistent haemoglobin adducts which are appropriate for the demonstration of exposure during the previous 4 months, the life-time of the erythrocytes. The results summarized in Table 4 indicate that the increase in the levels of aminobiphenyl adducts with haemoglobin caused by passive smoking was about 5 % to 20 % of that observed in smokers.

**Table 4.** Formation of 4-aminobiphenyl-haemoglobin adducts in smokers and non-smokers

| Exposed persons[a] | Determination of exposure | Adducts [pg/g haemoglobin] | | | References |
|---|---|---|---|---|---|
| | | S | PS | NS | |
| men 50 S, 50 NS, 15 PS (Italy) | questionnaire nicotine/cotinine in the urine | 112–175[b, d] 86–118[c, d] | 35[b] 34[c] | 30[b] 12[c] | Bartsch *et al.* 1990 |
| pregnant women 15 S, 40 NS, 9 PS (USA) | questionnaire nicotine in indoor air, "personal monitor" | 184[d] | 27.8[d] | 17.6 | Hammond *et al.* 1993 |

[a] S: smoker, PS: passive smoker, NS: non-smoker
[b] slow acetylators, [c] rapid acetylators,
[d] difference from value for non-smokers statistically significant

In another study of 73 pregnant non-smokers and 27 smokers, the formation of haemoglobin adducts with 3-aminobiphenyl and 4-aminobiphenyl and also with tobacco-specific nitrosamines was investigated. In this study the levels of haemoglobin adducts in the passive smokers were not increased. However, even in the smokers the levels of the 3-aminobiphenyl and 4-aminobiphenyl adducts were increased by a factor of only 2.0 to 2.7 (Richter *et al.* 1995).

In 9 pregnant passive smokers (determined by questioning and measurement of the levels of cotinine in plasma and urine), no increase in the levels of DNA adducts in the placenta were found in comparison with those in 10 women who were not exposed to ETS and with those in 11 smokers (Daube *et al.* 1997).

In 87 mothers and their children aged 2 to 5 years, cotinine concentrations and levels of albumin-PAH adducts were measured in the plasma. The smokers smoked on average 10 cigarettes per day and none smoked more than one packet per day (duration of exposure not specified). The highest cotinine concentrations and levels of PAH-adducts were found in the smokers (Table 5).

**Table 5.** PAH adduct formation and cotinine concentrations in female smokers, passive smokers, non-smokers and their children (Crawford *et al.* 1994)

| Women | Cotinine (ng/ml plasma) | PAH adducts (fmol/$\mu$g albumin) |
| --- | --- | --- |
| smokers | 170 | 0.8 |
| children of smokers | 4.14 | 0.35 |
| passive smokers | 1.64 | 0.49 |
| children of passive smokers | 0.87 | 0.18 |
| non-smokers | 0.96 | 0.31 |
| children of non-smokers | 0.25 | 0.18 |

The differences between the group of smokers and the other two groups (passive smokers and non-smokers) in the levels determined in both the mothers and the children were statistically significant. The concentration of PAH-albumin adducts in the passive smoking women was increased above that in the non-smokers but not significantly; in the children of these two groups of women, no differences in the PAH-adduct concentrations could be found (Crawford *et al.* 1994).

With $^{32}$P-postlabelling, DNA adducts (no other details) were found in the monocytes of 5 smokers who smoked 24 cigarettes in 8 hours. They were, however, no longer detectable 36 hours later. In 5 passive smokers, no DNA adducts could be detected. The authors were of the opinion that the method was not sensitive enough to detect the probably relatively low levels of DNA adducts expected in passive smokers (Holz *et al.* 1990).

In 50 passive smokers who were exposed to ETS for 70 hours weekly, the incidence of sister chromatid exchange (SCE) in lymphocytes was not different from that in 56 passive smokers who were exposed for only 5 hours per week. In this case too, the authors suggested that the SCE test is probably not sensitive enough to detect cytogenetic effects of ETS exposure (Gorgels *et al.* 1992).

## Carbon monoxide and carboxyhaemoglobin

The levels of carbon monoxide in exhaled air or carboxyhaemoglobin in blood are not good indicators of exposure to tobacco smoke because they are neither sensitive nor specific enough. The exposure to carbon monoxide from car exhaust gases, gas heaters and industrial incinerators have marked effects on the results of such studies. After 30 minutes exposure to air with high levels of tobacco smoke (no other details) increased levels of carbon monoxide in the exhaled air and carboxyhaemoglobin in blood were found in non-smokers (California EPA 1997).

In non-smokers exposed to tobacco smoke in closed rooms, the carboxyhaemoglobin levels in blood reached 2 % to 3 % (Adlkofer 1992, California EPA 1997, Scherer *et al.* 1992). The endogenous carbon monoxide production (Coburn *et al.* 1966) results in carboxyhaemoglobin levels of 0.4 % to 0.8 % in healthy non-smokers, whereby the inter-individual differences can be enormous (Werner 1978).

## Thioethers and mutagenic substances in urine

Reactive intermediates formed in normal metabolism are inactivated mainly by conjugation with glutathione and then, after several degradation steps, excreted as thioethers with the urine or the bile. Therefore thioethers are considered to be indicators of the formation of reactive and potentially mutagenic intermediates in the organism.

Non-smokers (number not specified) were exposed for 8 hours to various concentrations of tobacco smoke; in the first experiment the carbon monoxide concentration was 10 ml/m$^3$ and the particle concentration 1.0–1.5 mg/m$^3$, in the second the carbon monoxide concentration was 25 ml/m$^3$ and the particle concentration 3.0 to 4.0 mg/m$^3$. In the second experiment, smokers who smoked 20 cigarettes during the study were also included for comparison. The level of thioether excretion in the urine of passive smokers was about 44 % of that excreted by smokers. Determined from the thioether excretion in the urine, the non-smokers had "smoked" 9 cigarettes (Adlkofer 1992).

In another study in which 5 smokers each smoked 24 cigarettes during 8 hours in the presence of 5 non-smokers, the levels of thioethers excreted by the non-smokers were increased significantly by 16.5 µmol in 24 hours. The 8-hour average carbon monoxide concentration was 24 ml/m$^3$. A significant increase in the excretion of hydroxypropyl-mercapturic acid in the urine of the non-smokers was also demonstrated. The amount of hydroxypropyl-mercapturic acid excreted by the non-smokers was on average 70 % of that excreted by the smokers. Determined from the levels of hydroxypropyl-mercapturic acid excretion in the urine, the non-smokers had been exposed to the levels found in the mainstream smoke of 17 actively smoked cigarettes. The levels of hydroxyphenanthrenes and hydroxypyrenes in urine were of the same order of magnitude in smokers and non-smokers. No increase in the mutagenicity of the urine of non-smokers was observed in a test for mutagenicity in *Salmonella* strain TA98 in the presence of a metabolic activation system. All determinations were subject to very large intra-individual scatter and the background levels of phenanthrenes were very high (Scherer *et al.* 1992, 1996).

Six non-smokers were exposed for 8 hours to ETS produced by 4 smokers in the same room (88 cigarettes). The average carbon monoxide concentration was 19 ml/m$^3$. An increase in the mutagenicity of the urine of the passive smokers after the exposure was demonstrated in a mutagenicity test with the *Salmonella* strain TA98 in the presence of a metabolic activation system (Mohtashamipur *et al.* 1987a). Eight non-smokers were exposed to tobacco smoke from 157 cigarettes smoked by 10 smokers within 6 hours in the same room. The 12-hour urine samples revealed an increased mutagenicity in 6 non-smokers in a test with the *Salmonella* strain TA1538 and decreased mutagenicity in 2 non-smokers. In this publication there were no details of eating habits, alcohol consumption or intake of medicines (Bos *et al.* 1983).

27 non-smokers who had worked as waiters in restaurants were exposed to high concentrations of tobacco smoke for 40 hours per week (concentrations not specified). An increased mutagenicity of their urine was seen in a test with the *Salmonella* strain TA98, but the increase was not significant (Husgafvel-Pursiainen *et al.* 1987).

The increased levels of thioethers in urine and the increased mutagenicity of the urine demonstrated in numerous studies provide clear evidence for the intake or formation of reactive, potentially mutagenic intermediates in the body after exposure to ETS and thus for an increase in systemic effects.

# 4 Mechanism of Action of ETS

The effects which are essentially responsible for the vascular damage caused by sidestream smoke are thickening of the *tunica intima* and *tunica media* of vessel walls (see Section 6.4), endothelial damage, an increase in the tendency for thrombocyte aggregation and a reduction in the capacity for oxygen transport.

Ten non-smokers were exposed to ETS for 20 minutes. The tendency for thrombocyte aggregation, the number of endothelial cells, the nicotine concentration in plasma and the level of carboxyhaemoglobin increased (Davis *et al.* 1989).

In non-smokers and smokers exposed for 15 minutes to ETS, the inhibitory effect of prostaglandin $E_1$, $I_2$ and $D_2$ on thrombocyte aggregation was studied before, during, immediately after, and 20 and 60 minutes after the exposure. In the non-smokers the effects of prostaglandins $E_1$, $I_2$ and $D_2$ were reduced by a factor of 2 at the end of exposure; for $E_1$ and $I_2$ the effect was statistically significant, for $D_2$ not significant. These effects were no longer detectable 60 minutes after the end of exposure. In the smokers no significant effects of the prostaglandins on thrombocyte aggregation were detectable (Sinzinger and Kefalides 1982).

Similarly, in another study in which 13 smokers and 9 non-smokers were simultaneously exposed for 20 minutes to the smoke of 30 cigarettes, a significant decrease in the sensitivity of the thrombocytes to prostaglandin $I_2$ was observed. The sensitivity before the exposure was much lower in the smokers than in the non-smokers (Burghuber *et al.* 1986).

Eight non-smokers were exposed once or ten times for 60 minutes per day to the smoke of 30 cigarettes which were smoked by 8 smokers. Blood samples were taken before, immediately after, 20 and 60 minutes after, and 6 hours after the exposure. Changes in the parameters for thrombocyte adhesion, thrombocyte aggregation, thrombocyte migration, circulating aggregates, circulating endothelial cells and the prostaglandin system observed immediately after exposure were no longer detected 6 hours later. After a single exposure, the thrombocytes of the non-smokers reacted more sensitively than those of the smokers. During 5 days of exposure to smoke each day, the values for platelet sensitivity of the non-smokers became increasingly like those of the smokers whereas in the smoker group no significant changes were seen (Sinzinger and Virgolini 1989).

Passive smoking can result in an increase in the carboxyhaemoglobin level of 1 % to 3 %. The resulting lower oxygen-carrying capacity of the arterial blood can result in an inadequate supply of oxygen to the myocardium in patients with angina. This was confirmed by demonstrating that the duration of exercise required to induce angina was reduced in patients after inhalation of ETS (concentrations not determined) (Aronow 1978).

In healthy women exposed to ETS while they were exercising on a treadmill, an increase in the resting pulse, a significant reduction in oxygen uptake and a reduction in the time till exhaustion were observed. In addition heart rate, carbon dioxide concentration in the exhaled air and lactate concentration in the venous blood were increased. The reduced capacity for oxygen transport resulting from carboxyhaemoglobin formation and the increase in the blood lactate concentration resulted in overloading of aerobic metabolism and thus caused more rapid exhaustion during physical activity (McMurray *et al.* 1985).

The increased relative risk of developing lung cancer and tumours of the nasal cavity and nasal sinuses which is associated with exposure to ETS may be ascribed to the effects of carcinogenic components of ETS.


# 5 Acute Toxicity

The following components of sidestream smoke cause acute irritation of the respiratory tract: ammonia, acrolein, formaldehyde, nicotine, nitrogen oxides, phenol and sulfur dioxide. Some components such as acrolein, crotonaldehyde, formaldehyde and hydrogen cyanide impair mucociliary clearance (California EPA 1997).

After a 15-minute exposure to sidestream smoke (carbon monoxide level 44 to 48 ml/m$^3$), mild to severe eye irritation and increased odour perception were reported by all 11 healthy non-smokers, in 4 non-smokers moderate to severe irritation of the nose and throat and in 2 non-smokers rhinitis and coughing. In 10 ETS-sensitive persons, the symptoms were more severe and more frequent (Bascom *et al.* 1991).

A group of 24 non-smokers who suffered from mild asthma complained of eye irritation, irritation of the upper airways and a feeling of tightness in the chest after exposure to ETS for one hour. A slight reduction in the forced expiratory volume in one second (FEV$_1$) was not associated with an increase in airway resistance; 16 symptom-free non-smokers reported irritation of the eyes, nose and throat, and coughing. In the severity of the symptoms there was no significant difference between the groups with and without bronchial asthmatic symptoms (Jörres and Magnussen 1992).

A group of 12 test persons was exposed to sidestream smoke twice for 60 minutes with an interval of one week between the exposures; mucociliary clearance was more rapid in 6 of the persons, slower in 3 and unchanged in another 3 (Bascom *et al.* 1995).

10 hyper-reactive and 10 normally reactive test persons were exposed to sidestream smoke. In the hyper-reactive persons, a significant decrease in $FEV_1$, FVC (forced expiratory vital capacity) and $MEF_{50}$ (mid-expiratory flow at 50 % of FVC) was detected and shown to depend on the exposure level (determined as carbon monoxide concentration). In the healthy persons no changes were observed. At carbon monoxide concentrations of 2 $ml/m^3$ or more, the persons (no other details) complained of mild coughing, rough dry throats and a feeling of tightness in the chest (Danuser *et al.* 1993).

An increase in bronchial reactivity was seen in 10 of 31 smoke-sensitive persons with asthma and 8 of 39 smoke-sensitive persons without asthma after exposure to ETS for 6 hours (Menon *et al.* 1992).

# 6 Epidemiological Studies

Detailed descriptions of epidemiological studies of all end points and their results are to be found in US EPA (1993) and California EPA (1997). Results of case-control studies are given as the odds ratio and results of cohort studies as the relative risk (RR).

## 6.1 Lung cancer

Table 6 summarizes the results of cohort studies and case-control studies of the association between lung cancer and exposure to ETS. Results of studies published before 1997 were taken from the review by Hackshaw *et al.* (1997). More recent studies were added to the table and are described separately in the text. Studies in which the number of persons studied was too small, the documentation inadequate, potential confounders were not taken into account and other limiting factors reduced the meaningfulness of the results (see US EPA 1993) are indicated in Table 6. For each type of study, the results of the studies are presented first without differentiation according to the levels of exposure and then the dose-response relationship is described.

### 6.1.1 Cohort studies

In the three most useful studies (Cardenas *et al.* 1997, Garfinkel 1981, Hirayama 1984), the relative lung cancer risk for non-smoking women with partners who smoked was higher than that for those whose partner did not smoke. For non-smoking men whose wives smoked the relative risk was increased only in the study by Hirayama (1984) (see Table 6).

**Table 6.** Relative risk of lung cancer caused by passive smoking: epidemiological studies of the risk for life-long non-smokers whose partner smoked compared with that for those with non-smoking partners

| Study | Country | Cases ♀/♂ | Size of control group or cohort ♀/♂ | Relative risk (95 % confidence interval) ♀/♂ |
|---|---|---|---|---|
| **Cohort studies** | | | | |
| Butler 1988[1] | USA | 8 | 9199 | 2.02 (0.48–8.56) |
| Cardenas *et al.* 1997 | USA | 150/97 | 192084/96445 | 1.20 (0.80–1.60)/ 1.00 (0.60–1.80) |
| Garfinkel 1981 | USA | 153 | 176586 | 1.18 (0.90–1.54) |
| Hirayama 1984[2,3] | Japan | 200/64 | 91340/20225 | 1.45 (1.02–2.08)/ 2.25 (1.06–4.76) |
| **Case-control studies** | | | | |
| Akiba *et al.* 1986 | Japan | 94/19 | 270/110 | 1.52 (0.87–2.63)/ 2.10 (0.51–8.61) |
| Boffetta *et al.* 1998[4,5] | Europe | 650 | 1542 | 1.16 (0.93–1.44) |
| Brownson *et al.* 1987[1] | USA | 19 | 47 | 1.52 (0.39–5.96) |
| Brownson *et al.* 1992 | USA | 431 | 1166 | 0.97 (0.78–1.21) |
| Buffler *et al.* 1984[1] | USA | 41/11 | 196/90 | 0.80 (0.34–1.90)/ 0.51 (0.14–1.79) |
| Chan and Fung 1982[6,7] | Hong Kong | 84 | 139 | 0.75 (0.43–1.30) |
| Correa *et al.* 1983[1] | USA | 22 | 133 | 2.07 (0.81–5.25) |
| Du *et al.* 1993 | China | 75 | 128 | 1.19 (0.66–2.13) |
| Fontham *et al.* 1994 | USA | 651 | 1253 | 1.26 (1.04–1.54) |
| Gao *et al.* 1987 | China | 246 | 375 | 1.19 (0.82–1.73) |
| Garfinkel *et al.* 1985 | USA | 134 | 402 | 1.23 (0.81–1.87) |
| Geng *et al.* 1988[7] | China | 54 | 93 | 2.16 (1.08–4.29) |
| Hole *et al.* 1989[4,5] | Scotland | 7 | 7997 | 2.41 (0.45–12.83) |
| Humble *et al.* 1987[1] | USA | 20 | 162 | 2.34 (0.81–6.75) |
| Inoue and Hirayama 1988[1,7] | Japan | 22 | 47 | 2.55 (0.74–8.78) |
| Janerich *et al.* 1990[4,5] | USA | 188 | 191 | 0.75 (0.48–1.18) |
| Jöckel 1991[1] | Germany | 23/9 | 45/70 | 2.27 (0.75–6.82)/ 2.68 (0.58–12.36) |
| Jöckel *et al.* 1998[4,5,8] | Germany | | | |
| I | | 71 | 236 | 1.58 (0.83–2.98) |
| II | | 304 | 1423 | 0.93 (0.69–1.25) |
| Kabat *et al.* 1995 | USA | 67/39 | 173/98 | 1.10 (0.62–1.96)/ 1.63 (0.69–3.85) |
| Kabat and Wynder 1984[1] | USA | 24/12 | 25/12 | 0.79 (0.25–2.45)/ 1.00 (0.10–5.07) |
| Kalandidi *et al.* 1990 | Greece | 90 | 116 | 1.62 (0.90–2.91) |
| Koo *et al.* 1987 | Hong Kong | 86 | 136 | 1.55 (0.90–2.67) |
| Lam 1985 | Hong Kong | 60 | 144 | 2.01 (1.09–3.72) |
| Lam *et al.* 1987 | Hong Kong | 199 | 335 | 1.65 (1.16–2.35) |
| Lee *et al.* 1986[1] | England | 32/15 | 66/30 | 1.03 (0.41–2.55)/ 1.31 (0.38–4.52) |
| Liu *et al.* 1991[6,7] | China | 54 | 202 | 0.74 (0.32–1.69) |

**Table 6.** continued

| Study | Country | Cases ♀/♂ | Size of control group or cohort ♀/♂ | Relative risk (95 % confidence interval) ♀/♂ |
|---|---|---|---|---|
| Liu *et al.* 1993[1] | China | 38 | 69 | 1.66 (0.73–3.78) |
| Pershagen *et al.* 1987 | Sweden | 70 | 294 | 1.03 (0.61–1.74) |
| Shimzu *et al.* 1988 | Japan | 90 | 163 | 1.08 (0.64–1.82) |
| Sobue 1990 | Japan | 144 | 731 | 1.06 (0.74–1.52) |
| Stockwell *et al.* 1992 | USA | 210 | 301 | 1.60 (0.80–3.00) |
| Sun *et al.* 1996 | China | 230 | 230 | 1.16 (0.80–1.69) |
| Trichopolous *et al.* 1983[2] | Greece | 62 | 190 | 2.13 (1.19–3.83) |
| Wang *et al.* 1996 | China | 135 | 135 | 1.11 (0.67–1.84) |
| Wu *et al.* 1985[1] | USA | 29 | 62 | 1.20 (0.50–3.30) |
| Wu-Williams *et al.* 1990[2,7] | China | 417 | 602 | 0.79 (0.62–1.02) |
| Zaridze *et al.* 1995[6] | Russia | 162 | 285 | 1.66 (1.12–2.45) |
| Zaridze *et al.* 1998[5] | Russia | 189 | 358 | 1.53 (1.06–2.21) |
| Hackshaw *et al.* 1997, meta-analysis, 37 studies of women, 9 studies of men | | 4625/274 | 477924/117260 | 1.24 (1.13–1.36)/ 1.34 (0.97–1.84) |

[1] number of cases small (< 50)
[2] confounders not adequately taken into account
[3] 90 % confidence interval
[4] men and women
[5] not included in the meta-analysis of Hackshaw *et al.* (1997)
[6] documentation inadequate
[7] in US-EPA (1993) described as the least valid studies
[8] I: North Germany, II: West, South and East Germany


## Dose-response relationships

Demonstration of a causal relationship between ETS-exposure and the development of lung cancer requires the analysis of the dose-response relationship.

In the study by Garfinkel (1981), the relative risk was not increased with increasing exposure levels.

In the study by Hirayama (1984), the relative risks for the women were 1.36, 1.42, 1.58 and 1.91 (90 % confidence interval (CI): 1.34–2.71) when the husbands were ex-smokers, and when they smoked 1–14, 15–19 or more than 20 cigarettes per day, respectively (p value for trend test 0.002). Similarly for non-smoking men the relative risk increased with the exposure level: it was 2.14 when the wives smoked 1–19 cigarettes per day and 2.31 (90 % CI: 0.90–5.94) when they smoked more than 20 (p value for trend test: 0.02).

Potential confounders for lung cancer and misclassification of smokers were not allowed for, but the clear dose response relationship does speak for an effect of passive smoking.

In the study by Cardenas *et al.* (1997) a significant dose-response relationship was also found: in the exposure groups 1–19, 20–39 and > 40 cigarettes/day, the relative risks for the women exposed to ETS were 1.1, 1.2 and 1.9 (95 % CI: 1.0–3.6) (p value for trend test: 0.03). An increase in the relative risk with the length of time the couples had been married was not observed. Because the number of cases was too small, a similar analysis for the men exposed to ETS could not be carried out. This study is marked by almost complete data for causes of death (97%) and direct questioning of both partners as to their smoking habits. Numerous potential confounders such as previous lung diseases, occupational exposure to asbestos, eating habits and education were taken into account.

Taken together, the 3 cohort studies demonstrate an increased incidence of deaths from lung cancer associated with exposure to ETS. Significant is the increase in the study by Hirayama (1984). In this and in the study by Cardenas *et al.* (1997) there is also a significant dose-response relationship (see Table 7).

## 6.1.2 Case-control studies

### 6.1.2.1 ETS exposure by the partner

Without differentiation according to the level or duration of exposure to ETS (Table 6), increased relative risks for lung cancer were observed in most of the studies which can be evaluated and the increases were sometimes statistically significant (Fontham *et al.* 1994, Lam *et al.* 1987, Zaridze *et al.* 1998). The results of the studies by Fontham *et al.* 1994 and Lam *et al.* 1987 are of particular relevance for the reasons described below.

In the study by Fontham *et al.* (1994), all confounders known to date such as potential misclassification of smokers, occupational exposure to known carcinogens, eating habits and socio-economic status were taken into account. The smoking status was verified by means of cotinine determination to minimize the misclassification of smokers as non-smokers. The study by Lam *et al.* (1987) is characterized by good data for exposure (various sources of exposure) to ETS, valid classification of the smoking habits of the husband and compensation for potential confounders (education, place of birth, duration of residence in Hong Kong, marital status).

In some of the studies in which the lung cancer risk was increased after exposure to ETS, but not significantly (Gao *et al.* 1987, Jöckel *et al.* 1998, Kabat *et al.* 1995, Kalandidi *et al.* 1990, Koo *et al.* 1987, Liu *et al.* 1993, Stockwell *et al.* 1992) the number of cases was very small; the studies were, however, well-documented and took potential confounders into account. Thus the results of these studies support the conclusions of Fontham *et al.* (1994) and Lam *et al.* (1987).

In a European multicentre study with 650 cases and 1542 controls, the relative risk of developing lung cancer after exposure to ETS by the spouse was 1.16 and not significantly increased (Boffetta *et al.* 1998; Table 6).

Of the studies with negative results, that of Brownson *et al.* (1992) is characterized by good documentation and exposure data for home and workplace, and took into account potential confounders. In two other studies, although the relative risk of developing lung cancer for wives exposed to ETS by their smoking husbands was not increased, that for children (Janerich *et al.* 1990) or for women whose mothers or fathers-in-law smoked (Shimizu *et al.* 1988) was.

**Table 7.** Relative risk (RR) of lung cancer after exposure of life-long non-smokers to ETS (groups with highest level exposure) by a smoking partner (modified from Jinot and Bayard 1994)

| Reference | Exposure level | Power[1] | RR[2] | p value[3] |
|---|---|---|---|---|
| studies analysed in US-EPA (1993)[4]: | | | | |
| Akiba et al. 1986 | ≥ 30 cigarettes/day | 0.1 | 2.1 | 0.13 |
| Correa et al. 1983 | ≥ 41 packyears* | 0.06 | 3.2 | 0.005 |
| Fontham et al. 1991[5] | ≥ 80 packyears | no data | 1.32 | 0.21 |
| Gao et al. 1987 | ≥ 40 years | 0.33 | 1.7 | 0.02 |
| Garfinkel 1981 | ≥ 20 cigarettes/day | no data | 1.09 | 0.33 |
| Garfinkel et al. 1985 | ≥ 20 cigarettes/day | 0.21 | 2.05 | 0.01 |
| Geng et al. 1988 | ≥ 20 cigarettes/day | no data | 2.76 | < 0.00001 |
| Hirayama 1984 | ≥ 20 cigarettes/day | 0.13 | 1.91 | 0.00015 |
| Humble et al. 1987 | ≥ 21 cigarettes/day | no data | 1.09 | 0.46 |
| Inoue and Hirayama 1988 | ≥ 20 cigarettes/day | no data | 3.09 | 0.05 |
| Janerich et al. 1990 | ≥ 50 packyears | no data | 1.01 | 0.5 |
| Kalandidi et al. 1990 | ≥ 41 cigarettes/day | 0.06 | 1.57 | 0.16 |
| Koo et al. 1987 | ≥ 21 cigarettes/day | 0.11 | 1.18 | 0.36 |
| Lam et al. 1987 | ≥ 21 cigarettes/day | 0.16 | 2.05 | 0.02 |
| Pershagen et al. 1987 | ≥ 16 cigarettes/day | no data | 3.11 | 0.02 |
| Trichopoulos et al. 1983 | ≥ 21 cigarettes/day | 0.11 | 2.55 | 0.003 |
| Wu et al. 1985 | ≥ 31 years[6] | no data | 1.87 | no data |
| other studies: | | | | |
| Boffetta et al. 1998[4] | > 23 packyears | no data | 1.64 | < 0.05[7] |
| Brownson et al. 1992[4] | ≥ 40 packyears | no data | 1.3 | 0.025[8] |
| Cardenas et al. 1997 | ≥ 40 cigarettes/day | no data | 1.9 | 0.025[8] |
| Du et al. 1993 | > 20 cigarettes/day | no data | 1.62 | > 0.05[7] |
| Fontham et al. 1994[4] | ≥ 80 packyears | no data | 1.79 | 0.03[7] |
| Jöckel 1991 | > 90th percentile of the distribution | no data | 3.43 | 0.018[7] |
| Jöckel et al. 1998[9] | > 90th percentile of the distribution | no data | I: 1.51<br>II: 1.59 | > 0.05[7]<br>> 0.05[7] |
| Kabat et al. 1995 | > 10 cigarettes/day | no data | ♂ 7.48<br>♀ 1.06 | < 0.05[7]<br>< 0.05[7] |
| Liu et al. 1993 | ≥ 20 cigarettes/day | no data | 2.9 | 0.02[7] |
| Stockwell et al. 1992 | ≥ 40 years | no data | 2.2 | 0.025[8] |
| Zaridze et al. 1998 | > 15 years | no data | 1.42 | 0.07[7] |

* packyears is the number of packs of cigarettes smoked daily × years
[1] probability that the effect is significant (p < 0.05) , when the real relative risk is 1.5
[2] RR for cohort studies; OR for case-control studies
[3] one-sided p value test for RR = 1 against RR > 1, or trend test
[4] relative risk, corrected for misclassification of smokers
[5] relative risk for all types of tumour relative to two control groups
[6] years of exposure for adults (partner and workplace)
[7] two-sided p value test
[8] 95 % CI includes 1.0
[9] I: north Germany, II: west, south and east Germany

## Dose-response relationship

The increased risk of developing lung cancer after exposure to ETS by the partner becomes more apparent when the groups with the highest level exposure are examined: in most studies in which a dose-response relationship was investigated, the relative risk is significantly increased in this group (Table 7) or a statistically significant trend towards increase in the risk of lung cancer with exposure to ETS was found.

In the European multicentre study, the risk of developing lung cancer after exposure to ETS by the partner was highest in the group with the highest level exposure. Because the exposure was recorded differently in the different study centres, the number of cases available for evaluation of the dose-response relationship was reduced to about 550. For those variables which best describe the dose, "hours/day × years" and "packyears", there was a significant linear trend only for the variable "hours/day × years". Here the relative risks were also highest in the groups with the highest exposure levels (Boffetta *et al.* 1998; Table 8).

**Table 8.** Relative risk of lung cancer determined in the European multicentre study after exposure of life-long non-smokers to ETS by their partners (Boffetta *et al.* 1998)

| Exposure | Cases | Controls | RR[a] | 95 % CI | Trend[b] |
|---|---|---|---|---|---|
| [hours/day × years][c] | | | | | |
| 1–135 | 165 | 396 | 0.90 | 0.70–1.16 | |
| 136–223 | 44 | 81 | 1.20 | 0.78–1.85 | |
| > 223 | 41 | 53 | 1.80 | 1.12–2.90 | 0.02 |
| no data | 103 | 234 | | | |
| [packyears][d] | | | | | |
| 0.1–13.0 | 188 | 411 | 1.00 | 0.78–1.28 | |
| 13.1–23.0 | 36 | 83 | 0.89 | 0.57–1.39 | |
| > 23.0 | 42 | 55 | 1.64 | 1.04–2.59 | 0.09 |
| no data | 87 | 215 | | | |

[a] relative risk, adjusted for age and sex
[b] p value for test for linear trend
[c] in the presence of the non-smoking spouse
[d] years × packs/day

In two studies carried out in Germany from which some of the data is included in Boffetta *et al.* (1998), a non-significant increase in the relative risk with the level of exposure was seen for exposures caused exclusively by the partner; the number of cases with high level exposure was, however, small. The exposure was classed as "little or none" (dose below the 75th percentile of the distribution of cases and controls; 209 cases in the two studies), "relevant" (dose above the 75th percentile; 20 cases) or "high" (dose higher than the 90th percentile; 25 cases). Taking into account all sources of exposure (including workplace, see Section 6.1.2.2) the relative risk in the group with the highest level exposure was significantly increased (RR = 2.09, 95 % CI: 1.02–4.48, 18 cases) in one of the studies and increased but not significantly in the other (RR = 1.41, 95 % CI: 0.98–2.01, 57 cases) (Jöckel *et al.* 1998).

In particular in the light of the dose-response relationships, the available studies speak for a causal relationship between the development of lung cancer and exposure to ETS.

## 6.1.2.2 Exposure to ETS at the workplace

In some of the available studies, the exposure situation was described only for the current workplace, in others the dates and duration of the workplace exposure were not given at all (Brownson *et al.* 1992, Garfinkel *et al.* 1985, Kalandidi *et al.* 1990, Shimizu *et al.* 1988, Stockwell *et al.* 1992, Zaridze *et al.* 1998). In other studies, only exposure to ETS at the workplace during the previous three months was described (Koo *et al.* 1987) or the exposure at the previous four places of work (more than 1 year) (Kabat *et al.* 1995) was documented. In these studies the observed relative risks of lung cancer were sometimes increased but not significantly.

### Dose-response relationship

In two studies, no dose-response relationship was found (Kabat *et al.* 1995, Kalandidi *et al.* 1990). In studies which were more extensive than these two, more complete data for the exposure to ETS at the workplace were collected and the dose-response relationship was investigated (Boffetta *et al.* 1998, Fontham *et al.* 1994, Jöckel *et al.* 1998).

For women, who were exposed to ETS at work, the risk of lung cancer was significantly increased to 1.39 (95 % CI: 1.1–1.74). For women, who had been exposed to ETS for 1–15, 16–30 or more than 30 years, the relative risk of developing lung cancer increased significantly with the length of the exposure period: 1.30 (95 % CI: 1.01–1.67), 1.40 (95 % CI: 1.04–1.88) and 1.86 (95 % CI: 1.24–2.78), respectively (Fontham *et al.* 1994).

In the European multicentre study, the relative lung cancer risk after exposure to ETS at work was 1.17 (95 % CI: 0.94–1.45). A dose-response relationship was not seen when the data were analysed according to the duration of exposure but a significant trend was observed after analysis of weighted exposure, which better reflects the exposure dose (Table 9; Boffetta *et al.* 1998).

When the two studies from Germany were analysed together, the relative lung cancer risk for persons exposed to ETS only at work compared with persons with "little or no exposure" (definition see Section 6.1.2.1; 325 cases) was 1.63 (95 % CI: 0.94–2.84, 19 cases) for "not negligible" exposure and 1.95 (95 % CI: 1.11–3.42, 21 cases) for "high level" exposure (Jöckel *et al.* 1998).

In summary, the studies in which dose-response relationships were analysed revealed an increase in the relative risk of lung cancer associated with exposure to ETS at work and statistically significant increases in relative risk in those groups with the highest level exposure.

**Table 9.** Relative lung cancer risk after exposure of life-long non-smokers to ETS at work (Boffetta *et al.* 1998)

| Exposure | Cases | Controls | RR[a] | 95 % CI | Trend[b] |
|---|---|---|---|---|---|
| [years] | | | | | |
| none | 276 | 687 | 1.00 | – | |
| 0.1–29 | 278 | 634 | 1.15 | 0.91–1.44 | |
| 29.1–38 | 55 | 129 | 1.26 | 0.85–1.85 | |
| > 38 | 39 | 91 | 1.19 | 0.76–1.86 | 0.21 |
| no data | 2 | 1 | | | |
| | | | | | |
| weighted[c] | | | | | |
| none | 276 | 687 | 1.00 | – | |
| 1–16848 | 196 | 525 | 0.97 | 0.76–1.25 | |
| 16849–32448 | 47 | 105 | 1.41 | 0.93–2.12 | |
| > 32448 | 48 | 71 | 2.07 | 1.33–3.21 | <0.01 |
| no data | 83 | 154 | | | |

[a] relative risk, adjusted for age and sex
[b] trend: p value for linear trend test
[c] subjective estimation of ETS dose × hours exposure

## 6.1.3 Meta-analyses

The US EPA (1993) has carried out detailed evaluation of 31 studies (including 4 cohort studies) and has adjusted the results for misclassification of smokers (see Section 6.1.4). The relative risks associated with exposure to ETS by the partner which were found for different regions were: Greece (2.01), Hong Kong (1.48), Japan (1.41), USA (1.19), Western Europe (1.17) and China (0.95). When only the groups with the highest level exposure were analysed, the relative risks determined were 2.15, 1.68, 1.96, 1.38, 3.11 and 2.32 and the overall relative risk was 1.81 (90 % CI: 1.60–2.05).

In the meta-analysis of 37 case-control studies and cohort studies by Hackshaw *et al.* (1997) the relative lung cancer risks associated with ETS exposure were 1.24 (95 % CI: 1.13–1.36) for women, 1.34 (95 % CI: 0.97–1.84) for men. Overall for women and men together and including data from two other studies (Hole *et al.* 1989, Janerich *et al.* 1990) in which differentiation according to sex was not carried out, a relative risk of 1.23 (95 % CI: 1.13–1.34) was obtained. From the data from 16 studies of the relationship between the number of cigarettes smoked by the partner and lung cancer risk, a significant dose-response relationship was found. The lung cancer risk increases by 23 % (14 %–32 %) for each 10 cigarettes per day smoked by the partner. In 11 studies the relationship between the risk of lung cancer and the length of time the person had lived together with a smoker was studied. Here too the dose-response relationship was significant. The lung cancer risk increases by 11 % (4 %–17 %) for each 10 years of exposure (Hackshaw *et al.* 1997).

## 6.1.4 Potential bias

Erroneous classification of smokers or ex-smokers as non-smokers or false claims of smokers or ex-smokers to be non-smokers would lead to an overestimation of the risk of developing lung cancer because of exposure to ETS. In some publications (Lee 1988, 1987a, 1987b, Lee *et al.* 1986), in fact, the whole increase in relative risk is ascribed to such misclassification.

The reason for this is the assumption that it is more likely for a smoker to live together with a smoker than with a non-smoker. In that case it would also be more likely that a smoker with lung cancer misclassified as a non-smoker lived together with a smoking partner and this would in fact result in an overestimation of the risk associated with ETS. On the other hand, "random" misclassification of some of the smokers as non-smokers would not lead to an overestimation of the relative risk.

In the report of the US EPA (1993), this aspect of the problem was taken into account and an individual correction for this kind of misclassification carried out for all studies for which it was possible; in spite of the correction, the relative lung cancer risks associated with ETS exposure were still increased.

In a simpler approach it is taken into account that smokers live together with smokers about three times as often as with non-smokers. Assays of nicotine and cotinine demonstrate that 2 % of those who claim to be non-smokers are actually smokers. When these data are used for correction purposes, the relative risk revealed by the meta-analysis of Hackshaw *et al.* (1997) of 1.24 (1.13–1.36) is reduced to 1.18 (1.07–1.3). In addition, a reduced intake of fruit and vegetables increases the risk of developing lung cancer and smokers and non-smokers who live together with smokers eat less fruit and vegetables than do non-smokers without a partner who smokes. When these eating habits are also taken into account, the relative risk associated with ETS is reduced to 1.16 (1.04–1.03). However, factors which falsely reduce the determined relative risk, namely misclassification in the control group, must also be taken into account: cotinine can be detected in the urine of some of the non-smokers who live together with non-smokers, which means that these persons are exposed to ETS elsewhere. The associated lung cancer risk in the control group leads to a false reduction in the calculated relative risk. When this effect is taken into account, the relative risk is increased from 1.16 to 1.26 (1.06–1.47). Thus, because the potential sources of bias do not have large effects and tend to balance each other out, the non-adjusted relative risk of 1.24 (1.13–1.36) may be seen as a valid estimation of the real risk (Hackshaw *et al.* 1997).

Therefore it is very unlikely that the observed increase in relative lung cancer risk associated with ETS is caused by misclassification of smokers. This conclusion is not changed by the analyses of misclassification of smokers in southern Germany (Heller *et al.* 1998), because they have assumed levels of misclassification which are much too high (Keil *et al.* 1998).

Another source of error in the estimation of the relative lung cancer risk for passive smokers is the failure to take into account the proportion of ex-smokers among the partners classified as smokers. In most studies, exposure was given only as "ever smoked" or "never smoked", which means that the exposure of a passive smoker can have taken place a long time before the study when the partner is an ex-smoker. In ex-smokers the increased relative lung cancer risk has practically disappeared after 10–20 years (IARC 1986). The study of Nyberg *et al.* (1998) demonstrates that also for passive smokers with ex-smokers as partners the risk is smaller because of the time since exposure than it is for persons with active smokers as partners. This effect tends to result in an underestimation of the relative lung cancer risk for passive smokers with partners who are currently active smokers (Nyberg *et al.* 1998).

## 6.2 Other kinds of cancer

There are also studies of the association of exposure to ETS and the development of tumours of the nasal cavity, the nasal sinuses, bladder and cervix, which are certainly or probably induced by active smoking (California EPA 1997). They demonstrate a consistent association between the development of carcinomas of the nasal sinuses and ETS exposure (Buckley *et al.* 1981, Fukuda and Shibata 1990, Hirayama 1984, Zheng *et al.* 1993). For the development of cervix carcinoma after exposure to ETS, the relative risks were increased significantly (Sandler *et al.* 1985a, 1985b, Slattery *et al.* 1989) or not significantly (Hirayama 1981, Coker *et al.* 1992). The detection of DNA adducts in the cervical epithelium and of nicotine and cotinine in cervical mucous suggests that the association is biologically plausible (Slattery *et al.* 1989).

For other kinds of tumours such as bladder cancer (Burch *et al.* 1989, Kabat *et al.* 1986), brain tumours (Cordier *et al.* 1994, Filippini *et al.* 1994, Gold *et al.* 1993, Norman *et al.* 1996), renal cell carcinomas (Kreiger *et al.* 1993), mammary carcinomas, leukaemia, lymphomas and non-Hodgkin lymphomas the results are unclear and it cannot be decided whether or not there is an association with ETS exposure because either the numbers of persons studied were too small or the effects were seen only in some subgroups (California EPA 1997).

## 6.3 Other respiratory disorders

### 6.3.1 Children

In several cross-sectional studies, a significantly higher incidence of asthma or increased sensitivity of the airways was found in children whose parents included at least one smoker (Andrae *et al.* 1988, Burchfiel *et al.* 1986, Chinn and Rona 1991, Dekker *et al.* 1991, Dold *et al.* 1992, Evans *et al.* 1987, Forastière *et al.* 1992, Frischer *et al.* 1992, Goren and Hellman 1991, Gortmaker *et al.* 1982, Kühr *et al.* 1992, Murray and Morrison 1986, O'Connor *et al.* 1987, Weitzmann *et al.* 1990). In several epidemiological studies, increased relative risks of 1.4 to 4.8 for the development of respiratory diseases such as pneumonia and bronchitis were found for babies and small children (Fergusson and Horwood 1985, Hall *et al.* 1984, Harlap and Davies 1974, Leeder *et al.* 1976, Ogston *et al.* 1987, Pullan and Hey 1982, Rantakallio 1978).

In addition, in cohort studies bronchitis and asthma were seen especially in atopically predisposed children significantly more frequently if they were exposed to ETS at home (Arshad and Hilde 1992, Bisgaard *et al.* 1987, Halken *et al.* 1991, Martinez *et al.* 1992, Neuspiel *et al.* 1989, Rantakallio 1978, Taylor and Wadsworth 1987, Burr *et al.* 1989, Cogswell *et al.* 1987, Geller-Bernstein *et al.* 1987, Halken *et al.* 1992). The health of the children often improves later in life if the parents stop smoking in the presence of the children (Murray and Morrison 1993). In several of these studies a positive dose-response relationship was demonstrated between ETS exposure levels (number of cigarettes and number of smokers) and the development of the above-mentioned diseases (Jöckel and Knauth 1994).

Extensive studies have demonstrated an association between exposure to ETS and the occurrence of respiratory diseases such as coughing, bronchitis, throat infections with fever, and sinusitis, an increase in the frequency of these diseases, and adverse effects on lung function in school children (Charlton 1984, Dold *et al.* 1992, Meister 1990, Pershagen 1986, Samet *et al.* 1991, Ware *et al.* 1984, Weiss *et al.* 1983, Wolf-Ostermann *et al.* 1995). The presence of a dose-response relationship has been confirmed; thus the relative risk of coughing and bronchitis was increased by 13 % to 18 % after exposure to an average of 10 cigarettes per day. If the exposure level was twice as high, the relative risk increased by 27 % to 40 % (Meister 1990, Pershagen 1986, Somerville *et al.* 1988, Weiss *et al.* 1983).

There is good evidence, that babies whose mothers smoke are at higher risk of dying from sudden infant death syndrome before they are one year old. This association exists independently of all other known risk factors; the biological details are unknown (US EPA 1993).

The most recent meta-analyses of 300 epidemiological studies of the association between exposure to ETS and disorders in early childhood confirm the development of diseases of the lower respiratory tract, sudden infant death syndrome, asthma, asthmatic symptoms and inflammation of the middle ear (*otitis media*) (Anderson and Cook 1997, Cook and Strachan 1997, Strachan and Cook 1997, 1998a, 1998b, 1998c).

## 6.3.2 Adults

In the review by the US EPA (1993), 6 earlier studies are described in which ETS was shown to have slight adverse effects on the lung function of healthy non-smokers (reduction of mean $FEV_1$ by about 2.5 %) and to increase the incidence of respiratory symptoms by 30 % to 60 %. These studies are sometimes inconsistent, frequently do not demonstrate any dose-response relationships and do not contain sufficient controls to check for sources of bias (US EPA 1993).

In more recent studies described in the review by the California EPA (1997), exposure to ETS is more clearly associated with the development of respiratory diseases and symptoms and the development of impaired lung function in adults.

A study of indoor exposure of 164 persons, including many with severe to moderate asthma, revealed that the number of days with restricted activity increased with the level of exposure to ETS (RR 1.61, 95 % CI: 1.06–2.46). The relative risk of developing severe coughing and shortness of breath was not increased significantly. If the asthmatic non-smoker lived with a smoker, the daily moderate to severe episodes of shortness of breath were worse (RR 2.05, 95 % CI: 1.78–2.40) (Ostro *et al.* 1994).

In a 10-year prospective study of 3577 Seventh-day Adventists (men and women) the relative risk for the development of new cases of asthma was 1.45 (95 % CI: 1.21–1.80) if the persons were exposed to ETS at work, and was significantly associated with the duration of exposure. The relative risk was also increased significantly (no other details), when the analysis was restricted to the medically confirmed cases of asthma (30 of 49) (Greer *et al.* 1993).

Exposure of 4197 passive smokers to ETS increased the incidence of wheezing not associated with colds (RR 1.94, 95 % CI: 1.39–2.70), medically diagnosed asthma (RR 1.39, 95 % CI: 1.04–1.86), shortness of breath during physical activity (RR 1.45, 95 % CI: 1.20–1.76), acute (RR 1.59, 95 % CI: 1.17–2.15) and chronic bronchitis (RR 1.65, 95 % CI: 1.28–2.16). The incidence of allergic rhinitis was not increased. Taking the main sources of bias into account did not have an appreciable effect on the values of the relative risks. The incidence of respiratory symptoms increased with the duration of exposure (Leuenberger *et al.* 1994).

In other more recent studies reviewed by the California EPA (1997) there is also evidence for the association of exposure to ETS with the development of respiratory disorders and with adverse effects on lung function (Dayal *et al.* 1994, Jaakkola *et al.* 1995, Ng *et al.* 1993, Robbins *et al.* 1993, White *et al.* 1991, Xu and Li 1995). However, these results can be evaluated only with reservations because in some cases the findings were limited to certain conditions, for example, to persons exposed to ETS both as children and as adults (Robbins *et al.* 1993) or only in the age group up to 25 years (Jaakkola *et al.* 1995). The interpretation of other results (Dayal *et al.* 1994) is limited by incomplete documentation, for example, no data for demographic parameters or for the date of diagnosis (California EPA 1997).

The unambiguously documented association of passive smoking with the incidence of respiratory symptoms in children (Section 6.3.1) makes it plausible that ETS can cause the observed mild respiratory symptoms and lung function disorders also in adults, who are less sensitive than children.

## 6.4 Cardiovascular diseases

In a case-control study of 59 women with angiographically confirmed coronary heart disease and 126 control persons, it was shown that exposure to ETS either at home or at work was sufficient to cause a markedly increased risk of coronary heart disease (relative risk of 2.0 and 2.7, respectively). If the persons were exposed both at home and at work, the relative risk was much higher (4.2, 95 % CI 1.6–10.9). Positive and statistically significant dose-response relationships between ETS exposure level (number of smokers, amount of smoke, daily period of exposure to ETS) at home or at work and the development of coronary heart disease were found (He *et al.* 1994).

In other studies, a significant association between passive smoking and coronary heart disease or the risk of myocardial infarction was reported (Dobson *et al.* 1991, Kawachi *et al.* 1997).

The intimal-medial thickness of the walls of the carotid arteries increases consistently in the order: never-smokers without ETS exposure, never-smokers exposed to ETS, ex-smokers, currently active smokers. Even after correction for other potential effect factors in multivariate analyses, the effects remained (Howard *et al.* 1994). These results were confirmed in a follow-up study in which the increase in vessel wall thickness was shown to be 20 % higher in persons exposed to ETS (Howard *et al.* 1998).

The good agreement between the results of epidemiological studies of the effects of passive smoking on cardiovascular disease is documented in a number of reviews (Glantz and Parmley 1991, Hense 1995, Law *et al.* 1997, Wells 1994). In most of the studies, passive smoking was characterized in terms of the exposure by a smoking spouse. Many of the studies, however, were too small to yield significant risk estimates. Practically all the studies demonstrated increases between 20 % and 70 % in the relative risks for fatal and non-fatal cardiovascular diseases (Wells 1994). In the larger studies, a monotonic dose-response relationship without a clear threshold could be demonstrated. There was no appreciable difference between men and women. Even after control for confounders and taking into account a plausible level of misclassification, the effect of passive smoking remained significant.

The epidemiological data from prospective studies and case-control studies suggest that there is a causal relationship between exposure to ETS and death of non-smokers from coronary heart disease. For non-smokers exposed to ETS by the partner, the risk of dying of coronary heart disease is increased by a factor of 1.3 (California EPA 1997). The association is stronger for mortality than for morbidity including *angina pectoris*. Because similarly large relative risks result in much more heart disease than lung cancer, the absolute numbers of diseases of the cardiovascular system caused by passive smoking are about five to ten times the numbers of cases of bronchial carcinoma.


## 6.5 Evaluation of the epidemiological studies

The results of the epidemiological studies and the dose-response relationships which have been demonstrated indicate that exposure to ETS increases the relative risk of developing lung cancer. In extensive meta-analyses, a relative risk in the range of 1.2 to 1.3 was found, without differentiation according to duration or level of exposure. Exposure to ETS increases the risks of cancer of the nasal cavity or nasal sinuses and that of cardiovascular disease and the mortality from these diseases. These effects are plausible because they are also seen in smokers, they can be accounted for mechanistically and have been shown to yield dose-response relationships. Impairment of lung function, respiratory diseases and an increased incidence of asthma have been observed. In children, passive smoking causes pneumonia, bronchitis and bronchiolitis, and in asthmatic children severe and frequent attacks of asthma.

# 7 Results of Animal Studies

## 7.1 Effects on the respiratory passages

Lung function tests were carried out on rats which were 7–10 weeks old and which had been exposed before and after birth to sidestream smoke (total particle concentration: 1 mg/m$^3$). Studies of the isolated perfused lungs revealed a 24 % decrease in elasticity, and a 200 % greater hyper-reactivity after provocation with metacholine. The number of neuroendocrine cells was increased 22-fold after exposure during gestation and lactation. The reduced elasticity and hyper-reactivity are considered to be a result of the hyperplasia and persistence of neuroendocrine cells. These effects were not observed when the animals were exposed either only prenatally or only postnatally. In rats which were exposed to sidestream smoke only postnatally, the rate of cell division in the epithelial cells of the terminal bronchi but not in the proximal bronchi was reduced relative to that in controls (California EPA 1997).

Elasticity, pulmonary resistance and morphology were studied in isolated perfused lungs of 43-day old guinea pigs which had been exposed to sidestream smoke since they were 8 days old (total particle concentration: 1 mg/m$^3$). The elasticity of the lungs increased by 17 %, pulmonary morphology and resistance were unaffected. The reaction of the lungs to capsaicin, a specific C fibre agonist, was reduced after exposure, the reaction to substance P, which is released from C fibres, was unaffected. The authors concluded from these results that sidestream smoke can cause a reduced C fibre response which makes it possible for toxic and infectious substances to enter the lungs and to increase the incidence of respiratory effects (California EPA 1997).

Rats were exposed to sidestream smoke 6 hours per day, 5 days per week from day 2 of their lives for 8 to 15 weeks. The total particle concentration was given as 1 mg/m$^3$. In this study no changes in pulmonary resistance, elasticity, relative lung weights or pressure in the lung arteries were found (Witschi *et al.* 1997b).

## 7.2 Cardiovascular effects

Some animal studies have demonstrated, that short-term exposure to sidestream smoke promotes arteriosclerosis. Exposure of male rabbits for 10 weeks to a low (particle concentration: 4 mg/m$^3$, n = 16) or a high (particle concentration: 33 mg/m$^3$, n = 16) concentration of sidestream smoke accelerated the development of arteriosclerosis in the aorta and in the pulmonary arteries. The rabbits had been given a diet containing 0.3 % cholesterol (90 mg/kg body weight) for two weeks before the exposure (Zhu *et al.* 1993). In another study with male rabbits exposed to sidestream smoke, administration of the β-blocker metoprolol did not protect the animals from the development of arteriosclerosis. The authors concluded that the β-adrenergic receptor system is not involved in the mechanism of induction of arteriosclerosis by tobacco smoke (California EPA 1997).

In young hens exposed to sidestream smoke (particle concentration: 2.2–2.8 mg/m$^3$) for 16 weeks (6 hours/day, 5 days/week), the growth of arteriosclerosis plaques was accelerated. The authors state that similar concentrations of sidestream smoke would suffice to accelerate the development of arteriosclerosis plaques in man (Penn and Snyder 1993).

Sprague-Dawley rats were exposed to sidestream smoke 6 hours daily, 5 days per week, for 3 weeks (n = 21) or 6 weeks (n = 12). The serum lipid concentrations, carbon monoxide-Hb, plasma nicotine and cotinine concentrations and myocardial infarct size increased significantly and in a dose-dependent manner. The animals were exposed to carbon monoxide concentrations of 92 ml/m$^3$, total particles 60 mg/m$^3$ and nicotine 1.1 mg/m$^3$ (Zhu *et al.* 1994).

## 7.3 Reproductive toxicity

Rats were exposed to sidestream smoke for two hours daily during gestation. The body weights of the foetuses were reduced significantly (by 9 %) relative to the control values. The smoke used was not quantified or otherwise characterized (California EPA 1997).

In another study, rats were exposed to sidestream smoke for 6 hours daily on days 3 to 10 of gestation. No effects on the body weights of the foetuses were found. The number of litters, however, was significantly smaller than in the control group (California EPA 1997).

After exposure of pregnant rats to sidestream smoke (total particle concentration 1 mg/m$^3$) for 6 hours on each of days 3, 6 to 10, and 13 to 17 of gestation, a significant reduction (7 %) in the body weights of the foetuses relative to the control values was found. The body weights and food consumption of the dams, the number of foetuses and the number of implantations was unaffected (California EPA 1997).

From day 3 of gestation, pregnant Sprague-Dawley rats were exposed to sidestream smoke (particles: 1 mg/m$^3$, carbon monoxide 4.9 ml/m$^3$, nicotine 344 µg/m$^3$) for 4 hours daily, 7 days per week until the births, and the female progeny then exposed for 9 weeks postnatally. In two other groups the animals were exposed either just prenatally or just postnatally. No effects were detected in the pups exposed only prenatally. Postnatal exposure of the rats increased the mortality within the first 18 days of life from 14 % to 43 %. Both in the group exposed prenatally and postnatally and in that exposed only postnatally, significantly reduced body weights, an increase in relative brain weights, a significant reduction in the level of DNA in the brain and an increase in the protein:DNA ratio in the brain were observed (Gospe *et al.* 1996).

## 7.4 Genotoxicity

The mutagenicity of the urine of rats exposed to mainstream or sidestream smoke (smoking machine, 2–4 filter cigarettes) in a *Salmonella* mutagenicity test with strain TA1538 was increased (Mohtashamipur *et al.* 1984).

Male and female F344/N rats (number not specified) were exposed to cigarette smoke 6 hours daily, 5 days per week for 22 days either intermittently or continuously via the nose or continuously by whole-body exposure. In the first week the dose of particles to which the animals were exposed was 600 mg × hour/m$^3$, later 1200 mg × hour/m$^3$; 18 hours or 3 weeks after the last exposure the levels of DNA adducts in lung tissue were determined. A significant increase in the levels of DNA adducts was seen only in the 18-hour samples (Bond *et al.* 1989).

A significant increase in the levels of DNA adducts was observed in the lungs of female C57Bl and DBA mice (numbers not specified) which had been exposed nasally to sidestream smoke for 65–70 weeks (Gairola *et al.* 1993).

A micronucleus test carried out with groups of 3 NMRI mice exposed to the sidestream smoke of 1 to 4 cigarettes yielded positive results which were dependent on the exposure level (Mohtashamipur *et al.* 1987b).

Male SD rats were exposed 6 hours/day, 5 days/week for 14 days or 13 weeks to sidestream smoke (particle concentration: 0.1, 1, 10 mg/m$^3$). DNA-adduct formation and chromosomal aberrations were determined in the organs lung, heart, liver, larynx and bladder. The body weights of the exposed animals were not different from those of the controls. In the highest concentration group, mild and partially reversible hyperplasia of the nasal epithelium was seen. A significant increase in DNA-adduct formation was detected after 7, 14 and 28 days only in the highest concentration group in the lungs and heart, after 90 days also in the larynx. A significant increase in chromosomal aberrations was not seen in any of the concentration groups (Lee *et al.* 1992, 1993).


## 7.5 Carcinogenicity

Male A/J mice were exposed to ETS 6 hours/day, 5 days/week for 5 months. The concentrations in the exposure chamber averaged 87 mg/m$^3$ for total particles, 246 ml/m$^3$ carbon monoxide and 16 mg/m$^3$ nicotine. After 5 months the incidence of lung tumours was increased, but not significantly, from 11 % to 33 %. In a second group the follow-up period was 4 months during which time the mice were exposed to filtered air; 85 % of the exposed animals developed lung tumours compared with 38 % in the control group. The increase in tumour incidence was statistically significant. More than 80 % of the tumours were adenomas, the rest adenocarcinomas (Witschi *et al.* 1997a). In an earlier study in which AJ mice were exposed to ETS (total particle concentration 4 mg/m$^3$), the number of animals with lung tumours and the number of tumours per lung in the exposed group were comparable with those in the control group (Witschi *et al.* 1995).

Female NMRI mice were given doses of sidestream smoke condensate of 2.5, 5 and 7.5 mg (benzo[*a*]pyrene dose 0.65, 1.65, 2.5 µg/kg body weight) twice weekly during 3 months by application to the shaved dorsal skin. Survival of the treated animals was significantly reduced compared with that of the control animals. In the exposed animals the number of precancerous skin lesions was significantly increased in a dose-dependent manner. In the highest dose group there was also a 3-fold increase in the number of skin tumours; mammary adenocarcinomas, the spontaneous incidence of which is known to be very low in this mouse strain, were also found (Mohtashamipur *et al.* 1990).

## 7.6 Other effects

Rabbits were placed in a 50 litre incubator for 30 minutes and exposed to the smoke of 3 cigarettes smoked during this time. The animals were exposed either once, twice daily for two weeks, or twice daily for 8 weeks. In all the groups, mitochondrial respiration, oxidative phosphorylation and the activity of the cytochrome oxidase in heart cells were significantly reduced in comparison with the control values. The activities of NADPH oxidase and succinate oxidase were unaffected (Glantz and Parmley 1991).

A group of 12 male albino rats was exposed to sidestream smoke (total particle concentration: 1.2 mg/m$^3$) 2 hours daily for 60 days to investigate irritative effects on the conjunctiva. The thickness of the conjunctival epithelium was reduced and the basal epithelial cells became altered in form. The surface layer of the epithelium changed its form and became scaly. In addition, vacuolation of the epithelial cells was detected and macrophages and lymphocytes were found between the cells. The plasma thiocyanate concentrations increased from 3.7 µmol/l before the exposure to 119 mol/l after the exposure; the increase was significantly different from the control values (Avunduk *et al.* 1997).

# 8 Manifesto (MAK value/classification)

In most epidemiological studies, an increased relative risk of developing lung cancer was seen in persons exposed to ETS at home or at work. Of particular significance is the fact that the relative risks were generally largest in the groups with the highest exposure level and that an exposure-effect relationship was found. Likewise, meta-analyses of these studies revealed significantly increased relative risks and a significant exposure-effect relationship. Even after correction for misclassification of smokers as non-smokers, the increase in the relative risks remained statistically significant. Likewise for tumours of the nasal cavity and nasal sinuses, the relative risk has been shown to be increased in persons exposed to ETS.

ETS causes an increase in the incidence of lung tumours in a particularly sensitive strain of mouse, and the condensate of sidestream smoke is carcinogenic when applied epicutaneously to mice.

Taken together, the presence of carcinogenic and mutagenic substances in ETS, the demonstrated uptake of mutagenic substances from ETS by passive smokers, the exposure-effect relationship between ETS exposure levels and the incidence of lung cancer, and the results of the available carcinogenicity studies in animals meet the criteria for classification of ETS as carcinogenic for man. Passive smoking is therefore classified in Category 1 of Section III of the *List of MAK and BAT Values.*

# 9 References


Adlkofer FX (1992) Lungenkrebs durch Passivrauchen am Arbeitsplatz – ein eher theoretisches Problem. *Zentralbl Arbeitsmed 42*: 400–424

Akiba S, Kato H, Blot WJ (1986) Passive smoking and lung cancer among Japanese women. *Cancer Res 46*: 4804–4807

Anderson HR, Cook DG (1997) Passive smoking and sudden infant death syndrome: review of the epidemiological evidence. *Thorax 52*: 1003–1009

Andrae S, Axelson O, Björksten B, Fredriksson M, Kjellman N-IM (1988) Symptoms of bronchial hyperreactivity and asthma in relation to environmental factors. *Arch Dis Child 63*: 473–481

Aronow W (1978) Effect of passive smoking on angina pectoris. *N Engl J Med 299*: 21–24

Arshad SH, Hilde DW (1992) Effect of environmental factors on the development of allergic disorders in infancy. *J Allergy Clin Immunol 90*: 235–241

Avunduk AM, Avunduk MC, Evirgen O, Yardimci S, Tastan H, Güven C, Cetinkaya K (1997) Histopathological and ultrastructural examination of the rat conjunctiva after exposure to tobacco smoke. *Ophthalmologica 211*: 296–300

Bartsch H, Caporaso N, Coda M, Kadlubar F, Malaveille C, Skipper P, Talaska G, Tannenbaum SR, Vineis P (1990) Carcinogen hemoglobin adducts, urinary mutagenicity, and metabolic phenotype in active and passive cigarette smokers. *J Nat Cancer Inst 82*: 1826–1831

Bascom R, Kulle T, Kagey-Sobotka A, Proud D (1991) Upper respiratory tract environmental tobacco smoke sensitivity. *Am Rev Respir Dis 143*: 1304–1311

Bascom R, Kesavanathan J, Fitzgerald TK, Cheng KH, Swift DL (1995) Sidestream tobacco smoke exposure acutely alters human nasal mucociliary clearance. *Environ Health Perspect 103*: 1026–1030

Bisgaard H, Dalgaard P, Nyboe J (1987) Risk factors for wheezing during infancy. A study of 5953 infants. *Acta Paediatr Scand 76*: 719–726

Boffetta P, Agudo A, Ahrens W, Benhamou E, Benhamou S, Sarah CD, Fortes C, Gonzalez CA, Jöckel K-H, Krauss M, Kreienbrock L, Kreuzer M, Mendes A, Merletti F, Nyberg F, Pershagen G, Pohlabeln H, Riboli E, Simonato L, Schmid G, Whitley E, Tredaniel J, Wichmann HE, Winck C, Zambon P, Saracci R (1998) Multicenter case-control study of exposure to environmental tobacco smoke and lung cancer in Europe. *J Nat Cancer Inst 90*: 1440–1450

Bond JA, Chen BT, Griffith WC, Mauderly JL (1989) Inhaled cigarette smoke induces the formation of DNA adducts in lungs of rats. *Toxicol Appl Pharmacol 99*: 161–172

Bos RP, Theuws JLG, Henderson PTH (1983) Excretion of mutagens in human urine after passive smoking. *Cancer Lett 19*: 85–90

Brownson RC, Reif JS, Keefe TJ, Gerguson SW, Pritzl JA (1987) Risk factors for adenocarcinoma of the lung. *Am J Epidemiol 125*: 25–34

Brownson RC, Alavanja MC, Hock ET, Loy TS (1992) Passive smoking and lung cancer in non-smoking women. *Am J Public Health 82*: 1525–1530

Bryant MS, Vineis P, Skipper PL, Tannenbaum SR (1988) Hemoglobin adducts of aromatic amines. Associations with smoking status and type of tobacco. *Proc Nat Acad Sci USA 85*: 9788–9791

Buffler PA, Pickle LW, Mason TJ, Contant C (1984) The causes of lung cancer in Texas. in: Mizell M, Corres P (Eds) *Lung cancer causes and prevention*, Verlag Chemie International, New York, 83–99

Buckley JD, Doll R, Harris RWC, Vessy MP, Williams PT (1981) Case-control study of the husbands of women with dysplasia or carcinoma of the cervix uteri. *Lancet 2*: 1010–1014

Burch JD, Rohan TE, Howe GR, Risch HA, Hill GB, Steele R, Miller AB (1989) Risk of bladder cancer by source and type of tobacco exposure: a case-control study. *Int J Cancer 44*: 622–628

Burchfiel CM, Higgins MW, Keller JB, Howatt WF, Butler WJ, Higgins IT (1986) Passive smoking in childhood: respiratory conditions and pulmonary function in Tecumseh, Michigan. *Am Rev Respir Dis 133*: 966–973

Burghuber O, Punzengruber C, Sinzinger H, Haber P, Silberbauer K (1986) Platelet sensitivity to prostacyclin in smokers and non-smokers. *Chest 90*: 34–38

Burr M, Miskelly FG, Butland BK, Merrett TG, Vaughan-Williams E (1989) Environmental factors and symptoms in infants at high risk of allergy. *J Epidemiol Comm Health 43*: 125–132

Butler TL (1988) *The relationship of passive smoking to various health outcomes among seventh-day adventists in California.* University of California, Los Angeles, Doctoral thesis

California EPA (Environmental Protection Agency) (1997) *Health effects of exposure to environmental tobacco smoke.* Final report, Office of Environmental Health Hazard Assessment, Sacramento

Cardenas VM, Thun MJ, Austin H, Lally CA, Clark WS, Greenberg S, Heath jr CW (1997) Environmental tobacco smoke and lung cancer mortality in the American Cancer Society's cancer prevention study II. *Cancer Causes Control 8*: 57–64

Chan WC, Fung SC (1982) Lung cancer in non-smokers in Hong Kong. in: Grundmann E (Ed.) *Cancer Epidemiology*, Cancer Campaign 6, Gustav Fischer Verlag, Stuttgart, 199–202

Charlton A (1984) Children's coughs related to parental smoking. *Br Med J 288*: 1647–1649

Chinn S, Rona RJ (1991) Quantifying health aspects of passive smoking in British children ages 5–11 years. *J Epidemiol Comm Health 45*: 188–194

Coburn RF, Williams WJ, Kane PB (1966) Endogenous carbon monoxide production in patients with hemolytic anemia. *J Clin Invest 45*: 460–468

Cogswell JJ, Mitchell EB, Alexander J (1987) Parental smoking, breast feeding and respiratory infection in development of allergic diseases. *Arch Dis Child 62*: 338–344

Coker AL, Rosenberg AJ, McCann MF, Hulka BS (1992) Active and passive cigarette smoke exposure and cervical intraepithelial neoplasia. *Cancer Epidemiol Biomarkers Prev 1*: 349–356

Cook DG, Strachan DP (1997) Parental smoking and prevalence of respiratory symptoms and asthma in school age children. *Thorax 52*: 1081–1094

Cordier S, Iglesias MJ, Le Goaster C, Guyot MM, Mandereau L, Hemon D (1994) Incidence and risk factors for childhood brain tumors in the Ile de France. *Int J Cancer 59*: 776–782

Crawford FG, Mayer J, Santella RM, Cooper TB, Ottman R, Tsai W-Y, Simon-Cereijido, Wang M, Tang D, Perera FP (1994) Biomarkers of environmental tobacco smoke in preschool children and their mothers. *J Nat Cancer Inst 18*: 1398–1402

Cuzick J, Routledge N, Jenkins D, Garner C (1990) DNA adducts in different tissues of smokers and non-smokers. *Int J Cancer 45*: 673–678

Danuser B, Weber A, Lorenz-Hartmann A, Krueger H (1993) Effects of a bronchoprovocation challenge test with cigarette sidestream smoke on sensitive and healthy adults. *Chest 103*: 353–358

Daube H, Scherer G, Riedel K, Ruppert T, Tricker AR, Rosenbaum P, Adlkofer F (1997) DNA adducts in human placenta in relation to tobacco smoke exposure and plasma antioxidant status. *J Cancer Res Clin Oncol 123*: 141–151

Davis J, Shelton L, Watanabe I, Arnold J (1989) Passive smoking affects endothelium and platelets. *Arch Intern Med 149*: 386–389

Dayal HH, Khuder S, Sharrar R, Trieff N (1994) Passive smoking in obstructive respiratory diseases in an industrialized urban population. *Environ Res 65*: 161–171

Dekker C, Dales R, Barlett S, Brunekreff B, Zwanenburg H (1991) Childhood asthma and indoor environment. *Chest 100*: 922–926

Dobson AJ, Alexander HM, Heller RF, Lloyd DM (1991) Passive smoking and the risk of heart attack or coronary death. *Med J Aust 154*: 793–797

Dold S, Reitmeir P, Wjst M, Mutius EV (1992) Auswirkungen des Passivrauchens auf den kindlichen Respirationstrakt. *Monatszeitschr Kinderheilk 140*: 763–768

Du YX, Cha Q, Chen YZ, Wu JM (1993) Exposure to environmental tobacco smoke and female lung cancer in Guangzhou, China. *Proceedings of Indoor Air 1*: 511–516

Eliopoulos C, Klein J, Phan MK, Knie B, Greenwald M, Chitayat D, Koren G (1994) Hair concentration of nicotine and cotinine in women and their newborn infants. *J Am Med Assoc 271*: 621–623

Eliopoulos C, Klein J, Chitayat D, Greenwald M, Koren G (1996) Nicotine and cotinine in maternal and neonatal hair as markers of gestational smoking. *Clin Invest Med 19*: 231–242

Evans D, Levison MJ, Feldman CH, Clark MN, Wasilewski Y, Levin B, Mellins RB (1987) The impact of passive smoking on emergency room visits of urban children with asthma. *Am Rev Respir Dis 135*: 567–572

Falter B, Kutzer C, Richter E (1994) Biomonitoring of hemoglobin adducts: aromatic amines and tobacco-specific nitrosamines. *Clin Invest 72*: 364–371

Fergusson DM, Horwood LJ (1985) Parental smoking and respiratory illness during early childhood: a six-year longitudinal study. *Pediatr Pulmonol 1*: 99–106

Filippini G, Farionotti M, Lovicu G, Maisonneuve P, Boyle P (1994) Mothers' active and passive smoking during pregnancy and risk of brain tumours in children. *Int J Cancer 57*: 769–774

Fontham ETH, Correa P, Wu-Williams AH, Reynolds P, Greenberg RS, Buffler PA, Chen VW, Boyd P, Alterman T, Austin DF, Liff J, Greenberg SD (1991) Lung cancer in nonsmoking women: a multicenter case-control study. *Cancer Epidemiol Biomarkers Prev 1*: 35–43

Fontham ETH, Correa P, Reynolds P, Wu-Williams A, Buffler PA, Greenberg RS, Chen VW, Alterman T, Boyd P, Austin DF, Liff J (1994) Environmental tobacco smoke and lung cancer in nonsmoking women. A multicenter study. *J Am Med Assoc 271*: 1752–1759

Forastière F, Corbo GM, Michelozzi P, Pistelli R, Agabiti N, Brancato G, Ciappi G, Perucci CA (1992) Effects of environment and passive smoking on the respiratory health of children. *Int J Epidemiol 21*: 66–73

Frischer T, Kühr J, Meinart R, Karmaus W, Barth R, Herrmann-Kunz E, Urbanek KR (1992) Maternal smoking in early childhood: a risk factor for bronchial responsiveness to exercise in primary school children. *J Pediatr 121*: 17–22

Fukuda K, Shibata A (1990) Exposure-response relationships between woodworking, smoking or passive smoking, and squamous cell neoplasms of the maxillary sinus. *Cancer Causes Control 1*: 165–168

Gairola CG, Wu H, Gupta RC, Diana JN (1993) Mainstream and sidestream cigarette smoke-induced DNA adducts in C7B1 and DBA mice. *Environ Health Perspect 99*: 253–255

Gao Y-T, Blot WJ, Zheng W, Ershow AG, Hsu CW, Levin LI, Zhang R, Fraumeni Jr JF (1987) Lung cancer among Chinese women. *Int J Cancer 40*: 604–609

Garfinkel L (1981) Time trends in lung cancer mortality among nonsmokers and a note on passive smoking. *J Nat Cancer Inst 66*: 1061–1066

Garfinkel L, Auerbach O, Joubert L (1985) Involuntary smoking and lung cancer: a case-control study. *J Nat Cancer Inst 75*: 463–469

Geller-Bernstein G, Kenett R, Weisglass L, Tsur S, Lahav M, Levin S (1987) Atopic babies with wheezy bronchitis. *Allergy 42*: 85–91

Geng G, Liang ZH, Zhang AY, Wu GL (1988) On the relationship between smoking and female lung cancer. in: Aoki M, Hisamichi S, Tominaga S (Eds) *Smoking and Health*, Elsevier, Amsterdam, 483–486

Glantz SA, Parmley WW (1991) Passive smoking and heart disease. *Circulation 83*: 1–12

Gold EB, Leviton A, Lopez R, Gilles FH, Hedley-Whyte ET, Kolonel LN, Lyon JL, Swanson GM, Weiss NS *et al.* (1993) Parental smoking and risk of childhood brain tumors. *Am J Epidemiol 137*: 620–628

Goren AI, Hellman S (1991) Passive smoking among school-children in Israel. *Environ Health Perspect 96*: 203–211

Gorgels WJMJ, Van Poppel G, Jarvis MJ, Stenhuis W, Kok FJ (1992) Passive smoking and sister-chromatid exchanges in lymphocytes. *Mutat Res 279*: 233–238

Gortmaker SL, Klein-Walker D, Jacobs FH, Ruch-Ross H (1982) Parental smoking and the risk of childhood asthma. *Am J Public Health 72*: 574–579

Gospe SM, Zhou SS, Pinkerton KE (1996) Effects of environmental tobacco smoke exposure *in utero* and/or postnatally on brain development. *Pediatr Res 39*: 494–498

Greer JR, Abbey DE, Burchette RJ (1993) Asthma related to occupational and ambient air pollutants in nonsmokers. *J Occup Med 35*: 909–915

Hackshaw AK, Law MR, Wald NJ (1997) The accumulated evidence on lung cancer and environmental tobacco smoke. *Br Med J 315*: 980–988

Halken S, Host A, Husby S, Hansen LG, Osterballe O, Nyboe J (1991) Recurrent wheezing in relation to environmental risk factors in infancy. A prospective study of 276 infants. *Allergy 46*: 507–514

Halken S, Host A, Husby S, Hansen LG, Osterballe O (1992) Effect of an allergy prevention programme on incidence of atopic symptoms in infancy. A prospective study of 159 "high-risk" infants. *Allergy 47*: 545–553

Hall CB, Gala CL, Magill FB, Leddy JP (1984) Long-term prospective study in children after respiratory syncytial virus infection. *J Pediatr 105*: 358–364

Hammond SK, Coghlin J, Gann PH, Paul M, Taghizadeh K, Skipper PL, Tannenbaum SR (1993) Relationship between enviromental tobacco smoke exposure and carcinogen-hemoglobin adduct levels in nonsmokers. *J Nat Cancer Inst 85*: 474–478

Harlap S, Davies AM (1974) Infant admissions to hospital and maternal smoking. *Lancet 1*: 529–532

He Y, Lam TH, Si LS, Du RY, Jia GL, Huang JY, Zheng JS (1994) Passive smoking at work as a risk factor for coronary heart disease in Chinese women who have never smoked. *Br Med J 308*: 380–384

Hecht SS, Carmella SG, Murphy SE, Akekar S, Brunneman KD, Hoffmann D (1993) A tobacco-specific lung carcinogen in the urine of men exposed to cigarette smoke. *N Engl J Med 329*: 1543–1546

Heller W-D, Scherer G, Sennewald E, Adlkofer F (1998) Misclassification of smoking in a follow-up population study in southern Germany. *Clin Epidemiol 51*: 211–218

Hense HW (1995) Herz and Kreislauf. in: Wichmann HE, Schlipköter HW, Fülgraff G (Eds) *Handbuch der Umweltmedizin*, ecomed, Landsberg

Hirayama T (1981) Nonsmoking wives of heavy smokers have a higher risk of lung cancer: a study from Japan. *Br Med J 282*: 183–185

Hirayama T (1984) Cancer mortality in nonsmoking women with smoking husbands based on a large-scale cohort study in Japan. *Prev Med 13*: 680–690

Hoffmann D, Wynder EL (1994) Aktives and Passives Rauchen. in: Marquardt H, Schäfer SG (Eds) *Lehrbuch der Toxikologie*, Wissenschaftsverlag, Mannheim, 589–605

Hole DJ, Gills CR, Chopra C, Hawthorne VM (1989) Passive smoking and cardiorespiratory health in a general population in the west of Scotland. *Br Med J 299*: 423–427

Holz O, Krause T, Scherer G, Schmidt-Preuß U, Rüdiger HW (1990) $^{32}$P-postlabelling analysis of DNA adducts in monocytes of smokers and passive smokers. *Int Arch Occup Environ Health 62*: 299–303

Howard G, Burke GL, Szklo M, Tell GS, Eckfeldt J, Evans G, Heiss G (1994) Active and passive smoking are associated with increased carotid wall thickness. *Arch Intern Med 154*: 1277–1282

Howard G, Wagenknecht LE, Burke GL, Diez-Roux A, Evans GW, McGovern P, Nieto J, Tell GS (1998) Cigarette smoking and progression of atherosclerosis. *J Am Med Assoc 279*: 119–124

Humble CG, Samet JM, Pathak DR (1987) Marriage to a smoker and lung cancer risk. *Am J Public Health 77*: 598–602

Husgafvel-Pursiainen K, Sorsa M, Moller M, Benestad C (1986) Genotoxicity and polynuclear aromatic hydrocarbon analysis of environmental tobacco smoke samples from restaurants. *Mutagenesis 1*: 287–292

Husgafvel-Pursiainen K, Sorsa M, Engström K, Einistö P (1987) Passive smoking at work: biochemical and biological measures of exposure to environmental tobacco smoke. *Int Arch Occup Environ Health 59*: 337–345

IARC (International Agency for Research on Cancer) (1986) *IARC monographs on the evaluation of the carcinogenic risk of chemicals to humans: tobacco smoking*, Vol 38, IARC, Lyon

Inoue R, Hirayama T (1988) Passive smoking and lung cancer in women. in: Aoki M, Hisamichi S, Tominaga S (Eds) *Smoking and Health*, Elsevier, Amsterdam, 283–285

Jaakkola MS, Jaakkola JK, Becklake MR, Ernst P (1995) Passive smoking and evolution of lung function in young adults. An 8-year longitudinal study. *J Clin Epidemiol 48*: 317–327

Janerich DT, Thompson WD, Varela LR, Greenwald P, Chorost S, Tucci C, Zaman MB, Melamed M, Kiely M, McKneally MF (1990) Lung cancer and exposure to tobacco smoke in the household. *N Engl J Med 323*: 632–636

Jinot JJ, Bayard S (1994) Respiratory health effects of passive smoking: EPA's weight-of-evidence analysis. *J Clin Epidemiol 47*: 339–349

Jöckel K-H (1991) *Passivrauchen—Bewertung der epidemiologischen Befunde. VDI-Berichte 888*, VDI-Verlag, 517–535

Jöckel K-H, Knauth C (1994) Passivrauchen. in: Wichmann HE, Schlipköter HW, Fülgraff G (Eds) *Handbuch der Umweltmedizin*, ecomed, Landsberg

Jöckel K-H, Krauss M, Pohlabeln, Ahrens W, Kreuzer M, Kreienbrock L, Möhner M, Wichmann HE (1998) Lungenkrebsrisiko durch berufliche Exposition – Passivrauchen. in: Wichmann HE, Jöckel KH, Robra BP (Eds) *Fortschritte in der Epidemiologie, Lungenkrebsrisiko durch berufliche Exposition*, ecomed, Landsberg, 227–236

Jordanov JS (1990) Cotinine concentrations in amniotic fluid and urine of smoking, passive smoking and non-smoking pregnant women at term and in the urine of their neonates on 1st day of life. *Eur J Pediatr 149*: 734–737

Jörres R, Magnussen H (1992) Influence of short-term passive smoking on symptoms, lung mechanics and airway responsiveness in asthmatic subjects and healthy controls. *J Eur Respir 5*: 936–944

Kabat GC, Wynder EL (1984) Lung cancer in nonsmokers. *Cancer 53*: 1214–1221

Kabat GC, Dieck GS, Wynder EL (1986) Bladder cancer in nonsmokers. *Cancer 57*: 362–367

Kabat GC, Stellman SD, Wynder EL (1995) Relation between exposure to environmental tobacco smoke and lung cancer in lifetime nonsmokers. *Am J Epidemiol 142*: 141–148

Kalandidi A, Katsouyanni K, Voropoulou N, Bastas G, Saracci R, Trichopoulos D (1990) Passive smoking and diet in the etiology of lung cancer among non-smokers. *Cancer Causes Control 1*: 15–21

Kawachi I, Colditz GA, Speizer FE, Manson JAE, Stampfer MJ, Willett WC, Hennekens CH (1997) A prospective study of passive smoking and coronary heart disease. *Circulation 95*: 2374–2379

Keil U, Stieber J, Filipiak B, Löwel H, Wichmann HE (1998) On misclassification of smoking. Letter to the editor. *J Clin Epidemiol 52*: 91–93

Koo LC, Ho JH, Saw D, Ho CY (1987) Measurements of passive smoking and estimates of lung cancer risk among non-smoking Chinese females. *Int J Cancer 39*: 162–169

Kreiger N, Marrett LD, Dodds L, Hilditch S, Darlington GA (1993) Risk factors for renal cell carcinoma: results of a population-based case-control study. *Cancer Causes Control 4*: 101–110

Kühr J, Frischer T, Karmaus W, Meinert R, Barth R, Herrman-Kunz E, Forster J, Urbanek R (1992) Early childhood risk factors for sensitization at school age. *J Allergy Clin Immunol 90*: 358–363

Lam WK (1985) *A clinical and epidemiological study of carcinoma of the lung in Hong Kong.* University of Hong Kong, Doctoral thesis

Lam TH, Kung IT, Wong CM, Lam WK, Kleevens JW, Saw D, Hsu C, Seneviratne S, Lam SY, Lo KK (1987) Smoking, passive smoking and histological types in lung cancer in Hong Kong Chinese women. *Br J Cancer 56*: 673–678

Law MR, Morris JK, Wald NJ (1997) Environmental tobacco smoke exposure and ischaemic heart disease. An evaluation of the evidence. *Br Med J 315*: 973–980

Lee PN (1987a) Lung cancer and passive smoking: association of an artifact due to misclassification of smoking habits? *Toxicol Lett 35*: 157–162

Lee PN (1987b) Passive smoking and lung cancer association: A result of bias? *Hum Toxicol 6*: 517–524

Lee PN (1988) *Misclassification of smoking habits and passive smoking*, Springer Verlag, Berlin

Lee PN, Chamberlain J, Alderson MR (1986) Relationship of passive smoking to risk of lung cancer and other smoking-associated diseases. *Br J Cancer 54*: 97–105

Lee CKL, Brown BG, Reed EA, Rahn CA, Coggins CRE, Doolittle DJ, Hayes AW (1992) Fourteen-day inhalation study in rats, using aged and diluted sidestream smoke from a reference cigarette. *Fundam Appl Toxicol 19*: 141–146

Lee CKL, Brown BG, Reed EA, Coggins CRE, Doolittle DJ, Hayes AW (1993) Ninety-day inhalation study in rats, using aged and diluted sidestream smoke from a reference cigarette: DNA adducts and alveolar macrophage cytogenetics. *Fundam Appl Toxicol 20*: 393–401

Leeder SR, Corkhill R, Irwig LM, Holland WW, Colley JRT (1976) Influence of family factors on the incidence of lower respiratory illness during the first year of life. *Br J Prev Soc Med 30*: 203–212

Leuenberger PH, Schwartz J, Ackermann-Liebrich U, Blaser K, Bolognini G, Bongard JP, Brändli O, Braun P, Bron C, Brutsche M, Domenghetti G, Elsasser S, Guldimann P, Hollenstein C, Hufschmid P, Karrer W, Keller R, Keller-Wossidlo H, Künzli N, Lüthi JC, Martin BW, Medici T, Perruchoud AP, Radaelli A, Schindler C, Schoeni MH, Solari G, Tschopp JM, Villinger B, Wüthrich B, Zellweger JP, Zemp E (1994) Passive smoking exposure in adults and chronic respiratory symptoms (Sapaldia study). *Am J Respir Crit Care Med 150*: 1221–1228

Liu Z, He X, Chapman RS (1991) Smoking and other risk factors for lung cancer in Xuanwei, China. *Int J Epidemiol 20*: 26–31

Liu Q, Sasco AJ, Riboli E, Hu MX (1993) Indoor air pollution and lung cancer in Ghuangzhou, People's Republic of China. *Am J Epidemiol 137*: 145–154

Martinez FD, Cline M, Burrows B (1992) Increased incidence of asthma in children of smoking mothers. *Pediatrics 89*: 21–26

McMurray RG, Hicks LL, Thompson DL (1985) The effects of passive inhalation of cigarette smoke on exercise performance. *Eur J App Physiol 54*: 196–200

Meister R (1990) Allgemeine Umweltnoxen und Passivrauchen. *Pneumologie 44*: 378–386

Menon P, Rando RJ, Stankus RP, Salvaggio JE, Lehrer SB (1992) Passive cigarette smoke-challenge studies: increase in bronchial hyperreactivity. *J Allergy Clin Immunol 89*: 560–566

Mohtashamipur E, Norporth K, Heger M (1984) Urinary excretion of frameshift mutagens in rats caused by passive smoking. *J Cancer Res Clin Oncol 108*: 296–301

Mohtashamipur E, Müller G, Norpoth K, Endrikat M, Stücker W (1987a) Urinary excretion of mutagens in passive smokers. *Toxicol Lett 35*: 141–146

Mohtashamipur E, Norpoth K, Straeter H (1987b) Clastogenic effect of passive smoking on bone marrow polychromatic erythrocytes of NMRI mice. *Toxicol Lett 35*: 153–156

Mohtashamipur E, Mohtashamipur A, German P-G, Ernst H, Norpoth K, Mohr U (1990) Comparative carcinogenicity of cigarette mainstream and sidestream smoke condensates on the mouse skin. *J Cancer Res Clin Oncol 116*: 604–608

Murray AB, Morrison BJ (1986) The effects of cigarette smoke from the mother on bronchial responsiveness and severity of symptoms in children with asthma. *J Allergy Clin Immunol 77*: 575–581

Murray AB, Morrison BJ (1993) The decrease in severity of asthma in children of parents who smoke since the parents have been exposing them to less cigarette smoke. *J Allergy Clin Immunol 91*: 102–110

Neuspiel DR, Rush D, Butler NR, Golding J, Burr PE, Kurzon M (1989) Parental smoking and post-infancy wheezing in children: a prospective cohort study. *Am J Public Health 79*: 168–171

Ng TP, Hui KP, Tan WC (1993) Respiratory symptoms and lung function effects of domestic exposure to tobacco smoke and cooking by gas in non-smoking women in Singapore. *J Epidemiol Comm Health 47*: 454–458

NIOSH (National Institute for Occupational Safety and Health) (1991) *Environmental tobacco smoke in the workplace.* Publication No 91-108, NIOSH, Cincinnati

Norman MA, Holly EA, Ahn DK, Preston-Martin S, Müller BA, Bracci PM (1996) Prenatal exposure to tobacco smoke and childhood brain tumors: results from the United States West Coast childhood brain tumor study. *Cancer Epidemiol Biomarkers Prev 5*: 127–133

Nyberg F, Agrenius V, Svartengren K, Svensson C, Pershagen G (1998) Environmental tobacco smoke and lung cancer in nonsmokers – does time since exposure play a role? *Epidemiology 9*: 301–308

O'Connor GT, Weiss ST, Tager IB, Munoz A, Speizer FE (1987) The effect of passive smoking on pulmonary function and non-specific bronchial responsiveness in a population-based sample of children and young adults. *Am Rev Respir Dis 135*: 800–804

Ogston SA, Florey CV, Walker CVM (1987) Association of infant alimentary and respiratory illness with parental smoking and other environmental factors. *J Epidemiol Comm Health 41*: 21–25

Ostro BD, Lipsett MJ, Mann JM, Weiner M, Selner JS (1994) Indoor air pollution and asthma: results from a panel study. *Am Rev Respir Dis 149*: 1400–1496

Penn A, Snyder CA (1993) Inhalation of sidestream cigarette smoke accelerates development of arteriosclerotic plaques. *Circulation 88*: 1820–1825

Pershagen G (1986) Review of epidemiology to passive smoking. *Arch Toxicol, Suppl 9*: 63–73

Pershagen G, Hrubec Z, Svensson C (1987) Passive smoking and lung cancer in Swedish women. *Am J Epidemiol 1258*: 17–24

Pichini S, Altieri I, Pellegrini M, Pacifici R, Zuccaro P (1997a) Hair analysis for nicotine and cotinine: evaluation of extraction procedures, hair treatments, and development of reference material. *Forensic Sci 84*: 243–252

Pichini S, Altieri I, Pellegrini M, Pacifici R, Zuccaro P (1997b) The analysis of nicotine in infants' hair for measuring exposure to environmental tobacco smoke. *Forensic Sci 84*: 253–258

Pullan CR, Hey CN (1982) Wheezing, asthma, and pulmonary disfunction 10 years after infection with respiratory syncytial virus in infancy. *Br Med J 284*: 1665–1669

Rantakallio P (1978) Relationship of maternal smoking to morbidity and mortality of the child up to the age of five. *Acta Paediatr Scand 67*: 621–631

Richter E, Branner B, Kutzer C, Donhärl AME, Scherer G, Tricker AR, Heller W-D (1995) Comparison of biomarkers for exposure to environmental tobacco smoke: cotinine and haemoglobin adducts from aromatic amines and tobacco-specific nitrosamines in pregnant smoking and nonsmoking women. in: Maroni M (Ed.) *Healthy buildings '95. An international conference on healthy buildings in mild climate*, University of Milano, 599–640

Robbins AS, Abbey DE, Lebowitz MD (1993) Passive smoking and chronic respiratory disease symptoms in non-smoking adults. *Int J Epidemiol 22*: 809–817

Samet JM, Cain WS, Leaderer BP (1991) Environmental tobacco smoke. in: Samet JM, Spengler JD (Eds) *Indoor air pollution. A health perspective*, The Johns Hopkins University Press, Baltimore, 131–169

Sandler DP, Everson RB, Wilcox AJ (1985a) Passive smoking in adulthood and cancer risk. *Am J Epidemiol 121*: 37–48

Sandler DP, Everson RB, Wilcox AJ, Browder JP (1985b) Cancer risk in adulthood from early life exposure to parents smoking. *Am J Public Health 75*: 487–492

Scherer G, Conze C, Tricker AR, Adlkofer F (1992) Uptake of tobacco smoke constituents on exposure to environmental tobacco smoke (ETS). *Clin Invest 70*: 352–367

Scherer G, Doolittle DJ, Ruppert T, Meger-Kossien I, Riedel I, Tricker AR, Adlkofer F (1996) Urinary mutagenicity and thioethers in nonsmokers: role of environmental tobacco smoke (ETS) and diet. *Mutat Res 368*: 195–204

Shimizu H, Morishita M, Mizuno K, Masuda T, Ogura Y, Santo M, Nishimura M, Kunishima K, Karasawa K, Nishiwaki K, Yamamoto M, Hisamichi S, Tominaga S (1988) A case-control study of lung cancer in non-smoking women. *Tohoku J Exp Med 154*: 389–397

Siegel M (1993) Involuntary smoking in the restaurant workplace. A review of employee exposure and health effects. *J Am Med Assoc 270*: 490–493

Sinzinger H, Kefalides A (1982) Passive smoking severely decreases platelet sensitivity to anti-aggregatory prostaglandins. *Lancet 2*: 392–393

Sinzinger H, Virgolini I (1989) Are passive smokers at greater risk of thrombosis? *Wien Klin Wochenschr 20*: 684–688

Slattery ML, Robinson LM, Schuman KL, French TK, Abbott TM, Overall Jr JC, Gardner JW (1989) Cigarette smoking and exposure to passive smoke are risk factors for cervical cancer. *J Am Med Assoc 261*: 1593–1598

Sobue T (1990) Association of indoor air pollution and lifestyle with lung cancer in Osaka, Japan. *Int J Epidemiol 19*: 62–66

Somerville SM, Rona RJ, Chinn S (1988) Passive smoking and respiratory conditions in primary school children. *J Epidemiol Comm Health 42*: 105–110

Spengler JD (1991) Sources and concentrations of indoor air pollution. in: Samet JM, Spengler JD (Eds) *Indoor air pollution. A health perspective*, Johns Hopkins University Press, Baltimore, 33–67

Stockwell HG, Goldman AL, Lyman GH, Noss CI, Armstrong AW, Pinkham PA, Candelora EC, MR Brusa (1992) Environmental tobacco smoke and lung cancer risk in nonsmoking women. *J Nat Cancer Inst 84*: 1417–1422

Strachan DP, Cook DG (1997) Parental smoking and lower respiratory illness in infancy and early childhood. *Thorax 52*: 905–914

Strachan DP, Cook DG (1998a) Parental smoking, middle ear disease and adenotonsillectomy in children. *Thorax 53*: 50–56

Strachan DP, Cook DG (1998b) Parental smoking and allergic sensitisation in children. *Thorax 53*: 117–123

Strachan DP, Cook DG (1998c) Parental smoking and childhood asthma: longitudinal and case-control studies. *Thorax 53*: 204–212

Sun X-W, Dai X-D, Lin C-Y, Shi Y-B, Ma Y-Y, Li W (1996) Passive smoking and lung cancer among nonsmoking women in Harbin, China. International symposium on lifestyle factors and human lung cancer, China 1994. *Lung Cancer 14*: 237

Taylor B, Wadsworth J (1987) Maternal smoking during pregnancy and lower respiratory tract illness in early life. *Arch Dis Child 62*: 786–791

Trichopoulos D, Kalandidi A, Sparros L (1983) Lung cancer and passive smoking: conclusion of Greek study. *Lancet 2*: 677–678

US EPA (United States Environmental Protection Agency) (1993) *Respiratory health effects of passive smoking: lung cancer and other disorders*, Monograph 4, NIH Publication No 93-3605, National Institutes of Health, Bethesda, MD, USA

Waage H, Sisland T, Urdal P, Langard S (1992) Discrimination of smoking status by thiocyanate and cotinine in serum, and carbon monoxide in expired air. *Int J Epidemiol 21*: 488–493

Wang T-J, Zhou B-S, Shi J-P (1996) Lung cancer in nonsmoking Chinese women: a case-control study. International symposium on lifestyle factors and human lung cancer, China 1994. *Lung Cancer 14*: 93–98

Ware JH, Dockery DW, Spiro A, Speizer FE, Ferris BG (1984) Passive smoking, gas cooking, and respiratory health of children living in six cities. *Am Rev Respir Dis 129*: 366–374

Weiss SC, Tager IB, Schenker M, Speizer FE (1983) The health effects of involuntary smoking. *Am Rev Respir Dis 128*: 933–942

Weitzmann M, Gortmaker S, Klein Walker D, Sobol A (1990) Maternal smoking and childhood asthma. *Pediatrics 85*: 505–511

Wells AJ (1994) Passive smoking as a cause of heart disease. *J Am Coll Cardiol 24*: 546–554

Werner B (1978) Inter- and intra-individual variation of carbon monoxide production in young healthy males. *J Clin Lab Invest 38*: 199–202

White JR, Froeb HF, Kulik JA (1991) Respiratory illness in nonsmokers chronically exposed to tobacco smoke in the work place. *Chest 100*: 39–43

Witschi H, Oreffo VIC, Pinkerton KE (1995) Six month exposure of strain A/J mice to cigarette sidestream smoke: cell kinetics and lung tumor data. *Fundam Appl Toxicol 26*: 32–40

Witschi H, Espiritu I, Peake JL, Wu K, Maronpot RR, Pinkerton KE (1997a) The carcinogenicity of environmental tobacco smoke. *Carcinogenesis 18*: 575–586

Witschi H, Joad JP, Pinkerton KE (1997b) The toxicology of environmental tobacco smoke. *Annu Rev Pharmacol Toxicol 37*: 29–52

Wolf-Ostermann K, Luttmann H, Treiber-Klötzer C, Kreienbrock L, Wichmann HE (1995) Cohort study on respiratory diseases and lung function in schoolchildren in southwest Germany. Part 3: Influence of smoking and passive-smoking. *Zentralbl Hyg 197*: 459–488

Wu AH, Henderson BE, Pike MC, Yu MC (1985) Smoking and other risk factors for lung cancer in women. *J Nat Cancer Inst 74*: 747–751

Wu-Williams AH, Dai XD, Blot W, Xu ZY, Sun XW, Xiao HP, Stone BJ, Yu SF, Feng YP, Ershow AG, Sun J, Fraumeni Jr JF, Henderson BE (1990) Lung cancer among women in north-east China. *Br J Cancer 62*: 982–987

Xu X, Li B (1995) Exposure-response relationship between passive smoking and adult pulmonary function. *Am J Respir Crit Care Med 151*: 41–46

Zaridze DG, Zemlianaia GM, Aitakov ZN (1995) Contribution of out- and indoor air pollution to the etiology of lung cancer (Russian). *Vestn Ross Akad Med Nauk 4*: 6–10

Zaridze D, Maximovitch D, Zemlyanaya G, Aitakov ZN, Boffetta P (1998) Exposure to environmental tobacco smoke and risk of lung cancer in non-smoking women from Moscow, Russia. *Int J Cancer 73*: 335–338

Zheng W, McLaughlin JK, Chow WH, Chien HTC, Blot WJ (1993) Risk factors for cancers of nasal cavity and paranasal sinuses among white men in the United States. *Am J Epidemiol 138*: 965–972

Zhu B, Sun Y, Sievers RE, Isenberg WM, Glantz SA, Parmley WW (1993) Passive smoking increases experimental atherosclerosis in cholesterol-fed rabbits. *J Am Coll Cardiol 21*: 225–232

Zhu B, Sun Y, Sievers RE, Glantz SA, Parmley WW, Wolfe CL (1994) Exposure to environmental tobacco smoke increase myocardial infarct size in rats. *Circulation 89*: 1282–1290

completed 21.8.1998

# Styrene

| | |
|---|---|
| **MAK value (1987)** | **20 ml/m$^3$ (ppm) $\hat{=}$ 86 mg/m$^3$** |
| **Peak limitation (1987)** | **Category II, excursion factor 2** |
| **Absorption through the skin** | **–** |
| **Sensitization** | **–** |
| **Carcinogenicity (1998)** | **Category 5** |
| **Prenatal toxicity (1985)** | **Pregnancy risk group C** |
| **Germ cell mutagenicity** | **–** |
| **BAT value** | **600 mg mandelic acid plus phenyl glyoxylic acid/g creatinine urine sampling: end of exposure or end of shift; for long-term exposures after several shifts** |
| Synonyms | cinnamene<br>phenylethylene<br>styrol<br>vinylbenzene |
| Chemical name (CAS) | ethenylbenzene |
| CAS number | 100-42-5 |
| Structural formula | |
| Molecular formula | C$_8$H$_8$ |
| Molecular weight | 104.16 |
| Melting point * | –30.6°C |
| Boiling point * | 145°C at 1013 hPa |
| Vapour pressure at 20°C * | 7.3 hPa |

**1 ml/m$^3$ (ppm) $\hat{=}$ 4.33 mg/m$^3$**  
(calculated for 1013 hPa and 20°C)

**1 mg/m$^3$ $\hat{=}$ 0.23 ml/m$^3$ (ppm)**

---

* D'Ans – Lax Taschenbuch für Chemiker und Physiker Band I; Springer-Verlag 1967

*Essential MAK Value Documentations.* DFG, Deutsche Forschungsgemeinschaft
Copyright © 2006 WILEY-VCH Verlag GmbH & Co. KGaA, Weinheim
ISBN: 3-527-31394-X

The toxicology of styrene was reviewed by the Commission in 1987 (Henschler 1987). In the present English translation of the supplement from 1998, the genotoxic and carcinogenic potential of styrene is evaluated within the framework of the new system for classification of carcinogenic substances (1998 MAK documentation: see "Changes in the Classification of Carcinogenic Chemicals in the Work Area" in Volume 12 of the present series).

Since 1998 a series of publications have appeared dealing with the mode of action of styrene. These research activities, which are still ongoing, are centred on the observation that styrene was clearly carcinogenic only in the mouse where it led to lung tumours originating in the terminal bronchioles (Cruzan *et al.* 2001).

Because it was assumed that the metabolic intermediate styrene-7,8-oxide is most relevant for tumorigenesis, the lung burden of this substance in different species has been described in several toxicokinetic publications (Cohen *et al.* 2002, Sarangapani *et al.* 2002, Csanády *et al.* 2003). In the studies of Cohen *et al.* (2002) and Csanády *et al.* (2003) the difference between the lung burdens of styrene-7,8-oxide in the mouse and rat was concluded to be too small to explain why styrene produced lung tumours only in mice. This conclusion was also supported by the finding that the levels of alkylated DNA were extremely small and comparable in lungs and livers of rats and mice exposed to 690 mg/m$^3$ of $^{14}$C-styrene of high specific activity (Boogaard *et al.* 2000). In the light of these results and of the observation that an increase in cell replication rates was observed in the terminal bronchioles of mice after oral application or inhalative exposure to styrene but no effect whatsoever was seen in the lungs of rats exposed to styrene via inhalation (Green *et al.* 2001), it was hypothesized that other mechanisms not directly related to the lung burden of styrene-7,8-oxide and its alkylating potency are most relevant for lung tumorigenicity in the mouse (Filser *et al.* 2002). The authors found a drastic pulmonary glutathione depletion occurring only in lungs of styrene-exposed mice but not in lungs of rats. Explaining this observation in terms of species-specific pulmonary glutathione *S*-transferase activities and pulmonary glutathione turnover rates, Filser *et al.* (2002) speculated that glutathione depletion in the mouse lung could lead to cytotoxicity or apoptosis resulting in cell proliferation and finally in tumour formation. Since the amount of styrene-7,8-oxide formed from styrene in the human lung is orders of magnitudes smaller, if it is formed at all, (Nakajima *et al.* 1994, Carlson *et al.* 2000) and the glutathione *S*-transferase activity is also significantly less than that in the rodent lung (Oberste-Frielinghaus *et al.* 1999), it was concluded that such a mechanism of tumorigenicity should not be relevant in man.

Another attempt to explain the striking difference between mice and rats in the sensitivity to styrene is based on the following findings. Hynes *et al.* (1999) demonstrated that Clara-cell-enriched fractions of mouse lung metabolized a higher fraction of styrene to styrene-7,8-oxide than did corresponding fractions of rat lung. In mouse urine, products derived from phenylacetaldehyde formed in intermediary metabolism of styrene made up a higher percentage of metabolites than in rat urine (Cruzan *et al.* 2002). These authors state that metabolites derived from phenylacetaldehyde and from 4-vinylphenol, another intermediate of styrene metabolism, are present at lower levels in human urine than in mouse urine or are not detected. They also discuss the relevance of mouse pulmonary

CYP2F2 for the formation of styrene-7,8-oxide and toxic metabolites of 4-vinylphenol. In any case, an oxidative metabolic step is first required before pulmonary effects of styrene can develop in the mouse (Green *et al.* 2001). Since the level of pulmonary metabolism of styrene by CYP450 isozymes is at least 2 orders of magnitude smaller in man than in rodents (Oberste-Frielinghaus *et al.* 1999), an adverse effect on human lung after exposure of persons to a styrene concentration of 20 ml/m$^3$ is not probable.

# 1 Carcinogenicity

The currently available epidemiological studies of workers in the styrene producing and processing industries (reviewed in IARC 1994, Kolstad *et al.* 1995) have not yielded clear evidence of carcinogenic effects of styrene in man. The tumours reported most frequently were tumours of the lymphatic and haematopoietic systems. However, the results for an increase in the incidence of these tumours were inconsistent, especially in the studies carried out in the branch of industry with the highest concentrations of styrene. In addition, the observed tumour incidences did not correlate with the cumulative styrene exposure levels. In most of the studies the persons were also exposed to other substances such as 1,3-butadiene, ethylbenzene, dyes, benzene. The data for the carcinogenic effects of styrene in man were considered by the IARC as "inadequate evidence" (IARC 1994).

Carcinogenicity of styrene was investigated in 13 long-term animal studies. In all four studies with mice, styrene induced lung tumours, both after oral administration and following inhalation exposure (Cruzan *et al.* 2001, IARC 1994). In contrast, in nine studies with rats styrene was not carcinogenic (Cruzan *et al.* 1998, IARC 1994). Therefore the findings are considered as evidence of a carcinogenic effect in the mouse.

# 2 Genotoxicity

In most *in vitro* mutagenicity tests, positive results were obtained with styrene after addition of metabolic activation systems. Cytogenetic studies of experimental animals and exposed workers have yielded both negative and positive results (Henschler 1987, IARC 1994).

# 3 Mechanism of Action

Styrene is metabolized to the epoxide, styrene-7,8-oxide, which alkylates macromolecules *in vitro* and *in vivo* (IARC 1994, Osterman-Golkar *et al.* 1995). The epoxide is mutagenic *in vitro* and was carcinogenic in mouse and rat (IARC 1994).

# 4 Estimation of Internal Exposure

Studies were carried out in particular with the aim of quantifying exposure to styrene-7,8-oxide in the mouse, rat and man as a function of the administered dose of styrene (Csanády *et al.* 1994, Kessler *et al.* 1992, Korn *et al.* 1994, Morgan *et al.* 1993a, 1993b, 1993c, Osterman-Golkar *et al.* 1995, Pauwels *et al.* 1996). The parameters of internal exposure used were both the concentration of styrene-7,8-oxide in blood and that of its adducts with haemoglobin and DNA. After similar doses of styrene, in the low dose range the internal exposure to styrene-7,8-oxide in the mouse was two to three times that in the rat and with increasing dose the difference became even greater. The internal exposure levels of this metabolite in persons exposed to styrene concentrations up to $100 \text{ ml/m}^3$ are $^1/_5$ to $^1/_{20}$ of those in rodents (IARC 1994). For man a linear relationship was found between the styrene concentration in air and the styrene-7,8-oxide concentration in blood, whereby exposure to a styrene concentration of $20 \text{ ml/m}^3$ yielded at equilibrium a styrene-7,8-oxide concentration in venous blood of 1 µg/l. The detection limit was given as 0.9 µg/l (Korn *et al.* 1994).

The carcinogenic risk associated with styrene has been estimated on the basis of the internal exposure to styrene-7,8-oxide or the levels of its haemoglobin and DNA adducts which are formed, and taking into consideration the long-term animal studies which have been published to date (Csanády *et al.* 1995, Filser *et al.* 1993a). The determination of the internal exposure levels was based on extensive studies of the toxicokinetics of styrene and styrene-7,8-oxide in mouse, rat and man (Csanády *et al.* 1994, Filser *et al.* 1993b). For a 40-year exposure to styrene (vapour in air: $20 \text{ ml/m}^3$) at the workplace (8 h/day, 5 days/week, 48 weeks/year) the cancer risk was estimated to lie in the range between 1.7 and 7.5 per 100000 exposed persons. This risk is $^1/_{67}$ to $^1/_{880}$ of that estimated for a working lifetime exposure to a concentration of benzene in air of $1 \text{ ml/m}^3$ (TRK value) under workplace conditions (see MAK documentation for benzene (Henschler 1992)). It is also smaller than the unavoidable risk of about 1 per 10000 persons which has been estimated for endogenously formed ethylene oxide (see MAK documentation for ethylene in Volume 10 of the present series).

# 5 Manifesto (MAK value/classification)

In the organism, styrene is metabolized to styrene-7,8-oxide, a direct alkylating agent. Evidence for carcinogenic effects of styrene has been obtained from studies with mice. Consequently, styrene may be considered to have a genotoxic and carcinogenic potential to humans too. However, the risk for man may be evaluated.

The risk of developing cancer in the course of a lifetime as a result of 40 years occupational exposure to a concentration of styrene in air of 20 ml/m$^3$ is smaller than the unavoidable risk caused by endogenous ethylene oxide, for which a value of about 1/10000 has been estimated. It can therefore be considered to be very small indeed, although styrene acts by a genotoxic mechanism. Therefore styrene has been classified in Section III, Category 5 of the *List of MAK and BAT Values*. The previous MAK value of 20 ml/m$^3$ has been retained. Because the irritative and central nervous effects of styrene do not develop until the exposure concentration is more than twice the MAK value, the substance is classified in Peak limitation category II with an excursion factor of 2.

The internal exposure can be monitored by determination of haemoglobin adducts.


# 6 References


Boogaard PJ, deKloe KP, Wong BA, Sumner SCJ, Watson WP van Sittert NJ (2000) Quantification of DNA adducts formed in liver, lungs, and isolated lung cells of rats and mice exposed to C-14-styrene by nose-only inhalation. *Toxicol Sci 57*: 203–216

Carlson GP, Mantick NA, Powley MW (2000) Metabolism of styrene by human liver and lung. *J Toxicol Environ Health 59*: 591–595

Cohen JT, Carlson G, Charnley G, Coggon D, Delzell E, Graham JD, Greim H, Krewski D, Medinsky M, Monson R, Paustenbach D, Petersen B, Rappaport S, Rhomberg L, Ryan PB, Thompson K (2002) A comprehensive evaluation of the potential health risks associated with occupational and environmental exposure to styrene. *J Toxicol Environ Health B Crit Rev 5*: 1–265.

Cruzan G, Cushman JR, Andrews LS, Granville GC, Johnson KA, Hardy CJ, Coombs DW, Mullins PA, Brown WR (1998) Chronic toxicity/oncogenicity study of styrene in CD rats by inhalation exposure for 104 weeks. *Toxicol Sci 46*: 266–281

Cruzan G, Cushman JR, Andrews LS, Granville GC, Johnson KA, Bevan C, Hardy CJ, Coombs DW, Mullins PA, Brown WR (2001) Chronic toxicity/oncogenicity study of styrene in CD-1 mice by inhalation exposure for 104 weeks. *J Appl Toxicol 21*: 185–198

Cruzan G, Carlson GP, Johnson KA, Andrews LS, Banton MI, Bevan C, Cushman JR (2002) Styrene respiratory tract toxicity and mouse lung tumors are mediated by CYP2F-generated metabolites. *Regulary Toxicology and Pharmacology 35*: 308–319

Csanády GA, Mendrala AL, Nolan RJ, Filser JG (1994) A physiologic pharmacokinetic model for styrene and styrene-7,8-oxide in mouse, rat and man. *Arch Toxicol 68*: 143–157

Csanády GA, Kessler W, Filser JG (1995) Carcinogenic risk estimates for inhaled styrene based on the body burden of its metabolite styrene-7,8-oxide. *Naunyn-Schmiedeberg's Arch Pharmacol, Supp 351*: 122

Csanády GyA, Kessler W, Hoffmann HD, Filser JG (2003) A toxicokinetic model for styrene and its metabolite styrene-7,8-oxide in mouse, rat and human with special emphasis on the lung. *Toxicol Letters* (in press)

Filser JG, Kessler W, Csanády GA (1993a) Different approaches to estimate the carcinogenic risk of styrene based on animal studies. *The SIRC Review 3 (1)*: 54

Filser JG, Schwegler U, Csanády GA, Greim H, Kreuzer PE, Kessler W (1993b) Species-specific pharmacokinetics of styrene in rat and mouse. *Arch Toxicol 67*: 517–530

Filser JG, Kessler W, Csanády GA (2002) Estimation of a possible tumorigenic risk of styrene from daily intake via food and ambient air. *Toxicology Letters 126*: 1–18

Green T, Toghill A, Foster JR (2001) The role of cytochromes P-450 in styrene induced pulmonary toxicity and carcinogenicity. *Toxicology 169*: 107–117

Henschler D (Ed.) (1987) Styrol (Styrene) (German). in: *Toxikologisch-arbeitsmedizinische Begründungen von MAK-Werten*, 13th issue, VCH-Verlagsgesellschaft, Weinheim

Henschler D (Ed.) (1992) Benzol (Benzene) (German). in: *Toxikologisch-arbeitsmedizinische Begründungen von MAK-Werten*, 18th issue, VCH-Verlagsgesellschaft, Weinheim

Hynes DE, DeNicola DB, Carlson GP (1999) Metabolism of styrene by mouse and rat isolated lung cells. *Toxicol Sci 51*:195–201

IARC (International Agency for Research on Cancer) (1994) *Styrene. IARC monographs on the evaluation of carcinogenic risks to humans, Vol. 60*, IARC, Lyon

Kessler W, Jiang X, Filser JG (1992) Pharmacokinetics of styrene-7,8-oxide in the mouse and the rat. in: Kreuz R, Piekarski C (Eds) *32nd Annual Meeting of the German Society of Occupational Medicine*, Köln, 1992, Genter Verlag, Stuttgart, 622–626

Kolstad HA, Juel K, Olsen J, Lynge E (1995) Exposure to styrene and chronic health effects: mortality and incidence of solid cancers in the Danish reinforced plastics industry. *Occup Environ Med 52*: 320–327

Korn M, Gfrörer W, Filser JG, Kessler W (1994) Styrene-7,8-oxide in blood of workers exposed to styrene. *Arch Toxicol 68*: 524–527

Morgan DL, Mahler JF, O'Connor RW, Price Jr HC, Adkins Jr B (1993a) Styrene inhalation toxicity studies in mice. I Hepatotoxicity in B6C3F$_1$ mice. *Fundam Appl Toxicol 20*: 325–335

Morgan DL, Mahler JF, Dill JA, Price Jr HC, O'Connor RW, Adkins Jr B (1993b) Styrene inhalation toxicity studies in mice. II Sex differences in susceptibility of B6C3F$_1$ mice. *Fundam Appl Toxicol 21*: 317–325

Morgan DL, Mahler JF, Dill JA, Price Jr HC, O'Connor RW, Adkins Jr B (1993c) Styrene inhalation toxicity studies in mice. III Strain differences in susceptibility. *Fundam Appl Toxicol 21*: 326–333

Nakajima T, Elovaara E, Gonzalez FJ, Gelboin HV, Raunio H, Pelkonen O, Vainio H, Aoyama T (1994) Styrene metabolism by cDNA-expressed human hepatic and pulmonary cytochromes P450. *Chem Res Toxicol 7*: 891–896

Oberste-Frielinghaus H, Dhawan-Robl M, Pütz C, Csanády GyA, Baur C, Filser JG (1999) Metabolism of styrene and racemic styrene-7,8-oxide in microsomes and cytosol of lung obtained from mouse, cat and human. *Naunyn-Schmiedeberg's Arch Pharmacol (Suppl) 359*: R157

Osterman-Golkar S, Christakopoulos A, Zorec V, Svensson K (1995) Dosimetry of styrene 7,8-oxide in styrene- and styrene oxide-exposed mice and rats by quantification of haemoglobin adducts. *Chem-Biol Interact 95*: 79–87

Pauwels W, Vodičeka P, Severi M, Plná K, Veulemans H, Hemminki K (1996) Adduct formation on DNA and haemoglobin in mice intraperitoneally administered with styrene. *Carcinogenesis 17*: 2673–2680

Sarangapani R, Teeguarden JG, Cruzan G, Clewell HJ, Andersen ME (2002) Physiologically based pharmacokinetic modeling of styrene and styrene oxide respiratory tract dosimetry in rodents and humans. *Inhal Toxicol 14(8)*:789–834

completed 06.03.2003

# Trichloroethylene

| | |
|---|---|
| **Classification/MAK value:** | **see Section IIIA1**<br>**of the List of MAK and BAT Values** |
| **Classification dates from:** | **1996** |
| Synonyms: | 1-chloro-2,2-dichloroethylene<br>ethinyl trichloride<br>ethylene trichloride<br>tri |
| Chemical name (CAS): | trichloroethene |
| CAS number: | 79-01-6 |

Structural formula:

$$H_2C_2Cl\text{—}C=C\text{—}Cl$$

| | |
|---|---|
| Molecular formula: | $C_2HCl_3$ |
| Molecular weight: | 131.39 |
| Melting point: | $-86°C$ |
| Boiling point: | $87°C$ |
| Vapour pressure at 20°C: | 77 hPa |

**1 ml/m$^3$ (ppm) = 5.45 mg/m$^3$**     **1 mg/m$^3$ = 0.183 ml/m$^3$ (ppm)**


# 1 Toxic Effects and Modes of Action

Trichloroethylene has euphoretic effects which can cause habituation and psychic dependency. An overdose of the substance can result in paralysis of the respiratory and circulatory centres in the brain. Metabolism of trichloroethylene yields reactive intermediates which can produce liver and kidney damage.

A number of studies have demonstrated that purified, epoxide-free trichloroethylene is carcinogenic in experimental animals. Life-long administration of high doses of trichloroethylene to mice resulted in increased incidences of liver and lung tumours; in rats trichloroethylene produced kidney tumours. The mechanisms of the tumorigenic effects of trichloroethylene in animals have been elucidated in detail. Metabolites of the oxidative bioactivation pathway are responsible for the carcinogenic effects of trichloroethylene in the mouse liver and mouse lung. Because the mouse has a high capacity for

*Essential MAK Value Documentations.* DFG, Deutsche Forschungsgemeinschaft
Copyright © 2006 WILEY-VCH Verlag GmbH & Co. KGaA, Weinheim
ISBN: 3-527-31394-X

oxidative metabolism of trichloroethylene, because of the induction of peroxisome proliferation in the mouse liver and because of certain anatomical and biochemical characteristics of the mouse lung (large numbers of Clara cells with high cytochrome P450 activity) the carcinogenicity of trichloroethylene in these organs can be considered to be a species-specific finding. The glutathione-dependent toxification of trichloroethylene, which also takes place in man, is responsible for the induction of renal tumours in the rat; this metabolic pathway yields genotoxic and cytotoxic metabolites which have also been detected in man. Epidemiological studies have reported an increased incidence of renal cell tumours in workers exposed for many years to high levels of trichloroethylene. Together with the known mechanism of action of trichloroethylene, these studies demonstrate a causal relationship between occupational exposure to high concentrations of the substance and the development of kidney tumours.


# 2 Toxicokinetics and Metabolism

Trichloroethylene can be metabolized by both oxidative and reductive metabolic pathways. The kinetic data available to date, however, describe only the oxidative metabolism.


## 2.1 Comparative toxicokinetics in mouse, rat and man

It has been calculated that under resting conditions, 67 % of gaseous trichloroethylene inhaled into the alveolar region is absorbed by the rat (Filser and Deml 1989) and 70 % to 80 % by man (Bogen 1988). In the rat, mouse and man exposed to low trichloroethylene concentrations, metabolism is the main elimination route. It has been calculated that after exposure to concentrations below 50 ml/m$^3$, about 97 % of the inhaled trichloroethylene is metabolized by the rat (Filser and Deml 1989). Within 50 hours after a 6-hour exposure of rats and mice to 10 ml/m$^3$ 98 % and 99 % were metabolized, respectively (Stott *et al.* 1982). After oral administration, a trichloroethylene dose of 10 mg/kg body weight was metabolized almost completely by rats and mice; 1 % to 4 % of the dose was exhaled unchanged. From data obtained with exposed persons (Fernández *et al.* 1975, 1977, Monster *et al.* 1976, 1979, Nomiyama and Nomiyama 1971) it was calculated that after exposure to 320 ml/m$^3$ 81 % to 87 % of the absorbed trichloroethylene was metabolized, whereas after exposure to 54 to 140 ml/m$^3$ 89 % to 92 % was metabolized (Bogen 1988). This is evidence for saturability of trichloroethylene metabolism in man.

The capacity for oxidative metabolism of trichloroethylene is higher in the mouse than in the rat. Whereas in the mouse the rate of oxidative metabolism increased linearly with the oral trichloroethylene dose up to about 2000 mg/kg body weight (Prout *et al.* 1985) and saturation was first achieved at a dose of 2400 mg/kg body weight (Buben and O'Flaherty 1985), saturation is seen in the rat at doses as low as 1000 mg/kg body weight (Prout *et al.* 1985). The kinetics in the rat after inhalation of trichloroethylene reveal saturation of metabolism after exposure to concentrations above 100 ml/m$^3$ (Filser and Bolt 1981) whereas in the mouse no signs of saturation were seen even after exposure to 600 ml/m$^3$ (Stott *et al.* 1982). Per kilogram body weight, mice metabolize 2.6 times as much trichloroethylene as do rats (Stott *et al.* 1982). The relative proportions of the main metabolites are similar in rats and mice (Green and Prout 1985); therefore, after high doses of trichloroethylene mice are exposed to higher levels of trichloroacetic acid than are rats: after a trichloroethylene dose of 2000 mg/kg body weight the mouse was exposed to a seven times higher dose of trichloroacetic acid than was the rat (Prout *et al.* 1985). Under steady state conditions during exposure to a trichloroethylene concentration of 10 ml/m$^3$, the metabolic rate in µmol/hour and kg body weight was 13 in the mouse, 8 in the rat and 1.4 in man (Filser and Deml 1989).

The results from *in vitro* studies of oxidative conversion of trichloroethylene are in accordance with those obtained *in vivo*. After incubation with liver microsomes, trichloroethylene was metabolized mainly by the cytochrome P450 enzymes CYP2E1, CYP1A1 and CYP1A2. These enzymes have higher activities in the mouse than in the rat (Nakajima *et al.* 1992, 1993). The higher levels of reactive metabolites produced in the mouse in oxidative metabolism are probably responsible, at least in part, for the hepato-carcinogenic effects of trichloroethylene which have been seen repeatedly exclusively in the mouse (see Section 4.6, Carcinogenicity). *In vitro* human liver microsomes have been shown to have cytochrome P450 activities like those of the rat and lower than those of mouse microsomes (Guengerich 1991).

Current conceptions of the oxidative and reductive biotransformation of trichloroethylene are shown in Figure 1.

## 2.2 Oxidative metabolism

The oxidative metabolism of trichloroethylene by the mixed-function oxygenases in the liver of man and animals yields not only the main metabolites trichloroacetic acid and trichloroethanol or its glucuronic acid conjugate but also the following urinary metabolites (Dekant *et al.* 1984):

1. oxalic acid, formed perhaps by hydrolysis of the postulated epoxide intermediate, and dechlorination of the resulting unstable vicinal diol,
2. *N*-hydroxyacetylaminoethanol, as end product of the reaction with phosphatidyl ethanolamine,
3. dichloroacetic acid (detected to date only in the mouse) formed by rearrangement of the epoxide to dichloroacetyl chloride and subsequent hydrolysis.

After oral administration to mice of a dose of $^{14}$C-labelled trichloroethylene of 200 mg/kg body weight, up to 6 % of the dose was exhaled as [$^{14}$C]-CO$_2$; in the rat it was only 1.9 % (Dekant *et al.* 1984). This is in accordance with the fact that CO$_2$ is produced from dichloroacetic acid which has only been identified unequivocally in the mouse.

As the oxidative metabolic pathways become saturated at high trichloroethylene concentrations, it is to be expected that the proportion of the trichloroethylene which is metabolized via the second, glutathione-dependent, reductive metabolic pathway will increase with increasing trichloroethylene concentration.

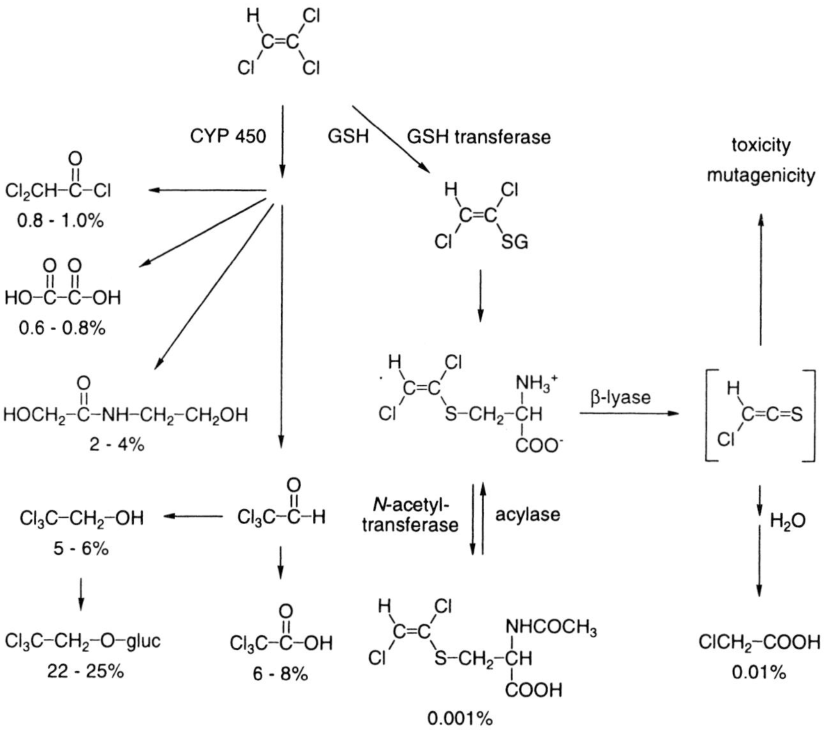

Figure 1. Metabolism of trichloroethylene in the rat. The percentages are proportions of the oral dose of 200 mg/kg body weight which were recovered in the urine (from Dekant *et al.* 1984)

## 2.3 Reductive metabolism

Another metabolite found in the urine of rodents exposed to trichloroethylene is the mercapturic acid 1,2-dichlorovinyl-*N*-acetylcysteine which, however, makes up only a very small fraction of the total metabolites (Commandeur and Vermeulen 1990, Dekant *et al.* 1986a). 2,2-Dichlorovinyl-*N*-acetylcysteine is also excreted by the rat in even lower concentrations (Commandeur and Vermeulen 1990). The formation of these mercapturic acids is considered to take place via enzymatic conjugation of trichloroethylene with glutathione in the liver in a reaction in which hydrogen chloride is cleaved from the molecule (Figure 1). Both the 1,2-dichloro and the 2,2-dichloro isomers can be formed. The dichlorovinylglutathione is transported to the kidneys where it is broken down stepwise by cleavage of glutamic acid and glycine to yield dichlorovinylcysteine. Dichlorovinylcysteine is concentrated in the proximal tubule cells by the efficient transport system for amino acid derivatives. In this part of the nephron some of the cysteine conjugate is acetylated to yield the urinary metabolite, the mercapturic acid. However, 1,2-dichloro-vinylcysteine and, to a lesser extent, the 2,2-isomer (Commandeur *et al.* 1991) are also substrates for the β-lyases of the renal tubule cells. The β-lyase cleavage of 1,2-dichloro-vinylcysteine yields an unstable thiol intermediate which eliminates hydrogen chloride spontaneously to form a reactive thioketene which can react with cellular nucleophiles. Reaction of the thioketene with water and subsequent hydrolysis of the thiocarboxylic acid yields monochloroacetic acid which has also been detected in the urine of rats exposed to trichloroethylene (Prout *et al.* 1985, Vamvakas *et al.* 1993a).

1,2-Dichlorovinylcysteine has been shown to be nephrotoxic in all animal species studied to date apart from the bovine calf. This organ-specific effect and the β-lyase-dependent cytotoxicity and mutagenicity are considered to provide evidence that metabolism to 1,2-dichlorovinylcysteine is required for the nephrotoxic and nephrocarcinogenic effects of trichloroethylene in long-term animal studies (see Section 4.6, Carcinogenicity). The low concentrations of the mercapturic acids in the urine do not speak against a role for the glutathione-dependent bioactivation in the kidney-specific effects of trichloroethylene because the amount of dichlorovinylcysteine available for acetylation is reduced by the β-lyase-mediated metabolism. In addition, the amount excreted depends on the state of the equilibrium between acetylation and deacetylation of the dichlorovinylcysteine. Acetylation and deacetylation rates differ for 1,2-dichlorovinyl-cysteine and 2,2-dichlorovinylcysteine (Commandeur *et al.* 1991). Therefore, the urine concentration of dichlorovinyl-*N*-acetylcysteine reflects neither the absolute level of glutathione conjugation in the liver, that is, the dichlorovinylcysteine concentration in the kidney, nor the amount of substance metabolized via the β-lyase pathway to the ultimate mutagenic metabolite.

That the β-lyase pathway is also of significance in man has been demonstrated by the detection of β-lyase in the human kidney (Lash *et al.* 1990, Nelson *et al.* 1988) and by the demonstration that 1,2-dichlorovinylcysteine has toxic effects on human kidney cells in culture (Chen *et al.* 1990). In human cells *in vitro* the β-lyase activity is about 10 % of that in rat cells (Nelson *et al.* 1988).

Dichlorovinyl-*N*-acetylcysteine was also detected, without differentiation of the 1,2-isomer and 2,2-isomer, in the urine of four workers exposed to trichloroethylene. The trichloroethylene concentration in the workplace air was not known (Birner *et al.* 1993).

After simultaneous exposure of three male test persons and four Wistar rats to trichloroethylene concentrations of 40, 80 and 160 ml/m$^3$ for 6 hours, the oxidative and glutathione-dependent metabolites were quantified in the 48-hour urine (Bernauer *et al.* 1996). The ratio of oxidative to reductive metabolites was similar in man and rat: expressed relative to the creatinine concentration, 1000 to 10000 times more oxidative metabolites than reductive metabolites (sum of 1,2-dichlorovinyl-*N*-acetylcysteine and 2,2-dichlorovinyl-*N*-acetylcysteine) were excreted. The test persons excreted a total of up to 0.43 µmole dichlorovinyl-*N*-acetylcysteine isomers. The concentrations of the two isomers in urine were similar. The rats, in contrast, excreted 3 to 4 times as much of the 2,2-isomer than of the more mutagenic and toxic 1,2-isomer (see Section 4.5.2, *In vitro* genotoxicity of the metabolites). Because of the various enzymes and transport mechanisms involved, any attempt to derive the concentrations of the substances in the kidney cells from the amounts of the isomers excreted must be viewed with reservations. The significance of this study lies rather in the fact that the genotoxic glutathione-dependent metabolites were detected in human urine after exposures to trichloroethylene concentrations below 50 ml/m$^3$, which was the MAK value for trichloroethylene until 1996.

The acute intoxication of a 17-year old youth who swallowed about 70 ml trichloroethylene with suicidal intent has been described (Brüning *et al.* 1996a). The highest concentration of trichloroethylene in blood was 4.05 mg/l, found 13 hours after the substance had been ingested. The level of dichlorovinyl-*N*-acetylcysteine excreted in the urine increased continuously until day 3 when it reached 1.25 nmol/mg creatinine. The excretion of 2,2,2-trichloroethanol also increased continuously to 12500 nmol/mg creatinine on day 3. The concentration of trichloroacetic acid, on the other hand, was highest on day 2 (3400 nmol/mg creatinine) and had fallen off markedly on day 3 (410 nmol/mg creatinine). The reported kinetics are not in line with the known half-lives of the metabolites; it is conceivable that the excretion pattern was affected by the multiplicity of intensive medical procedures which were carried out, such as forced diuresis and administration of various medicines. However the results do demonstrate that trichloroethylene can be metabolized both oxidatively and reductively in man.

It is also conceivable that quantitative differences between man, mouse and rat in the formation and metabolism of 1,2-dichlorovinylcysteine and 2,2-dichlorovinylcysteine contribute to the observed species differences in renal tumour formation because the two isomers differ in their genotoxic and nephrotoxic effects (Commandeur *et al.* 1991).

1,2-Dichlorovinylcysteine was considered to be the substance responsible for the severe bone marrow damage and aplastic anaemia which developed in calves given a soya meal feed which had been extracted with trichloroethylene (McKinney *et al.* 1959).


# 3 Effects in Man


## 3.1 Acute toxicity

Acute intoxication of persons exposed occupationally or otherwise to trichloroethylene has been described in several reports (Stüber 1931, von Oettingen 1964) which are of little significance for the assessment of the effects of chronic exposure to the substance because they generally do not include analytical data for exposure levels.

The symptoms which have been observed during short-term inhalation exposure of test persons are shown in Table 1.

**Table 1.** Effects of short-term inhalation of trichloroethylene

| Trichloroethylene concentration $(ml/m^3)$ | Symptoms | References |
|---|---|---|
| 20 | odour just detectable | Weitbrecht 1957 |
| 160 | eye irritation | Stewart *et al.* 1962 |
| 200 | after a period of hours to days: slight subjective impairment of performance | Stewart *et al.* 1970 |
| 279–372 | 30 min: tolerable | Lehmann and Schmidt-Kehl 1936 |
| 229–384 | after several hours: prenarcotic state | Nomura 1962 |
| 1280 | after 6 min: mucosal irritation, prenarcotic state | Lehmann and Schmidt-Kehl 1936 |
| 2500–6000 | complete narcosis | Rehrmann 1953, Longley and Jones 1963 |

In experiments in which persons were exposed briefly to trichloroethylene, psycho-motor performance was said to be impaired even at very low concentrations (Ettema *et al.* 1975, Nakaaki *et al.* 1973, Salvini *et al.* 1971). However, the data are contradictory. The reason for the different results and, more particularly, for differences in the evaluation of the results is likely to be that unspecific factors were not adequately taken into account. One very critical study (Stewart *et al.* 1974) has demonstrated that no significant changes in the results of performance tests are detectable in persons inhaling a trichloroethylene concentration of $110 \, ml/m^3$ for several hours.

An evaluation of a collective of 288 acute occupational intoxications (McCarthy and Jones 1983) confirmed that adverse effects on central nervous functions is the most frequent symptom: unconsciousness was reported in 125 cases, prenarcotic symptoms without unconsciousness in 128 cases. In 76 cases the persons suffered from nausea and vomiting and in 55 cases from reversible lung function disorders. Hepatic and cardiac symptoms were unusual with 5 and 3 cases, respectively, and nephrotoxic effects of short-term exposure to high levels of trichloroethylene were not described in this review. According to one case-report (David *et al.* 1989) a worker employed in degreasing metal developed acute renal failure as a result of allergic interstitial nephritis with secondary tubular necrosis. Since the exposure levels were not stated in the report, a causal role of trichloroethylene has not been demonstrated.

In the course of an acute intoxication of a 17-year old youth who swallowed about 70 ml trichloroethylene with suicidal intent, the patient developed a high temperature, tremor, generalized hyperkinesia, sinus tachycardia and then lost consciousness about 5 hours after ingestion of the trichloroethylene (Brüning *et al.* 1996a). After 5 days artificial respiration under narcosis with forced hyperventilation and diuresis, the patient regained consciousness. Whereas the blood tests revealed no evidence of liver or kidney damage, on day 5 the excretion of low molecular weight proteins ($\alpha_1$-microglobulin and $\beta_2$-microglobulin) and $\beta$-*N*-acetylglucosaminidase ($\beta$-NAG) in the urine was increased by a factor of five to six relative to the values obtained immediately after ingestion of trichloroethylene when the patient was admitted to the intensive care unit. Although the total protein level was not increased, the changes in these parameters are evidence of tubule damage. In a follow-up examination 5 months after the ingestion of the trichloroethylene the concentrations of $\alpha_1$-microglobulin and $\beta_2$-microglobulin and $\beta$-NAG were still increased by a factor of three to four but all the values appeared to be declining.

## 3.2 Chronic toxicity

In several occupational medical studies (see MAK documentation 1976: Henschler 1976) subjective symptoms were described repeatedly in workers exposed to trichloroethylene. The symptoms were mainly vegetative nervous and central nervous changes and mild neurological disorders or visus disorders which belong in the category of psycho-organic syndromes. Occasionally liver damage and cardiac arrhythmias were reported in these early studies. The symptoms developed only after long-term exposure to high concentrations. The subjective symptoms have also been documented more recently. Attempts to make objective records of the psychic changes with performance tests and self-assessment scales and by measuring nerve conduction velocities were, however, not successful (Triebig *et al.* 1977a, 1977b, 1982, Triebig and Grobe 1987).

The protein pattern in the urine of nephrectomized patients with renal cell tumours was studied by SDS polyacrylamide gradient gel electrophoresis (SDS-PAGE), a method which can distinguish between tubular, glomerular or mixed-type kidney damage (Brüning *et al.* 1996b). One of the groups studied comprised 17 patients who had been exposed to trichloroethylene at work before the operation (average exposure period 15.2 years, average latency period between beginning of exposure and tumour diagnosis 30.4 years). The other group comprised 35 nephrectomized patients whose occupational anamnesis did not indicate exposure to trichloroethylene. In both groups the date of the operation was on average five years before the SDS-PAGE investigation. In 14 of 17 exposed patients (82.3 %) the protein pattern indicated more or less severe tubular damage, in 3 cases (17.7 %) mixed-type tubular-glomerular damage. In the patients who had not been exposed the protein pattern was normal in 18 of the 35 (51.4 %) cases. In 12 of these patients (34.4 %) there was evidence of tubular damage, in 4 cases (11.4 %) evidence of mixed type tubular-glomerular damage and in one case (2.8 %) evidence of glomerular damage. These results are evidence that long-term exposure to trichloroethylene can induce chronic nephrotoxicity.

The cranial nerve damage (*polyneuritis cranialis*) resulting from occupational exposure to trichloroethylene is not an effect of trichloroethylene itself but of dichloroacetylene which is produced in the presence of alkali or at high temperatures by cleavage of hydrogen chloride from trichloroethylene (Greim *et al.* 1984, Henschler *et al.* 1970, Spencer and Schaumburg 1985).

## 3.4 Genotoxicity

A number of cytogenetic studies carried out with workers exposed to trichloroethylene indicate that the substance has effects on chromosomes. In the cultured lymphocytes from 9 of a total of 28 exposed workers (average exposure concentration 75 ml/m$^3$, 5 hours daily) the incidence of hypoploidy was pathologically increased (Konietzko *et al.* 1978) and correlated with the trichloroethylene concentrations in the workplace air. In addition, sister chromatid exchange was observed in lymphocytes from 6 of the workers (Gu *et al.* 1981) and various structural chromosomal aberrations such as breaks, translocations, insertions and gaps in another collective of 15 workers (Rasmussen *et al.* 1988). In contrast, a similar study (Nagaya *et al.* 1989) in which the average trichloroethylene concentrations in the workplace air were only 30 ml/m$^3$ yielded negative results.

The available data do not permit any conclusive statement as to the genotoxicity of trichloroethylene in man, especially in the light of the very small cohort sizes and the lack of data for the purity of the trichloroethylene.

## 3.5 Carcinogenicity

The cancer mortality and incidence in collectives of individuals exposed to trichloroethylene has been recorded in several epidemiological studies during the last two decades. Here only those studies are included in which persons were exposed only to trichloroethylene, although the exposure concentrations could not be quantified in every case.

In a cohort of 518 workers exposed to trichloroethylene, no increase in the incidence of tumours could be detected (Axelson *et al.* 1978). The duration of exposure and the period of time between the start of exposure and the qualifying date were mostly less than ten years. In a more recent study by the same authors the original collective was extended to a total of 1670 exposed persons (Axelson *et al.* 1994). This update also did not reveal significant increases in cancer incidence or cancer mortality. The trichloroacetic acid levels in the urine at the time of exposure were less than 100 mg/ml (BAT value) in about 80 % of the collective; this corresponds to an average workplace exposure level of about 30 to 50 ml/m$^3$. Even in the past, the concentrations were probably not much higher because 30 ml/m$^3$ has long been the maximum workplace concentration permitted for trichloroethylene in Sweden (Axelson *et al.* 1994).

Likewise, in a group of 2117 exposed workers no association between exposure to trichloroethylene and cancer could be detected (Tola *et al.* 1980). However, the time period between the beginning of exposure and the qualifying date was at most 13 years and about half of the persons included were younger than 40 years at the end of the study period. The exposure levels must have been low, in most cases below 50 ml/m$^3$, because the trichloroacetic acid concentrations in the urine at the time of exposure were less than 100 mg/l in about 90 % of the cases.

Studies of 188 exposed persons (Barret *et al.* 1984) and 2646 exposed persons (Shindell and Ulrich 1985) also yielded no evidence of a carcinogenic potential of trichloroethylene. Whereas the exposure levels in the study by Barret *et al.* (1984) can be considered to be high—50 % of cases were exposed to trichloroethylene levels of 150 ml/m$^3$ or more—the average age of the collective of 41 years was too low and the average exposure time of 7 years too short for the detection of a carcinogenic effect. Shindell and Ulrich (1985) did not comment on the level of exposure of their cohort. The collectives studied by Tola *et al.* (1980) and by Shindell and Ulrich (1985) both showed clearly a healthy worker effect.

A large study investigated 4733 workers who were exposed to trichloroethylene in an aircraft maintenance facility (ENSR Health Sciences 1990).The exposure period was as much as 30 years in some cases; measurements of the levels of trichloroethylene in the workplace air or of trichloroacetic acid in the urine were not available. The exposure levels were allocated on the basis of the work carried out as follows: high for work at or in the immediate vicinity of degreasing machines, medium for work at larger distances from the degreasing machines but involving occasional contact with trichloroethylene and low for work away from degreasing machines. About 80 % of the 4733 exposed persons belonged to the groups with medium or low exposure. In the whole group there were 125 observed deaths from cancer compared with 159.7 expected deaths, that is, the mortality from malignant tumours was significantly reduced (standardized mortality ratio, SMR 78). Analysis of changes in cancer mortality during the period of employment revealed a tendency to increase within the three groups: employed for (1) less than 10 years, (2) 10 to 20 years and (3) more than 20 years. Even in the longest exposed group, however, the SMR of 81.6 % was smaller than in the control collective. Non-significant increases in mortality from liver and bile duct tumours were detected in all three groups. The number of deaths from renal tumours was also increased non-significantly with one case after medium-term and two cases after long-term exposure. On the other hand, analysis of causes of death revealed a significantly increased mortality from benign tumours with 6 observed cases where 2.02 would have been expected. Of these 6 persons, 3 died with benign brain tumours; the localization of the other tumours was not stated in this publication. Because of the short exposure periods (3 cases < 2 years, 2 cases < 10 years), the authors considered it unlikely that the trichloroethylene exposure could be causally related to the development of the benign tumours.

The largest cohort study to date was carried out by the US National Cancer Institute in an aircraft servicing plant in which the mortality of 6929 workers exposed to trichloroethylene was analysed for the period between 1952 and 1982 (Spirtas *et al.* 1991, Stewart *et al.* 1991). On the basis of the trichloroethylene concentrations in the air of the relevant workplace, the exposure period per work shift and the total period of employment, the cumulative exposure of the persons in the cohort was calculated and expressed in units (< 5 units, 5–25 units, > 25 units). The cancer mortality ratios were lower for the exposed workers than for the control group (general population of the region) and no significant association with the trichloroethylene exposure was detected for any specific kind of cancer.

A study carried out in Finland included 3974 workers of whom 3089 had been exposed to trichloroethylene, 849 to tetrachloroethylene and 271 to 1,1,1-trichloroethane between 1965 and 1983. The average trichloroacetic acid concentrations in the urine of the persons exposed to trichloroethylene (8.3 mg/l in women and 6.3 mg/l in men) suggest that the exposure concentrations were low. For comparison, the BAT value for trichloroethylene was 100 mg trichloroacetic acid per litre urine and corresponded to a MAK value for trichloroethylene of 50 ml/m$^3$. The cancer incidence in this cohort was registered until the end of 1992 and compared with the incidence in the general population of Finland. The cohort was divided into three groups on the basis of the observation period in years: (1) 0 to 9 years, (2) 10 to 19 years and (3) 20 or more years. In the group exposed to trichloroethylene with an observation period of more than 20 years the overall incidence of cancer was slightly increased (standardized incidence ratio, SIR = 1.57); if the data for all the persons exposed to trichloroethylene were analysed, however, there was no increase in the total cancer incidence. Of the specific cancers, and also only for the group with an observation period of more than 20 years, the incidence was increased for liver tumours (SIR = 6.07), prostate tumours (SIR = 3.57), tumours of the lympho-haematopoietic system (SIR = 2.98) and stomach tumours (SIR = 2.98) (Anttila *et al.* 1995).

In their assessment of the carcinogenicity of trichloroethylene in man, in February 1995 the IARC (1995) combined the data from the three large cohort studies from the USA (Spirtas *et al.* 1991), Sweden (Axelson *et al.* 1994) and Finland (Anttila *et al.* 1995). The analysis revealed increased risks for tumours at the following locations: in the three studies a total of 23 liver and bile duct tumours were observed where 12.97 would be expected and 27 non-Hodgkin lymphomas where 18.82 would be expected. In two of the studies (Anttila *et al.* 1995, Spirtas *et al.* 1991) the liver tumours were registered separately from the bile duct tumours and 7 liver tumours were found where 4 would be expected. On the basis of these data the IARC came to the conclusion that "there is limited evidence in humans for the carcinogenicity of trichloroethylene".

In a German cardboard factory a retrospective cohort study of 169 workers who had been exposed to trichloroethylene for at least one year between 1956 and 1975 was carried out (Henschler *et al.* 1995). The reason for the study was the appearance of renal tumours in four persons employed at this factory. Because it was known that trichloroethylene causes renal tumours in experimental animals, the study aimed to establish whether trichloroethylene was responsible for the renal tumours in these workers. The study was carried out without knowledge of whether or not the affected persons had been exposed to trichloroethylene. Of the total of 870 persons who had been employed in the factory, 183 were male workers exposed only or mainly to trichloroethylene. Of the exposed persons, one refused to take part in the study, 8 could not be located and 5 were not able to take part because of their poor state of health (Lammert 1995). Although this could result in a slight increase in the calculated risk, the proportion of the exposed persons available for the study, 92.3 %, is similar to that available in the other epidemiological studies of cohorts exposed to trichloroethylene. The control group comprised all the employees of the factory who fulfilled the following selection criteria: age, sex and physical activity like those in the exposed group, no exposure to trichloroethylene and employment in the same period of time. The earliest start of employment in the factory was 1918, the earliest exposure and thus beginning of the observation period was 1956. Thus for a total of 5188 person years, the average observation period was 30.7 years. The persons exposed to trichloroethylene and the controls were subjected to ultrasonography which, however, did not reveal any additional cases of renal cancer. Not only the diagnosis of renal cancer (morbidity and mortality) but also mortality from a number of other disorders was recorded (Ulm 1995) including tumours of the lungs and brain, lymphatic, haematopoietic and cardiovascular disorders. During the course of the study an additional case of renal cell cancer was detected. Between the exposed and the control persons there were no differences in factors which predispose to renal cancer such as overweight, smoking habits, alcohol consumption and ingestion of diuretics. Trichloroethylene was introduced into the factory in 1956 as a cleaning and degreasing agent and remained the main organic solvent used until 1975. From 1967 chlorine-free and fluorine-free hydrocarbons, hexachlorobenzene and 1,1,1-trichloroethane were also used but only occasionally and in very small quantities. A causal involvement of the highly nephrocarcinogenic dichloroacetylene, which could have been produced from trichloroethylene during cleaning of hot metal surfaces, is considered unlikely because dichloroacetylene is unstable in the presence of oxygen. In addition, the typical symptoms of dichloroacetylene intoxication such as irritation of the mucosa of the eyes and nose, numbness around the mouth and damage to cranial nerves (Henschler *et al.* 1970) were not observed in any of the exposed workers. On the basis of workplace descriptions and detailed interviews with workers who had been employed in the factory for many years and who complained of regularly occurring headaches, dizziness and confusion, and on the basis of data for the annual consumption of trichloroethylene, it was concluded that the concentrations of trichloroethylene had been very high in the cardboard machine area and high in the metal-working shop and electrical workshop. The reported symptoms are seen in persons exposed occupationally or in experimental studies to trichloroethylene concentrations above 200 ml/m$^3$ (Stopps and McLaughlin 1970, Torkelson and Rowe 1981). Data for trichloroethylene concentrations in the workplace air or of trichloroethylene metabolites in the urine of the exposed persons were not available. The average period of exposure was 17.8 years and the median total observation period 34 years. The numbers of observed renal tumours in the exposed and control groups were compared with the expected incidence from the Danish cancer registry. The diagnosis "renal cell cancer" was made for four of the workers exposed to trichloroethylene; two of these cases were non-smokers, one ex-smoker since 1972 and one smoker (Lammert 1995). Another non-smoker exposed to trichloroethylene had a carcinoma of the renal pelvis. This tumour type is histologically different from the renal cell tumour but, since only the combined incidence of these two tumour types is given in the Danish cancer registry, this case was also included in the statistical analysis. The SIR relative to the Danish cancer registry was 7.97 and was highly significant. In the control group, no cases of renal cancer were diagnosed. Direct comparison of the incidence in the exposed and non-exposed groups without the case of renal pelvis carcinoma revealed a significantly increased risk in the exposed group. Thus, all the different kinds of analysis applied in this study demonstrated a significant increase of the incidence of renal cancer in the group of persons exposed to trichloroethylene. After the study had been completed, two further cases of kidney cancer were diagnosed in routine examinations in the group of persons exposed to trichloroethylene, one was a renal cell tumour in a non-smoker, the other a renal pelvis carcinoma in an ex-smoker. In Table 2 the workplace descriptions, duration of exposure and latency period until diagnosis of the tumour are shown for the 7 cases of renal tumours. The subdivision of the persons into groups of occupations presumed to differ in the level of exposure did not demonstrate dependence of the tumour development on exposure levels. However, for this purpose the number of cases was too small and it was not possible to characterize the exposure of the workers precisely enough, especially with respect to changing workplaces. In the group of exposed persons and the control group there were one ($p < 0.05$, compared with the registry data) and three ($p < 0.05$) persons with brain tumours, respectively. For these no association with exposure to a chemical substance could be demonstrated and no plausible explanation found.

**Table 2.** Workplace description, beginning and duration of exposure and the latency period until diagnosis of the tumour in the seven cases of renal tumours among 169 workers exposed to trichloroethylene in a German cardboard factory (data from Henschler *et al.* 1995).

| Case no. | Work place | Beginning of exposure (year) | Duration of exposure (years) | Date of tumour diagnosis | Latency period (years) | Tumour |
|---|---|---|---|---|---|---|
| 1 | 2 | 1956 | 19.4 | 1974 | 18 | renal cell carcinoma |
| 2 | 3 | 1961 | 14.2 | 1980 | 19 | renal cell carcinoma |
| 3 | 1 | 1956 | 19.4 | 1990 | 34 | renal cell carcinoma |
| 4 | 1 | 1956 | 13.0 | 1990 | 34 | renal pelvis carcinoma |
| 5 | 3 | 1972 | 3.0 | 1991 | 19 | renal cell carcinoma |
| 6 | 1 | 1957 | 19.3 | 1993 | 36 | renal cell carcinoma |
| 7 | 3 | 1956 | 19.4 | 1993 | 37 | renal pelvis carcinoma |
| | | | ∅15.2 | | ∅28.1 | |

workplace description:    1: near the cardboard machines
2: in the metal-working shop
3: in the electrical workshop

In the other cohort studies published to date, no association between exposure to trichloroethylene and the occurrence of renal cell carcinomas was detected. The study by Henschler *et al.* (1995) differs from the others in that the cohort was exposed almost exclusively to very high, although not exactly quantifiable levels of trichloroethylene. However, poorly quantifiable exposure is also a feature of the other cohort studies in which data for concentrations of trichloroethylene in the workplace air are not available. In the study by Henschler *et al.* the observation period was long enough and the average age of the persons studied was high. In most of the cohort studies published to date, renal cancer mortality and not morbidity was recorded. Morbidity is, however, a more sensitive parameter than mortality: because of improved diagnostic methods, the chance of survival after an operation for this kind of cancer has been high since the 1950s and since then, therefore, the incidence of renal cancer has increased more steeply than mortality from renal cancer (Coleman *et al.* 1993). At the end of the 1980s the incidence of renal cancer in males in the Scandinavian countries was about 20/100000 (Coleman *et al.* 1993) and the mortality about 5/100000 and in Germany 6/100000 (La Vecchia *et al.* 1992). Incidence data are not available for Germany. For the calculation of the expected incidence of renal cancer in the study by Henschler *et al.* (1995) data from the Danish cancer registry of 1960 were used because at that time more recent data were not available. At that time, however, the incidence of renal cancer was about the same as the mortality from renal cancer (Ulm 1995). Therefore the expected incidence given in the study by Henschler *et al.* (1995) is about the same size as the expected mortality. In the other two incidence studies (Axelson *et al.* 1994, Anttila *et al.* 1995) the persons studied were exposed to much lower levels of trichloroethylene than those studied by Henschler *et al.* (1995) and this accounts for the fact that even here no association could be demonstrated between the incidence of renal cancer and trichloroethylene exposure.

The relationship between trichloroethylene exposure and certain cancers has also been investigated in a number of case-control studies.

From a cancer registry 95 patients were identified in whom primary liver cell carcinoma had been diagnosed in the years 1951 to 1977 and who lived near a trichloroethylene producing factory. The personnel records of this factory demonstrated that none of the liver carcinoma patients had ever worked there (Paddle 1983).

The results of a case control study of 59 patients who were nephrectomized in the urological department of a German hospital have been made available to the Commission; in this study the effects of occupational exposure especially to trichloroethylene and tetrachloroethylene on the occurrence of renal cell tumours was investigated (Vamvakas *et al.* 1998). The study group included all patients (n = 59) for whom the diagnosis renal cell tumour had been made at the histopathological examination after nephrectomy between December 1987 and May 1992. The control group comprised 84 traffic accident patients who were treated in the same clinic. To exclude renal tumours, all members of the control group were subjected to abdominal ultrasonography. In the region served by the clinic there are many small factories of the metal-working and electrical industries. Evaluation of workplaces and exposure was carried out by means of questionnaires and personal interviews; in addition, information was obtained from works physicians and occupational hygienists as well as from the technical health and safety authorities. Of the nephrectomy cases, 20 (34 %) stated that they had been exposed to trichloroethylene; none reported exposure to tetrachloroethylene. Five members of the control group had been exposed occupationally to trichloroethylene (6 %) and two to tetrachloroethylene (2 %). After allowance for the factors age, sex, smoking habits, body mass index (body weight divided by the square of the height), blood pressure and consumption of diuretics by means of logistic regression, an odds ratio for the combined exposure to trichloroethylene and tetrachloroethylene of 13.42 (95 % confidence interval 3.50–51.39) was obtained. Since no exposure to tetrachloroethylene had been reported in the group of nephrectomized patients, this highly significant odds ratio is evidence for an association between exposure to trichloroethylene and the occurrence of renal cell carcinomas. The average period of exposure to trichloroethylene in the group of nephrectomized patients was 19 years, clearly longer than the 5-year average in the control group. On the basis of duration and frequency of exposure and the workplace description (size of open vats containing trichloroethylene, temperature of the trichloroethylene, etc.) the persons were classified in one of three categories: high, medium and low level exposure. Of the 20 exposed patients with renal tumours, 8 were in the high-level exposure group, 10 in the medium and 2 in the low. The corresponding numbers for the 7 exposed persons in the control group were 2, 3 and 2. Since trichloroethylene was for decades the main cleaning and degreasing agent in the metal-working industry, the groups were also studied for employment in metal-working independent of any information about exposure. This comparison yielded an odds ratio of 1.95 (95 % confidence interval 0.72–5.41) which is very close to the odds ratio of 1.6 given in the study by Mandel *et al.* (1995) for the occurrence of renal cancer among employees in the metal-working industry. In the study by Vamvakas *et al.* (1998) not only exposure to trichloroethylene but also to asbestos, cadmium, gasoline (petrol) and other petroleum products were studied as these too are associated with the occurrence of renal cell tumours. Only two of the nephrectomized patients reported brief exposure to asbestos. In addition, the effect of the age difference between the nephrectomized group (62 years) and the control group (53 years) on the association was analysed by means of logistic regression and stratification into three age groups. Both analyses confirmed the association between exposure to trichloroethylene and the occurrence of renal cell tumours. Thus this case-control study yields more evidence that long-term occupational exposure to high concentrations of trichloroethylene can lead to an increase in the incidence of renal cell tumours.

Analysis of 329 cases of colon carcinoma from the Swedish cancer registry and comparison with 623 control persons revealed an increased risk for 10 women employed in dry cleaning businesses where trichloroethylene was used (odds ratio 7.4) but not for general occupational exposure to trichloroethylene (odds ratio 1.5) (Fredriksson *et al.* 1989). This study cannot be considered to be very meaningful because of the small number of exposed persons.

There are a number of other small case-control studies, mostly available only as abstracts without adequate data, which did not reveal any association between cancer and exposure to trichloroethylene. They are not listed here (see IARC 1995).

# 4 Animal Experiments and *in vitro* Studies

## 4.1 Acute toxicity

The $LD_{50}$ and $LC_{50}$ values available for trichloroethylene are shown in Table 3; they demonstrate that the substance is of low acute toxicity after oral administration and inhalation.

Deaths caused by trichloroethylene doses in the region of the $LD_{50}$ are a result of CNS depression caused by the narcotic effects of the substance and the associated paralysis of the respiratory and circulatory centres in the brain. In addition, single doses of trichloroethylene cause sensitization of the heart to epinephrine which results in cardiac arrhythmia. This effect was seen in dogs exposed to trichloroethylene concentrations of 5000 or 10000 ml/m$^3$ for 10 minutes (Reinhardt *et al.* 1973) and in rabbits exposed to 6000 ml/m$^3$ for 1 hour (White and Carlson 1981).

Trichloroethylene doses in the range of the $LD_{50}$ ($\geq$ 2000 mg/kg body weight) produced damage in the Clara cells in the mouse lung (Forkert and Birch 1989, Scott *et al.* 1988). This species-specific and cell-specific effect was more marked after exposure by inhalation. After a single 6-hour exposure of mice to trichloroethylene concentrations between 200 and 1000 ml/m$^3$, concentration-dependent damage was seen in the Clara cells; exposure of rats to 1000 ml/m$^3$ for 6 hours had no effect on the lungs (Odum *et al.* 1992).

Unlike trichloroethylene itself, which is of very low renal toxicity, the urinary metabolite, 1,2-dichlorovinylcysteine, is highly and selectively nephrotoxic. A single 1,2-dichlorovinylcysteine dose of 5 mg/kg body weight or more, administered orally or intraperitoneally, induced marked functional and histopathological changes in the proximal tubules of the kidneys of rats, mice and rabbits (Beuter *et al.* 1989, Darnerud *et al.* 1988, 1989, 1991, Hassall *et al.* 1983, Terracini and Parker 1965, Wolfgang *et al.* 1989).

In cultured kidney cells it has been demonstrated that 2,2-dichlorovinylcysteine and its mercapturic acid are less cytotoxic than the corresponding 1,2-isomers (Commandeur *et al.* 1991). There are no other studies of the cytotoxicity of the 2,2-isomers.

**Table 3.** Acute toxicity of trichloroethylene. LC$_{50}$ values after exposure of animals by inhalation and LD$_{50}$ values after oral administration

| Species | | LC$_{50}$ (ml/m$^3$) | Duration of exposure (h) | References |
|---|---|---|---|---|
| rat | | 26300 | 1 | Vernot *et al.* 1977 |
| rat | | 12500 | 4 | Siegel *et al.* 1971 |
| mouse | | 41122 | 0.33 | Aviado *et al.* 1976 |
| mouse | | 49000 | 0.5 | Vernot *et al.* 1977 |
| mouse | | 8450 | 4 | Kylin *et al.* 1972 |
| | | LD$_{50}$ (mg/kg body weight) | | |
| rat | | 7183 | | Smyth *et al.* 1969 |
| mouse | | 2400 | | Tucker *et al.* 1982 |
| mouse | | 2850 | | Aviado *et al.* 1976 |
| mouse | ♂ | 2402 | | Tucker *et al.* 1982 |
| mouse | ♀ | 2443 | | Tucker *et al.* 1982 |
| cat | | 5864 | | NIOSH 1984 |
| rabbit | | 7330 | | NIOSH 1984 |

## 4.2 Subacute, subchronic and chronic toxicity

### 4.2.1 Liver, kidney

Table 4 summarizes the effects of trichloroethylene on the liver and kidneys which have been reported in recent studies. When all the available data for the subchronic and chronic toxicity of trichloroethylene are reviewed, the mouse liver is seen to be the most sensitive organ.

**Table 4.** Subchronic and chronic toxicity of trichloroethylene

| Species | Dose or concentration; duration; administration route | **Dose or concentration:** findings | References |
| --- | --- | --- | --- |
| rat | 150 ml/m$^3$; 24 h/day, 30 days | **150:** liver weights increased by 20–30 %, kidney weights slightly increased | Kjellstrand *et al.* 1981a |
| | 50, 200, 800 ml/m$^3$; 24 h/day, 7 weeks | **from 50:** kidney weights increased in a dose-dependent manner (no other details in the abstract), functional kidney disorders (glucosuria, altered creatinine clearance) **800:** liver weights increased, pathological values for blood liver parameters | Nomiyama *et al.* 1986 |
| | 1100 mg/kg body weight; 5 days/week, 3 weeks; gavage | **1100:** liver weights and DNA synthesis in the liver increased, no histopathological changes in the liver, no effects on the kidneys · | Stott *et al.* 1982 |
| | 1500 mg/kg body weight; 10 days; gavage | **1500:** liver weights increased by 30 %, no effects on DNA synthesis, peroxisome levels or cell proliferation | Elcombe *et al.* 1985 |
| mouse | 150 ml/m$^3$; 24 h/day, 30 days | **150:** liver weights increased by 60–80 %, kidney weights slightly increased | Kjellstrand *et al.* 1981a |
| | 37–300 ml/m$^3$; 24 h/day, 30 days | **from 37:** liver weights increased in a dose-dependent manner, vacuolation/enlargement of hepatocytes, kidney weights slightly increased, spleen weights decreased | Kjellstrand *et al.* 1981a, 1983a, 1983b |
| | 1500 mg/kg body weight; 10 days; gavage | **1500:** liver weights increased by 75 %, DNA synthesis, peroxisome and cell proliferation increased 5-fold | Elcombe *et al.* 1985 |
| | 24, 240 mg/kg body weight; 14 days; drinking water | **24:** NOEL **240:** liver weights increased, no effects on the kidneys | Tucker *et al.* 1982 |
| | 500, 1200, 2400 mg/kg body weight; 5 days/week, 3 weeks; gavage | **from 500:** liver weights and DNA synthesis in the liver increased in a dose-dependent manner, hepatocellular swelling/necrosis, no effects on the kidneys | Stott *et al.* 1982 |
| | 18.4; 216.7; 393; 660 mg/kg body weight; 180 days; drinking water | **18.4:** NOEL **216.7:** increased liver weights in the ♂ mouse **393:** ♂: increased liver weights **660:** ♂/♀: increased liver weights, increased kidney weights, proteinuria, ketonuria | Tucker *et al.* 1982 |
| Mongolian gerbil | 150 ml/m$^3$; 24 h/day, 30 days | **150:** liver weights increased by 20–30 %, kidney weights increased by 10–20 % | Kjellstrand *et al.* 1981a |

NOEL: no observed effect level

In contrast, the trichloroethylene metabolite, 1,2-dichlorovinylcysteine, is a powerful and selective nephrotoxin also after medium-term and long-term administration. Cell necrosis, cytomegaly and karyomegaly, pyknosis, multiple nucleoli and atrophy and lumen dilation were seen in the proximal tubules of the mouse kidney after administration of daily 1,2-dichlorovinylcysteine doses of 1 to 22 mg/kg body weight with the drinking water during a period of 4 to 37 weeks; toxic effects on the liver or lungs were not detected (Jaffe *et al.* 1984).

### 4.2.2 Central nervous system

Rats which were exposed to a trichloroethylene concentration of 100 ml/m$^3$, 6 hours daily on 5 days per week, were less active than the control animals after 1.5 weeks (Silvermann and Williams 1975). Mongolian gerbils also showed slight behavioural disorders during continuous exposure to 150 ml/m$^3$ for 10 to 15 weeks (Kjellstrand *et al.* 1981b). Biochemical studies demonstrated changes in the fatty acid composition of the phospholipids in the brains of rats exposed continuously to a trichloroethylene concentration of 320 ml/m$^3$ for 30 or 90 days (Kyrklund *et al.* 1986).

### 4.2.3 Immune system

In a study of the subacute toxicity of trichloroethylene, mice were given doses of 24 or 240 mg/kg body weight by gavage daily during a period of 14 days. The treatment produced an impairment of the cell-mediated immune response to sheep erythrocytes but no effect on the humoral immune response (Tucker *et al.* 1982). After administration of trichloroethylene in the drinking water in average daily doses between 18 and 800 mg/kg body weight for periods of 4 or 6 months, impairment of the cell-mediated immune response and of the colony-forming ability of bone marrow cells was detected in the female mice given doses of 18 mg/kg body weight or more for 4 months. The humoral immune response was first significantly reduced after 6 months at a dose of 800 mg/kg body weight. In the male animals, in contrast, no significant changes were detected (Sanders *et al.* 1982). Changes in the thymus weights in rats exposed to a trichloroethylene concentration of 800 ml/m$^3$ for 14 days (Nomiyama *et al.* 1986) could not be taken into consideration because the documentation was inadequate.

### 4.2.4 Blood count

Exposure of dogs to trichloroethylene concentrations between 200 and 2000 ml/m$^3$ for 1 hour or to 700 ml/m$^3$ for 4 hours resulted in a concentration-dependent significant reduction in the leukocyte count while the erythrocyte and thrombocyte counts were unaffected. After oral administration of trichloroethylene doses of 400 mg/kg body weight to mice daily during 4 or 6 months, a significant decrease in leukocyte and erythrocyte counts was detected (Tucker *et al.* 1982).

## 4.3 Effects on skin and mucous membranes

Skin and mucosal tolerance of trichloroethylene was studied after single applications of the substance to the scarified or untreated skin of the rabbit and after instillation into the rabbit eye (Duprat *et al.* 1976). The effects of trichloroethylene were described as slightly irritating in the rabbit eye and highly irritating on rabbit skin.

## 4.4 Reproductive and developmental toxicity

Trichloroethylene has been tested for prenatal toxicity in inhalation studies with mice, rats and rabbits. Mice were exposed from day 6 to day 15 of gestation for 7 hours daily to a trichloroethylene concentration of 300 ml/m$^3$, which was not toxic for the dams. No prenatal toxic effects were detected (Leong *et al.* 1975, Schwetz *et al.* 1975). Pregnant Sprague-Dawley rats, for which 300 ml/m$^3$ was also not maternally toxic, were exposed under the same experimental conditions. Evidence of prenatal toxic effects was not found in this study either (Leong *et al.* 1975, Schwetz *et al.* 1975).

Likewise, in a similar study with rats exposed to a trichloroethylene concentration of 300 ml/m$^3$, no evidence of prenatal toxicity was found (Bell 1977).

In another experiment with a different exposure pattern, Sprague-Dawley rats were exposed to a trichloroethylene concentration of 500 ml/m$^3$ for 7 hours daily, 5 days per week for 3 weeks before mating and on days 0 to 18 or 6 to 18 of gestation. This concentration was not toxic for the dams. Under these conditions as well, no prenatal toxicity was detected (Beliles *et al.* 1980).

Likewise, exposure of Long-Evans rats to the much higher trichloroethylene concentration of 1800 ml/m$^3$ before and during the entire gestation period did not produce maternal toxicity. However signs of slight prenatal toxicity were seen in the form of incomplete ossification of the sternum (Dorfmüller *et al.* 1979).

New Zealand White rabbits were exposed to a trichloroethylene concentration of 500 ml/m$^3$ for 7 hours daily, 5 days per week on days 0 to 21 or 7 to 21 of gestation; another group was also exposed for 3 weeks before mating. This concentration was not toxic for the dams. In the group exposed only during gestation there was a non-significant increase in the incidence of hydrocephalus in the foetuses. In the group which was also exposed before mating, this anomaly was not detected. Since hydrocephalus also occurs spontaneously in this strain of rabbits the findings were not considered to reflect prenatal toxic effects of trichloroethylene. In an analogous study with Sprague-Dawley rats neither prenatal toxic nor teratogenic effects were found (Hardin *et al.* 1981, NIOSH 1980).

Another publication (Healy *et al.* 1982) describes a study in which 32 Wistar rats were exposed to a trichloroethylene concentration of 100 ml/m$^3$, 4 hours daily from day 8 to day 21 of gestation; the progeny differed significantly from the controls for three of the nine parameters investigated. Firstly, there was an increased number of dams in which all the embryos were resorbed during a very early phase of gestation. This finding has not been reported in any other study to date, not even in those in which the animals were exposed to much higher concentrations. The suggestion made by the authors that trichloroethylene could be particularly toxic for Wistar embryos in the early phase of gestation has not been confirmed in any other study. Secondly, the foetuses exposed to trichloroethylene (average weight 2.93 g) were slightly lighter than the controls (3.21 g); the difference was statistically significant. It was considered to be a sign of prenatal toxicity because (thirdly) the incidence of absent or bipartite centres of ossification in the sternum was also higher in the exposed group than in the controls. The only other information given on this point in the publication was that 70 % of the skeletons examined were normal in the control group and 50 % in the exposed group. Maternal toxicity was not mentioned. The findings must be seen as minimal differences between exposed and control animals which were not found even in this form in the studies described above with exposures to trichloroethylene concentrations of 300 and 500 ml/m$^3$. Even the exposures to the much higher concentration of 1800 ml/m$^3$ (Dorfmüller *et al.* 1979) yielded merely evidence for slight foetotoxicity of trichloroethylene.

The effects of oral administration of purified trichloroethylene on the fertility of F344 rats were investigated in a 2-generation study (NTP 1986). Trichloroethylene was mixed with the diet to provide daily doses of 0, 75, 150 or 300 mg/kg body weight from day 7 before mating until the birth of the $F_2$ generation. No significant adverse effects on mating behaviour or fertility, no sperm anomalies or histopathological effects on the sex organs were found. Marginal effects of the highest dose on the testis/epididymis weights in the $F_0$ and $F_1$ generations and on the numbers of surviving pups per litter were considered to be results of general toxicity.

Analogous 2-generation studies with CD-1 mice (NTP 1985) given daily trichloroethylene doses of 0, 180, 375 or 750 mg/kg body weight according to the same dose scheme also revealed no significant adverse effects on the reproductive system but maternal effects in the highest dose group. Sperm motility was reduced in the highest dose group in the $F_0$ generation by 45 % and in the $F_1$ generation by 18 %. In the $F_1$ generation the survival *in utero* and survival until weaning were reduced.

Gavage of male Long-Evans rats with trichloroethylene in corn oil in doses of 0, 10, 100 or 1000 mg/kg body weight on 5 days weekly for 6 weeks did not produce toxic effects on the sperm. During the first week of treatment in the 1000 mg/kg group, adverse effects on mating behaviour were seen, possibly as a result of the depressive effects of trichloroethylene on the central nervous system (Zenick *et al.* 1984).

Likewise, in female Long-Evans rats which were given trichloroethylene doses between 0 and 1000 mg/kg body weight daily for 2 weeks before mating, during mating and during the entire gestation period, no adverse effects on reproduction were detected (Manson *et al.* 1984). In the highest dose group the treatment resulted in general toxicity with reduced weight gains and increased mortality of the dams and reduced survival of the new-born pups.

In contrast to the studies described above in which neither inhalation of trichloroethylene nor oral administration of the substance produced teratogenic effects in the various animal species, in one study the administration of trichloroethylene in the drinking water to Sprague-Dawley rats resulted in cardiac malformations in the offspring (Dawson *et al.* 1993). Trichloroethylene was administered in the drinking water (1.5 or 1100 ml/m$^3$ water) for 2 months before mating and during the whole gestation period or only before mating or only during gestation. Signs of maternal or foetal toxicity were not seen in any of the treated groups. On the other hand, in the progeny of the animals exposed before and during gestation there was a significant increase in the incidence of cardiac malformations: in 8.2 % of the foetuses of dams given trichloroethylene at a concentration of 1.5 ml/m$^3$ in the drinking water, various malformations of the heart and large vessels were seen; in the 1100 ml/m$^3$ group the incidence was 9.2 % whereas in the untreated control group it was only 3 %. In the high concentration group the incidence of heart malformations was also increased (10.5 %) when trichloroethylene was administered only during gestation. The other groups were not different from the controls. Although the microscopic methods and the nature of the malformations were described very precisely in the publication, many important questions remain unanswered. Firstly, the doses were not given and cannot be calculated exactly because there are no data for the body weights of the dams. Secondly, the daily water consumption, calculated from the data given in the publication for the total amount of trichloroethylene ingested per animal, varies widely from group to group, from 14 to 160 ml. Thirdly, the incidence of cardiac malformations is not in accordance with the absence of any foetotoxicity. The viability of the new-born pups was not investigated. Finally, this study is alone in the literature in reporting such a finding.

## 4.5 Genotoxicity

### 4.5.1 In vitro

Many of the early genotoxicity studies tested technical grade trichloroethylene products which were stabilized with epoxides. Since then the genotoxic effects of the oxiranes which were used, 1,2-epoxybutane and epichlorohydrin, have been demonstrated unequivocally *in vitro* and *in vivo* (Henschler *et al.* 1977, McGregor *et al.* 1989). The present discussion does not include studies in which technical grade trichloroethylene was tested or those in which the purity of the substance was not specified.

In the bacterial test systems with *Salmonella typhimurium* TA98, TA100, TA1535, TA1537 (Baden *et al.* 1979, NTP 1988, Waskell 1978) and *Escherichia coli* K12 (Greim *et al.* 1975) purified trichloroethylene was not mutagenic without metabolic activation. In the presence of liver homogenates prepared from rats or mice pretreated with Aroclor or phenobarbital, purified trichloroethylene tested in closed systems demonstrated in some studies a just detectable mutagenicity (doubling of the spontaneous revertant frequency) in bacteria (Baden *et al.* 1979, Bartsch *et al.* 1979, Greim *et al.* 1975, Simmon *et al.* 1977). In more recent studies, however, no mutagenic effects of purified trichloroethylene in the presence of S9 mix from liver have been found (McGregor *et al.* 1989, NTP 1988, Shimada *et al.* 1985).

Studies of the induction of gene mutations in yeast (*Saccharomyces cerevisiae*) and fungi (*Aspergillus nidulans*) yielded similar results: not significant to weak mutagenicity in the presence of liver S9 mix (Bronzetti *et al.* 1978, Callen *et al.* 1980, Crebelli *et al.* 1985, Crebelli and Carere 1989, Shahin and von Borstel 1977).

In *A. nidulans* trichloroethylene induced unequivocal aneuploidy (Crebelli *et al.* 1985) whereas in CHO cells (a cell line derived from Chinese hamster ovary cells) it did not induce chromosomal aberrations in the presence or absence of a metabolic activation system (NTP 1988).

In freshly isolated hepatocytes from rats which had not been pretreated, trichloroethylene did not induce UDS (unscheduled DNA synthesis) (Shimada *et al.* 1985). Other authors have reported the induction of UDS in hepatocytes isolated from rats which had been pretreated with liver enzyme inducers (Costa and Ivanetich 1984). The method used in this study for detecting UDS, however, yields only qualitative information and is not adequately validated. In human lymphocytes in the presence of S9 mix, trichloroethylene induced DNA repair but only with concentrations which also produced severe general cytotoxicity (Perocco and Prodi 1981).

In studies with L5178Y mouse lymphoma cells (NTP 1988) trichloroethylene induced a doubling of the control mutation rate only after metabolic activation with liver S9 mix from male rats pretreated with Aroclor. The weak mutagenic effect was not dose-dependent and was observed only with the highest trichloroethylene concentration which also had marked adverse effects on cell growth (reduction to 15 % to 20 % of the control value).

Trichloroethylene yielded weak positive results in the morphological cell transformation test with embryo cells from the Syrian hamster (Price *et al.* 1978).

A critical assessment of the numerous *in vitro* genotoxicity tests which have been carried out demonstrates that in some studies trichloroethylene shows a weak mutagenic effect (at most a doubling of the control values) only at high concentrations which are also cytotoxic.


### 4.5.2 *In vitro* genotoxicity of the metabolites

The toxic and carcinogenic effects observed *in vivo* with trichloroethylene are not induced by the substance itself but by the metabolites formed by the oxidative and glutathione-dependent bioactivation in the organism. The complex transport and bioactivation cascades in the organism of the experimental animal can often not be simulated in the *in vitro* test systems. Therefore the mutagenicity of the main metabolites (prepared by chemical synthesis) must be investigated separately.

Trichloroethanol and trichloroacetic acid were not mutagenic in various strains of *S. typhimurium* with or without addition of a metabolic activation system (Bignami *et al.* 1980, Moriya *et al.* 1983). Trichloroethanol induced weak gene mutation in *A. nidulans* and was a potent inducer of chromosomal nondisjunction in the same system. Trichloroacetic and dichloroacetic acids, on the other hand, did not cause an increase in DNA single strand breaks in cultured primary rat hepatocytes (Bignami *et al.* 1980).

Unlike the metabolites of the oxidative metabolism of trichloroethylene, 1,2-dichlorovinylcysteine was highly mutagenic in *S. typhimurium* both in the presence and absence of a metabolic activation system (Dekant *et al.* 1986b, Green and Odum 1985). 2,2-Dichlorovinylcysteine and its mercapturic acid proved to be less genotoxic in *S. typhimurium* than the corresponding 1,2-isomers (Commandeur *et al.* 1991). Other studies of the genotoxicity of the 2,2-isomers are not available at present. Studies of the cleavage of 1,2-dichlorovinylcysteine in bacterial cell homogenates revealed high $\beta$-lyase activities in the *S. typhimurium* strains. The mercapturic acid, 1,2-dichlorovinyl-*N*-acetylcysteine, was also mutagenic in *S. typhimurium* in the presence of rat kidney cytosol which contains high deacetylase activity. Inhibition of bacterial $\beta$-lyase inhibited the mutagenicity of 1,2-dichlorovinylcysteine and that of its mercapturic acid (Vamvakas *et al.* 1987). In the presence of rat kidney S9 mix with high $\gamma$-glutamyltranspeptidase activity, 1,2-dichlorovinylglutathione was also a potent mutagen in *S. typhimurium* (Vamvakas *et al.* 1988a). The mutagenicity of the glutathione conjugate could be prevented by inhibition of the $\gamma$-glutamyltranspeptidase or the $\beta$-lyase.

These results have demonstrated clearly that the glutathione-dependent bioactivation of trichloroethylene produces metabolites with genotoxic effects *in vitro*. It must be admitted, however, that the hepatorenal glutathione-dependent activation of trichloroethylene has not yet been simulated in *in vitro* test systems as it has for the structurally related nephrocarcinogenic haloalkenes, hexachlorobutadiene and tetrachloroethylene, which are also bioactivated via the mercapturic acid metabolic pathway (Vamvakas *et al.* 1988b, 1989a). The reason for this is that trichloroethylene is in comparison a poor substrate for the glutathione transferases so that the yield of the glutathione conjugate and therefore of 1,2-dichlorovinylcysteine *in vitro* is not sufficient for the induction of a detectable mutagenic effect.

In cultured kidney cells (LLC-PK$_1$) 1,2-dichlorovinylcysteine induced a slight increase in DNA repair (Vamvakas *et al.* 1989b) and also calcium efflux from the mitochondria which led to activation of calcium-dependent endonucleases. A result of the activation of this cascade was the induction of DNA double strand breaks and subsequent poly(ADP-ribosyl)ation of nuclear proteins. 1,2-Dichlorovinylcysteine also induced the expression of c-fos and c-myc in LLC-PK$_1$ cells (Vamvakas *et al.* 1990, 1992, 1993b).

### 4.5.3 *In vivo*

Oral administration of purified trichloroethylene in doses of 375 to 3000 mg/kg body weight to CD-1 and B6C3F$_1$ mice led to a significant and dose-dependent increase in the incidence of micronuclei in bone marrow cells (Duprat and Gradiski 1981; Sbrana *et al.* 1985). On the other hand, pure trichloroethylene did not induce an increase in structural chromosomal aberrations in the bone marrow cells of B6C3F$_1$ mice either after oral administration (once, 1200 mg/kg body weight) or after inhalation (600 ml/m$^3$, 7 hours/day, 5 days/week for 10 weeks) (Sbrana *et al.* 1985).

In another study, male C57BL/6 mice and CD rats were exposed to trichloroethylene concentrations of 0, 5, 500 or 5000 ml/m$^3$ for 6 hours, and then 18 hours later tissue samples were taken for cytogenetic examination (Kligerman *et al.* 1994). The tests for induction of sister chromatid exchange, chromosomal aberrations and micronuclei in peripheral lymphocytes from rat blood and splenocytes from the mouse yielded negative results. The frequency of micronuclei in polychromatic bone marrow erythrocytes was also determined. In the mouse the results were negative but in the rat there was a concentration-dependent increase which reached four times the control value at 5000 ml/m$^3$.

This concentration was, however, also cytotoxic which was seen in the reduction of the percentage of polychromatic bone marrow erythrocytes relative to the normochromatic erythrocytes, from 60 % in the untreated animals to about 35 % in the exposed group. In the same study, rats were also exposed to trichloroethylene concentrations of 0, 5, 50 and 500 ml/m$^3$, 6 hours daily on four consecutive days. This repeated exposure did not lead to a significant increase in micronucleus frequency relative to that in untreated animals. The result after repeated inhalation is, however, dubious as the micronucleus frequency in the control animals was unusually high.

In a host-mediated assay, commercially available trichloroethylene of unspecified purity yielded positive results (Bronzetti *et al.* 1978) but purified, epoxide-free trichloroethylene negative results (Rossi *et al.* 1983). Trichloroethylene (131.4 mg/kg body weight, i.p.) produced positive results in the coat colour spot test in mice (Fahrig 1977); whether or not epoxide stabilizers were present in the preparation was not reported. In contrast, no mutagenic effects were seen in the dominant lethal test after administration of trichloroethylene concentrations of 50 to 450 ml/m$^3$ by inhalation (Slacic-Erben *et al.* 1980). Administration of trichloroethylene by gavage to B6C3F$_1$ and CD-1 mice and to F344 rats did not cause induction of DNA repair in the liver (Doolittle *et al.* 1987, Mirsalis *et al.* 1989). However, the rate of DNA synthesis was increased in the two strains of mice but not in the F344 rats. This is in line with the species-specific hepatocarcinogenicity of trichloroethylene in the mouse.

Epoxide-free trichloroethylene (528 to 1310 mg/kg body weight, i.p.) caused significant induction of DNA single strand breaks in the liver and kidneys of the mouse whereas effects in the lung were slight or not detectable (Walles 1986). DNA single strand breaks were detected 1, 2 and 4 but not 8 hours after oral administration of epoxide-free trichloroethylene (92 to 3930 mg/kg body weight) to rats and mice; the effects in the mouse liver were clearly more marked than those in the rat liver (Nelson and Bull 1988). No DNA single strand breaks could be detected in the rat liver 6 hours after oral administration of a trichloroethylene dose of 2000 mg/kg body weight (Parchman and Magee 1982). In this study no assays were carried out at shorter intervals after administration; this could be the reason for the apparently contradictory results of the two studies.

Thus purified trichloroethylene induces *in vivo* micronuclei and DNA single strand breaks, effects which can also be produced by cytotoxic processes such as the secondary formation of reactive oxygen species. Purified trichloroethylene yielded negative results in systems which detect gene mutations (host-mediated assay). In one study (Nelson and Bull 1988) the induction of DNA single strand breaks in the rat liver after administration of trichloroacetic acid or dichloroacetic acid was observed whereas other similar studies yielded negative results (Parchman and Magee 1982).

Intraperitoneal administration of 1,2-dichlorovinylcysteine resulted in a marked increase in DNA single strand breaks in the rabbit renal cortex and DNA double strand breaks in the rat renal cortex (Jaffe *et al.* 1985, McLaren *et al.* 1994).

## 4.5.4 Binding to protein and DNA

Metabolites of the oxidative and the glutathione-dependent bioactivation of trichloroethylene can readily bind to proteins both *in vitro* and *in vivo* (Allemand *et al.* 1978, Banerjee and Van Duuren 1978, Bolt and Filser 1977, Forkert and Birch 1989, Miller and Guengerich 1983, Uehleke and Poplawski-Tabarelli 1977, Vamvakas *et al.* 1987, Van Duuren and Banerjee 1976). In a more recent study it could be shown that after administration of $^{14}$C-trichloroethylene to rats and mice 30 % to 75 % of the radioactivity measured in the proteins was derived from metabolic incorporation of $^{14}$C into the amino acid pool and thus into protein synthesis; the residual 25 % to 70 % of the measured radioactivity could be ascribed to covalent binding of trichloroethylene metabolites to proteins (Stevens *et al.* 1992).

The binding of radioactivity from $^{14}$C-trichloroethylene to protein in the liver and kidneys of male F344 rats and B6C3F$_1$ mice was studied after oral administration of trichloroethylene (20, 45, 100, 400 and 1000 mg/kg body weight) and of the oxidative metabolites trichloroacetate (20 mg/kg body weight) and dichloroacetate (5 mg/kg body weight) and the glutathione-dependent metabolite dichlorovinylcysteine (5 and 25 mg/kg body weight) (Eyre *et al.* 1995a, 1995b). After the trichloroethylene treatment the total protein binding in the liver of the rats was comparable with that in the mice whereas the total protein binding in the rat kidneys was 1.7 times that in the mouse kidneys. In this study it was also attempted to quantify the protein binding of oxidative and reductive metabolites separately. It was shown that after administration of a trichloroethylene dose of 1000 mg/kg body weight, the protein binding associated with dichlorovinylcysteine in the mouse kidney was twice that found in the rat kidney. Likewise, after administration of a dichlorovinylcysteine dose of 25 mg/kg body weight the protein binding in the mouse kidney was about 12 times that in the rat kidney. Further studies from the same laboratory showed that after oral administration of a trichloroethylene dose of 1000 mg/kg body weight, the maximum level of dichlorovinylcysteine which accumulated in the rat kidney was 6 times that which accumulated (or was formed) in the mouse kidney. However, since at the same time the mouse kidney bioactivated 12 times as much dichlorovinylcysteine as did the rat kidney, the total glutathione-dependent toxification of trichloroethylene in the mouse kidney was twice that in the rat kidney. Tested for the induction of cell proliferation after oral administration of dichlorovinylcysteine, the mouse kidney also proved to be more sensitive than the rat kidney: a dichlorovinylcysteine dose of 25 mg/kg body weight was required to produce a significant increase in cell proliferation in the rat kidney whereas a dose of 1 mg/kg body weight was sufficient to do the same in the mouse kidney. It must be pointed out, however, that the oral administration of a bolus dose of dichlorovinylcysteine in no way imitates the kinetics of formation of this metabolite during long-term inhalation of trichloroethylene and so results obtained by this method can have only limited relevance for assessing the chronic toxicity and carcinogenicity of the parent substance.

In the presence of liver microsomes or rat hepatocytes, marked covalent binding of trichloroethylene to calf thymus DNA was demonstrated (Banerjee and Van Duuren 1978, Bergman 1983, DiRenzo *et al.* 1982, Mazzullo *et al.* 1992, Miller and Guengerich 1983). *In vivo*, on the other hand, covalent binding of trichloroethylene metabolites to

DNA has not been demonstrated unequivocally. After administration of [14]C-trichloro-ethylene in isolated studies, low levels of radioactivity were detected in the DNA in various organs, e.g., liver, lungs, kidneys, spleen, pancreas, testes and brain. In most cases, however, the radioactivity originated either from protein impurities or from the incorporation of C-fragments from the metabolic breakdown of radioactively labelled trichloroethylene into guanine and cytosine and not from DNA adducts with trichloroethylene metabolites (Bergman 1983, Mazzullo *et al.* 1992, Stott *et al.* 1982).

## 4.6 Carcinogenicity

Epichlorohydrin, one of the epoxides used frequently in the past to stabilize trichloroethylene, induced sarcomas in mice after subcutaneous injection (Van Duuren *et al.* 1974) and nasal carcinomas in rats after adminstration by inhalation (Simmon 1977). In addition, forestomach tumours were induced in male and female Swiss mice after oral administration of epichlorohydrin or of epoxybutane, which was also often used as a stabilizer (Henschler *et al.* 1984). In the study by Henschler *et al.* (1984) not only trichloroethylene stabilized with epoxides but also epoxide-free trichloroethylene was tested. Because of marked toxicity the initial dose of 2400 mg/kg body weight had to be halved from week 40 and administration discontinued in week 68 (instead of week 104). The pathological examination after 106 weeks revealed no evidence of carcinogenicity in the mice treated with epoxide-free trichloroethylene. In spite of its methodic shortcomings, this comparative study produced important evidence as to the activity of the epoxide stabilizers in the carcinogenicity studies with technical grade trichloroethylene products.

A review of the long-term carcinogenicity studies carried out with trichloroethylene, giving the purity of the trichloroethylene products tested, is shown in Table 5.

In an inhalation study, male and female Wistar rats, NMRI mice and Syrian hamsters were exposed to purified trichloroethylene in concentrations of 0, 100 and 500 ml/m$^3$. With the exception of a significant but not concentration-dependent increase in the incidence of malignant lymphomas in the female NMRI mice, no evidence of carcinogenicity was found. As the incidence of lymphomas was also high in the control mice, it was suggested that this result could be a strain-specific finding (Henschler *et al.* 1980).

With B6C3F$_1$ and Swiss mice and Sprague-Dawley rats, long-term inhalation studies were carried out with epoxide-free trichloroethylene in concentrations of 100, 300 and 600 ml/m$^3$ (Maltoni *et al.* 1988). Survival and body weight development in the treated groups were not significantly different from the control values. After an exposure period of 78 weeks and a total observation period of 154 weeks, in the male Swiss mice there was a significant increase in the incidence of hepatomas in the high concentration group and of lung tumours in the high and medium concentration groups. In the high concentration group the number of lung tumours was also significantly increased in the female B6C3F$_1$ mice. The total number of tumours was increased significantly in the female animals of all exposed groups. After an exposure period of 104 weeks and a total observation period of 150 weeks, a significant concentration-dependent increase in testis tumours (Leydig cell tumours) and a not significant and not concentration-dependent increase in lymphosarcomas was seen in male Sprague-Dawley rats. Exposure to a trichloroethylene concentration of 600 ml/m$^3$ also produced a slight, statistically not significant increase in the incidence of adenocarcinomas of the renal tubules in male Sprague-Dawley rats (control 0/135, 600 ml/m$^3$ 4/130). The authors considered this result to be "borderline evidence". In 77 % of the male rats of the 600 ml/m$^3$ group and in 16.9 % of the 300 ml/m$^3$ group, cytokaryomegaly was detected in the proximal renal tubules. The lesion, considered to be a precursor of the renal cell tumour, was found neither in the control group nor in the treated female rats. In addition an adenocarcinoma of the renal tubules was found in one of the female animals of the 600 ml/m$^3$ group (control 0/145, 600 ml/m$^3$ 1/130).

In another long-term study, trichloroethylene (dissolved in corn oil) was administered to male and female F344/N rats and B6C3F$_1$ mice for 103 weeks. The animals were killed in weeks 103 to 107. The rats were given trichloroethylene doses of 500 or 1000 mg/kg body weight, the mice 1000 mg/kg body weight (NTP 1990). The survival of the male rats of both dose groups and that of the treated male mice was significantly lower than that of the control animals. The average body weights of the treated rats of both sexes and that of the treated male mice were lower than that of the controls; no weight differences were detected in the female mice. Toxic nephropathy (cytomegaly) was diagnosed in more than 95 % of the treated rats and mice. The lesions were most severe in the male rats of the high dose group; none of the control animals developed kidney damage. Adenocarcinomas of the renal tubules were detected in 3 male rats of the high dose group, all in animals which survived until the end of the study. The numbers of tubule adenocarcinomas at the end of the study were 0/33 in the control group, 0/20 in the low dose group and 3/16 (19 %) in the high dose group. Calculated on the basis of the number of surviving animals, the increase in incidence was thus significant. Renal cell tumours are very rare in F344/N rats; the incidence in historical controls is 3/748 (0.4 %). In addition, in the 500 mg/kg group of male rats one renal pelvis carcinoma and two tubule adenomas were detected and in the 1000 mg/kg body weight group one renal pelvis carcinoma. In the control groups there were no renal cell carcinomas or renal pelvis carcinomas; only one benign renal pelvis papilloma was detected in a control animal. In the females of the 1000 mg/kg group one renal cell carcinoma was detected. The kidney tumours in the male rats were considered in the final report to be an equivocal finding because the survival in the two treatment groups was significantly reduced: 40 % in the 500 mg/kg group, 32 % in the 1000 mg/kg group compared with 70 % in the control group. Thus the doses selected were clearly higher than the MTD. In the same study trichloroethylene produced an increase in the incidence of hepatocellular carcinomas in male (control 8/48, 1000 mg/kg 31/50) and female (control 2/48, 1000 mg/kg 13/49) mice. The incidence of hepatocellular adenomas was also increased in males (control 7/48, 1000 mg/kg 14/50) and females (control 4/48, 1000 mg/kg 16/49).

**Table 5.** Carcinogenicity studies with trichloroethylene

| Author: | Henschler *et al.* 1980 |
| Substance: | trichloroethylene, stabilized with triethanolamine |
| Species: | mouse (NMRI), 30 ♂, 30 ♀ |
| Administration: | inhalation in whole animal exposure chamber |
| Concentration: | 100, 500 ml/m$^3$ |
| Duration: | 18 months, 6 h/day, 5 days/week, necropsy after 30 months |
| Toxicity: | survival of both sexes decreased in a dose-dependent manner, renal toxicity not specified |

| | | 0 | 100 ml/m$^3$ | 500 ml/m$^3$ |
|---|---|---|---|---|
| **Tumours:** | | | | |
| malignant lymphoma | ♂ | 7/30 (23 %) | 7/29 (24 %) | 6/30 (20 %) |
| | ♀ | 9/29 (31 %) | 17/30 (57 %) | 18/28 (64 %) |
| kidney | | | | |
| cystadenoma | ♂ | 4/30 | 1/29 | 1/30 |
| | ♀ | 0/29 | 0/30 | 0/28 |
| adenoma | ♂ | 0/30 | 1/29 | 0/30 |
| | ♀ | 0/29 | 0/30 | 0/28 |
| adenocarcinoma | ♂ | 1/30 | 0/29 | 0/30 |
| | ♀ | 0/29 | 0/30 | 0/28 |

| Author: | Henschler *et al.* 1980 |
| Substance: | trichloroethylene, stabilized with triethanolamine |
| Species: | hamster (Syrian), 30 ♂, 30 ♀ |
| Administration: | inhalation in animal exposure chamber |
| Concentration: | 100, 500 ml/m$^3$ |
| Duration: | 18 months, 6 h/day, 5 days/week, necropsy after 30 months |
| Toxicity: | no effects on survival |
| Tumours: | no significant increase in tumour incidence |

| Author: | Henschler *et al.* 1980 |
| Substance: | trichloroethylene, stabilized with triethanolamine |
| Species: | rat (Wistar), 30 ♂, 30 ♀ |
| Administration: | inhalation in animal exposure chamber |
| Concentration: | 100, 500 ml/m$^3$ |
| Duration: | 18 months, 6 h/day, 5 days/week, necropsy after 36 months |
| Toxicity: | no effects on survival |
| Tumours: | no significant increases in tumour incidence, renal toxicity not specified |

| | | 0 | 100 ml/m$^3$ | 500 ml/m$^3$ |
|---|---|---|---|---|
| **Tumours, kidney:** | | | | |
| cystadenoma | ♂ | 2/29 | 0/30 | 1/30 |
| | ♀ | 0/28 | 0/30 | 0/30 |
| adenoma | ♂ | 0/29 | 1/30 | 1/30 |
| | ♀ | 0/28 | 0/30 | 0/30 |
| angiosarcoma | ♂ | 1/29 | 0/30 | 0/30 |
| | ♀ | 0/28 | 0/30 | 0/30 |
| adenocarcinoma | ♂ | 0/29 | 0/30 | 1/30 |
| | ♀ | 0/28 | 0/30 | 0/30 |

**Table 5.** continued

| Author: | Fukuda *et al.* 1983 |
| Substance: | trichloroethylene, stabilized with 0.019 % epichlorohydrin |
| Species: | mouse (ICR), 49–50 ♀ |
| Administration: | inhalation in animal exposure chamber |
| Concentration: | 50, 150, 450 ml/m$^3$ |
| Duration: | 104 weeks, 7 h/day, 5 days/week, necropsy after 107 weeks |
| Toxicity: | no effects on survival |

| | 0 | 50 ml/m$^3$ | 150 ml/m$^3$ | 450 ml/m$^3$ |
|---|---|---|---|---|
| **Tumours:** | | | | |
| lung adenocarcinoma | 1/49 (2 %) | 3/50 (6 %) | 8/50 (16 %)[a] | 7/46 (15 %)[a] |

[a] $p < 0.05$ (Fisher exact test)

| Author: | Fukuda *et al.* 1983 |
| Substance: | trichloroethylene, stabilized with 0.019 % epichlorohydrin |
| Species: | rat (Sprague-Dawley), 49–51 ♀ |
| Administration: | inhalation in animal exposure chamber |
| Concentration: | 50, 150, 450 ml/m$^3$ |
| Duration: | 104 weeks, 7 h/day, 5 days/week, necropsy after 107 weeks |
| Toxicity: | survival of the control animals significantly less than that of the exposed animals |
| Tumours: | none |

| Author: | Maltoni *et al.* 1988 |
| Substance: | trichloroethylene, epoxide-free, stabilized with butylhydroxytoluene |
| Species: | rat (Sprague-Dawley), 130 ♂, 130 ♀, controls 135 ♂ and 135 ♀ |
| Administration: | inhalation in animal exposure chamber |
| Concentration: | 100, 300, 600 ml/m$^3$ |
| Duration: | 104 weeks, 5 days/week, 7 h/day, observation until death |
| Toxicity: | survival and body weight development not different from control values |

| | | 0 | 100 ml/m$^3$ | 300 ml/m$^3$ | 600 ml/m$^3$ |
|---|---|---|---|---|---|
| **Findings** | | | | | |
| testis tumours (Leydig cell tumours) | ♂ | 4.4 % | 12.3 % | 23.1 % | 23.8 % |
| lymphosarcoma | ♂ | 0.7 % | 3.8 % | 3.1 % | 1.5 % |
| | ♀ | 0 % | 3.1 % | 0.8 % | 0.8 % |
| cytokaryomegaly in | ♂ | 0 % | 0 % | 16.9 % | 77 % |
| proximal tubules | ♀ | 0 % | 0 % | 0 % | 0 % |
| tubule adenocarcinoma | ♂ | 0 % | 0 % | 0 % | 3.1 % |
| | ♀ | 0 % | 0 % | 0 % | 0.7 % |

**Table 5.** continued

| Author: | Maltoni *et al.* 1988 |
| --- | --- |
| Substance: | trichloroethylene, epoxide-free, stabilized with butylhydroxytoluene |
| Species: | mouse (Swiss and B6C3F$_1$), 90 ♂, 90 ♀ |
| Administration: | inhalation in animal exposure chamber |
| Concentration: | 100, 300, 600 ml/m$^3$ |
| Duration: | 78 weeks, 5 days/week, 7 h/day, observation until death |
| Toxicity: | survival and body weight development not different from control values |

| | | 0 | 100 ml/m$^3$ | 300 ml/m$^3$ | 600 ml/m$^3$ |
| --- | --- | --- | --- | --- | --- |
| **Findings** | | | | | |
| **B6C3F$_1$ mice** | | | | | |
| hepatoma | ♂ | 1.1 % | 1.1 % | 3.3 % | 6.7 % |
| | ♀ | 3.3 % | 4.4 % | 4.4 % | 10.0 %[a] |
| lung tumours | ♂ | 2.2 % | 2.2 % | 3.3 % | 1.1 % |
| | ♀ | 4.4 % | 6.7 % | 7.8 % | 16.7 %[b] |
| **Swiss mice** | | | | | |
| hepatoma | ♂ | 4.4 % | 2.2 % | 8.9 % | 14.4 %[b] |
| | ♀ | 0 % | 0 % | 0 % | 1.1 % |
| lung tumours | ♂ | 11.1 % | 12.2 % | 25.5 %[b] | 30.0 %[c] |
| | ♀ | 16.7 % | 16.7 % | 14.4 % | 22.2 % |

[a] $p < 0.01$ for ♂ and ♀ together
[b] $p < 0.05$
[c] $p < 0.01$

| Author: | NCI 1976 |
| --- | --- |
| Substance: | trichloroethylene, stabilized with epichlorohydrin and 1,2-epoxybutane |
| Species: | rat (Osborne-Mendel), 50 ♂, 50 ♀ |
| Administration: | gavage, corn oil vehicle |
| Dose: | 550, 1110 mg/kg body weight |
| Duration: | 78 weeks, 5 days/week, necropsy after 110 weeks |
| Toxicity: | survival significantly decreased |

| | | 0 | | 550 mg/kg body weight | | 1110 mg/kg body weight | |
| --- | --- | --- | --- | --- | --- | --- | --- |
| survivors after | ♂ | 3/20 | (15 %) | 8/50 | (16 %) | 3/50 | (6 %) |
| 110 weeks | ♀ | 8/20 | (40 %) | 7/48 | (15 %) | 13/50 | (26 %) |
| overall tumour incidence | ♂ | 5/25 | (20 %) | 7/50 | (14 %) | 5/50 | (10 %) |
| | ♀ | 7/20 | (35 %) | 2/48 | (4 %) | 12/50 | (24 %) |

**Table 5.** continued

| Author: | NCI 1976 |
|---|---|
| Substance: | trichloroethylene, stabilized with epichlorohydrin and 1,2-epoxybutane |
| Species: | mouse (B6C3F$_1$), 50 ♂, 50 ♀ |
| Administration: | gavage, corn oil vehicle |
| Dose: | ♂: 1170, 2340 mg/kg body weight, |
|  | ♀: 870, 1740 mg/kg body weight |
| Duration: | 78 weeks, 5 days/week, necropsy after 90 weeks |
| Toxicity: | survival of exposed ♀ significantly decreased |

|  |  | 0 | 1170/870 mg/kg body weight | 2340/1740 mg/kg body weight |
|---|---|---|---|---|
| survivors after | ♂ | 8/20 (40 %) | 36/50 (72 %) | 22/50 (44 %) |
| 90 weeks | ♀ | 20/20 (100 %) | 42/50 (84 %) | 39/47 (83 %) |
| hepatocellular carcinoma | ♂ | 1/20 (5 %) | 26/50 (52 %)[a] | 31/48 (65 %)[a] |
|  | ♀ | 0/20 (0 %) | 4/50 (8 %)[a] | 11/47 (23 %)[a] |

[a] p < 0.01

| Author: | Henschler *et al.* 1984 |
|---|---|
| Substance: | trichloroethylene, stabilized with triethanolamine or epichlorohydrin or 1,2-epoxybutane or epichlorohydrin and 1,2-epoxybutane |
| Species: | mouse (ICR/Ha Swiss), 50 ♂, 50 ♀ |
| Administration: | gavage, corn oil vehicle |
| Dose: | weeks 1–35: 2400 mg/kg body weight, |
|  | weeks 35–40: no treatment, |
|  | weeks 41–68: 1200 mg/kg body weight |
| Duration: | 68 weeks, 5 days/week, necropsy after 90 weeks |
| Toxicity: | survival of exposed animals decreased, because of toxicity, administration was discontinued during the study and the dose reduced |
| Tumours: | trichloroethylene stablized with triethanolamine: no significant increases in tumour incidence; trichloroethylene with all other stabilizers: forestomach carcinomas in ♂ and ♀ animals significantly increased |

|  |  | 0 | trichloroethylene stablized with triethanolamine |
|---|---|---|---|
| Tumours, kidney |  |  |  |
| cystadenoma | ♂ | 1/50 | 1/50 |
|  | ♀ | 0/50 | 4/50 |
| adenocarcinoma | ♂ | 0/50 | 1/50 |
|  | ♀ | 0/50 | 0/50 |
| carcinosarcoma | ♂ | 1/50 | 0/50 |
|  | ♀ | 1/50 | 0/50 |

**Table 5.** continued

| Author: | NTP 1988 |
|---|---|
| Substance: | trichloroethylene, stabilized with amine base |
| Species: | rat (ACI, August, Marshall, Osborne-Mendel), of each strain 50 ♂, 50 ♀ per group |
| Administration: | gavage, corn oil vehicle |
| Dose: | 500, 1000 mg/kg body weight |
| Duration: | 103 weeks, 5 days/week, necropsy after 103–107 weeks |
| Toxicity: | toxic nephropathy and central nervous system toxicity, reduced survival, high incidence of accidental deaths |

|  |  | 0 (untreated) | | 0 (vehicle control) | | 500 mg/kg body weight | 1000 mg/kg body weight |
|---|---|---|---|---|---|---|---|
| **ACI rats** | | | | | | | |
| survivors after | ♂ | 39/50 | (78 %) | 38/50 | (76 %) | 19/50[a] (38 %) | 11/50[a] (22 %) |
| 103 weeks | ♀ | 37/50 | (74 %) | 35/50 | (70 %) | 20/50 (40 %) | 19/50[a] (38 %) |
| accidental deaths | ♂ | 0 | · | 0 | | 11 | 18 |
| | ♀ | 0 | | 2 | | 14 | 12 |
| Findings: | | | | | | | |
| renal cytomegaly | ♂ | | | 0/50 | | 40/49 (82 %) | 48/49 (98 %) |
| | ♀ | | | 0/48 | | 43/47 (91 %) | 42/43 (98 %) |
| toxic nephropathy | ♂ | | | 0/50 | | 18/49 (37 %) | 18/49 (37 %) |
| | ♀ | | | 0/48 | | 21/47 (45 %) | 19/43 (44 %) |
| tubule adenoma | ♂ | 0/49 | | 0/50 | | 0/49 | 0/49 |
| | ♀ | 0/49 | | 0/48 | | 2/47 (4 %) | 0/43 |
| tubule adenocarcinoma | ♂ | 0/49 | | 0/50 | | 1/49 (2 %) | 0/49 |
| | ♀ | 0/49 | | 0/50 | | 1/47 (2 %) | 1/43 (2 %) |
| **August rats** | | | | | | | |
| survivors after | ♂ | 24/50 | (48 %) | 21/50 | (42 %) | 13/50 (26 %) | 16/50 (32 %) |
| 104 weeks | ♀ | 26/50 | (52 %) | 23/50 | (46 %) | 26/50 (52 %) | 25/50 (50 %) |
| accidental deaths | ♂ | 1 | | 6 | | 12 | 11 |
| | ♀ | 0 | | 1 | | _6 | 13 |
| Findings: | | | | | | | |
| renal cytomegaly | ♂ | | | 0/50 | | 46/50 (92 %) | 46/49 (94 %) |
| | ♀ | | | 0/49 | | 46/48 (96 %) | 50/50 (100 %) |
| toxic nephropathy | ♂ | | | 0/50 | | 10/50 (20 %) | 31/49 (63 %) |
| | ♀ | | | 0/49 | | 8/48 (17 %) | 29/50 (58 %) |
| tubule adenoma | ♂ | 0/50 | | 0/50 | | 1/50 (2 %) | 1/49 (2 %) |
| | ♀ | 0/50 | | 1/49 | (2 %) | 2/48 (4 %) | 0/50 |
| tubule adenocarcinoma | ♂ | 0/50 | | 0/50 | | 1/50 (2 %) | 0/49 |
| | ♀ | 0/50 | | 0/49 | | 2/48 (4 %) | 0/50 |
| **Marshall rats** | | | | | | | |
| survivors after | ♂ | 32/50 | (64 %) | 26/50 | (52 %) | 12/50[a] (24 %) | 7/50 (14 %) |
| 102–103 weeks | ♀ | 31/50 | (62 %) | 30/50 | (60 %) | 13/50[a] (26 %) | 9/50[a] (18 %) |
| accidental deaths | ♂ | 1 | | 2 | | 12 | 25 |
| | ♀ | 1 | | 3 | | 14 | 18 |
| Findings: | | | | | | | |
| renal cytomegaly | ♂ | | | 0/49 | | 48/50 (96 %) | 47/47 (100 %) |
| | ♀ | | | 0/50 | | 46/48 (96 %) | 43/44 (98 %) |
| toxic nephropathy | ♂ | | | 0/49 | | 18/50 (36 %) | 23/47 (49 %) |
| | ♀ | | | 0/50 | | 30/48 (63 %) | 30/44 (68 %) |

**Table 5.** continued

| | | | | | | | | |
|---|---|---|---|---|---|---|---|---|
| tubule adenoma | ♂ | 2/49 | (4 %) | 0/49 | | 1/50 | (2 %) | 0/47 |
| | ♀ | 1/49 | (2 %) | 1/50 | (2 %) | 1/48 | (2 %) | 0/44 |
| tubule adenocarcinoma | ♂ | 0/49 | | 0/49 | | 0/50 | | 1/47 (2 %) |
| | ♀ | 0/49 | | 0/50 | | 1/48 | (2 %) | 1/44 (2 %) |
| testis tumours | ♂ | 16/46 | (35 %) | 17/46 | (37 %) | 21/48 | (44 %) | 32/48 (65 %) |
| **Osborne-Mendel rats** | | | | | | | | |
| survivors after | ♂ | 21/50 | (42 %) | 22/50 | (44 %) | 17/50 | (34 %) | 14/50 (28 %) |
| 104 weeks | ♀ | 19/50 | (38 %) | 18/50 | (36 %) | 10/50 | (20 %) | 7/50[a] (14 %) |
| accidental deaths | ♂ | 0 | | 1 | | 6 | | 7 |
| | ♀ | 0 | | 8 | | 6 | | 6 |
| Findings: | | | | | | | | |
| renal cytomegaly | ♂ | | | 0/50 | | 48/50 | (96 %) | 49/50 (98 %) |
| | ♀ | | | 0/50 | | 49/50 | (98 %) | 49/49 (100 %) |
| toxic nephropathy | ♂ | | | 0/50 | | 39/50 | (78 %) | 35/50 (70 %) |
| | ♀ | | | 0/50 | | 30/50 | (60 %) | 39/49 (80 %) |
| tubule adenoma | ♂ | 0/50 | | .0/50 | | 6/50 | (12 %) | 1/50 (2 %) |
| | ♀ | 1/50 | (2 %) | 0/50 | | 0/50 | | 1/49 (2 %) |
| tubule adenocarcinoma | ♂ | 0/50 | | 0/50 | | 0/50 | | 1/50 (2 %) |
| | ♀ | 0/50 | | 0/50 | | 0/50 | | 0/49 |

[a] $p < 0.05$

| | |
|---|---|
| Author: | NTP 1990 |
| Substance: | trichloroethylene, stabilized with amine base |
| Species: | rat (F344/N), 50 ♂, 50 ♀ |
| Administration: | gavage, corn oil vehicle |
| Dose: | 500, 1000 mg/kg body weight |
| Duration: | 103 weeks, 5 days/week, necropsy after 103–107 weeks |
| Toxicity: | survival of ♂ reduced, body weights of all treated animals reduced, toxic nephropathy (cytomegaly) in more than 95 % of treated animals, most marked in ♂ of the high dose group; 20 % of ♂ in the high dose group died after gavage errors |

| | | 0 (vehicle control) | | 500 mg/kg body weight | | 1000 mg/kg body weight | |
|---|---|---|---|---|---|---|---|
| survivors after | ♂ | 35/50 | (70 %) | 20/50 | (40 %) | 16/50 | (32 %) |
| 103–107 weeks | ♀ | 37/50 | (74 %) | 33/50 | (66 %) | 6/50 | (52 %) |
| Findings: | | | | | | | |
| toxic nephropathy | ♂ | 0/48 | | 48/49 | (98 %) | 48/49 | (98 %) |
| | ♀ | 0/50 | | 48/49 | (98 %) | 48/48 | (100 %) |
| tubule adenoma | ♂ | 0/48 | | 2/49 | (4 %) | 0/49 | |
| | ♀ | 0/50 | | 0/49 | | 0/48 | |
| tubule adenocarcinoma | ♂ | 0/48 | | 0/49 | | 3/49 [a] | (6 %) |
| | ♀ | 0/50 | | 0/49 | | 1/48 | (2 %) |
| renal pelvis carcinoma | ♂ | 0/48 | | 1/49 | (2 %) | 1/49 | (2 %) |
| | ♀ | 0/50 | | 0/49 | | 0/48 | |
| renal pelvis papilloma | ♂ | 1/48 | (2 %) | 0/49 | | 0/49 | |
| | ♀ | 0/50 | | 0/49 | | 0/48 | |

[a] expressed in terms of the surviving animals (n = 16) significantly increased relative to the control values ($p < 0.05$)

**Table 5.** continued

| Author: | NTP 1990 |
| Substance: | trichloroethylene, stabilized with amine base |
| Species: | mouse (B6C3F$_1$), 50 ♂, 50 ♀ |
| Administration: | gavage, corn oil vehicle |
| Dose: | 1000 mg/kg body weight |
| Duration: | 103 weeks, 5 days/week, necropsy after 103–107 weeks |
| Toxicity: | survival and body weights of ♂ significantly reduced, ♀ not different from controls, toxic nephropathy (cytomegaly) in more than 95 % of treated animals |

|  |  | 0 (vehicle control) | 1000 mg/kg body weight |
|---|---|---|---|
| survivors after | ♂ | 33/50 (66 %) | 16/50 (32 %) |
| 103–107 weeks | ♀ | 32/50 (64 %) | 23/50 (46 %) |
| Findings: |  |  |  |
| hepatocellular carcinoma | ♂ | 8/48 (17 %) | 31/50 (62 %) |
|  | ♀ | 2/48 (4 %) | 13/49 (26 %) |
| hepatocellular adenoma | ♂ | 7/48 (15 %) | 4/50 (28 %) |
|  | ♀ | 4/48 (8 %) | 16/49 (33 %) |

| Author: | Van Duuren *et al.* 1979 |
| Substance: | trichloroethylene (purity n.s.) |
| Species: | mouse (ICR/Ha Swiss) 30 ♀ |
| Administration: | subcutaneous, vehicle trioctanoin |
| Dose: | 0.5 mg/application |
| Duration: | 88 weeks, once weekly |
| Toxicity: | n.s. |
| Tumours: | none |

| Author: | Van Duuren *et al.* 1979 |
| Substance: | trichloroethylene (purity n.s.) |
| Species: | mouse (ICR/Ha Swiss) 30 ♀ |
| Administration: | epicutaneous, vehicle acetone, not occlusive |
| Dose: | 1 mg/application |
| Duration: | 88 weeks, 3 times weekly |
| Toxicity: | n.s. |
| Tumours: | none |

In another NTP study (NTP 1988) epichlorohydrin-free trichloroethylene dissolved in corn oil was administered to male and female ACI, August, Marshall and Osborne-Mendel rats by gavage of doses of 500 or 1000 mg/kg body weight. In 7 of the total of 16 treated groups, survival was significantly reduced relative to the control values. Toxic nephropathy in up to 80 % of the treated animals and central nervous system toxicity (sedation, tremor, spasms and loss of consciousness) also showed that the doses selected were too high. The carcinogenicity findings demonstrated that trichloroethylene was a weak nephrocarcinogen for male Osborne-Mendel rats (tubule adenomas: 500 mg/kg 6/50, 1000 mg/kg 1/50) female ACI rats (tubule adenocarcinomas: 500 mg/kg 1/47, 1000 mg/kg 1/43) and female August rats (tubule adenocarcinomas: 500 mg/kg 2/48, 1000 mg/kg 0/50). In the Marshall rats of the 1000 mg/kg group the incidence of testis tumours was significantly increased: 500 mg/kg 21/48, 1000 mg/kg 32/48, control 17/46. Because of the marked toxicity, the reduced survival and the high level of accidental deaths among the treated animals the study was considered inadequate for the assessment of the carcinogenicity of trichloroethylene.

The studies of both acute and subchronic toxicity in animals and the genotoxicity studies *in vitro* support the hypothesis that 1,2-dichlorovinylcysteine is responsible for the kidney-specific effects of trichloroethylene. To date, 1,2-dichlorovinylcysteine has not been tested in a long-term study. However, a number of initiation-promotion studies have been carried out for the mouse kidney (Meadows *et al.* 1988). According to these authors, 1,2-dichlorovinylcysteine administered with the drinking water is not a complete carcinogen but could possibly have promoting effects. However, the exclusion of an initiating effect for 1,2-dichlorovinylcysteine cannot be accepted without reservations because of considerable weaknesses in this study. For the initiation experiments dichloro-vinylcysteine was administered for 4 weeks in the drinking water and the promotor $\beta$-cyclodextrin for only 7 days subcutaneously; a positive control was not included. The increased incidence of tumours seen in the promotion study with the initiator dimethyl-nitrosamine after 14 weeks administration of dichlorovinylcysteine could, therefore, equally well have been caused by a genotoxic effect of dichlorovinylcysteine itself. The study cannot be used for the assessment of an initiating or promoting potential of dichlorovinylcysteine.

The observation that mice given trichloroethylene excrete dichlorovinyl-*N*-acetyl-cysteine like the rats and develop kidney damage but no kidney tumours can at present not be accounted for satisfactorily. It is conceivable that this observation and the observed differences in renal toxicity of dichlorovinylcysteine in the mouse and rat could also be a result of different transport mechanisms and repair capabilities in the relevant tissues. It is also important that the urine concentration of dichlorovinyl-*N*-acetylcysteine is not a measure of the total level of reductive metabolism and does not reflect the level of activity of the toxifying $\beta$-lyase pathway. At present, however, no better marker for this metabolic process is available. In addition, it is not known which is the rate-limiting step of reductive metabolism; it is conceivable that the excretion of dichlorovinyl-*N*-acetylcysteine must be seen as a detoxication step.


# 5 Manifesto (MAK value, classification)


The nephrocarcinogenic effects of trichloroethylene in man were demonstrated because of an increase in the incidence of kidney cancer in a collective of workers exposed con-tinuously for many years to very high concentrations of trichloroethylene and confirmed in a case-control study. A causal relationship between the observed renal cell tumours and the trichloroethylene exposure is suggested by the fact that the cohort studied was exposed for a long period (average exposure period 18–19 years) exclusively to high concentrations of trichloroethylene. In addition, most of the tumours identified in the cohort of workers were histopathologically identical with the renal cell tumours found in rats exposed to trichloroethylene; this kind of cancer occurs extremely rarely as a spontaneous tumour in the strain of rats used in these studies. Finally, the toxification pathway which is responsible for the nephrocarcinogenicity is identical in the rat and man and yields genotoxic metabolites. The findings of the epidemiological studies must, however, be seen in context because the exposure concentrations were very much higher than the MAK value of 50 ml/m$^3$ which was valid until 1996. Under the high level exposure conditions, saturation of oxidative metabolism and an increase in the reductive metabolic processes is to be expected. For exposures in the range of the no longer valid MAK value, a very much smaller renal cancer risk can be assumed, as has been confirmed by the study of Axelson *et al.* (1994).

In the light of the new epidemiological findings of an increased incidence of renal cell tumours in exposed workers, the nephrocarcinogenicity in the rat and the known molecular mechanism of the organ specificity, trichloroethylene is classified in Category IIIA1 in the "List of MAK and BAT Values"; the MAK value and the classification in Pregnancy risk group C have been withdrawn.

# 6 References


Allemand H, Pessayre D, Descatoire V, Degott C, Feldmann G, Benhamou J-P (1978) Metabolic activation of trichloroethylene into a chemically reactive metabolite toxic to the liver. *J Pharmacol Exp Ther 204*: 714–723

Anttila A, Pukkala E, Salmén M, Hernberg S, Hemminki K (1995) Cancer incidence among Finnish workers exposed to halogenated hydrocarbons. *J Occup Environ Med 37*: 797–806

Aviado DM, Zakhari S, Simaan JA, Ulsamer AG (1976) In: Goldberg L (Ed.) *Methyl chloroform and trichloroethylene in the environment*, CRC Press, Cleveland, OH, 47–89

Axelson O, Andersson K, Hogstedt C, Holmberg B, Molina G, de Verdier A (1978) A cohort study on trichloroethylene exposure and cancer mortality. *J Occup Med 20*: 194–196

Axelson O, Seldén A, Andersson K, Hogstedt C (1994) Updated and expanded Swedish cohort study on trichloroethylene and cancer risk. *J Occup Med 36*: 556–562

Baden JM, Kelley MJ, Mazze RI, Simmon VF (1979) Mutagenicity of inhalation anesthetics: nitrous oxide, cyclopropane, trichloroethylene and divinyl ether. *Br J Anaesth 51*: 417–421

Banerjee S, Van Duuren BL (1978) Covalent binding of the carcinogen trichloroethylene to hepatic microsomal proteins and to endogenous DNA in vitro. *Cancer Res 38*: 776–780

Barret L, Faure J, Guilland B, Chomat D, Didier B, Debru JL (1984) Trichloroethylene occupational exposure: Elements for better prevention. *Int Arch Occup Environ Health 53*: 283–289

Bartsch H, Malaveille C, Barbin A, Planche G (1979) Mutagenic and alkylating metabolites of haloethylenes, chlorobutadienes and dichlorobutenes produced by rodent or human liver tissues. *Arch Toxicol 41*: 249–277

Beliles RP, Brusick DJ, Meeler FJ (1980) *Teratogenic/mutagenic risk of workplace contaminants: trichloroethylene, perchloroethylene and carbon disulfide.* US DHEW, Contract No. 210-77-0047, p 234

Bell Z (1977) *Teratogenicity study via inhalation with trichlor 132, trichloroethylene, in albino rats.* EPA-No-600/8-22-006, EPA, Washington, DC, USA

Bergman K (1983) Interactions of trichloroethylene with DNA in vitro and with RNA and DNA of various mouse tissues in vivo. *Arch Toxicol 54*: 181–193

Bernauer U, Birner G, Dekant W, Henschler D (1996) Biotransformation of trichloroethene: dose-dependent excretion of 2,2,2-trichloro-metabolites and mercapturic acids in rats and humans after inhalation. *Arch Toxicol 70*: 338–346

Beuter W, Cojocel C, Müller W, Donaubauer HH, Mayer D (1989) Peroxidative damage and nephrotoxicity of dichlorovinylcysteine in mice. *J Appl Toxicol 9*: 181–186

Bignami M, Conti G, Crebelli R, Misuraca F, Puglia AM, Randazzo R, Sciandrello G, Carere A (1980) Mutagenicity of halogenated aliphatic hydrocarbons in Salmonella typhimurium, Streptomyces coelicolor and Aspergillus nidulans. *Chem Biol Interact 30*: 9–23

Birner G, Vamvakas S, Dekant W, Henschler D (1993) Nephrotoxic and genotoxic N-acetyl-S-dichlorovinyl-L-cysteine is a urinary metabolite after occupational 1,1,2-trichloroethene exposure in humans: Implications for the risk of trichloroethene exposure. *Environ Health Perspect 99*: 281–284

Bogen KT (1988) Pharmacokinetics for regulatory risk analysis: The case of trichloroethylene. *Regul Toxicol Pharmacol 8*: 447–466

Bolt HM, Filser JG (1977) Irreversible binding of chlorinated ethylenes to macromolecules. *Environ Health Perspect 21*: 107–112

Bronzetti G, Zeiger E, Frezza D (1978) Genetic activity of trichloroethylene in yeast. *J Environ Pathol Toxicol 1*: 411–418

Brüning T, Golka K, Makropoulos V, Bolt HM (1996b) Preexistence of chronic tubular damage in cases of renal cell cancer after long and high exposure to trichloroethylene. *Arch Toxicol 70*: 259–260

Brüning T, Vamvakas S, Makropoulos V, Birner G (1996a) A case of acute intoxication by trichloroethene: clinical symptoms, toxicokinetics and renal damage. *Arch Pharmacol 353, Suppl*: R109

Buben JA, O'Flaherty EJ (1985) Delineation of the role of metabolism in the hepatotoxicity of trichloroethylene and perchloroethylene: a dose-effect study. *Toxicol Appl Pharmacol 78*: 105–122

Callen DF, Wolf CR, Philpot RM (1980) Cytochrome P-450 mediated genetic activity and cytotoxicity of seven halogenated aliphatic hydrocarbons in Sacharomyces cerevisiae. *Mutat Res 77*: 55–63

Chen JC, Stevens JL, Trifillis AL, Jones TW (1990) Renal cysteine conjugate β-lyase-mediated toxicity studied with primary cultures of human proximal tubular cells. *Toxicol Appl Pharmacol 103*: 463–473

Coleman MP, Esteve J, Damiecki P, Arslan A, Renard H (1993) *Trends in cancer incidence and mortality*. IARC Scientific Publication 121, IARC, Lyon

Commandeur JNM, Vermeulen NPE (1990) Identification of N-acetyl(1,1-dichlorovinyl)- and N-acetyl(1,2-dichlorovinyl)-L-cysteine as two regioisomeric mercapturic acids of trichloroethylene in the rat. *Chem Res Toxicol 3*: 212–218

Commandeur JNM, Boogaard J, Mulder GJ, Vermeulen NPE (1991) Mutagenicity and cytotoxicity of two regioisomeric mercapturic acids and cysteine S-conjugates of trichloroethylene. *Arch Toxicol 65*: 373–380

Costa AK, Ivanetich KM (1984) Chlorinated ethylenes: their metabolism and effect on DNA repair in rat hepatocytes. *Carcinogenesis 5*: 1629–1636

Crebelli R, Conti G, Conti L, Carere A (1985) Mutagenicity of trichloroethylene, trichloroethanol and chloral hydrate in Aspergillus nidulans. *Mutat Res 155*: 105–111

Crebelli R, Carere A (1989) Genetic toxicology of 1,1,2-trichloroethylene. *Mutat Res 221*: 11–37

Darnerud PO, Brandt I, Feil VJ, Bakke JE (1988) S-(1,2-Dichloro-($^{14}$C)vinyl)-L-cysteine (DCVC) in the mouse kidney: correlation between tissue-binding and toxicity. *Toxicol Appl Pharmacol 95*: 423–434

Darnerud PO, Brandt I, Feil VJ, Bakke JE (1989) Dichlorovinyl cysteine (DCVC) in the mouse kidney: tissue-binding and toxicity after glutathione depletion and probenecid treatment. *Arch Toxicol 63*: 345–350

Darnerud PO, Gustafson A-L, Törnwall U, Feil VJ (1991) Age- and sex-dependent dichlorovinyl cysteine (DCVC) accumulation and toxicity in the mouse kidney: relation to development of organic anion transport and β-lyase activity. *Pharmacol Toxicol 68*: 104–109

David NJ, Wolman R, Milne FJ, van Niererk I (1989) Acute renal failure due to trichloroethylene poisoning. *Br J Ind Med 46*: 347–349

Dawson BV, Johnson PD, Goldberg SJ, Ulreich JB (1993) Cardiac teratogenesis of halogenated hydrocarbon-contaminated drinking water. *J Am Coll Cardiol 21*: 1466–1472

Dekant W, Metzler M, Henschler D (1984) Novel metabolites of trichloroethylene through dechlorination reactions in rats, mice and humans. *Biochem Pharmacol 33*: 2021–2027

Dekant W, Metzler M, Henschler D (1986a) Identification of S-1,2-dichlorovinyl-N-acetyl-cysteine as a urinary metabolite of trichloroethylene: A possible explanation for its nephrocarcinogenicity in male rats. *Biochem Pharmacol 35*: 2455–2458

Dekant W, Vamvakas S, Berthold K, Schmidt S, Wild D, Henschler D (1986b) Bacterial β-lyase mediated cleavage and mutagenicity of cysteine conjugates derived from the nephrocarcinogenic alkenes trichloroethylene, tetrachloroethylene and hexachlorobutadiene. *Chem Biol Interact 60*: 31–45

DiRenzo AB, Gandolfi AJ, Sipes IG (1982) Microsomal bioactivation and covalent binding of aliphatic halides to DNA. *Toxicol Lett 11*: 243–252

Doolittle DJ, Müller G, Scribner HE (1987) The in vivo-in vitro hepatocyte assay for assessing DNA repair and DNA replication: studies in the CD-1 mouse. *Food Chem Toxicol 25*: 399–405

Dorfmüller MA, Henne SP, York RG, Bornschein RL, Manson JM (1979) Evaluation of teratogenicity and behavioural toxicity with inhalation exposure of maternal rats to trichloroethylene. *Toxicology 14*: 153–166

Duprat P, Delsaut L, Gradiski D (1976) Pouvoir irritant des principaux solvants chlorés aliphatiques sur la peau et le muqueuses oculaires du lapin. *Eur J Toxicol 9*: 171–177

Duprat P, Gradiski D (1981) Cytogenetic effect of trichloroethylene in the mouse as evaluated by the micronucleus test. *IRCS Med Sci 8*: 182

Elcombe CR, Rose MS, Pratt IS (1985) Biochemical, histological, and ultrastructural changes in rat and mouse liver following the administration of trichloroethylene: possible relevance to species differences in hepatocarcinogenicity. *Toxicol Appl Pharmacol 79*: 365–376

ENSR Health Sciences (1990) *Historical prospective mortality study of Hughes Aircraft employees at Air Force Plant #44*. Document Number 8701-356-001, ENSR, Alameda, CA

Ettema JH, Kleerekoper L, Duba WC (1975) *Staub-Reinhalt. Luft 35*: 409

Eyre RJ, Stevens DK, Parker JC, Bull RJ (1995a) Acid-labile adducts to protein can be used as indicators of the S-conjugate pathway of trichloroethene metabolism. *J Toxicol Environ Health 46*: 443–464

Eyre RJ, Stevens DK, Parker JC, Bull RJ (1995b) Renal activation of trichloroethene and S-(1,2-dichlorovinyl)-L-cysteine and cell proliferative responses in the kidneys of F344 rats and B6C3F1 mice. *J Toxicol Environ Health 46*: 465–481

Fahrig R (1977) The mammalian spot test (Fellfleckentest) with mice. *Toxicology 38*: 87–98

Fernández JG, Humbert BE, Droz PO, Caperos JR (1975) Exposition au trichloroéthylène: Bilan de l'absorption, de l'excrétion et du métabolisme sur des sujets humains. *Arch Mal Prof Med Trav Secur Soc 36*: 397–407

Fernández JG, Droz PO, Humbert BE, Caperos JR (1977) Trichloroethylene exposure. Simulation of uptake, excretion, and metabolism using a mathematical model. *Br J Ind Med 34*: 43–45

Filser JG, Bolt HM (1981) Inhalation pharmacokinetics based on gas uptake studies. I. Improvement of kinetic models. *Arch Toxicol 47*: 279–292

Filser JG, Deml E (1989) Pharmakokinetik und Metabolismus leichtflüchtiger chlorierter Verbindungen am Beispiel von Trichlorethylen, Perchlorethylen und 1,1,1-Trichlorethan (Methylchloroform). In: Verein Deutscher Ingenieure (Ed.) *Halogenierte organische Verbindungen in der Umwelt. Herkunft, Messung, Wirkung, Abhilfemaßnahmen, Band II*, VDI Berichte Nr 745, VDI-Verlag, Düsseldorf, 679–711

Forkert PG, Birch DW (1989) Pulmonary toxicity of trichloroethylene in mice. Covalent binding and morphological manifestations. *Drug Metab Dispos 17*: 106–113

Fredriksson M, Bengtsson NO, Hardell L, Axelson O (1989) Colon cancer, physical activity, and occupational exposures. *Cancer 63*: 1838–1842

Fukuda K, Takemoto K, Tsuruta H (1983) Inhalation carcinogenicity of trichloroethylene in mice and rats. *Ind Health 21*: 243–254

Green T, Odum J (1985) Structure/activity studies of the nephrotoxic and mutagenic action of cysteine conjugates of chloro- and fluoroalkenes. *Chem Biol Interact 54*: 15–31

Green T, Prout MS (1985) Species differences in response to trichloroethylene. II. Biotransformation in rats and mice. *Toxicol Appl Pharmacol 79*: 401–411

Greim H, Bonse G, Radwan Z, Reichert D, Henschler D (1975) Mutagenicity in vitro and potential carcinogenicity of chlorinated ethylenes as a function of metabolic oxirane formation. *Biochem Pharmacol 24*: 2013–2017

Greim H, Wolff T, Höfler M, Lahaniatis M (1984) Formation of dichloroacetylene from trichloroethylene in the presence of alkaline material - possible cause of intoxication after abundant use of chloroethylene-containing solvents. *Arch Toxicol 56*: 74–77

Gu ZW, Sele B, Jalbert B (1981) Effects of trichloroethylene and its metabolites on the rate of sister chromatid exchange. In vivo and in vitro study on the human lymphocytes. *Ann Genet 24*: 105–106

Guengerich FP (1991) Oxidation of toxic and carcinogenic chemicals by human cytochrome P-450 enzymes. *Chem Res Toxicol 4*: 391–407

Hardin BD, Bond GP, Sikov MR, Andrew FD, Beliles RP, Niemeier RW (1981) Testing of selected workplace chemicals for teratogenic potential. *Scand J Work Environ Health 7, Suppl 4*: 66–75

Hassall CD, Gandolfi AJ, Brendel K (1983) Correlation of the in vivo and in vitro renal toxicity of S-(1,2-dichlorovinyl)-L-cysteine. *Drug Chem Toxicol 6*: 507–520

Healy TE, Poole TR, Hooper A (1982) Rat fetal development and maternal exposure to trichloroethylene at 100 ppm. *Br J Anaesth 54*: 337–341

Henschler D, Broser F, Hopf HC (1970) "Polyneuritis cranialis" durch Vergiftung mit chlorierten Acetylenen beim Umgang mit Vinylidenchlorid-Copolymeren. *Arch Toxicol 26*: 62–69

Henschler D (Ed.) (1976) Trichloräthylen, in *Gesundheitsschädliche Arbeitsstoffe. Toxikologisch-arbeitsmedizinische Begründungen von MAK-Werten*, 5th issue, VCH-Verlagsgesellschaft, Weinheim

Henschler D, Eder E, Neudecker T, Metzler M (1977) Carcinogenicity of trichloroethylene: Fact or artifact? *Arch Toxicol 37*: 233–236

Henschler D, Romen W, Elsässer HM, Reichert D, Eder E, Radwan Z (1980) Carcinogenicity study of trichloroethylene by longterm inhalation in three animal species. *Arch Toxicol 43*: 237–248

Henschler D, Elsässer H, Romen W, Eder E (1984) Carcinogenicity study of trichloroethylene, with and without epoxide stabilizers, in mice. *J Cancer Res Clin Oncol 107*: 149–156

Henschler D, Vamvakas S, Lammert M, Dekant W, Kraus B, Thomas B, Ulm K (1995) Increased incidence of renal cell tumors in a cohort of cardboard workers exposed to trichloroethene. *Arch Toxicol 69*: 291–299

IARC (International Agency for Research on Cancer) (1995) *Dry cleaning, some chlorinated solvents and other industrial chemicals, IARC monographs, Vol. 63*, IARC, Lyon

Jaffe DR, Gandolfi AJ, Nagle RB (1984) Chronic toxicity of S-(trans-1,2-dichlorovinyl)-L-cysteine in mice. *J Appl Toxicol 4*: 315–319

Jaffe DR, Hassall CD, Gandolfi AJ, Brendel K (1985) Production of DNA single strand breaks in renal tissue after exposure to 1,2-dichlorovinylcysteine. *Toxicology 35*: 25–33

Kjellstrand P, Kanje M, Mansson L, Bjerkemo M, Mortensen I, Lanke J, Holmquist B (1981a) Trichloroethylene: effects on body and organ weights in mice, rats and gerbils. *Toxicology 21*: 105–115

Kjellstrand P, Bjerkemo M, Mortensen I, Mansson L, Lanke J, Holmquist B (1981b) Effects of longterm exposure to trichloroethylene on the behaviour of Mongolian gerbils (Meriones unguiculatus). *J Toxicol Environ Health 8*: 787–793

Kjellstrand P, Holmquist B, Alm P, Kanje M, Romare S, Jonsson I, Mansson L, Bjerkemo M (1983a) Trichloroethylene: further studies on the effects on body and organ weights and plasma butyryl-cholinesterase activity in mice. *Acta Pharmacol Toxicol 53*: 375–384

Kjellstrand P, Holmquist B, Mandahl N, Bjerkemo M (1983b) Effects of continuous trichloroethylene inhalation on different strains of mice. *Acta Pharmacol Toxicol 53*: 369–374

Kligerman AD, Bryant MF, Doerr CL, Erexson GL, Evansky PA, Kwanyuen P, McGee JK (1994) Inhalation studies of the genotoxicity of trichloroethylene to rodents. *Mutat Res 322*: 87–96

Konietzko H, Haberlandt W, Heilbronner H, Reill G, Weichardt H (1978) Cytogenetische Untersuchungen an Trichloräthylen-Arbeitern. *Arch Toxicol 40*: 201–206

Kylin B, Reichard H, Sümegi I, Yllner S (1972) Hepatotoxic effect of tri- and tetrachloroethylene on mice. *Nature (London) 193*: 395–398

Kyrklund T, Kjellstrand P, Haglid KG (1986) Fatty acid changes in rat brain ethanolamine phosphoglycerides during and following chronic exposure to trichloroethylene. *Toxicol Appl Pharmacol 85*: 145–153

La Vecchia C, Levi F, Lucchini F, Negri E (1992) Descriptive epidemiology of kidney cancer in Europe. *J Nephrol 5*: 37–43

Lammert M (1995) personal communication to the Commission

Lash LH, Nelson RM, Dyke RA, Anders MW (1990) Purification and characterization of human kidney cytosolic cysteine conjugate β-lyase activity. *Drug Metab Dispos 18*: 50–54

Leong BKJ, Schwetz BA, Gehring PJ (1975) Embryo and fetotoxicity of inhaled trichloroethylene, perchloroethylene, methylchloroform and methylene chloride in mice and rats. *Toxicol appl Pharmacol 33*: 136, Abstract No. 134, 14th SOT Meeting

Longley EO, Jones R (1963) Acute trichloroethylene narcosis. *Arch environm Hlth 7*: 249–252

Maltoni C, Lefemine G, Cotti G, Perino G (1988) Long-term carcinogenicity bioassays on trichloroethylene administered by inhalation to Sprague-Dawley rats and Swiss and B6C3F1 mice. *Ann N Y Acad Sci 534*: 316–342

Mandel JS, McLaughlin JK, Schlehofer B, Mellemgaard A, Helmert U, Lindblad P, McCredie M, Adam H-O (1995) International renal-cell cancer study. IV. Occupation. *Int J Cancer 61*: 601–605

Manson JM, Hustert K, Richdale M, Smith MK (1984) Effects of oral exposure to trichloroethylene on female reproductive function. *Toxicology 32*: 229–242

Mazzullo M, Bartoli S, Bonora B, Collacci A, Lattanzi G, Niero A, Silingardi P, Grilli S (1992) In vivo and in vitro interaction of trichloroethylene with macromolecules from various organs of rat and mouse. *Res Commun Chem Pathol Pharmacol 76*: 192–208

McCarthy TB, Jones RD (1983) Industrial gassing poisonings due to trichloroethylene, perchloroethylene, and 1,1,1-trichloroethane 1961–80. *Br J Ind Med 40*: 450–455

McGregor DB, Reynolds DM, Zeiger E (1989) Conditions affecting the mutagenicity of trichloroethylene in Salmonella. *Environ Mol Mutagen 13*: 197–202

McKinney LL, Picken JCJ, Weakley FB, Eldridge AC, Campbell RE, Cowan JC, Biester HE (1959) Possible toxic factor of trichloroethylene-extracted soybean oil meal. *J Am Chem Soc 81*: 909–915

McLaren J, Boulikas T, Vamvakas S (1994) Induction of poly(ADP-ribosyl)ation in the kidney after in vivo application of renal carcinogens. *Toxicology 88*: 101–112

Meadows SD, Gandolfi AJ, Nagle RB, Shively JW (1988) Enhancement of DMN-induced kidney tumors by 1,2-dichlorovinylcysteine in Swiss-Webster mice. *Drug Chem Toxicol 11*: 307–318

Miller RE, Guengerich FP (1983) Metabolism of trichloroethylene in isolated hepatocytes, microsomes and reconstituted enzyme systems containing cytochrome P-450. *Cancer Res 43*: 1145–1152

Mirsalis JC, Tyson CK, Steinmetz KL, Loh EK, Hamilton CM, Bakke JP, Spalding JW (1989) Measurement of unscheduled DNA synthesis and S-phase synthesis in rodent hepatocytes following in vivo treatment: testing of 24 compounds. *Environ Mol Mutagen 14*: 155–164

Monster AC, Boersma G, Duba WC (1976) Pharmacokinetics of trichloroethylene in volunteers, influence of workload and exposure concentration. *Int Arch Occup Environ Health 38*: 87–102

Monster AC, Boersma G, Duba WC (1979) Kinetics of trichloroethylene in repeated exposure of volunteers. *Int Arch Occup Environ Health 42*: 283–292

Moriya M, Ohta T, Watanabe K, Miyatawa T, Kato K, Shirazu Y (1983) Further mutagenicity studies on pesticides in bacterial reversion assay systems. *Mutat Res 116*: 185–216

Nagaya T, Ishikawa N, Hata H (1989) Sister-chromatid exchanges in lymphocytes of workers exposed to trichloroethylene. *Mutat Res 22*: 279–282

Nakaaki K, Onishi N, Iida H, Kimotsuki K, Fukabori S (1973) Experimental study on the effect of exposure to trichloroethylene in man (Japan.). *Rodo Kagaku 49*: 499–563, cited in *Chemical Abstracts 80*: 56210n (1974)

Nakajima T, Wang R-S, Elovaara E, Park SS, Gelboin HV, Vainio H (1993) Cytochrome P450-related differences between rats and mice in the metabolism of benzene, toluene and trichloroethylene in liver microsomes. *Biochem Pharmacol 45*: 1079–1085

Nakajima T, Wang RS, Elovaara E, Park SS, Gelboin HV, Vainio H (1992) A comparative study on the contribution of cytochrome P450 isoenzymes to metabolism of benzene, toluene and trichloroethylene in rat liver. *Biochem Pharmacol 43*: 251–257

NCI (National Cancer Institute) (1976) *Carcinogenesis Bioassay of Trichloroethylene.* NCI Technical Report 2, US Department of Health, Education, and Welfare, Public Health Service, NCI, National Institutes of Health, Bethesda, MD

Nelson MA, Bull RJ (1988) Induction of strand breaks in DNA by trichloroethylene and metabolites in rat and mouse liver in vivo. *Toxicol Appl Pharmacol 94*: 45–54

Nelson R, Van Dyke R, Powis G, Lash L, Anders M (1988) Characterization of human liver and kidney cysteine conjugate β-lyase activity. *FASEB J 2*: A1355

NIOSH (National Institute of Occupational Safety and Health) (1980) *Teratogenic-mutagenic risk of workplace contaminants: trichloroethylene, perchloroethylene and carbon disulfide.* Litton Bionetics Inc, Report No 210-77-0047, US Dept Health Educat Welfare, NIOSH, Cincinnati, OH

NIOSH (1984) KX4550000 Ethylene, trichloro-. *Registry of toxic effects of chemical substances (RTECS)*, 906–907

Nomiyama K, Nomiyama H (1971) Metabolism of trichloroethylene in humans: sex difference in urinary excretion of trichloroacetic acid and trichloroethanol. *Int Arch Arbeitsmed 28*: 37–48

Nomiyama K, Nomiyama H, Arai H (1986) Reevaluation of subchronic toxicity of trichloroethylene. *Toxicol Lett 31*: 225–231

Nomura S (1962) Health hazards in workers exposed to trichloroethylene vapour. I: Trichloroethylene poisoning in an electroplating plant. *Kumamoto med J 15*: 29–37

NTP (National Toxicology Program) (1985) *Trichloroethylene: Reproduction and fertility assessment in CD-1 mice when administered in the feed.* NTP-86-068, US Department of Health and Human Services, National Institute of Environmental Health Sciences, Research Triangle Park, NC

NTP (1986) *Trichloroethylene: Reproduction and fertility assessment in F344 rats when administered in the feed.* Final Report. NTP-86-085, US Department of Health and Human Services, National Institute of Environmental Health Sciences, Research Triangle Park, NC

NTP (1988) *Toxicology and carcinogenesis studies of trichloroethylene in four strains of rats (ACI, August, Marshall, Osborne-Mendel) (gavage studies).* Technical Report 273, US Department of Health and Human Services, Public Health Service, National Institutes of Health, Research Triangle Park, NC

NTP (1990) *Carcinogenesis studies of trichloroethylene (without epichlorhydrin) in F344/N rats and B6C3F1 mice (gavage studies).* Technical Report 243, US Department of Health and Human Services, Public Health Service, National Institutes of Health, Research Triangle Park, NC

Odum J, Foster JR, Green T (1992) A mechanism for the development of Clara cell lesions in the mouse lung after exposure to trichloroethylene. *Chem Biol Interact 83*: 135–153

Paddle GM (1983) Incidence of liver cancer and trichloroethylene manufacture: joint study by industry and a cancer registry. *Br Med J 286*: 846

Parchman LG, Magee PN (1982) Metabolism of [$^{14}$C]trichloroethylene to $^{14}CO_2$ and interaction of a metabolite with liver DNA in rats and mice. *J Toxicol Environ Health 9*: 797–813

Perocco P, Prodi G (1981) DNA damage by haloalkanes in human lymphocytes cultured in vitro. *Cancer Lett 13*: 213–218

Price PJ, Hassett CM, Mansfield JI (1978) Transforming activities of trichloroethylene and proposed industrial alternatives. *In Vitro 14*: 290–293

Prout MS, Provan WM, Green T (1985) Species differences in response to trichloroethylene. *Toxicol Appl Pharmacol 79*: 389–400

Rasmussen K, Sabroe S, Wohlert M, Ingerslo HJ, Kappel B, Nielsen J (1988) A genotoxic study of metal workers exposed to trichloroethylene. Sperm parameters and chromosome aberrations in lymphocytes. *Int Arch Occup Environ Health 60*: 419–423

Rehrmann A (1953) Thesis (Habilitation), University of Hamburg

Reinhardt CF, Mullin LS, Maxfield ME (1973) Epinephrine-induced cardiac arrythmia potential of some common industrial solvents. *J Occup Med 15*: 953–955

Rossi AM, Migliore L, Barale R, Loprieno N (1983) In vivo and in vitro mutagenicity studies of a possible carcinogen, trichloroethylene, and its two stabilizers, epichlorohydrin and 1,2-epoxybutane. *Teratogen Carcinogen Mutagen 3*: 75–87

Salvini M, Binaschi S, Riva M (1971) Evaluation of the psychophysiological functions in humans exposed to trichloroethylene. *Brit J ind Med 28*: 293–295

Sanders VM, Tucker AN, White K, Kauffmann BM, Hallett P, Carchman RA, Borzelleca JF, Munson AE (1982) Humoral and cell-mediated immune status in mice exposed to trichloroethylene in the drinking water. *Toxicol Appl Pharmacol 62*: 385–368

Sbrana I, Lascialfari D, Loprieno N (1985) TCE induces micronuclei but not chromosomal aberrations in mouse bone marrow cells. *Fourth International Conference on Environmental Mutagens, Stockholm*, 4.–28.06., Abstracts, 163

Schwetz BA, Leong BKJ, Gehring PJ (1975) The effect of maternally inhaled trichloroethylene, perchloroethylene, methyl chloroform, and methylene chloride on embryonal and fetal development in mice and rats. *Toxicol Appl Pharmacol 32*: 84–96

Scott JE, Forkert PG, Oulton M, Rasmusson MG, Temple S, Fraser MO, Whitefield S (1988) Pulmonary toxicity of trichloroethylene: induction of changes in surfactant phospholipids and phospholipase A2 activity in the mouse lung. *Exp Mol Pharmacol 49*: 141–150

Shahin MM, von Borstel RC (1977) Mutagenic and lethal effects of α-benzene hexachloride, dibutyl phthalate and trichloroethylene in Sacharomyces cerevisiae. *Mutat Res 48*: 173–180

Shimada T, Swanson AF, Leber P, Williams GM (1985) Activities of chlorinated ethane and ethylene compounds in the Salmonella/rat microsome mutagenesis and rat hepatocyte/DNA repair assays under vapor phase exposure conditions. *Cell Biol Toxicol 1*: 159–179

Shindell S, Ulrich S (1985) A cohort study of employees of a manufacturing plant using trichloroethylene. *J Occup Med 27*: 577–579

Siegel J, Jones RA, Coon RA, Lyon JP (1971) Effects on experimental animals of acute, repeated and continuous inhalation exposures to dichloroacetylene mixtures. *Toxicol Appl Pharmacol 18*: 168–174

Silvermann AP, Williams H (1975) Behaviour of rats exposed to trichloroethylene vapour. *Br J Ind Med 32*: 308–315

Simmon VF (1977) Structural correlations of mutagenic and carcinogenic alkyl halides. In: Asher JM, Zervos C (Eds) *Symposium on structural correlates in carcinogenesis and mutagenesis*, Office of Science, US Food and Drug Administration, Rockville, MD, 163–171

Simmon VF, Kauhanen K, Tardiff RG (1977) Mutagenicity of chemicals identified in drinking water. *Dev Toxicol Environ Sci 2*: 249–258

Slacic-Erben R, Roll R, Franke G, Uehleke H (1980) Trichloroethylene vapours do not produce dominant lethal mutations in mice. *Arch Toxicol 4*: 37–44

Smyth jr HF, Carpenter CP, Weil CS, Pozzani UC, Striegel JA, Nycum JS (1969) Range-finding toxicity data. List VII. *Am Ind Hyg Assoc J 30*: 470–476

Spencer PS, Schaumburg HH (1985) Organic solvent neurotoxicity: facts and research needs. *Scand J Work Environ Health 11*: 53–60

Spirtas R, Stewart PA, Lee JS, Marano DE, Forbes CD, Grauman DJ, Pettigrew HM, Blair A, Hoover RN, Cohen JL (1991) Retrospective cohort mortality study of workers at an aircraft maintenance facility. I. Epidemiological results. *Br J Ind Med 48*: 515–530

Stevens DK, Eyre RJ, Bull RJ (1992) Adduction of hemoglobin and albumin in vivo by metabolites of trichloroethylene, trichloroacetate, and dichloroacetate in rats and mice. *Fundam Appl Toxicol 19*: 336–342

Stewart PA, Lee JS, Marano DE, Spirtas R, Forbes CD, Blair A (1991) Retrospective cohort mortality study of workers at an aircraft maintenance facility. II. Exposures and their assessment. *Br J Ind Med 48*: 531–537

Stewart RD, Gay HH, Erley DS, Hake CL, Peterson JE (1962) Observations on the concentrations of trichloroethylene in blood and expired air following exposures of humans. *Amer. Ind. Hyg. Ass. J. 23*: 167–170

Stewart RD, Dodd HC, Gay HH, Erley DS (1970) Experimental human exposure to trichloroethylene. *Arch environm Hlth 20*: 64–71

Stewart RD, Hake CL, Lebrun AJ, Kalbfleisch JH, Newton PE, Peterson JE, Cohen HH, Struble R, Busch KA (1974) Effects of Trichloroethylene on Behavioral Performance Capabilities in Behavioral Toxicology. US Department of Health, Education and Welfare, Public Health Service, *NIOSH Publication No. 74 126, p 96*

Stopps GJ, McLaughlin M (1970) Physiological tests in human subjects exposed to solvent vapors. *Am Ind Hyg Assoc J 28*: 43–50

Stott WT, Quast JF, Watanabe PG (1982) The pharmacokinetics and macromolecular interactions of trichloroethylene in mice and rats. *Toxicol Appl Pharmacol 62*: 137–151

Stüber K (1931) Gesundheitsschädigungen bei der gewerblichen Verwendung des Trichloräthylens und die Möglichkeiten ihrer Verhütung. *Arch Gewerbepath Gewerbehyg 2*: 398–456

Terracini B, Parker VH (1965) A pathological study on the toxicity of S-dichlorovinyl-L-cysteine. *Food Cosmet Toxicol 3*: 67–74

Tola S, Vilhunen R, Järvinen E, Korkala ml (1980) A cohort study on workers exposed to trichloroethylene. *J Occup Med 22*: 737–740

Torkelson TR, Rowe VK (1981) Halogenated aliphatic hydrocarbons containing chlorine, bromine and iodine. In: Clayton GD, Clayton FE (Eds) *Patty's industrial hygiene and toxicology*, John Wiley & Sons, New York, 3353–3360

Triebig G, Grobe T (1987) Toxische Enzephalopathie durch chronische Lösemittelexposition als Berufskrankheit. *Arbeitsmed Sozialmed Praeventivmed 22*: 222–228

Triebig G, Lehrl S, Kinzel W, Erzigkeit H, Galster JV, Schaller K-H (1977a) Psychopathometrische Ergebnisse von Verlaufsstudien an Trichloräthylen-exponierten Personen. *Zentralbl Bakteriol Parasitenk Infektionskr Hyg, Abt 1, Orig, Reihe B 164*: 314–327

Triebig G, Schaller K-H, Erzigkeit H, Valentin H (1977b) Biochemische Untersuchungen und psychologische Studien an chronisch Trichloräthylen-belasteten Personen unter Berücksichtigung expositionsfreier Intervalle. *Int Arch Occup Environ Health 38*: 149–162

Triebig G, Trautner P, Weltle D, Saure E, Valentin H (1982) Untersuchungen zur Neurotoxizität von Arbeitsstoffen. III. Messung der motorischen und sensorischen Nervenleitgeschwindigkeit bei beruflich Trichloräthylen-belasteten Personen. *Int Arch Occup Environ Health 51*: 25–34

Tucker AN, Sanders VM, Barnes DW, Bradshaw TJ, White jr KL, Sain LE, Borzelleca JF, Munson AE (1982) Toxicology of trichloroethylene in the mouse. *Toxicol Appl Pharmacol 62*: 351–357

Uehleke H, Poplawski-Tabarelli S (1977) Irreversible binding of $^{14}$C-labelled trichloroethylene to mice liver constituents in vivo and in vitro. *Arch Toxicol 37*: 289–294

Ulm K (1995) personal communication to the Commission, 24.03.1995

Vamvakas S, Dekant W, Berthold K, Schmidt S, Wild D, Henschler D (1987) Enzymatic transformation of mercapturic acids derived from halogenated alkenes to reactive and mutagenic intermediates. *Biochem Pharmacol 36*: 2741–2748

Vamvakas S, Elfarra AA, Dekant W, Henschler D, Anders MW (1988a) Mutagenicity of amino acid and glutathione S-conjugates in the Ames test. *Mutat Res 206*: 83–90

Vamvakas S, Kordowich FJ, Dekant W, Neudecker T, Henschler D (1988b) Mutagenicity of hexachloro-1,3-butadiene and its S-conjugates in the Ames test—Role of activation by the mercapturic acid pathway in its nephrocarcinogenicity. *Carcinogenesis 9*: 907–910

Vamvakas S, Herkenhoff M, Dekant W, Henschler D (1989a) Mutagenicity of tetrachloroethylene in the Ames-test—Metabolic activation by conjugation with glutathione. *J Biochem Toxicol 4*: 21–27

Vamvakas S, Dekant W, Henschler D (1989b) Assessment of unscheduled DNA synthesis in a cultured line of renal epithelial cells exposed to cysteine S-conjugates of haloalkenes and haloalkanes. *Mutat Res 222*: 329–335

Vamvakas S, Sharma VK, Shen S-S, Anders MW (1990) Perturbations of intracellular calcium distribution in kidney cells by nephrotoxic haloalkenyl cysteine S-conjugates. *Mol Pharmacol 38*: 455–461

Vamvakas S, Bittner D, Dekant W, Anders MW (1992) Events that precede and that follow S-(1,2-dichlorovinyl)-L-cysteine-induced release of mitochondrial $Ca^{2+}$ and their association with cytotoxicity to renal cells. *Biochem Pharmacol 44*: 1131–1138

Vamvakas S, Dekant W, Henschler D (1993a) Nephrocarcinogenicity of haloalkenes and alkynes. In: Anders MW, Dekant W, Henschler D, Oberleithner H, Silbernagl S (Eds) *Renal disposition and nephrotoxicity of xenobiotics*, Academic Press Inc, San Diego, 323–342

Vamvakas S, Bittner D, Köster U (1993b) Enhanced expression of the protooncogenes c-myc and c-fos in normal and malignant growth. *Toxicol Lett 67*: 161–172

Vamvakas S, Brüning T, Thomas B, Lammert M, Baumüller A, Bolt HM, Dekant W, Birner G, Henschler D, Ulm K (1998) Nephrocarcinogenicity of trichloroethene: a case-control study in nephrectomized patients. (submitted for publication)

Van Duuren BL, Goldschmidt BM, Katz C, Seidman I, Paul JS (1974) Carcinogenic activity of alkylating agents. *J Nat Cancer Inst 53*: 695–700

Van Duuren BL, Banerjee S (1976) Covalent interaction of the carcinogen trichloroethylene in rat hepatic microsomes. *Cancer Res 36*: 2419–2422

Van Duuren BL, Goldschmidt BM, Loewengart G, Smith AC, Melchionne S, Seidman I, Roth D (1979) Carcinogenicity of halogenated and aliphatic hydrocarbons in mice. *J Nat Cancer Inst 63*: 1433–1439

Vernot EH, MacEwen JD, Haun CC, Kinkead ER (1977) Acute toxicity and skin corrosion data for some organic and inorganic compounds and aqueous solutions. *Toxicol Appl Pharmacol 42*: 417–423

von Oettingen WF (1964) *The Halogenated Hydrocarbons of Industrial and Toxicological Importance*, Elsevier Publishing Co., Amsterdam, p 240

Walles SA (1986) Induction of single-strand breaks in DNA of mice by trichloroethylene and tetrachloroethylene. *Toxicol Lett 31*: 31–35

Waskell L (1978) Study on the mutagenicity of anesthetics and their metabolites. *Mutat Res 57*: 141–153

Weitbrecht U (1957) Beurteilung der Trichlorethylen-Gefährdung in Betrieb. *Zbl Arbeitsmed 7*: 55–58

White JF, Carlson GP (1981) Epinephrine-induced cardiac arrhythmias in rabbits exposed to trichloroethylene: role of trichloroethylene metabolites. *Toxicol Appl Pharmacol 60*: 458–465

Wolfgang GHI, Gandolfi AJ, Stevens JL, Brendel K (1989) In vitro and in vivo nephrotoxicity of the L and D isomers of S-(1,2-dichlorovinyl)-cysteine. *Toxicology 58*: 33–42

Zenick H, Blackburn K, Hope E, Richdale N, Smith MK (1984) Effects of trichloroethylene exposure on male reproductive function in rats. *Toxicology 31*: 237–250

completed 17.06.1996

# Contents for Volumes 1–21


6-12® **V** 345
abachi **XIII** 288, **XVIII** 283
absolute ethanol **XII** 129
Abstensil **V** 165
Abstinyl® **V** 165
*Acacia melanoxylon* **XIII** 285, 291–293
acetaldehyde **III** 1–9
acetene **X** 91
acetic acid 2-butoxyethyl ester **VI** 53
acetic acid butyl ester **XIX** 79
acetic acid *sec*-butyl ester **XIX** 89
acetic acid *tert*-butyl ester **XIX** 93
acetic acid 1,1-dimethylethyl ester **XIX** 93
acetic acid ethenyl ester **V** 229
acetic acid 2-ethoxyethyl ester **VI** 213
acetic acid ethyl ester **XII** 167
acetic acid isobutyl ester **XIX** 211
acetic acid 2-methoxy-1-methylethyl ester **V** 217
acetic acid methyl ester **XVIII** 191
acetic acid 1-methylpropyl ester **XIX** 89
acetic acid 2-methylpropyl ester **XIX** 211
acetic acid pentyl ester **XI** 211
acetic acid 2-propoxyethyl ester **XII** 187
acetic acid vinyl ester **V** 229, **XXI** 271
acetic anhydride **XIII** 43–46
acetic ether **XII** 167
acetic oxide **XIII** 43
acetic peroxide **VII** 229
acetone **VII** 1–8
acetonitrile **XIX** 1–41
1-acetoxyethylene **V** 229
2-acetoxypentane **XI** 211
acetyl hydroperoxide **VII** 229
acetyl oxide **XIII** 43
aclarubicin **I** 5
acquinite **VI** 105
acraldehyde **XVI** 1
acrolein **I** 41–43, 60, **XVI** 1–33
acrylaldehyde **XVI** 1
acrylamide **III** 11–21
acrylic acid *n*-butyl ester **V** 5, **XII** 57, **XVI** 35
acrylic acid ethyl ester **VI** 217, **XVI** 41
acrylic acid 2-ethylhexyl ester **XVI** 47
acrylic acid 2-hydroxyethyl ester **XVI** 89
acrylic acid hydroxypropyl ester **XVI** 95
acrylic acid methyl ester **VI** 253, **XVI** 177
acrylic acid monoester with propanediol **XVI** 95

acrylic acid pentaerythritol triester **XVI** 193
acrylic acid polymer, neutralized, cross-linked **XV** 1–29
acrylic acid 1,1,1-(trihydroxymethyl)propane triester **XVI** 201
acrylic aldehyde **XVI** 1
acrylic amide **III** 11
Acticide® 45 **XVI** 263
actinolite **II** 96, 184, 185
actinomycin D **I** 5
adriamycin **I** 5
AEPD® **IX** 223
aeropur® **I** 125
aerosols **XII** 271–292, **XVI** 289
afara **XIII** 288
African acajou **XVIII** 239
African afzelia **XIII** 285
African black walnut **XIV** 299
African blackwood **XIII** 286, 305
AfricaN'cherry' **XIII** 288
African ebony **XIII** 286, **XIV** 287
African mahogany **XIII** 285, 286, 287, **XVIII** 239
African maple **XIII** 288, **XVIII** 283
African satinwood **XIII** 286, **XIV** 291
African whitewood **XIII** 288, **XVIII** 283
*Afrormosia elata* **XIII** 287
*Afzelia* spp. **XIII** 285
AGE **VII** 9
alabaster **II** 117
alcohol **XII** 129
allyl alcohol **XV** 31–40
allyl chloride **XVIII** 1–18
allyl 2,3-epoxypropyl ether **VII** 9
allyl glycidyl ether **VII** 9–16
1-allyloxy-2,3-epoxypropane **VII** 9
allyl trichloride **IX** 171
altretamine **I** 2
aluminium **II** 69–93
aluminium hydroxide **II** 69–93
aluminium oxide **II** 69–93, **VIII** 141–338
Amazon mahogany **XVIII** 253
American arborvitae **XIII** 288
American black walnut **XIII** 286
American mahogany **XIII** 287, **XVIII** 253
American red oak **XVIII** 247
American walnut **XIII** 286
*Amerimnum ebenus* **XIII** 285, 299
Amine 220® **V** 373



*Essential MAK Value Documentations*. DFG, Deutsche Forschungsgemeinschaft
Copyright © 2006 WILEY-VCH Verlag GmbH & Co. KGaA, Weinheim
ISBN: 3-527-31394-X

Amine CS-1246 **IX** 275
aminic acid **XIX** 169
2-aminoaniline **XIII** 215
3-aminoaniline **VI** 287
4-aminoaniline **VI** 311
*m*-aminoaniline **VI** 287
*o*-aminoaniline **XIII** 215
*p*-aminoaniline **VI** 311
2-aminoanisole **X** 1
*o*-aminoanisole **X** 1
aminobenzene **VI** 17
4-(4-aminobenzyl)aniline **VII** 37
4-aminobiphenyl **I** 257–259
6-aminocaproic acid lactam **IV** 65
1-amino-2-chlorobenzene **III** 31
1-amino-3-chlorobenzene **III** 37
1-amino-4-chlorobenzene **III** 45
*m*-aminochlorobenzene **III** 37
1-amino-3-chloro-6-methylbenzene **VI** 143
2-amino-4-chlorotoluene **VI** 143
2-amino-5-chlorotoluene **VI** 127
3-amino-*p*-cresol methyl ether **IV** 135
*m*-amino-*p*-cresol methyl ether **IV** 135
aminodiglycol **IX** 215
amino-dimethyl-benzene isomers **XIX** 299
2-aminodimethylethanol **IX** 229
2-aminoethanol **XII** 15–35
2-(2-aminoethoxy)ethanol **IX** 215–222
2-aminoethoxyethanol **IX** 215
2-amino-6-ethoxynaphthalene **VII** 17
6-amino-2-ethoxynaphthalene **VII** 17–19
*β*-aminoethyl alcohol **XII** 15
3-amino-9-ethylcarbazole **V** 1–3
3-amino-*N*-ethylcarbazole **V** 1
2-amino-2-ethyl-1,3-propanediol **IX** 223–227
Aminoform **V** 355
aminoglutethimide **I** 2
6-aminohexanoic acid cyclic lactam **IV** 65
2-aminoisobutanol **IX** 229
*beta*-aminoisobutanol **IX** 229
*alpha*-aminoisopropylalcohol **IX** 237
aminomethane **VII** 145
1-amino-2-methoxybenzene **X** 1
1-amino-2-methoxy-5-methylbenzene **IV** 135
3-amino-4-methoxytoluene **IV** 135
2-amino-4-methylanisole **IV** 135
1-amino-2-methylbenzene **III** 307
1-amino-4-methylbenzene **III** 323
2-amino-1-methylbenzene **III** 307
4-amino-1-methylbenzene **III** 323
1-amino-2-methyl-5-nitrobenzene **VI** 271
2-amino-2-methyl-1-propanol **IX** 229–236
amino-methyl-toluene isomers **XIX** 299
6-aminonaphthol ether **VII** 17

4-amino-2-nitroaniline **IV** 295, **XIII** 207
4-amino-2-nitrophenol **IV** 289
2-amino-4-nitrotoluene **I** 167, **VI** 271, **XXI** 3
*p*-aminophenyl ether **VI** 277
4-aminophenyl ether **VI** 277
1-amino-2-propanol **IX** 237–244
2-aminotoluene **III** 307
4-aminotoluene **III** 323
*o*-aminotoluene **III** 307
*p*-aminotoluene **III** 323
3-amino-*p*-toluidine **VI** 339
5-amino-*o*-toluidine **VI** 339
aminotriazole **XVIII** 19
2-amino-1,3,4-triazole **IV** 1
3-amino-1,2,4-triazole **IV** 1
3-amino-1*H*-1,2,4-triazole **IV** 1
1-amino-2,4,5-trimethylbenzene **IV** 335
aminoxylene isomers **XIX** 299
1,2-aminozophenylene **II** 231
amitrole **IV** 1–10, **XVIII** 19–34
Ammoform **V** 355
ammonia **I** 40, **VI** 1–16, **XIII** 47–48
ammonium molybdate **XVIII** 199
amosite **II** 95–116, 188
AMP® **IX** 229
AMS **XV** 137
1-amyl acetate **XI** 211
*n*-amyl acetate **XI** 211
*sec*-amyl acetate **XI** 211
*tert*-amyl acetate **XI** 211
*α*-amylase **XI** 1–3
amylcarbinol **IX** 283
anaesthetic ether **XIII** 149
anatase **II** 199
anhydrite **II** 117
anhydrous hydrobromic acid **XIII** 187
aniline **I** 40, **III** 32, 42, 50, 151, 152, 159,
    160, **IV** 132, **VI** 17–36, **XXI** 3
anilinomethane **VI** 263
*Aningeria* spp. **XIII** 285
aningré **XIII** 285
2-anisidine **X** 1
*o*-anisidine **X** 1–13, **XXI** 3
*p*-anisidine **XXI** 3
anone **X** 35
anprolene **V** 181
Antabuse® **V** 165
Antadix **V** 165
anthophyllite **II** 96, 182, 183, 187
anthracite dust **XVIII** 107
apyonine auramine base **IV** 12
aqua fortis **III** 233
aquinite **VI** 105
arborvitae **XIII** 288, **XVIII** 263

arsenic acid **XXI** 49
arsenic and its inorganic compounds **XXI** 49–106
arsenic compounds **II** 48, 136
arsenic pentoxide **XXI** 49
arsenic trihydride **XV** 41
arsenic trioxide **XXI** 49
arsenic(lll) acid **XXI** 49
arsenic(lll) oxide **XXI** 49
arsenic(V) acid **XXI** 49
arsenic(V) oxide **XXI** 49
arsenous acid **XXI** 49
arsine **XV** 41–50
artificial almond oil **XVII** 13
artificial oil of ants **XVIII** 189
Arubren CP® **III** 81
asbestos **I** 28, **II** 48, 95–116, 136, 181–198, **VIII** 141–338
ASP 47 **XIII** 237
*Aspalatus ebenus* **XIII** 285, 299
asphalt **XVII** 37–118
Asulgan® K **V** 289
asymmetrical dichloroethylene **VIII** 109
atmospheric pressure **XIV** 11–16
attapulgite **VIII** 141–338
*Aucoumea klaineana* **XIII** 285
auramine **I** 250, **IV** 11–25
auramine base **IV** 11–25
Australian blackwood **XIII** 285, 291
Australian silk(y) oak **XIII** 286, **XIV** 295
avodiré **XIII** 288
ayan **XIII** 286, **XIV** 291
ayous **XVIII** 283
1-aza-2-cycloheptanone **IV** 65
5-azacytidine **I** 4
1-aza-3,7-dioxa-5-ethylbicyclo[3.3.0]octane **IX** 275
azathioprine **I** 4
azimethylene **XIII** 141
azimidobenzene **II** 231
aziminobenzene **II** 231
azotic acid **III** 233
BADGE **XIX** 43
barium chromate **III** 101–122, **VII** 33
basic yellow 2 **IV** 11
basic zinc chromate **XV** 289
battery acid **XV** 165
Bayer E393 **XIII** 237
Baywood **XIII** 287, **XVIII** 253
α-BCH **V** 193
β-BCH **V** 193
BCNU **I** 2
beech wood dust **IV** 363, **XVIII** 221
Beisugi **XVIII** 263

Belize mahogany **XVIII** 253
benzal chloride **VI** 79
benzaldehyde **XVII** 13–36
benzenamine **VI** 17
benzene azimide **II** 231
benzenecarbonyl chloride **VI** 80
benzene chloride **XII** 103
1,2-benzenediamine **XIII** 215
1,3-benzenediamine **VI** 287
1,4-benzenediamine **VI** 311, **XIV** 137
*m*-benzenediamine **VI** 287
*o*-benzenediamine **XIII** 215
*p*-benzenediamine **VI** 311
benzene-1,2-dicarboxylic acid **VII** 241
benzene-1,3-dicarboxylic acid **VII** 241
benzene-1,4-dicarboxylic acid **VII** 241
1,2-benzenedicarboxylic acid anhydride **VII** 247
1,2-benzenedicarboxylic acid di-2-propenyl ester **IX** 11
1,3-benzenediol **XX** 239
1,4-benzenediol **X** 113
*p*-benzenediol **X** 113
α-benzene hexachloride **V** 193
β-benzene hexachloride **V** 193
benzenyl chloride **VI** 80
benzenyl trichloride **VI** 80
benzidine **I** 258–259
1,2-benzisothiazol-3(2*H*)-one **II** 221–230
benzohydroquinone **X** 113
benzoic acid chloride **VI** 80
benzoic aldehyde **XVII** 13
benzoic trichloride **VI** 80
benzoisotriazole **II** 231
benzo[*a*]pyrene **I** 41–43, 58–60, 112, 180
1,2,3-benzotriazole **II** 231
1*H*-benzotriazole **II** 231–238
1*H*-benzotriazole, methyl **II** 295
benzotrichloride **VI** 80
2*H*-3,1-benzoxazine-2,4-[1*H*]-dione **XIII** 97
benzoyl chloride **VI** 80
(benzoyloxy)tributylstannane **I** 315
benzoyl peroxide **III** 249–256
benztriazole **II** 231
benzyl alcohol mono(poly)hemiformal **II** 239–247
benzyl disulfide **II** 283
α-(benzyldithio)toluene **II** 283
benzylene chloride **VI** 79
benzyl hemiformal **II** 239
benzyl hydroxymethyl ether **II** 239
benzylidene chloride **VI** 79
benzylidyne chloride **VI** 80
benzyl oxy methanol **II** 239

benzyl trichloride **VI** 80
bertrandite **XXI** 107
beryl **XXI** 107
beryllium **III** 23–29
beryllium acetate **XXI** 107
beryllium and its inorganic compounds **XXI** 107–160
beryllium carbonate **XXI** 107
beryllium chloride **XXI** 107
beryllium citrate **XXI** 107
beryllium compounds **III** 23–29
beryllium fluoride **XXI** 107
beryllium hydroxide **XXI** 107
beryllium nitrate **XXI** 107
beryllium oxide **XXI** 107
beryllium silicate **XXI** 107
beryllium stearate **XXI** 107
beryllium sulfate **XXI** 107
bété **XIII** 287, **XIV** 299
*N,N'*-bianiline **IX** 83
bianisidine **V** 153
bicarburetted hydrogen **X** 91
1,2-bichloroethane **III** 137
bicyclopentadiene **V** 125
biethylene **XV** 51
big-leaf mahogany **XIII** 287
binitrobenzene **I** 147
Bioban® CS-1246 **IX** 275
Bioban® P-1487 **IX** 295
[1,1'-biphenyl]-2-ol **II** 299
2-biphenylol **II** 299
*o*-biphenylol **II** 299
[1,1'-biphenyl]-2-ol sodium salt **II** 299
2-biphenylol sodium salt **II** 299
(2-biphenyloxy)sodium **II** 299
[1,1'-biphenyl]-3,3',4,4'-tetramine **III** 131
3,3',4,4'-biphenyltetramine **III** 131
bis(4-amino-3-chlorophenyl)methane **VII** 193
bis(4-aminophenyl)ether **VI** 277
bis(4-aminophenyl)methane **VII** 37
bis(*p*-aminophenyl)methane **VII** 37
bis(4-aminophenyl)sulfide **IV** 331
bis(*p*-aminophenyl)sulfide **IV** 331
2,6-bis(*tert*-butyl)phenol **V** 341
bis[(3-carboxyacryloyl)oxy]dioctylstannane diisooctyl ester **VII** 93
bis(*β*-chloroethyl)methylamine **I** 211
*N,N*-bis(2-chloroethyl)-*N*-methylamine **I** 211
bischloroethyl nitrosourea (BCNU) **I** 2
bis(2-chloroethyl)sulfide **IV** 27
bis(*β*-chloroethyl)sulfide **IV** 27–43
biscyclopentadiene **V** 125
bis(3,5-dichloro-2-hydroxyphenyl) sulfide **IX** 245

bis((diethylamino)thioxomethyl)disulfide **V** 165
bis(diethylthiocarbamoyl) disulfide **V** 165
bis(diethylthiocarbamyl) disulfide **V** 165
bis(*N,N*-diethylthiocarbamyl) disulfide **V** 165
4,4'-bis(dimethylamino)diphenylmethane **I** 249
*p,p'*-bis(dimethylamino)diphenylmethane **I** 249
bis(*p*-(dimethylamino)phenyl)methane **I** 249
bis(*p*-dimethylaminophenyl)methyleneimine **IV** 11
2,6-bis(1,1-dimethylethyl)phenol **V** 341
bis(dimethylthiocarbamoyl) disulfide **XV** 163
bis(dimethylthiocarbamyl) disulfide **XV** 163
1,3-bis(2,3-epoxypropoxy)benzene **VII** 69
*m*-bis(2,3-epoxypropoxy)benzene **VII** 69
2,2-bis(4-(2,3-epoxypropoxy)phenyl)propane **XIX** 43
*m*-bis(glycidyloxy)benzene **VII** 69
bis(4-glycidyloxyphenyl)dimethylmethane **XIX** 43
2,2-bis(*p*-glycidyloxyphenyl)propane **XIX** 43
bis(2-hydroxy-3,5-dichlorophenyl) sulfide **IX** 245
bis(2-hydroxyethyl) ether **X** 73
bis(2-hydroxyethyl)methylamine **IX** 291
bis(hydroxymethyl)acetylene **XV** 71
1,3-bis(hydroxymethyl)urea **V** 289–293
*N,N'*-bis(hydroxymethyl)urea **V** 289
bis(4-hydroxyphenyl)dimethylmethane diglycidyl ether **XIX** 43
2,2-bis(4-hydroxyphenyl)propane **XIII** 49
2,2-bis(4-hydroxyphenyl)propane diglycidyl ether **XIX** 43
bis(1,4-isocyanatophenyl)methane **VIII** 65
bis(*p*-isocyanatophenyl)methane **VIII** 65
bis(isooctyloxymaleoyloxy)dioctylstannane **VII** 93
bis(2-methoxyethyl)ether **IX** 41
bisphenol A **XIII** 49–87
bisphenol A diglycidyl ether **XIX** 43–78
bis(phenylmethyl)disulfide **II** 283
bis(tri-*n*-butyltin)oxide **I** 315
bithionol **IX** 245–262
bithionol sulfide **IX** 245
4,4'-bi-*o*-toluidine **V** 153
bitumen **XVII** 37–118
bituminous coal dust **XVIII** 107
bivinyl **XV** 51
black afara **XVIII** 259
black coal dust **XVIII** 107
black ebony **XIII** 286
black sucupira **XIII** 285, 295

black walnut **XIII** 286
black wattle **XIII** 285, 291
Bladafum® **XIII** 237
bleomycin **I** 5
BNPD **II** 249
body burden and hyperbaric pressure **XIV**
   11–16
Bombay blackwood **XIII** 285, 301
boracic acid **V** 295
boric acid **V** 295–321
boric anhydride **XIII** 89
boric oxide **XIII** 89
boroethane **XIV** 83
Borofax® **V** 295
boron fluoride **XIII** 93
boron oxide **XIII** 89–91
boron sesquioxide **XIII** 89
boron trifluoride **XIII** 93–95
boron trioxide **XIII** 89
*Bowdichia nitida* **XIII** 285, 295–297
Brazilian mahogany **XVIII** 253
Brazilian rosewood **XIII** 286, 309
brilliant oil yellow **IV** 11
broadleaf mahogany **XIII** 287, **XVIII** 253
bromic ether **VII** 115
2-bromo-2-(bromomethyl)glutaronitrile **XIV**
   87
2-bromo-2-(bromomethyl)pentanedinitrile
   **XIV** 87
bromoethane **VII** 115
bromofluoroform **VI** 37
bromoform **VII** 319
bromomethane **VII** 155
2-bromo-2-nitro-1,3-propanediol **II** 249–270
$\beta$-bromo-$\beta$-nitrotrimethylene glycol **II** 249
bromotrifluoromethane **VI** 37–45
Bronopol® **II** 249
brookite **II** 199
brown ebony **XIII** 285, 299
brucite **VIII** 141–338
*Brya ebenus* **XIII** 285, 299–300
busulfan **I** 2
$\alpha,\gamma$-butadiene **XV** 51
1,3-butadiene **XV** 51–70
butadiene diamer **XIV** 185
butane **XX** 1–9
*n*-butane **XX** 1
butanesulfone **IV** 45
$\delta$-butane sultone **IV** 45
1,4-butane sultone **IV** 45–49
2,4-butane sultone **IV** 45–49
butanethiol **XXI** 161–170
1-butanol **XIX** 99
2-butanol **XIX** 117

2-butanone **XII** 37–56
2-butanone peroxide **III** 250, 252, 254, 255
*cis*-butenedioic anhydride **IV** 275
1-butene oxide **V** 13
1,2-butene oxide **V** 13
3-buten-2-one **IX** 91
butoxydiethylene glycol **VII** 59
butoxydiglycol **VII** 59
1-*n*-butoxy-2,3-epoxypropane **IV** 51
1-*tert*-butoxy-2,3-epoxypropane **IV** 51
2-butoxyethanol **V** 210, **VI** 47–52
*n*-butoxyethanol **VI** 47
2-butoxyethanol acetate **VI** 53
2-(2-butoxyethoxy)ethanol **VII** 59
2-[2-(2-butoxyethoxy)ethoxy]ethanol **IX** 315
2-butoxyethyl acetate **VI** 53–55
butoxymethyl oxirane **IV** 51
*tert*-butoxymethyl oxirane **IV** 51
butoxytriethylene glycol **IX** 315
butoxytriglycol **IX** 315
buttercup yellow **XV** 289
2-butyl acetate **XIX** 89
*n*-butyl acetate **XIX** 79–88
*sec*-butyl acetate **XIX** 89–92
*tert*-butyl acetate **XIX** 93–97
*n*-butyl acrylate **V** 5–12, **XII** 57–62, **XVI** 35–
   40
*n*-butyl alcohol **XIX** 99–115
*sec*-butyl alcohol **XIX** 117–124
*tert*-butyl alcohol **XIX** 125–140
butyl carbamic acid 3-iodo-2-propynyl ester
   **XVI** 247
butyl carbitol **VII** 59
Butyl Cellosolve® **VI** 47
Butyl Cellosolve® acetate **VI** 53
*O*-butyl diethylene glycol **VII** 59
butyl diglycol **VII** 59
butyl dioxitol **VII** 59
1,2-butylene oxide **V** 13–18
$\alpha$-butylene oxide **V** 13
1,4-butylene sulfone **IV** 45
*O*-butyl ethylene glycol **VI** 47
*n*-butyl glycidyl ether **IV** 51–64
*tert*-butyl glycidyl ether **IV** 51–64
butyl glycol **VI** 47
butyl glycol acetate **VI** 53
*tert*-butyl hydroperoxide **III** 250, 253, 254
butyl-3-iodo-2-propynylcarbamate **XVI** 247
*N*-butyl mercaptan **XXI** 161
*tert*-butyl methyl ether **XVII** 119–145
*tert*-butyl peracetate **III** 250
*tert*-butyl peroxyacetate **III** 250
4-*tert*-butylphenol **XI** 5
*p*-*tert*-butyl phenol **XI** 5–25

466    *Contents for Volumes 1–21*

butyl 2-propenoate **V** 5, **XII** 57
*n*-butyl triethylene glycol **IX** 315
*n*-butyl triglycol **IX** 315
butynediol **XV** 71–74
2-butyne-1,4-diol **XV** 71
2-butyne-2,4-diol **XVI** 233
cadmium **I** 40, **V** 19–50
cadmium compounds **V** 19–50, **VII** 21
cadmium sulfide **VII** 21–25
calcium carbimide **V** 51
calcium cyanamide **V** 51–64
calcium molybdate **XVIII** 199
calcium sodium metaphosphate **VIII** 141–
   338
calcium sulfate **II** 117–128, **VIII** 141–338
Californian redwoood **XIII** 287
*Calocedrus decurrens* **XIII** 285, **XIV** 275–
   278
CAM **IX** 263
Cameroon ebony **XIV** 287
camphechlor **XIX** 281
camphochlor **XIX** 281
6-caprolactam **IV** 65
*ε*-caprolactam **IV** 65–78
caproyl alcohol **IX** 283
caprylic alcohol **XX** 227
captan **I** 292, 295
Carbamol® **V** 289
carbanil **XVII** 267
carbazotic acid **XVII** 273
carbinamine **VII** 145
carbinol **XVI** 143
carbomethene **XX** 191
carbon bichloride **III** 271
carbon bisulfide **XII** 63, **XXI** 171
carbon black **XVIII** 35–80
carbon dichloride **III** 271
carbon disulfide **XII** 63–79, **XXI** 171–185
carbon monoxide **I** 40, 42–43, 109, 111, 119,
   **IV** 79-95, 174, 188
carbon oxide **IV** 79
carbonic oxide **IV** 79
4,4′-carbonimidoylbis(*N,N*-dimethylbenzen-
   amine) **IV** 11, 12
4,4′-carbonimidoylbis[*N,N*-dimethylbenzen-
   amine] monochloride **IV** 11
carbon sulfide **XII** 63, **XXI** 171
carbon tetrachloride **XVIII** 81–106
Carbowax® **X** 247
*N*-carboxyanthranilic anhydride **XIII** 97–98
Caribbean mahogany **XIII** 287, **XVIII** 253
carmustine **I** 2
caustic soda **XII** 195
caviuna **XIII** 287, 321

CCNU **I** 2
CDM **VI** 57
cedar of Lebanon **XIII** 285
*Cedrus* spp. **XIII** 285
Cellosolve **VI** 205
cellosolve acetate **VI** 213
ceramic fibres **VIII** 141–338
cereal flour dusts **XIII** 99–100
Cereclor® **III** 81
Ceylonese ebony **XIII** 286, **XIV** 287
CFC 12 **V** 109
CFC 13 **I** 69
CFC 21 **V** 119
CFC 31 **V** 75
CFC 112 **I** 297, **III** 366
CFC 142b **I** 65
'cherry mahogany' **XIII** 288
chlorallylene **XVIII** 1
chlorambucil **I** 2
chlordimeform **VI** 57–78
chlorfenamidine **VI** 57
chlorinated camphene **XIX** 281
chlorinated hydrocarbon waxes **III** 81
chlorinated naphthalenes **XIII** 101–110
chlorinated paraffins **III** 81, **VII** 27–32
chlorinated paraffin waxes **III** 81
*α*-chlorinated toluenes **VI** 79–103
chlormethine **I** 211
chloroacetaldehyde **XII** 81–102
2-chloroacetaldehyde **XII** 81
chloroacetaldehyde monomer **XII** 81
chloroacetamide-*N*-methylol **IX** 263–267
chloroacetic acid methyl ester **IX** 1–8
2-chloroacrylonitrile **X** 15–19
4-chloro-2-aminotoluene **VI** 143
5-chloro-2-aminotoluene **VI** 127
2-chloroaniline **III** 31
3-chloroaniline **III** 37
4-chloroaniline **III** 45
*m*-chloroaniline **III** 37–43, **XXI** 3
*o*-chloroaniline **III** 31–35, 42, **XXI** 3
*p*-chloroaniline **III** 42, 45–61, **IV** 132, **XXI** 3
chlorobenzal **VI** 79
*α*-chlorobenzaldehyde **VI** 80
2-chlorobenzenamine **III** 31
3-chlorobenzenamine **III** 37
4-chlorobenzenamine **III** 45
chlorobenzene **XII** 103–127
chlorobenzol **XII** 103
*p*-chlorobenzotrichloride **X** 21–30
chlorocamphene **XIX** 281
2-chloro-*N*-(2-chloroethyl)-*N*-methyl-
   ethanamine **I** 211
1-chloro-2-(*β*-chloroethylthio)ethane **IV** 27

*p*-chlorocresol **II** 271
*p*-chloro-*m*-cresol **II** 271–282
chlorodeoxyglycerol **V** 83
1-chloro-2,2-dichloroethylene **X** 201
chlorodifluoroethane **I** 65
1-chloro-1,1-difluoroethane **I** 65–67
chlorodifluoromethane **I** 69, **III** 63–71
2-chloro-2-(difluoromethoxy)-1,1,1-trifluoro-
ethane **VII** 127
2-chloro-1-(difluoromethoxy)-1,1,2-trifluoro-
ethane **IX** 51
1-chloro-2,3-dihydroxypropane **V** 83
3-chloro-1,2-dihydroxypropane **V** 83
1-chloro-2,4-dinitrobenzene **XIII** 111–115,
**XXI** 3
2-chloro-1-ethanal **XII** 81
chloroethane **III** 73–79
2-chloroethanol **V** 65–73
*β*-chloroethanol **V** 65
chloroethene **V** 66, 241
chloroethene homopolymer **II** 149
2-chloroethyl alcohol **V** 65
1-(2-chloroethyl)-3-cyclohexyl-1-nitrosourea
**I** 2
chloroethylene **V** 241
chloroethylidene fluoride **I** 65
chlorofluoromethane **III** 63, 69, 70, **V** 75–77
chloroform **IV** 173, **XIV** 19–58
*N*-chloroformylmorpholine **V** 79–81
chlorohydric acid **VI** 231
*α*-chlorohydrin **V** 83–98
2-chloro-*N*-(hydroxymethyl)acetamide **IX**
263
6-chloro-3-hydroxytoluene **II** 271
*γ*-chloroisobutylene **IV** 97
chloromethane **I** 69, **VII** 173
3-chloro-6-methylaniline **VI** 143
4-chloro-2-methylaniline **VI** 127
5-chloro-2-methylaniline **VI** 143
4-chloro-2-methylbenzenamine **VI** 127
5-chloro-2-methylbenzenamine **VI** 143
(chloromethyl)benzene **VI** 79
4-chloro-2-methylbenzeneamine **VI** 127
5-chloro-2-methyl-2,3-dihydroisothiazol-
3-one **V** 323–339
5-chloro-2-methyl-4-isothiazolin-3(2*H*)-one
**V** 323
5-chloro-2-methyl-3-isothiazolone **V** 323
4-chloro-3-methylphenol **II** 271
*N'*-(4-chloro-2-methylphenyl)-*N,N*-dimethyl-
methanimidamide **VI** 57
3-chloro-2-methylpropene **IV** 97–106
3-chloro-2-methyl-1-propene **IV** 97

1-chloro-2-nitrobenzene **IV** 107
1-chloro-3-nitrobenzene **IV** 115
1-chloro-4-nitrobenzene **IV** 121
2-chloro-1-nitrobenzene **IV** 107
3-chloro-1-nitrobenzene **IV** 115
4-chloro-1-nitrobenzene **IV** 121
*m*-chloronitrobenzene **IV** 107, 115–120, 121,
**XXI** 3
*o*-chloronitrobenzene **IV** 107–114, 115, 121,
**XXI** 3
*p*-chloronitrobenzene **IV** 107, 115, 121–133,
**XXI** 3
1-chloro-1-nitropropane **XI** 27–29
chloroparaffins **III** 81–100, **VII** 27
chlorophenamidine **VI** 57
2-chlorophenylamine **III** 31
3-chlorophenylamine **III** 37
4-chlorophenylamine **III** 45
*m*-chlorophenylamine **III** 37
*o*-chlorophenylamine **III** 31
*p*-chlorophenylamine **III** 45
*p*-chlorophenyl chloride **IV** 141, **XX** 33
chlorophenylmethane **VI** 79
*Chlorophora excelsa* **XIV** 279–285
*Chlorophora* spp. **XIII** 285
chloropicrin **VI** 105–111
chloroprene **III** 205
1-chloro-2,3-propanediol **V** 83
3-chloro-1,2-propanediol **V** 83
3-chloro-1-propene **XVIII** 1
2-chloro-2-propenenitrile **X** 15
3-chloropropylene **XVIII** 1
3-chloropropylene glycol **V** 83
chlorothalonil **VI** 113–126
*α*-chlorotoluene **VI** 79
*ω*-chlorotoluene **VI** 79
4-chloro-2-toluidine **VI** 127
4-chloro-*o*-toluidine **VI** 127–141, 144, **XXI** 3
5-chloro-*o*-toluidine **VI** 143–144, **XXI** 3
*N'*-(4-chloro-*o*-tolyl)-*N,N*-dimethyl-
formamidine **VI** 57
1-chloro-4-(trichloromethyl)benzene **X** 21
1-chloro-2,2,2-trifluoroethyldifluoromethyl
ether **VII** 127
2-chloro-1,1,2-trifluoroethyldifluoromethyl
ether **IX** 51
chlorotrifluoromethane **I** 69–70
Chlorowax® **III** 81
chlorphenamidine **VI** 57
chlorpromazine **IX** 9–10
chlorvinphos **IV** 201
chromates **III** 101–122
chromic acid zinc salt **XV** 289

chromium compounds **III** 101–122
chromium(VI) compounds **II** 47, 48, **III** 101–122, **VII** 33–35
chromium potassium zinc oxide **XV** 289
chrysotile **I** 28, **II** 95–116, 182, 184, **VIII** 141–338
C.I. 23060 **V** 99
C.I. 24110 **V** 139
C.I. 37105 **VI** 271
C.I. 37230 **V** 153
C.I. 41000 **IV** 11
C.I. 76000 **VI** 17
C.I. 76025 **VI** 287
C.I. 76035 **VI** 339
C.I. 76050 **VI** 145
C.I. 76060 **VI** 311
C.I. 76070 **IV** 295, **XIII** 207
C.I. 76555 **IV** 289
C.I. 77947 **XVIII** 305
C.I. azoic red 83 **IV** 135
cigarette smoke **I** 39–62, **XIII** 3–39
cinnamene **XX** 285
C.I. oxidation base 22 **IV** 295, **XIII** 207
C.I. pigment 4 **XVIII** 305
C.I. pigment yellow 36 **XV** 289
cisplatin **I** 6
citric acid **XVI** 209–230
citron yellow **XV** 289
classification of carcinogenic chemicals **XII** 3–12
Cloparin® **III** 81
coal mine dust **XVIII** 107–156
coastal redwood **XIII** 287
cobalt **III** 123–130, **X** 31–34
cobalt compounds **III** 123–130, **X** 31–34
Cobratec® 99 **II** 231
cocobolo **XIII** 286, 313
cocuswood **XIII** 285, 299
colloidal mercury **XV** 81
colophony **XI** 239
common oak **XVIII** 247
Compound 469 **VII** 127
Contralin **V** 165
coromandel **XIII** 286, **XIV** 287
*p*-cresidine **IV** 135–139
*o*-cresol **XIV** 59
*m*-cresol **XIV** 59
*p*-cresol **XIV** 59
cresol (all isomers) **XIV** 59–82
*o*-cresylic acid **XIV** 59
*m*-cresylic acid **XIV** 59
*p*-cresylic acid **XIV** 59
cristobalite **II** 182, **XIV** 205
crocidolite **II** 95–116, **VIII** 141–338

Cronetal **V** 165
cross-linked polyacrylates **XV** 1
crystalline silicon dioxide **XIV** 205–271
Cuban mahogany **XIII** 287, **XVIII** 253
cumene **XIII** 117–128
cumene hydroperoxide **III** 250, 253–255
cumol **XIII** 117
cyanamide, calcium salt (1:1) **V** 51
cyanoacrylates **XIII** 201
2-cyanoacrylic acid ethyl ester **I** 233, **XIII** 201
2-cyanoacrylic acid methyl ester **I** 233, **XIII** 201
cyanomethane **XIX** 1
2-cyano-2-propenoic acid ethyl ester **I** 233
2-cyano-2-propenoic acid methyl ester **I** 233
cyclohexane **XIII** 129–140
cyclohexanone **X** 35–51
cyclohexanone iso-oxime **IV** 65
cyclohexenylethylene **XIV** 185
cyclohexyl ketone **X** 35
1,3-cyclopentadiene dimer **V** 125
cyclophosphamide **I** 2
Cystamin **V** 355
Cystogen **V** 355
cytarabine **I** 4
2,4-D **IV** 191, **XI** 61
2,4-DAA **VI** 145
dacarbazine **I** 6
DACPM **VII** 193
dactinomycin **I** 5
*Dalbergia latifolia* **XIII** 285, 301–303
*Dalbergia melanoxylon* **XIII** 286, 305–307
*Dalbergia nigra* **XIII** 286, 309–311
*Dalbergia retusa* **XIII** 286, 313–316
*Dalbergia stevensonii* **XIII** 286, 317–319
DAPM **VII** 37
daunomycin **I** 5
daunorubicin hydrochloride **I** 5
dawsonite **VIII** 141–338
DBDCB **XIV** 87
1,2-DCE **III** 137
DDVP **IV** 201
decyl 9-octadecenoate **IX** 269
decyl oleate **IX** 269–274
demeton **XIX** 141–149
demeton-O **XIX** 141
demeton-S **XIX** 141
deodar cedar **XIII** 285
DGA® **IX** 215
DGEBA **XIX** 43
DGEBPA **XIX** 43
diallyl phthalate **IX** 11–19
diamide **I** 171

diamide monohydrate **XIII** 181
diamine **I** 171
2,4-diamineanisole **VI** 145
2,4-diaminoanisole **VI** 145–156, **XXI** 3
*m*-diaminoanisole **VI** 145
2,4-diaminoanisole base **VI** 145
2,4-diaminoanisole sulfate **VI** 145–156
1,2-diaminobenzene **XIII** 215, **XXI** 3
1,3-diaminobenzene **VI** 287, **XXI** 3
1,4-diaminobenzene **VI** 311, **XXI** 3
*m*-diaminobenzene **VI** 287
*o*-diaminobenzene **XIII** 215
*p*-diaminobenzene **VI** 311
3,3′-diaminobenzidine **III** 131–135
3,3′-diaminobenzidine tetrahydrochloride **III** 131–135
4,4′-diamino-3,3′-bichlorobiphenyl **V** 99
di(4-amino-3-chlorophenyl)methane **VII** 193
4,4′-diamino-3,3′-dichlorodiphenyl **V** 99
4,4′-diamino-3,3′-dichlorodiphenylmethane **VII** 193
4,4′-diamino-3,3′-dimethoxybiphenyl **V** 139
4,4′-diamino-3,3′-dimethylbiphenyl **V** 153
4,4′-diaminodiphenyl ether **VI** 277
4,4′-diaminodiphenylmethane **VII** 37–57
*p,p*′-diaminodiphenylmethane **VII** 37
4,4′-diaminodiphenyl oxide **VI** 277
4,4′-diaminodiphenylsulfide **IV** 331
*p,p*′-diaminodiphenylsulfide **IV** 331
diaminoditolyl **V** 153
1,3-diamino-4-methoxybenzene **VI** 145
2,4-diamino-1-methoxybenzene **VI** 145
1,3-diamino-4-methylbenzene **VI** 339
2,4-diamino-1-methylbenzene **VI** 339
1,4-diamino-2-nitrobenzene **IV** 295, **XIII** 207
di(4-aminophenyl)methane **VII** 37
di(*p*-aminophenyl)sulfide **IV** 331
2,4-diaminotoluene **VI** 339
dianilinomethane **VII** 37
dianisidine **V** 139
diatomaceous earth **II** 157
1,4-diazacyclohexane **IX** 303, **XII** 177
diazinon **XI** 31–59
diazomethane **XIII** 141–148
dibenzoyl peroxide **III** 249–256
dibenzyl disulfide **II** 283–286
diborane **XIV** 83–85
diborane(6) **XIV** 83
diboron hexahydride **XIV** 83
[2,7-dibromo-9-(*o*-carboxyphenyl)-6-hydroxy-3-oxo-3*H*-xanthen-4-yl] hydroxymercury disodium salt **XV** 75
1,2-dibromo-2,4-dicyanobutane **XIV** 87–90

(2′,7′-dibromo-3′,6′-dihydroxy-3-oxo-spiro-[isobenzofuran-1(3H),9′-[9H]-xanthen]-4′-yl)hydroxymercury disodium salt **XV** 75
1,2-dibromoethane **III** 144
dibromohydroxymercurifluorescein disodium salt **XV** 75
di-*tert*-butyl hydroperoxide **III** 254
2,6-di-*tert*-butylphenol **V** 341–344
dicarboxybenzene **VII** 241
dichloroacetylene **VI** 157–163
1,2-dichlorobenzene **I** 71–80, **XX** 11–32
1,3-dichlorobenzene **I** 83–89
1,4-dichlorobenzene **I** 71, 73, 83, 196, 347, **IV** 141–171, **XX** 33–92
*m*-dichlorobenzene **I** 83
*o*-dichlorobenzene **I** 71, **XX** 11
*p*-dichlorobenzene **I** 71, 73, 83, 196, 347, **IV** 141, **XX** 33
3,3′-dichlorobenzidine **V** 99–107
*o,o*′-dichlorobenzidine **V** 99
3,3′-dichloro(1,1′-biphenyl)-4,4′-diamine **V** 99
3,3′-dichlorobiphenyl-4,4′-diamine **V** 99
3,3′-dichloro-4,4′-biphenylenediamine **V** 99
3,3′-dichloro-4,4′-diaminobiphenyl **V** 99
3,3′-dichloro-4,4′-diaminodiphenylmethane **VII** 193
2,2′-dichlorodiethylether **V** 66, 71, 72
2,2′-dichlorodiethyl sulfide **IV** 27
*β,β*′-dichlorodiethyl sulfide **IV** 27
dichlorodifluoromethane **I** 336, 341, **V** 109–117
dichlorodioctylstannane **VII** 91
1,2-dichloroethane **III** 137–147
*α,β*-dichloroethane **III** 137
1,1-dichloroethene **V** 66, 72, **VIII** 109
2,2-dichloroethenyl dimethyl phosphate **IV** 201
2,2-dichloroethenyl phosphoric acid dimethyl ester **IV** 201
1,1-dichloroethylene **VIII** 109
di(2-chloroethyl)methylamine **I** 211
1,2-dichloroethylmethyl ether **I** 91
*α,β*-dichloroethylmethyl ether **I** 91
di-2-chloroethyl sulfide **IV** 27
dichloroethyne **VI** 157
dichlorofluoromethane **I** 69, **V** 119–124
*α*-dichlorohydrin **I** 95
*sym*-dichloroisopropyl alcohol **I** 95
dichloromethane **IV** 173–190, **XVII** 147–161
1,2-dichloromethoxyethane **I** 91–93
(dichloromethyl)benzene **VI** 79
2,2′-dichloro-*N*-methyldiethylamine **I** 211

dichloromonofluoromethane **V** 119
dichloronaphthalenes **XIII** 101
2,4-dichlorophenoxyacetic acid **IV** 191–200,
    **XI** 61–104
2,4-dichlorophenoxyacetic acid salts and
    esters **XI** 61
dichlorophos **IV** 201
1,2-dichloropropane **IX** 21–39
1,3-dichloro-2-propanol **I** 95–98
*α,α*-dichlorotoluene **VI** 79
2,2-dichloro-1,1,1-trifluoroethane **X** 53–71
2,2-dichlorovinyl dimethyl phosphate **IV** 201
2,2-dichlorovinyl phosphoric acid dimethyl
    ester **IV** 201
dichlorovos **IV** 201
dichlorvos **IV** 201–216
dicumyl hydroperoxide **III** 252
dicumyl peroxide **III** 251
1,3-dicyanotetrachlorobenzene **VI** 113
dicyclopentadiene **V** 125–133
diesel engine emissions **I** 101–120, **II** 136
diethanolmethylamine **IX** 291
diethylamine **I** 25–28, 32–34
(diethylamino)ethane **XIII** 267
2-diethylaminoethanol **XIV** 91–100
*β*-diethylamino ethyl alcohol **XIV** 91
diethylcarbinol acetate **XI** 211
1,4-diethylenediamine **IX** 303, **XII** 177
1,4-diethylene dioxide **XX** 105
diethylene ether **XX** 105
diethylene glycol **X** 73–90
diethylene glycol *n*-butyl ether **VII** 59
diethylene glycol dimethyl ether **IX** 41–50
diethylene glycol monobutyl ether **VII** 59–67
1,4-diethylene oxide **XX** 105
*N,N*-diethylethanamine **XIII** 267
diethyl ether **I** 125, **XIII** 149–160
*O,O*-diethyl-*O*-(2-(ethylthio)ethyl)-
    phosphorthioate **XIX** 141
*O,O*-diethyl-*S*-(2-(ethylthio)ethyl)-
    phosphorthioate **XIX** 141
diethylmethylmethane **IV** 269
diethyl monosulfate **XX** 93
diethylnitrosamine **I** 23–35, 41, 60, 261–263,
    268, 286
diethyl oxide **XIII** 149
diethyl sulfate **XX** 93–104
Diflamoll® TP **II** 321
difluorochloromethane **III** 63
difluorodichloromethane **V** 109
1,1-difluoroethene **V** 135
1,1-difluoroethylene **V** 135–137
difluoromonochloroethane **I** 65
difluoromonochloromethane **III** 63

1,2-difluoro-1,1,2,2-tetrachloroethane **I** 297
1,3-diformyl propane **VIII** 45, **XVI** 59
diglycidyl ether **IV** 59
diglycidyl ether of bisphenol A **XIX** 43
1,3-diglycidyloxybenzene **VII** 69
diglycidyl resorcinol ether **VII** 69–74
diglycolamine **IX** 215
Diglycolamine® Agent **IX** 215
diglycol monobutyl ether **VII** 59
diglyme **IX** 41
1,3-dihydro-1,3-dioxo-isobenzofuran **VII** 247
dihydrooxirene **V** 181
1,4-dihydroxybenzene **X** 113
*m*-dihydroxybenzene **XX** 239
4,4′-dihydroxydiphenylpropane **XIII** 49
*p,p′*-dihydroxydiphenylpropane **XIII** 49
1,2-dihydroxyethane **IV** 225
2,2′-dihydroxyethyl ether **X** 73
*β,β′*-dihydroxyisopropyl chloride **V** 83
2,4-dihydroxy-2-methylpentane **XVI** 233
*N,N′*-dihydroxymethylurea **V** 289
2,3-dihydroxypropyl chloride **V** 83
2,2′-dihydroxy-3,3′,5,5′-tetrachloro-
    diphenylsulfide **IX** 245
diisobutylketone **XVIII** 157–164
4,4′-diisocyanatodiphenylmethane **VIII** 65
*p,p′*-diisocyanatodiphenylmethane **VIII** 65
2,4-diisocyanato-1-methylbenzene **XX** 291
2,6-diisocyanato-1-methylbenzene **XX** 291
diisooctyl[(dioctylstannylene)dithio]diacetate
    **VII** 92
diisopropyl ether **XXI** 187–193
diisopropyl peroxydicarbonate **III** 250
dilauroyl peroxide **III** 250–255
dimethanol urea **V** 289
3,3′-dimethoxybenzidine **V** 139–152
3,3′-dimethoxy-[1,1′-biphenyl]-4,4′-diamine
    **V** 139
3,3′-dimethoxy-4,4′-diaminobiphenyl **V** 139
dimethoxyphosphine oxide **I** 133
dimethylamine **I** 25–28, 32–34, **VII** 75–89
*N,N*-dimethylaminobenzene **III** 149
4,4′-dimethylaminobenzophenonimide **IV** 11
dimethylaminosulfonyl chloride **I** 143
2,4-dimethylaniline **XXI** 3
2,6-dimethylaniline **XXI** 3
*N,N*-dimethylaniline **III** 149–161, **XXI** 3
dimethylaniline isomers **XIX** 299
*N,N*-dimethylbenzenamine **III** 149
dimethylbenzenamine isomers **XIX** 299
dimethylbenzene **V** 263, **XV** 257
3,3′-dimethylbenzidine **V** 153–164
*α,α*-dimethylbenzyl hydroperoxide **III** 250,
    253–255

3,3′-dimethyl-[1,1′-biphenyl]-4,4′-diamine  **V** 153

3,3′-dimethylbiphenyl-4,4′-diamine  **V** 153

2,2-dimethylbutane  **IV** 269

2,3-dimethylbutane  **IV** 269

*N,N*-dimethylcarbamoyl chloride  **I** 143, 145

dimethyl 2,2-dichlorovinyl phosphate  **IV** 201

*O,O*-dimethyl-*O*-(2,2-dichlorovinyl)-phosphate  **IV** 201

3,3′-dimethyldiphenyl-4,4′-diamine  **V** 153

dimethyldithiocarbamic acid iron salt  **XIX** 163

dimethylene oxide  **V** 181

dimethyl ether  **I** 125–131

[(1,1-dimethylethoxy)methyl]oxirane  **IV** 51

4-(1,1-dimethylethyl)phenol  **XI** 5

dimethylformaldehyde  **VII** 1

dimethylformamide  **VIII** 1–44

*N,N*-dimethylformamide  **VIII** 1

2,6-dimethyl-4-heptanone  **XVIII** 157

1,2-dimethylhydrazine  **I** 176

dimethylketal  **VII** 1

dimethyl ketone  **VII** 1

*N,N*-dimethylmethanamide  **VIII** 1

dimethyl monosulfate  **IV** 217

dimethylnitromethane  **III** 241

dimethylnitrosamine  **I** 23–35, 41–44, 59, 261–264, 286

1,3-dimethylolurea  **V** 289

*N,N′*-dimethylolurea  **V** 289

*N,N*-dimethylphenylamine  **III** 149

dimethylphenylamine isomers  **XIX** 299

dimethyl phosphite  **I** 133

dimethyl phosphonate  **I** 133

1,1-dimethylpropyl acetate  **XI** 211

dimethylpropylmethane  **IV** 269

*N,N*-dimethylsulfamoyl chloride  **I** 143–145

*N,N*-dimethylsulfamyl chloride  **I** 143

dimethyl sulfate  **IV** 48, 217–223

dimethyl sulfoxide  **III** 163–171

dinitrobenzene (all isomers)  **I** 147–167, **XXI** 3

2,4-dinitro-1-chlorobenzene  **XIII** 111

4,6-dinitro-*o*-cresol  **XIX** 151–161, **XXI** 3

dinitrogen monoxide  **IX** 115

dinitrophenylmethane  **VI** 165

dinitrotoluenes  **I** 167, 361, **VI** 165–198

*ar,ar*-dinitrotoluene  **VI** 165

2,3-dinitrotoluene  **VI** 165

2,4-dinitrotoluene  **VI** 165, **XXI** 3

2,5-dinitrotoluene  **VI** 165

2,6-dinitrotoluene  **VI** 165, **XXI** 3

3,4-dinitrotoluene  **VI** 165

3,5-dinitrotoluene  **VI** 165

dioctyldichlorostannane  **VII** 91

2,2-dioctyl-1,3,2-dioxastannepin-4,7-dione  **VII** 93

1,3-dioctyl-1,3-dioxodistannoxane  **VII** 93

dioctyloxostannane  **VII** 91

dioctylstannium dichloride  **VII** 91

4,4′-[(dioctylstannylene)bis(oxy)]bis(4-oxo-2-butenoic acid) diisooctyl ester  **VII** 93

2,2′-[(dioctylstannylene)-bis(thio)]bis(acetic acid)bis(2-ethylhexyl) ester  **VII** 92

2,2′-[(dioctylstannylene)bis(thio)]bis(acetic acid) diisooctyl ester  **VII** 92

dioctylstannylene maleate  **VII** 93

di-*n*-octyltin bis(2-ethylhexyl mercaptoacetate)  **VII** 92

di-*n*-octyltin bis(2-ethylhexyl thioglycolate)  **VII** 92

di-*n*-octyltin bis(isooctyl maleate)  **VII** 93

dioctyltin-*S,S′*-bis(isooctyl mercaptoacetate)  **VII** 92

di-*n*-octyltin bis(isooctyl thioglycolate)  **VII** 92

di-*n*-octyltin compounds  **VII** 91–114

di-*n*-octyltin dichloride  **VII** 91

dioctyltin di(isooctyl thioglycolate)  **VII** 92

di-*n*-octyltin dithioglycolic acid 2-ethylhexyl ester  **VII** 92

di-*n*-octyltin-2-ethylhexyl-dimercapto-ethanoate  **VII** 92

di-*n*-octyltin maleate  **VII** 93

di-*n*-octyltin oxide  **VII** 91

Diolane  **XVI** 233

*Diospyros* spp.  **XIII** 286, **XIV** 287–290

1,4-dioxacyclohexane  **XX** 105

1,4-dioxane  **XX** 105–133

*p*-dioxane  **XX** 105

2,5-dioxo-dihydrofuran  **IV** 275

1,3-dioxophthalan  **VII** 247

dioxyethylene ether  **XX** 105

dipentene  **I** 185

1,4-diphenyl-2,3-dithiabutane  **II** 283

1,2-diphenylhydrazine  **IX** 83

*N,N′*-diphenylhydrazine  **IX** 83

4,4′-diphenylmethanediamine  **VII** 37

diphenylmethane-4,4′-diisocyanate  **VIII** 65

diphenylmethane-*p,p′*-diisocyanate  **VIII** 65

di(phenylmethyl)disulfide  **II** 283

*o*-diphenylol  **II** 299

diphenylolpropane  **XIII** 49

dipropylamine  **I** 25

dipropylene glycol monomethyl ether  **VI** 199–204

dipropyl methane  **XI** 165

disodium 2′,7′-dibromo-4′-(hydroxymercury)fluorescein **XV** 75
*Distemonanthus benthamianus* **XIII** 286,
**XIV** 291–293
disulfiram **V** 165–180
disulfuram **V** 165
2,6-ditertiary butyl phenol **V** 341
dithio **XIII** 237
1,1-dithiobis(*N*,*N*-diethylthioformamide) **V**
165
dithiocarbamates **XV** 147–149
dithiocarbonic anhydride **XII** 63, **XXI** 171
dithione **XIII** 237
dithiophos **XIII** 237
4,4′-di-*o*-toluidine **V** 153
divinyl **XV** 51
DMF **VIII** 1
DMFA **VIII** 1
DMSO **III** 163
DNCB **XIII** 111
DNOC **XIX** 151
DNT **VI** 165
dolomite **II** 182
Dominican mahogany **XVIII** 253
DOTC **VII** 91
DOTO **VII** 91
DOTTG **VII** 92
douka **XVIII** 277
doussié **XIII** 285
Dowanol® TBH **IX** 315
Dowicide® **II** 299
doxorubicin **I** 5
dry-zone mahogany **XIII** 287
*Dumoria africana* **XVIII** 277
*Dumoria heckelii* **XIII** 288, **XVIII** 277
Durmast oak **XVIII** 247
dusts **II** 3–46, 47–57, **XI** 281–301, **XII** 239,
271, **XVI** 287–315
Dymel A® **I** 125
eastern white cedar **XIII** 288, **XVIII** 263
East Indian ebony **XIV** 287
East Indian rosewood **XIII** 285, 301
EDAO **IX** 275, **XVI** 231
elayl **X** 91
endrin **XVIII** 165–188
enflurane **IX** 51–68
*Entandrophragma* spp. **XVIII** 229–233
environmental tobacco smoke **XIII** 3
epirubicin **I** 5
epirubicin hydrochloride **I** 5
1,2-epoxy-3-allyoxypropane **VII** 9
1,2-epoxybutane **V** 13
1,2-epoxy-4-(epoxyethyl)cyclohexane **I** 391
1,2-epoxyethane **V** 181

1-(epoxyethyl)-3,4-epoxycyclohexane **I** 391
3-(epoxyethyl)-7-oxabicyclo(4.1.0)heptane **I**
391
4-(epoxyethyl)-7-oxabicyclo(4.1.0)heptane **I**
391
3-(1,2-epoxyethyl)-7-oxabicyclo(4.1.0)-
heptane **I** 391
4-(1,2-epoxyethyl)-7-oxabicyclo(4.1.0)-
heptane **I** 391
1,2-epoxy-3-isopropoxypropane **VII** 141
1,2-epoxy-3-phenoxypropane **IV** 305
1,2-epoxypropane **V** 221
2,3-epoxypropane **V** 221
2,3-epoxy-1-propanol **XX** 179
2,3-epoxypropyl phenyl ether **IV** 305
(2,3-epoxypropyl)trimethylammonium
chloride **IV** 247
erionite **VIII** 141–338
erythrene **XV** 51
Esperal® **V** 165
essence of mirbane **XIX** 227
ESTOL IDCO 3667 **XVI** 257
estramustine **I** 2
Etabus **V** 165
ethanal **III** 1
ethane dichloride **III** 137
1,2-ethanediol **IV** 225
ethanenitrile **XIX** 1
ethaneperoxoic acid **VII** 229
ethanethiol **XXI** 195–203
ethanoic acid ethenyl ester **V** 229
ethanol **XII** 129–165
ethanolamine **XII** 15
Ethanox 701® **V** 341
ethene **X** 91
ethene oxide **V** 181
ethenone **XX** 191
ethenyl acetate **V** 229
ethenylbenzene **XX** 285
9-ethenyl-9*H*-carbazole **XIV** 181
4-ethenyl-1-cyclohexene **XIV** 185
ethenyl ethanoate **V** 229
1-ethenyl-2-pyrrolidinone **V** 249
ether **XIII** 149
ethinyl trichloride **X** 201
ethohexadiol **V** 345
2-ethoxy-6-aminonaphthalene **VII** 17
ethoxyethane **XIII** 149
2-ethoxyethanol **I** 205, 209, **V** 210, **VI** 205–
211, **XI** 105–119
2-ethoxyethanol acetate **VI** 213, **XI** 105
2-ethoxyethyl acetate **VI** 213–216, **XI** 105–
119
ethyl acetate **XII** 167–176

ethyl acrylate **VI** 217–229, **XVI** 41–46
ethyl alcohol **XII** 129
ethyl aldehyde **III** 1
ethylamine **I** 28, 34
ethyl bromide **VII** 115–120
ethylcellosolve **VI** 205
ethyl chloride **III** 73
ethyl 2-cyanoacrylate **I** 233–245, **XIII** 201–206
ethyl $\alpha$-cyanoacrylate **I** 233, **XIII** 201
7$\alpha$-ethyldihydro-1*H*,3*H*,5*H*-oxazolo-[3,4-*c*]oxazole **IX** 275, **XVI** 231
10-ethyl-4,4-dioctyl-7-oxo-8-oxa-3,5-dithia-4-stannatetradecanoic acid 2-ethylhexyl ester **VII** 92
5-ethyl-3,7-dioxa-1-azabicyclo[3.3.0]octane **IX** 275–281, **XVI** 231–232
ethyldithiurame **V** 165
ethylene **X** 91–107
ethylene carboxamide **III** 11
ethylene chloride **III** 137
ethylene chlorohydrin **V** 65
ethylene dichloride **III** 137
ethylene glycol **IV** 225–245
ethylene glycol *n*-butyl ether **VI** 47
ethylene glycol chlorohydrin **V** 65
ethylene glycol dimethacrylate **XIII** 161–163
ethylene glycol ethyl ether **VI** 205
ethylene glycol ethyl ether acetate **VI** 213
ethylene glycol isopropyl ether **V** 207
ethylene glycol methacrylate **XIII** 193
ethylene glycol methyl ether **VI** 239
ethylene glycol methyl ether acetate **VI** 249
ethylene glycol monoacrylate **XVI** 89
ethylene glycol monobutyl ether **VI** 47
ethylene glycol monobutyl ether acetate **VI** 53
ethylene glycol monoethyl ether **VI** 205
ethylene glycol monoethyl ether acetate **VI** 213
ethylene glycol monoisopropyl ether **V** 207
ethylene glycol monomethacrylate **XIII** 193
ethylene glycol monomethyl ether **VI** 239
ethylene glycol monomethyl ether acetate **VI** 249
ethylene glycol mono-*n*-propyl ether **XII** 179
ethylene glycol monopropyl ether acetate **XII** 187
ethylene monochloride **V** 241
ethylene oxide **V** 181–192, **XIII** 165–168
1-ethyleneoxy-3,4-epoxycyclohexane **I** 391
ethylene tetrachloride **III** 271
ethylene thiourea **XI** 121–163
ethylene trichloride **X** 201

ethyl ether **XIII** 149
ethyl-ethylene oxide **V** 13
ethyl formate **XIX** 181
ethyl glycol acetate **VI** 213
2-ethylhexane-1,3-diol **V** 345
2-ethyl-1,3-hexanediol **V** 345–353
2-ethylhexanol **XX** 135–178
2-ethyl-1-hexanol **XX** 135
2-ethylhexyl acrylate **XVI** 47–50
ethyl hexylene glycol **V** 345
2-ethylhexyl propenoate **XVI** 47
2-ethyl-2-(hydroxymethyl)-1,3-propanediol triacrylate **XVI** 201
ethyl mercaptan **XXI** 195
ethyl(2-mercaptobenzoato-*S*)mercury sodium salt **XV** 249
ethyl methacrylate **XVI** 51–58
ethyl $\alpha$-methacrylate **XVI** 51
ethyl methyl ketone **XII** 37
ethyl 2-methyl-2-propenoate **XVI** 51
4,4′-(2-ethyl-2-nitro-1,3-propanediyl)-bismorpholine **IX** 295–301
4,4′-(2-ethyl-2-nitro-trimethylene)-dimorpholine **IX** 295
ethyl orthosilicate **III** 299
ethyl oxide **XIII** 149
ethyl oxirane **V** 13
ethyl propenate **VI** 217
ethyl-2-propenoate **VI** 217
2-ethyl-3-propyl-1,3-propanediol **V** 345
ethyl silicate **III** 299
ethyl sulfate **XX** 93
ethyl sulfhydrate **XXI** 195
ethyl thiram **V** 165
Ethyl Thiurad **V** 165
ethyltrimethylmethane **IV** 269
etoposide **I** 6
ETS **XIII** 3
European walnut **XIII** 286
excursion limits **I** 15–22
exhaust fumes **I** 101–120
Exhoran® **V** 165
exposure peaks **I** 15–22
fast dark blue base R **V** 153
F-13B1 **VI** 37
F 134a **XIII** 251
FC 11 **I** 335, **III** 366
FC 12 **I** 336, 341, **III** 366, **V** 109
FC 13 **I** 69
FC 21 **V** 119
FC 22 **III** 63
FC 31 **V** 75
FC 112 **I** 297
FC 113 **III** 365, 366

FC 123 **X** 53
FC 142b **I** 65
ferbam **XIX** 163–168
fermentation butyl alcohol **XIX** 217
ferric dimethyldithiocarbamate **XIX** 163
ferric oxide **II** 135
ferrous oxide **II** 135
fibrous dust **VIII** 141–338
fine dust **II** 53
flint **II** 157
flowers of zinc **XVIII** 305
Fluorocarbon 134a **XIII** 251
fluorochloroform **I** 335
fluorodichloromethane **V** 119
fluorotrichloromethane **I** 335
5-fluorouracil **I** 4
flutamide **I** 6
FLX-0012® **III** 81
formaldehyde **I** 41–42, 44, 60, **II** 240, 250,
    287, 331, **III** 1–8, 173–189, **IV** 136,
    **XVII** 163–201
formalin **III** 173
formic acid **XIX** 169–180
formic acid dimethylamide **VIII** 1
formic acid ethyl ester **XIX** 181–186
Formin **V** 355
*N*-formyldimethylamine **VIII** 1
2-formylfuran **IX** 69
formylic acid **XIX** 169
fraké **XIII** 288
framiré **XVIII** 259
Freon® 13B1 **VI** 37
Freon® FE 1301 **VI** 37
fumes **II** 59–67, **XII** 271
Fundal® **VI** 57
Fungicide E® **XVI** 263
2-furaldehyde **XVIII** 189
2-furanaldehyde **IX** 69
2-furancarbinol **VII** 121, **XIX** 187
2-furancarbonal **IX** 69
2-furancarboxaldehyde **IX** 69, **XVIII** 189
2,5-furandione **IV** 275
2-furanmethanol **VII** 121, **XIX** 187
furfural **IX** 69–79, **XVIII** 189–190
furfural alcohol **VII** 121
furfuralcohol **XIX** 187
furfuraldehyde **IX** 69
furfuryl alcohol **VII** 121–125, **XIX** 187–188
furyl alcohol **VII** 121
2-furylaldehyde **IX** 69
2-furylcarbinol **VII** 121, **XIX** 187
2-furylmethanal **IX** 69
2-furylmethanol **VII** 121
fused silica **II** 157

Gaboon ebony **XIV** 287
Gaboon mahogany **XIII** 285
Galecron® **VI** 57
gasoline engine emissions **I** 109–120
gedu nohor **XVIII** 229
general threshold limit value for dust **II** 3–46,
    **XII** 239–270
germ cell mutagens **I** 9–13, **XVII** 3–9
giant arborvitae **XIII** 288, **XVIII** 263
giant cedar **XIII** 288, **XVIII** 263
glass fibres **VIII** 141–338
glass wool **VIII** 141–338
glauramine **IV** 12
glutaral **VIII** 45, **XVI** 59
glutaraldehyde **VIII** 45–64, **XVI** 59–64
glutardialdehyde **VIII** 45, **XVI** 59
glutaric dialdehyde **VIII** 45, **XVI** 59
glycerin α-monochlorohydrin **V** 83
glycerol chlorohydrin **V** 83
*sym*-glycerol dichlorohydrin **I** 95
glycerol-α,γ-dichlorohydrin **I** 95
glycerol trichlorohydrin **IX** 171
glyceryl monothioglycolate **IX** 81–82
glyceryl trichlorohydrin **IX** 171
glycidol **XX** 179–190
glycidyl allyl ether **VII** 9
glycidyl butyl ether **IV** 51
glycidyl isopropyl ether **VII** 141
1-*n*-glycidyloxybutane **IV** 51
glycidyl phenyl ether **IV** 305
glycidyl trimethylammonium chloride **IV**
    247–255
glycol **IV** 225
glycol butyl ether **VI** 47
glycol chlorohydrin **V** 65
glycol dichloride **III** 137
glycol ether DB **VII** 59
glycol ethyl ether **VI** 205
glycol methacrylate **XIII** 193
glycol methyl ether **VI** 239
glycol methyl ether acetate **VI** 249
glycol monobutyl ether acetate **VI** 53
glycolmonochlorohydrin **V** 65
glycol monoethyl ether **VI** 205
glycol monoethyl ether acetate **VI** 213
glycol monomethyl ether **VI** 239
glycol monomethyl ether acetate **VI** 249
G-MAC® **IV** 247
gold 'teak' **XIII** 287
*Gonystylus bancanus* **XIII** 286, **XVIII** 235–
    237
Grand Bassam mahogany **XIII** 287, **XVIII**
    239
graphite **II** 129–134

green ebony **XIII** 285, 299
*Grevillea robusta* **XIII** 286, **XIV** 295–297
Grotan® BK **II** 331
Grotan® HD **IX** 263
gypsum **II** 117, **VIII** 141–338
haematite **II** 136
halloysite **VIII** 141–338
Halon® 1301 **VI** 37
hard coal dust **XVIII** 107
hard metal **III** 123, 124
HCB **XVI** 65
α-HCH **V** 193
β-HCH **V** 193
γ-HCH **XVI** 113
2-(8-heptadecenyl)-4,5-dihydro-1-*H*-imid-
    azole-1-ethanol **V** 373
2-(8-heptadecenyl)-2-imidazoline-1-ethanol
    **V** 373
2-heptadecenyl-2-imidazoline-1-ethanol **V**
    373
*n*-heptane **XI** 165–176
*Heritiera utilis* **XIII** 288
hexabutyldistannoxane **I** 315
hexachlorobenzene **XVI** 65–88
α-1,2,3,4,5,6-hexachlorocyclohexane **V** 193
β-1,2,3,4,5,6-hexachlorocyclohexane **V** 193
1α,2α,3β,4α,5α,6β-hexachlorocyclohexane
    **XVI** 113
1α,2α,3β,4α,5β,6β-hexachlorocyclohexane **V**
    193
1α,2β,3α,4β,5α,6β-hexachlorocyclohexane **V**
    193
α-hexachlorocyclohexane **V** 193–206
β-hexachlorocyclohexane **V** 193–206
γ-1,2,3,4,5,6-hexachlorocyclohexane **XVI**
    113
1,2,3,4,10,10-hexachloro-6,7-epoxy-
    1,4,4a,5,6,7,8,8a-octahydro-*endo,endo*-
    1,4:5,8-dimethanonaphthalene **XVIII** 165
hexachloronaphthalenes **XIII** 101
(1aα2β,2aβ,3α,6α,6aβ,7β,7aα)-3,4,5,6,9,9-
    hexachloro-1a,2,2a,3,6,6a,7,7a-octahydro-
    2,7:3,6-dimethanonaphth[2,3-*b*]oxirene
    **XVIII** 165
hexahydro-2*H*-azepin-2-one **IV** 65
hexahydro-2*H*-azepin-7-one **IV** 65
hexahydrobenzene **XIII** 129
hexahydro-1,4-diazine **IX** 303, **XII** 177
hexahydrophthalic anhydride **X** 109–111
hexahydropyrazine **IX** 303, **XII** 177
hexahydro-1,3,5-tris(2-hydroxyethyl)-
    s-triazine **II** 331
hexamethylene **XIII** 129
hexamethyleneamine **V** 355

hexamethylenetetramine **V** 355–372
hexamethylmelamine **I** 2
hexamine **V** 355
hexanaphthene **XIII** 129
*n*-hexane **IV** 257–268, **XIV** 101–126
hexane (all isomers except *n*-hexane) **IV**
    269–273
6-hexanelactam **IV** 65
1-hexanol **IX** 283–290
*n*-hexanol **IX** 283
hexone **XIII** 169–180
*n*-hexyl alcohol **IX** 283
hexylene glycol **XVI** 233–246
*Heyderia decurrens* **XIII** 285, **XIV** 275
HFC-134a **XIII** 251
HMT **V** 355
HMTA **V** 355
HN2 **I** 211
Honduras mahogany **XIII** 287, **XVIII** 253
Honduras rosewood **XIII** 286, 317
Hordaflex® **III** 81
hydrazine **I** 40, 171–182, **XIII** 181–186
hydrazine anhydrous **XIII** 181
hydrazine hydrate **XIII** 181–186
hydrazine monohydrate **XIII** 181
hydrazine salts **XIII** 181–186
hydrazinobenzene **XI** 225
hydrazobenzene **IX** 83–89
hydrobromic acid **XIII** 187
hydrobromic ether **VII** 115
hydrochloric acid **VI** 231
hydrochloric ether **III** 73
hydrocyanic acid **I** 42, **XIX** 189
hydrogen arsenide **XV** 41
hydrogen bromide **XIII** 187–191
hydrogen carboxylic acid **XIX** 169
hydrogen chloride **VI** 231–238
hydrogen chloride gas **VI** 231
hydrogen cyanide **XIX** 189–210
hydrogen nitrate **III** 233
hydrogen peroxide **III** 249, 252, 253
hydrogen sulfate **XV** 165
α-hydro-ω-hydroxypoly(oxy-1,2-ethanediyl)
    **X** 247
α-hydro-ω-hydroxypoly[oxy(methyl-1,2-
    ethandiyl)] **X** 271
hydroquinone **III** 195, **X** 113–145
2-hydroxybiphenyl **II** 299
*o*-hydroxybiphenyl **II** 299
2-hydroxybutane **XIX** 117
4-hydroxy-2-butane sulfonic acid **IV** 45
1-hydroxy-4-*tert*-butylbenzene **XI** 5
hydroxycarbamide **I** 7
2-hydroxyethyl acrylate **XVI** 89–93

2-hydroxyethyl methacrylate **XIII** 193–199
2-hydroxy-1,3-di-*tert*-butylbenzene **V** 341
2-hydroxy-3,5-dichlorophenyl sulfide **IX** 245
2-hydroxydiphenyl **II** 299
hydroxy ether **VI** 205
2-(2-hydroxyethoxy)ethylamine **IX** 215
2,2′-hydroxyethoxyethylamine **IX** 215
2-hydroxyethylamine **XII** 15
1-hydroxyethyl-2-heptadecenylglyoxalidine **V** 373
1-(2-hydroxyethyl)-2-heptadecenylglyoxalidine **V** 373
1-(2-hydroxyethyl)-2-(8-heptadecenyl)-2-imidazoline **V** 373
1-(2-hydroxyethyl)-2-*n*-heptadecenyl-2-imidazoline **V** 373
1-hydroxyethyl-2-heptadecenyl-imidazoline **V** 373–375
*β*-hydroxyethyl isopropyl ether **V** 207
2-(*N*-2-hydroxyethyl-*N*-methylamino)ethanol **IX** 291
*N*-(2-hydroxyethyl)-3-methyl-2-quinoxaline-carboxamide 1,4-dioxide **IX** 151
1-hydroxy hexane **IX** 283
1-hydroxyisopropyl-(2-methoxypropyl)ether **VI** 200
1-hydroxy-1′-methoxydiisopropylether **VI** 200
2-hydroxy-2′-methoxydi-*n*-propylether **VI** 200
1-hydroxy-2-methylbenzene **XIV** 59
1-hydroxy-3-methylbenzene **XIV** 59
1-hydroxy-4-methylbenzene **XIV** 59
hydroxymethylene benzyl ether **II** 239
hydroxymethylene mono(poly)oxymethylene benzyl ether **II** 239
2-hydroxymethylfuran **VII** 121, **XIX** 187
2-hydroxymethyl-*n*-heptan-4-ol **V** 345
2-(hydroxymethyl)-2-nitro-1,3-propanediol **II** 287–293
1-hydroxymethylpropane **XIX** 217
4-hydroxy-3-nitroaniline **IV** 289
hydroxy octyl oxostannane **VII** 93
2-hydroxypropanamine **IX** 237
3-hydroxy-1-propanesulfonic acid sulfone **IV** 313
3-hydroxy-1-propanesulfonic acid sultone **IV** 313
*α*-hydroxypropanetricarboxylic acid **XVI** 209
2-hydroxy-1,2,3-propanetricarboxylic acid **XVI** 209
hydroxypropyl acrylate (all isomers) **XVI** 95–104
1-hydroxy-2-propyl acrylate **XVI** 95

1-hydroxy-3-propyl acrylate **XVI** 95
2-hydroxy-1-propyl acrylate **XVI** 95
2-hydroxypropylamine **IX** 237
3-hydroxypropylene oxide **XX** 179
2-hydroxypropyl methacrylate **XVI** 105–111
2-hydroxypropyl-(1-methoxyisopropyl)ether **VI** 200
2-hydroxypropyl 2-methyl-2-propenoate **XVI** 105
1-hydroxy-2-(1*H*)-pyridinethione sodium salt **X** 287
2-hydroxytoluene **XIV** 59
3-hydroxytoluene **XIV** 59
4-hydroxytoluene **XIV** 59
1-hydroxy-2,4,5-trichlorobenzene **XII** 209
2-hydroxytriethylamine **XIV** 91
hydroxyurea **I** 7
hyperbaric pressure and body burden **XIV** 11–16
hyponitrous acid anhydride **IX** 115
hytrol O **X** 35
idigbo **XVIII** 259
ifosfamide **I** 2
IGE **VII** 141
2-imidazolidinethione **XI** 121
4,4′-imidocarbonyl-bis(*N*,*N*-dimethylaniline) **IV** 11
4,4′-imidocarbonyl-bis(*N*,*N*-dimethylbenzenamine) **IV** 11
inactive limonene **I** 185
incense cedar **XIII** 285, **XIV** 275
Indian ebony **XIII** 286, **XIV** 287
Indian rosewood **XIII** 285, 301
inhalable dust **XII** 239
inhalation kinetics **XIV** 3–10
inorganic fibres **VIII** 141–338
inorganic lead compounds **XVII** 203
inorganic tin compounds **XIV** 149–165
iodomethane **VII** 219
3-iodo-2-propynylbutylcarbamate **XVI** 247–256
3-iodo-2-propynyl-*N*-butylcarbamate **XVI** 247
3-iodo-2-propynylcarbamic acid butyl ester **XVI** 247
Ionox 99® **V** 341
ipe **XIII** 288, **XIV** 311
ipé peroba **XIII** 287, **XIV** 307
iroko **XIII** 285, **XIV** 279
iron oxides **II** 135–144
isatoic acid anhydride **XIII** 97
isatoic anhydride **XIII** 97
isoamyl acetate **XI** 211
1,3-isobenzofurandione **VII** 247

isobutane **XX** 1
isobutanol-2-amine **IX** 229
isobutenyl chloride **IV** 97
isobutyl acetate **XIX** 211–215
isobutyl alcohol **XIX** 217–226
isobutyltrimethylmethane **I** 347
isocyanatobenzene **XVII** 267
isocyanic acid methylene di-*p*-phenylene ester
   **VIII** 65
isocyanic acid polymethylenepolyphenylene
   ester **VIII** 65
isodecyl oleate **XVI** 257–261
isoflurane **VII** 127–140
isohexane **IV** 269
Isol **XVI** 233
isonitropropane **III** 241
isooctane **I** 347
isoparaffins **IV** 269–273
isopentyl acetate **XI** 211
isophorone **I** 196
isophosphamid **I** 2
isophthalic acid **VII** 241
isopropanolamine **IX** 237
isopropenylbenzene **XV** 137
4-isopropenyl-1-methyl-1-cyclohexene **I** 185
2-isopropoxyethanol **V** 207-211
(isopropoxymethyl)oxirane **VII** 141
2-isopropoxypropane **XXI** 187
isopropylacetone **XIII** 169
isopropylamine **I** 34
isopropylbenzene **XIII** 117
isopropylcarbinol **XIX** 217
isopropyl cellosolve **V** 207
isopropyl ether **XXI** 187
isopropyl ethylene glycol ether **V** 207
isopropyl glycidyl ether **VII** 141–144
isopropyl glycol **V** 207
4,4'-isopropylidenediphenol **XIII** 49
4,4'-isopropylidenediphenol diglycidyl ether
   **XIX** 43
3-isopropyloxypropylene oxide **VII** 141
isosystox **XIX** 141
isothiourea **I** 301
'jacaranda' **XIII** 287, 321
Jamaica mahogany **XVIII** 253
jeweler's rouge **II** 135
*Juglans* spp. **XIII** 286
kambala **XIII** 285, **XIV** 279
Kathon® **V** 323
Kathon® 893 MW **XVI** 263
Kathon® biocide **V** 323
Kaurit® S **V** 289
ketene **XX** 191–196
ketohexamethylene **X** 35

ketone propane **VII** 1
*β*-ketopropane **VII** 1
Khaya mahagony **XIII** 286–287, **XVIII** 239
*Khaya* spp. **XIII** 286–287, **XVIII** 239–246
kieselguhr **II** 157
kosipo **XVIII** 229
lapacho **XIII** 288, **XIV** 311
Larvacide 100 **VI** 105
laughing gas **IX** 115
lead **XVII** 203–244
lead chromate **III** 101–122, **VII** 33
lead compounds **XVII** 203
lead molybdate **XVIII** 199
lead tetraethyl **XV** 223
lead tetramethyl **XV** 237
*Libocedrus decurrens* **XIII** 285, **XIV** 275
limba **XIII** 288
limitation of exposure peaks **I** 15–22
limonene **I** 185–200, 347
lindane **XVI** 113–141
*α*-lindane **V** 193
*β*-lindane **V** 193
lithium hydride **III** 191–193
loadstone **II** 135
lodestone **II** 135
longleaf pine **XIII** 287
long-term threshold values **XI** 281
Lorol C 6 **IX** 283
low cristobalite **XIV** 205
low quartz **XIV** 205
low tridymite **XIV** 205
Lubrimet® P600, P900 **X** 271
lusamba **XIII** 288
lye **XII** 195
Macassar ebony **XIII** 286, **XIV** 287
*Machaerium scleroxylon* **XIII** 287, 321–323
macore **XVIII** 277
Mafu® **IV** 201
magnesia **II** 145
magnesium oxide **II** 145–148
magnesium oxide sulfate **VIII** 141–338
magnetite **II** 136
mahogany **XIII** 287
mainstream smoke **XIII** 3
makoré **XIII** 288, **XVIII** 277
maleic acid anhydride **IV** 275, **XI** 177
maleic anhydride **IV** 275–287, **XI** 177–178
manganese **XII** 293–328
manganese compounds **XII** 293
man-made mineral fibres **VIII** 141–338
*Mansonia altissima* **XIII** 287, **XIV** 299–305
maritime pine **XIII** 287
MBOCA **VII** 193
MDEA **IX** 291

MDI **VIII** 65
mechlorethamine **I** 211
mecrylate **I** 233
MEK **XII** 37
melphalan **I** 3
*p*-mentha-1,8-diene **I** 185
1,8(9)-*p*-menthadiene **I** 185
merbromin **XV** 75–79
mercaptoethane **XXI** 195
2-mercaptoimidazoline **XI** 121
mercaptomethane **XX** 217
mercaptophos **XIX** 141
6-mercaptopurine **I** 4
mercurialin **VII** 145
mercurochrome **XV** 75
Mercurothiolate **XV** 249
mercury and inorganic mercury compounds **XV** 81–122
mercury, organic compounds **XV** 123–136
mesitylene **IV** 341
metallic mercury **XV** 81
metal-working fluids **II** 207–220, **IX** 195–213, **XVI** 207–286, **XX** 197–215
metaphenylenediamine **VI** 287
methacrylic acid ethyl ester **XVI** 51
methacrylic acid 2-hydroxyethyl ester **XIII** 193
methacrylic acid 2-hydroxypropyl ester **XVI** 105
methacrylic acid methyl ester **XVI** 181
2-methallyl chloride **IV** 97
*β*-methallyl chloride **IV** 97
methanal **III** 173, **XVII** 163
methanamine **VII** 145
methane base **I** 249
methane dichloride **IV** 173, **XVII** 147
methanethiol **XX** 217
methanoic acid **XIX** 169
methanol **XVI** 143–175
methenamine **V** 355
methotrexate **I** 3
4-methoxy-2-aminoanisole **IV** 135
2-methoxyaniline **X** 1, **XXI** 3
4-methoxyaniline **XXI** 3
*o*-methoxyaniline **X** 1
2-methoxybenzenamine **X** 1
4-methoxy-1,3-benzenediamine **VI** 145
2-methoxyethanol **V** 210, **VI** 239–248
2-methoxyethanol acetate **VI** 249
2-methoxyethyl acetate **VI** 249–251
methoxyhydroxyethane **VI** 239
2-methoxy-5-methylaniline **IV** 135
2-methoxy-5-methylbenzenamine **IV** 135
(2-methoxymethylethoxy)propanol **VI** 199

2-(2-methoxy-1-methylethoxy)-1-propanol **VI** 200
1-(2-methoxy-1-methylethoxy)-2-propanol **VI** 200
2-(2-methoxy-2-methylethoxy)-1-propanol **VI** 200
1-(2-methoxy-2-methylethoxy)-2-propanol **VI** 200
2-methoxy-2-methylpropane **XVII** 119
1-methoxy-2-nitrobenzene **IX** 103
*o*-methoxyphenylamine **X** 1
*p*-methoxy-*m*-phenylenediamine **VI** 145
4-methoxy-*m*-phenylenediamine **VI** 145
1-methoxy-2-propanol **V** 213–216, **XIV** 127–129
1-methoxypropan-2-ol **I** 204, **V** 213
2-methoxy-1-propanol **I** 203, 207
1-methoxy-2-propanol acetate **V** 217
2-methoxy-1-propanol acetate **I** 203, 207
1-(1-methoxy-2-propoxy)-2-propanol **VI** 200
1-(2-methoxy-1-propoxy)-2-propanol **VI** 200
2-(1-methoxy-2-propoxy)-1-propanol **VI** 200
2-(2-methoxy-1-propoxy)-1-propanol **VI** 200
1-methoxypropyl-2-acetate **V** 217–220
2-methoxypropyl-1-acetate **I** 207–209
4-methoxy-*m*-toluidine **IV** 135
methural **V** 289
methyl acetate **XVIII** 191–196
methyl acrylate **VI** 253–262, **XVI** 177–180
2-methylacrylic acid methyl ester **III** 195
methyl alcohol **XVI** 143
methyl aldehyde **III** 173, **XVII** 163
methyl allyl chloride **IV** 97
methylamine **I** 28, 34, **VII** 145–153
1-methyl-2-aminobenzene **III** 307
2-methyl-1-aminobenzene **III** 307
(methylamino)benzene **VI** 263
*N*-methylaminobenzene **VI** 263
*N*-methylaminodiglycol **IX** 291
1-methyl-2-aminoethanol **IX** 237
*N*-methylaniline **I** 25, **VI** 263–270, **XXI** 3
2-methylaniline **III** 307
4-methylaniline **III** 323
*p*-methylaniline **III** 323
5-methyl-*o*-anisidine **IV** 135
*p*-methylanisidine **IV** 135
1-methylazacyclopentan-2-one **X** 147
2-methylbenzenamine **III** 307
4-methylbenzenamine **III** 323
4-methyl-1,3-benzenamine **VI** 339
*N*-methylbenzenamine **VI** 263
methylbenzene **VII** 257
*p*-methylbenzeneamine **III** 323
methylbenzotriazole **II** 295

methyl-1*H*-benzotriazole **II** 295–298
*N*-methyl-bis(2-chloroethyl)amine **I** 211–230
methylbis(2-hydroxyethyl)amine **IX** 291
methyl bromide **VII** 155–172
2-methyl-1-butanol acetate **XI** 211
2-methyl-2-butanol acetate **XI** 211
3-methyl-1-butanol acetate **XI** 211
1-methylbutyl acetate **XI** 211
2-methylbutyl acetate **XI** 211
3-methylbutyl acetate **XI** 211
*β*-methylbutyl acetate **XI** 211
methyl-*tert*-butyl ether **XVII** 119
methyl carbinol **XII** 129
Methyl Cellosolve® **VI** 239
Methyl Cellosolve® acetate **VI** 249
methyl chloride **VII** 173–191
methyl chloroacetate **IX** 1
2-methyl-4-chloroaniline **VI** 127
3-methyl-4-chlorophenol **II** 271
methyl cyanide **XIX** 1
methyl 2-cyanoacrylate **I** 233–245, **XIII** 201–206
methyl *α*-cyanoacrylate **I** 233, **XIII** 201
methyldibromoglutaronitrile **XIV** 87
*N*-methyl-2,2′-dichlorodiethylamine **I** 211
methyldiethanolamine **IX** 291–294
*N*-methyldiethanolamine **IX** 291
*N*-methyldiethanolimine **IX** 291
2-methyl-2,3-dihydroisothiazol-3-one **V** 323–339
methyldinitrobenzene **VI** 165
1-methyl-2,3-dinitrobenzene **VI** 165
1-methyl-2,4-dinitrobenzene **VI** 165
1-methyl-3,4-dinitrobenzene **VI** 165
1-methyl-3,5-dinitrobenzene **VI** 165
2-methyl-1,3-dinitrobenzene **VI** 165
2-methyl-1,4-dinitrobenzene **VI** 165
4-methyl-1,2-dinitrobenzene **VI** 165
2-methyl-4,6-dinitrophenol **XIX** 151
methylene acetone **IX** 91
methylene base **I** 249
methylene bichloride **IV** 173, **XVII** 147
4,4′-methylenebis(aniline) **VII** 37
4,4′-methylenebis(benzenamine) **VII** 37
4,4′-methylene(bis)chloroaniline **VII** 193
4,4′-methylenebis(2-chloroaniline) **VII** 193–218
*p,p*′-methylenebis(*alpha*-chloroaniline) **VII** 193
4,4′-methylenebis(2-chlorobenzenamine) **VII** 193
4,4′-methylene-bis(*N,N*-dimethylaniline) **I** 249-256, **IV** 15
4,4′-methylene-bis(*N,N*-dimethyl)benzenamine **I** 249
1,1′-methylenebis(4-isocyanatobenzene) **VIII** 65
4,4′-methylene-bis(2-methylaniline) **XVIII** 197–198
methylenebis(4-phenylene isocyanate) **VIII** 65
methylenebis(*p*-phenylene isocyanate) **VIII** 65
methylene bisphenyl isocyanate **VIII** 65
methylene chloride **IV** 173, **XVII** 147
4,4′-methylenedianiline **VII** 37
*p,p*′-methylenedianiline **VII** 37
methylene dichloride **IV** 173, **XVII** 147
methylenedi-4-phenylene diisocyanate **VIII** 65
methylenedi-*p*-phenylene diisocyanate **VIII** 65
methylene di(phenylene isocyanate) **VIII** 65
4,4′-methylenediphenylene isocyanate **VIII** 65
4,4′-methylenediphenyl diisocyanate **VIII** 65
4,4′-methylene diphenyl isocyanate **VIII** 65, **XIV** 131–133
methylene glycol **III** 173
methylene oxide **III** 173, **XVII** 163
(1-methylethenyl)-benzene **XV** 137
methyl ether **I** 125
methyl ethoxol **VI** 239
2-(1-methylethoxy)ethanol **V** 207
[(1-methylethoxy)methyl]oxirane **VII** 141
(1-methylethyl)benzene **XIII** 117
methyl ethyl carbinol **XIX** 117
methylethylene oxide **V** 221
4,4′-(1-methylethylidene)bisphenol **XIII** 49
2,2′-[(1-methylethylidene)bis(4,1-phenylene-oxymethylene)]bis-oxirane **XIX** 43
methyl ethyl ketone **XII** 37
methyl ethyl ketone peroxide **III** 250, 252, 254, 255
methyl glycol **VI** 239
methyl glycol acetate **VI** 249
methyl glycol monoacetate **VI** 249
2,2′-(methylimino)bisethanol **IX** 291
2,2′-(methylimino)diethanol **IX** 291
*N*-methyliminodiethanol **IX** 291
*N*-methyl-2,2′-iminodiethanol **IX** 291
methyl iodide **VII** 219–228
methyl isobutyl ketone **XIII** 169
1-methyl-4-isopropenyl-1-cyclohexene **I** 185
2-methyl-4-isothiazolin-3(2*H*)-one **V** 323
2-methyl-3(2*H*)-isothiazolone **V** 323

methyl ketone **VII** 1
*N*-methyl-2-ketopyrrolidine **X** 147
*N*-methyl lost **I** 211
methyl mercaptan **XX** 217–226
methyl methacrylate **III** 195–231, **XVI** 181–191
methyl *α*-methacrylate **III** 195
*N*-methylmethanamine **VII** 75
methyl methylacrylate **III** 195, **XVI** 181
1-methyl-4-(1-methylethenyl)cyclohexene **I** 185
methyl 2-methyl-2-propenoate **III** 195, **XVI** 181
methyl monochloroacetate **IX** 1
2-methyl-5-nitroaniline **VI** 271
6-methyl-3-nitroaniline **VI** 271
2-methyl-5-nitrobenzenamine **VI** 271
1-methyl-2-nitrobenzene **VIII** 97
2-methyl-1-nitrobenzene **VIII** 97
*o*-methylnitrobenzene **VIII** 97
2-methyl-5-nitrobenzeneamine **VI** 271
(*Z*)-(2-methylnonyl)-9-octadecenoate **XVI** 257
*N*-methyl-*N*-oleoyl-aminoacetic acid **XVI** 281
4-methyl-3-oxapentan-1-ol **V** 207
3-methyl-1,2-oxathiolane 2,2-dioxide **IV** 45
methyloxirane **V** 221
*N*-methyl-*N*-(1-oxo-9-octadecenyl)glycine **XVI** 281
*N*-methyl-2-oxypyrrolidine **X** 147
2-methylpentane **IV** 269
3-methylpentane **IV** 269
2-methyl-2,4-pentanediol **XVI** 233
4-methyl-2-pentanone **XIII** 169
2-methylphenol **XIV** 59
3-methylphenol **XIV** 59
4-methylphenol **XIV** 59
*N*-methylphenylamine **VI** 263
4-methyl-*m*-phenylenediamine **VI** 339
1-methyl-1-phenylethylene **XV** 137
2-methylpropane **XX** 1
1-methyl-1,3-propane sultone **IV** 45
methyl propenate **VI** 253
methyl-2-propenoate **VI** 253
2-methyl-2-propenoic acid ethyl ester **XVI** 51
2-methyl-2-propenoic acid 2-hydroxyethyl ester **XIII** 193
2-methyl-2-propenoic acid 2-hydroxypropyl ester **XVI** 105
2-methyl-2-propenoic acid methyl ester **III** 195, **XVI** 181
1-methyl-1-propanol **XIX** 117
2-methyl-1-propanol **XIX** 217

2-methyl-2-propanol **XIX** 125
2-methyl-1-propyl acetate **XIX** 211
1-methyl propyl alcohol **XIX** 117
*beta*-methylpropyl ethanoate **XIX** 211
*N*-methyl-2-pyrrolidinone **X** 147
1-methyl-2-pyrrolidinone **X** 147
1-methyl-5-pyrrolidinone **X** 147
*N*-methylpyrrolidone **X** 147
1-methyl-2-pyrrolidone **X** 147
*N*-methyl-2-pyrrolidone **X** 147–170
*α*-methyl styrene **XV** 137–140
methyl sulfhydrate **XX** 217
methylsulfinylmethane **III** 163
methyl sulfoxide **III** 163
methyltetrahydrophthalic anhydride **X** 109–111
*N*-methyl-*N*,2,4,6-tetranitroaniline **XI** 179–185, **XXI** 3
*N*-methyl-*N*,2,4,6-tetranitrobenzenamine **XI** 179
methyl toluene **V** 263
methyltoluidine isomers **XIX** 299
methyl tribromide **VII** 319
1-methyl-2,4,6-trinitrobenzene **I** 359
2-methyl-1,3,5-trinitrobenzene **I** 359
methyl vinyl ketone **IX** 91–101
Michler's base **I** 249
Michler's hydride **I** 249
Michler's ketone **IV** 12, 13, 20, 21
Michler's methane **I** 249
*Microberlinia* spp **XIII** 287
*Mimusops africana* **XVIII** 277
*Mimusops heckelii* **XIII** 288, **XVIII** 277
mineral fibres **II** 185
MIPA **IX** 237
mists **XII** 271
mithramycin **I** 5
mitobronitol **I** 7
mitomycin **I** 5
mitoxantrone **I** 5
MMA **III** 195, **XVI** 181
MME **III** 195
4-MMPD **VI** 145
MOCA **VII** 193
molybdenite **XVIII** 199
molybdenum **XVIII** 199–219
molybdenum compounds **XVIII** 199–219
molybdenum dichloride **XVIII** 199
molybdenum dioxide **XVIII** 199
molybdenum dioxide **XVIII** 199
molybdenum disulfide **XVIII** 199
molybdenum pentachloride **XVIII** 199
molybdenum trichloride **XVIII** 199
molybdenum trioxide **XVIII** 199

molybdic anhydride **XVIII** 199
molybdophosphoric acid **XVIII** 199
monobromoethane **VII** 115
monobromomethane **VII** 155
monobutyl glycol ether **VI** 47
monochloroacetaldehyde **XII** 81
monochloroacetic acid methyl ester **IX** 1
monochlorobenzene **XII** 103
monochlorodifluoromethane **III** 63
monochloroethane **III** 73
2-monochloroethanol **V** 65
monochloroethene **V** 241
monochloroethylene **V** 241
monochlorofluoromethane **V** 75
$\alpha$-monochlorohydrin **V** 83
monochloromethane **VII** 173
monochloromonofluoroethane **V** 75
monochloronaphthalenes **XIII** 101
monochloropropanediol **V** 83
monochlorotrifluoromethane **I** 69
monocyclic aromatic amino and nitro
    compounds **XXI** 3–45
monoethanolamine **XII** 15
monoethyleneglycol **IV** 225
mono-isopropanolamine **IX** 237
monomethylamine **VII** 145
monomethylaniline **VI** 263
monomethyl glycol acetate **VI** 249
mononitroethane **XIX** 245
mononitromethane **XIX** 251
mono-*n*-octyltin compounds **VII** 91–114
monooctyltin hydroxide **VII** 93
mono-*n*-octyltin oxide **VII** 93
monooctyltin thioglycolate **VII** 94
mono-*n*-octyltin trichloride **VII** 93
mono-*n*-octyltin tris(2-ethylhexyl thio-
    glycolate) **VII** 94
mono-*n*-octyltin tris(2-ethylhexylmercapto-
    acetate) **VII** 94
mono-*n*-octyltin tris(isooctylthioglycolate)
    **VII** 94
mopidamol **I** 4
morpholine **I** 25, 28, 33–34
4-morpholinecarbonyl chloride **V** 79
morpholinylcarbamoyl chloride **V** 79
morpholinylcarbonyl chloride **V** 79
MOTC **VII** 93
MOTO **VII** 93
MOTTG **VII** 94
*Morus excelsa* **XIV** 279
muriatic acid **VI** 231
mustard gas **IV** 27
mustargen **I** 211
myleran **I** 2–3

Mystox WFA® **II** 299
nadone **X** 35
naphthalene **XI** 187–210
2-naphthylamine **I** 258–259
Natriphene® **II** 299
Natrium-Pyrion® **X** 287
natural isobutyl acetate **XIX** 211
natural rubber latex (*Hevea* species) **XV** 141–
    143
NCI-C56417 **V** 295
2-NDB **XIII** 207
nemalite **VIII** 141–338
neohexane **IV** 269
niangon **XIII** 288
Nicaragua mahogany **XVIII** 253
nickel compounds **I** 40, **II** 47, 48, 136
nicotine **I** 28, 40, 42–44, 59
Nigerian ebony **XIV** 287
Nigerian satinwood **XIII** 286, **XIV** 291
Nigerian walnut **XIV** 299
nimustine **I** 3
nitramine **XI** 179
nitric acid **III** 233–240
nitrite **I** 27–34, 59
2-nitro-4-aminophenol **I** 167, **IV** 289–293,
    **XXI** 3
*o*-nitro-*p*-aminophenol **IV** 289
4-nitro-2-aminotoluene **VI** 271
nitroaniline **I** 149–150, 166
4-nitroaniline **XXI** 3
2-nitroanisole **IX** 103–114, **XXI** 3
*o*-nitroanisole **IX** 103
nitrobenzene **XIX** 227–243, **XXI** 3
2-nitro-1,4-benzenediamine **IV** 295, **XIII** 207
nitrobenzol **XIX** 227
4-nitrobiphenyl **I** 257–259
4-nitro-1,1'-biphenyl **I** 257
*p*-nitrobiphenyl **I** 257
4-(2-nitrobutyl)morpholine **IX** 295–301
*N*-(2-nitrobutyl)morpholine **IX** 295
nitrocarbol **XIX** 251
*o*-nitrochlorobenzene **IV** 107
*m*-nitrochlorobenzene **IV** 115
*p*-nitrochlorobenzene **IV** 121
nitrochloroform **VI** 105
2-nitro-1,4-diaminobenzene **IV** 295, **XIII**
    207
4-nitrodiphenyl **I** 257
*p*-nitrodiphenyl **I** 257
nitroethane **XIX** 245–250
nitrogen dioxide **XXI** 205–260
nitrogen mustard **I** 211
nitrogen oxide **IX** 115
nitrogen oxides **I** 28, 40, 42–44, 58

nitrogen peroxide **XXI** 205
nitro-isobutylglycerol **II** 287
nitroisopropane **III** 241
nitromethane **XIX** 251–260
3-nitro-6-methylaniline **VI** 271
1-nitro-2-methylbenzene **VIII** 97
2-nitro-1,4-phenylenediamine **IV** 295, **XIII** 207
2-nitro-*p*-phenylenediamine **IV** 295–303, **XIII** 207–210, **XXI** 3
*o*-nitro-*p*-phenylenediamine **IV** 295, **XIII** 207
*o*-nitrophenyl methyl ether **IX** 103
1-nitropropane **XIII** 211–213
2-nitropropane **III** 241–247
nitropyrenes **I** 112, 118, 258
*N*-nitrosamines **I** 23–35, 42, 58–59, 261–286, **II** 250
nitrosation of amines **I** 23–35
*N*-nitrosodi-*n*-butylamine **I** 24, 29, 261–262, 274–276, 286
*N*-nitrosodiethanolamine **I** 27, 29, 261–262, 277–278, 286
*N*-nitrosodiethylamine **I** 23–35, 41, 60, 261–262, 268–271, 286
*N*-nitrosodiisopropylamine **I** 261–262, 274, 286
*N*-nitrosodimethylamine **I** 3–35, 41–44, 59, 261–267, 286
*N*-nitrosodiphenylamine **I** 27
*N*-nitrosodi-*n*-propylamine **I** 261–262, 271–273, 286
*N*-nitrosoethylphenylamine **I** 261–262, 280, 286
*N*-nitrosomethylethylamine **I** 261–262, 267, 286
*N*-nitrosomethylphenylamine **I** 29, 261–262, 279, 286
*N*-nitrosomorpholine **I** 29, 34, 261–262, 280–282, 286
*p*-nitrosophenol **XIV** 135
4-nitrosophenol **XIV** 135–136
*N*-nitrosopiperidine **I** 261–262, 282–284, 286
*N*-nitrosopyrrolidine **I** 40, 261–262, 284–285, 286
2-nitrotoluene **VIII** 97–108, **XXI** 3
3-nitrotoluene **XXI** 3
4-nitrotoluene **XXI** 3
*o*-nitrotoluene **VIII** 97
5-nitro-*o*-toluidine **VI** 271–275, **XXI** 3
nitrotrichloromethane **VI** 105
nitrous oxide **IX** 115–150
nitroxanthic acid **XVII** 273
NO$_x$ **I** 28

normal butyl thioalcohol **XXI** 161
northern red oak **XVIII** 247
northern white cedar **XIII** 288, **XVIII** 263
Noxal **V** 165
1-NP **XIII** 211
2-NP **XIII** 207
2-NPPD **XIII** 207
Nuran® **IV** 201
oak wood dust **IV** 363, **XIII** 287, **XVIII** 221
obeche **XIII** 288, **XVIII** 283
octachlorocamphene **XIX** 281
octachloronaphthalene **XIII** 101
(*Z*)-9-octadecenoic acid **XVII** 245
9-octadecenoic acid (*Z*) decyl ester **IX** 269
(*Z*)-9-octadecenoic acid isodecyl ester **XVI** 257
1-octanol **XX** 227–237
Octhilinone **XVI** 263
*n*-octyl alcohol **XX** 227
octylene glycol **V** 345
2-octyl-4-isothiazolin-3-one **XVI** 263–280
*N*-*n*-octyl-3-isothiazolone **XVI** 263
2-*n*-octyl-3-isothiazolone **XVI** 263
2-octyl-3(2*H*)-isothiazolone **XVI** 263
2,2′,2″-[(octylstannylidyne)tris(thio)]tris-(acetic acid)tri-isooctyl ester **VII** 95
2,2′,2″-[(octylstannylidyne)tris(thio)]tris-(acetic acid)tris(2-ethylhexyl) ester **VII** 95
[(octylstannylidyne)trithio]tri(acetic acid)-tris(2-ethylhexyl ester) **VII** 94
octyltin trichloride **VII** 93
octyltin tris(isooctyl thioglycolate) **VII** 94
octyltrichlorostannane **VII** 93
octyltris(2-ethylhexyloxycarbonylmethylthio)-stannane **VII** 94
oil of mirbane **XIX** 227
oil of turpentine **XIV** 173, **XVII** 315
oil of vitriol **XV** 165
OKO® **IV** 201
okoumé **XIII** 285
olaquinox **IX** 151–169
olefiant gas **X** 91
oleic acid **XVII** 245–266
oleic acid decyl ester **IX** 269
oleic acid isodecyl ester **XVI** 257
oleic sarcosine **XVI** 281
oleoyl *N*-methylaminoacetic acid **XVI** 281
oleoyl *N*-methylglycine **XVI** 281
oleoyl sarcosine **XVI** 281–286
*N*-oleoyl sarcosine **XVI** 281
Onyxide® 500 **II** 249
opal **II** 157
organic fibres **VIII** 205

orthoboric acid  **V** 295
oxacyclopropane  **V** 181
3-oxa-1-hepatanol  **VI** 47
oxane  **V** 181
3-oxapentane-1,5-diol  **X** 73
1,2-oxathiane 2,2-dioxide  **IV** 45
1,2-oxathiolane 2,2-dioxide  **IV** 313
Oxazolidine E  **IX** 275
oxidation base 25  **IV** 289
*α,β*-oxidoethane  **V** 181
oxiran  **V** 181
oxirane  **V** 181, **XIII** 165
oxirane methane ammonium-*N,N,N*-trimethyl
    chloride  **IV** 247
oxiranemethanol  **XX** 179
3-oxiranyl-7-oxabicyclo[4.1.0]heptane  **I** 391
Oxitol  **VI** 205
3-oxo-1,2-benzisothiazoline  **II** 221
2-oxobutane  **XII** 37
oxodioctylstannane  **VII** 91
2-oxo-hexamethylenimine  **IV** 65
oxomethane  **III** 173, **XVII** 163
oxybis(4-aminobenzene)  **VI** 277
4,4′-oxybisaniline  **VI** 277
4,4′-oxybisbenzenamine  **VI** 277
1,1′-oxybisethane  **XIII** 149
oxybismethane  **I** 125
1,1′-oxybis(2-methoxyethane)  **IX** 41
2,2′-oxybispropane  **XXI** 187
4,4′-oxydianiline  **VI** 277–286
2,2-oxydiethanol  **X** 73
4,4′-oxydiphenylamine  **VI** 277
oxyfume  **V** 181
oxymethurea  **V** 289
oxymethylene  **III** 173, **XVII** 163
ozone  **X** 171–199
Pacific red cedar  **XIII** 288, **XVIII** 263
PAH  **I** 42, 58–59, 111, 112
palygorskite  **VIII** 141–338
pâo ferro  **XIII** 287, 321
paraphenylenediamine  **VI** 311
*Paratecoma peroba*  **XIII** 287, **XIV** 307–309
Parmetol® F85  **XVI** 263
Paroil®  **III** 81
passive smoking at work  **I** 39–62, **XIII** 3–39
PCM  **I** 291
PCMC  **II** 271
pedunculate oak  **XVIII** 247
pencil cedar  **XIII** 285, **XIV** 275
pentachloronaphthalenes  **XIII** 101
pentachlorophenol  **III** 261–270
pentaerythritol triacrylate  **XVI** 193–199
1,5-pentanedial  **VIII** 45, **XVI** 59
1,5-pentanedione  **VIII** 45

3-pentanol acetate  **XI** 211
2-pentanol acetate  **XI** 211
pentyl acetate  **XI** 211–223
1-pentyl acetate  **XI** 211
3-pentyl acetate  **XI** 211
pentylcarbinol  **IX** 283
per  **III** 271
peracetic acid  **VII** 229–239
perchlorobenzene  **XVI** 65
perchloroethylene  **III** 271
perchloromercaptan  **I** 291
perchloromethane  **XVIII** 81
perchloromethylmercaptan  **I** 291–295
periclase  **II** 145
*Pericopsis elata*  **XIII** 287
peroba do(s) campo(s)  **XIII** 287, **XIV** 307
peroxides  **III** 249–260
peroxoacetic acid  **VII** 229
peroxyacetic acid  **III** 250, 253, 254, **VII** 229
petrol engine emissions  **I** 109–120
3-phenoxy-1,2-epoxypropane  **IV** 305
(phenoxymethyl)oxirane  **IV** 305
phenoxypropene oxide  **IV** 305
phenoxypropylene oxide  **IV** 305
phenylamine  **VI** 17
phenyl carbimide  **XVII** 267
phenyl chloroform  **VI** 80
2,2′-[1,3-phenylenebis(oxymethylene)]-
    bisoxirane  **VII** 69
1,2-phenylenediamine  **XIII** 215
1,3-phenylenediamine  **VI** 287
1,4-phenylenediamine  **VI** 311
*m*-phenylenediamine  **VI** 287–299, **XXI** 3
*o*-phenylenediamine  **XIII** 215–235, **XXI** 3
*p*-phenylenediamine  **VI** 311–337, **XIV** 137–
    141, **XXI** 3
*p*-phenylenediamine compounds  **XV** 151–
    153
phenyl 2,3-epoxypropyl ether  **IV** 305
phenylethylene  **XX** 285
phenyl glycidyl ether  **IV** 59, 305–311
phenylhydrazine  **XI** 225–234
phenyl isocyanate  **XVII** 267–272
phenyl methanal  **XVII** 13
phenylmethane  **VII** 257
(phenylmethoxy)methanol  **II** 239
[(phenylmethoxy)methoxy]methanol  **II** 239
[[[(phenylmethoxy)methoxy]methoxy]-
    methanol  **II** 239
*N*-phenylmethylamine  **VI** 263
1-phenyl-1-methylethylene  **XV** 137
phenyl mono(poly)methoxy methanol  **II** 239
4-phenylnitrobenzene  **I** 257
*p*-phenylnitrobenzene  **I** 257

2-phenylphenol **II** 299
*o*-phenylphenol **II** 299–316
*o*-phenylphenol sodium salt **II** 299
2-phenylpropane **XIII** 117
2-phenylpropene **XV** 137
phenyl trichloromethane **VI** 80
philosopher's wool **XVIII** 305
phosphomolybdic acid **XVIII** 199
phosphonic acid dimethyl ester **I** 133
phosphoric acid 2,2-dichloroethenyl dimethyl
    ester **IV** 201
phosphoric acid 2,2-dichlorovinyl dimethyl
    ester **IV** 201
phosphoric acid tributyl ester **XVII** 285
phosphoric acid triphenyl ester **II** 321
phosphorothioic acid *O*,*O*-diethyl *O*-[2-
    (ethylthio)ethyl] ester **XIX** 141
phosphorothioic acid *O*,*O*-diethyl *O*-[6-
    methyl-2-(1-methylethyl)-4-pyrimidinyl]
    ester **XI** 31
phosphorous acid dimethyl ester **I** 133
1,3-phthalandione **VII** 247
phthalic acid **VII** 241–246
phthalic acid anhydride **VII** 247, **XI** 235
phthalic acid diallyl ester **IX** 11
phthalic anhydride **VII** 247–255, **XI** 235–237
physical activity and inhalation kinetics **XIV**
    3–10
Pic-clor **VI** 105
Picfume® **VI** 105
picric acid **XVII** 273–284
Picride **VI** 105
picronitric acid **XVII** 273
picrylmethylnitramine **XI** 179
picrylnitromethylamine **XI** 179
pimelic ketone **X** 35
*Pinus* spp. **XIII** 287
piperazidine **IX** 303, **XII** 177
piperazine **IX** 303–313, **XII** 177–178
piperidine **I** 25
pitch pine **XIII** 287
plaster **II** 117
plaster of Paris **II** 117
plicamycin **I** 5
Pluracol® P410, P1010, P2010, P4010 **X** 271
Pluriol® P600, P2000 **X** 271
PMDI **VIII** 65–95, **XIV** 131
PNB **I** 257
polychlorinated camphenes **XIX** 281
polychlorocamphene **XIX** 281
polyethylene glycol **X** 247–270
polyethylene oxide **X** 247
polyglycol **X** 247

Polyglycol P425, P1200, P2000, P4000 **X**
    271
"polymeric MDI" **VIII** 65–95, **XIV** 131
polyoxyethylene **X** 247
poly(propane-1,2-diol) **X** 271
polypropylene glycol **X** 271–285
polypropylene oxide **X** 271
Poly-Solv® TB **IX** 315
polyvinyl chloride **II** 149–156
portland cement **XI** 303–333
potassium citrate **XVI** 209
potassium cyanide **XIX** 189
potassium titanates **VIII** 141–338
potassium zinc chromate **XV** 289
PPG **X** 271
Preventol® CMK **II** 271
Preventol® D 2 **II** 239
Preventol® O Extra **II** 299
Preventol® ON Extra **II** 299
procarbazine **I** 7
1,2-propanediol-1-acrylate **XVI** 95
1,3-propanediol-1-acrylate **XVI** 95
1,2-propanediol-2-acrylate **XVI** 95
1,3-propane sultone **IV** 46, 313–321
2-propanone **VII** 1
propargyl alcohol **XXI** 261–269
2-propenal **XVI** 1
2-propenamide **III** 11
propene oxide **V** 221
2-propenoic acid butyl ester **V** 5, **XII** 57,
    **XVI** 35
2-propenoic acid ethyl ester **VI** 217, **XVI** 41
2-propenoic acid 2-ethylhexyl ester **XVI** 47
2-propenoic acid 2-ethyl-2-[[(1-oxo-2-propen-
    yl)oxy]methyl]-1,3-propane-diyl ester
    **XVI** 201
2-propenoic acid homopolymer, sodium salt
    **XV** 1
2-propenoic acid 2-hydroxyethyl ester **XVI**
    89
2-propenoic acid 2-hydroxy-1-methylethyl
    ester **XVI** 95
2-propenoic acid 2-(hydroxymethyl)-2-
    [[(1-oxo-2-propenyl)oxy]methyl]-
    1,3-propanediyl ester **XVI** 193
2-propenoic acid 2-hydroxypropyl ester **XVI**
    95
2-propenoic acid 3-hydroxypropyl ester **XVI**
    95
2-propenoic acid hydroxypropyl ester **XVI** 95
2-propenoic acid methyl ester **VI** 253, **XVI**
    177
1-propenol-3 **XV** 31

2-propen-1-ol **XV** 31
[(2-propenyloxy)methyl]oxirane **VII** 9
2-propoxyethanol **XII** 179–186
2-propoxyethanol acetate **XII** 187
2-propoxyethyl acetate **XII** 187–193
propyl carbinol **XIX** 99
propyl cellosolve **XII** 179
propylene chloride **IX** 21
propylene dichloride **IX** 21
propylene epoxide **V** 221
propylene glycol 1-methyl ether **V** 213, **XIV** 127
propylene glycol 2-methyl ether **I** 203
propylene glycol 1-methyl ether 2-acetate **V** 217
propylene glycol 2-methyl ether 1-acetate **I** 207
propylene glycol monoacrylate **XVI** 95
propylene glycol monomethyl ether **V** 213
1,2-propylene oxide **V** 221–228
*n*-propyl glycol **XII** 179
Protectol® DMU **V** 289
Proxel® **II** 221
pruno **XIII** 287, **XIV** 299
pseudothiourea **I** 301
PVC **II** 149–156
pyrazine hexahydride **IX** 303, **XII** 177
2-pyridinethiol 1-oxide sodium salt **X** 287
pyroacetic acid **VII** 1
pyroacetic ether **VII** 1
pyrrolylene **XV** 51
QUAB 151® **IV** 247
quartz **I** 113, **II** 129–134, 182–195, **XIV** 205–271
quartz glass **II** 157
*Quercus* spp **XIII** 287, **XVIII** 247–252
quinone monoxime **XIV** 135
quinone oxime **XIV** 135
R 11 **I** 335, **III** 366
R 12 **V** 109
R 13 **I** 69
R 21 **V** 119
R 22 **III** 63
R 31 **V** 75
R 112 **I** 297
R 113 **III** 365, 366
R 142b **I** 65
radon **II** 136
ramin **XIII** 286, **XVIII** 235
RDGE **VII** 69
red cedar **XIII** 288, **XVIII** 263
red oak **XVIII** 247
red oxide of zinc **XVIII** 305
red pine **XIII** 287

reduced Michler's ketone **I** 249
Remol TRF® **II** 299
Reomet® SBT 75 **II** 317
resorcin **XX** 239
resorcinol **XX** 239–273
resorcinol bis(2,3-epoxypropyl) ether **VII** 69
resorcinol diglycidyl ether **VII** 69
resorcinyl diglycidyl ether **VII** 69
respirable dust **XII** 239
Rewopon® IM OA **V** 373
RH 893 HQ Technical **XVI** 263
RH 893 T **XVI** 263
Ribeclor® **III** 81
rock wool **VIII** 141–338
rosewood **XIII** 286, 313
rosin **XI** 239–242
Ro-sulfiram® **V** 165
rotenone **XIX** 261–271
rubber components **XV** 145–164
Rutgers 612® **V** 345
rutile **II** 199
rye flour dust **XIII** 99–100
samba **XVIII** 283
*Samba scleroxylon* **XIII** 288
Santos rosewood **XIII** 287, 321
sapele **XVIII** 229
sapeli **XVIII** 229
sapupira (da mata) **XIII** 285, 295
Sarkosyl® O **XVI** 281
Sarkosyl® OT **XVI** 281
Scots pine **XIII** 287
selenite **II** 117
sepiolite **VIII** 141–338
*Sequoia sempervirens* **XIII** 287
serpentine **II** 182, 183
sessile oak **XVIII** 247
shinglewood **XIII** 288, **XVIII** 263
short-term exposures **I** 15–22
sidestream smoke **XIII** 3
silica, amorphous **II** 157–179
silica, crystalline **XIV** 205–271
silica fume **II** 160, 165
silica gel **II** 157
silica glass **II** 157
silicic acid tetraethyl ester **III** 299
silicon carbide **VIII** 141–338
silicon dioxide, crystalline **XIV** 205–271
silk(y) oak **XIII** 286
silver oak **XIII** 286, **XIV** 295
sipo **XVIII** 229
Skane® M-8 **XVI** 263
Skane® M-8 HQ **XVI** 263
slag wool **VIII** 141–338
soapstone **II** 181

486    *Contents for Volumes 1–21*

soda lye **XII** 195
sodium arsenate **XXI** 49
sodium arsenite **XXI** 49
sodium azide **XX** 275–284
sodium (1,1′-biphenyl)-2-olate **II** 299
sodium 2-biphenylolate **II** 299
sodium citrate **XVI** 209
sodium cyanide **XIX** 189
sodium ethylmercurithiosalicylate **XV** 249
sodium hydrate **XII** 195
sodium hydroxide **XII** 195–207
sodium 2-hydroxybiphenyl **II** 299
sodium 1-hydroxy-2-(1*H*)-pyridinethione **X** 287
sodium molybdate **XVIII** 199
Sodium Omadine® **X** 287
sodium *o*-phenylphenate **II** 299
sodium *o*-phenylphenol **II** 299–316
sodium *o*-phenylphenolate **II** 299
sodium *o*-phenylphenoxide **II** 299
sodium pyridinethione **X** 287
sodium pyrithione **X** 287–311
solvent yellow 34 **IV** 11
South American mahogany **XIII** 287
Southern silk(y) oak **XIII** 286, **XIV** 295
spanon **VI** 57
spartalite **XVIII** 305
spinel **III** 128
spirit of hartshorn **VI** 1
spirits of salt **VI** 231
spirits of turpentine **XIV** 173, **XVII** 315
spirits of wine **XII** 129
steatite **II** 181
sterlingite **XVIII** 305
Stopetyl **V** 165
Stopmold B® **II** 299
strychnidin-10-one **XIX** 273
strychnine **XIX** 273–279
styrene **XX** 285–290
styrol **XX** 285
succinic acid peroxide **III** 254
sucupira **XIII** 285, 295
sulfate of lime **II** 117
sulfate turpentine **XIV** 173, **XVII** 315
sulfinylbismethane **III** 163
sulfotep(p) **XIII** 237
sulfuric acid **XV** 165–222
sulfuric acid calcium salt **II** 117
sulfuric acid diethyl ester **XX** 93
sulfuric acid dimethyl ester **IV** 217
sulfuric ether **XIII** 149
sulfur mustard **IV** 27
sulourea **I** 301

γ-sultone **IV** 45
*Swietenia* spp. **XIII** 287, **XVIII** 253–258
systox **XIX** 141
2,4,5-T **XI** 243
*Tabebuia* spp. **XIII** 288, **XIV** 311–315
talc **II** 181–198
tantalum **XVI** 293–296
*Tarrietia utilis* **XIII** 288
Tasmanian blackwood **XIII** 285, 291
TBP **IX** 245
2,4-TDI **XX** 291
2,6-TDI **XX** 291
teak **XIII** 288, **XIV** 317
*Tecoma peroba* **XIII** 287
*Tectona grandis* **XIII** 288, **XIV** 317–323
TEDP **XIII** 237–249
tegafur **I** 4
TEL **XV** 223
teniposide **I** 6
terephthalic acid **VII** 241
*Terminalia ivorensis* **XVIII** 259–262
*Terminalia superba* **XIII** 288, **XVIII** 259–262
3,3′,4,4′-tetraaminobiphenyl **III** 131
1,3,5,7-tetraazaadamantane **V** 355
1,3,5,7-tetraazatricyclo[3.3.1.1$^{3,7}$]decane **V** 355
2,4,5,6-tetrachloro-1,3-benzene-dicarbonitrile **VI** 113
2,4,5,6-tetrachloro-3-cyanobenzonitrile **VI** 113
1,1,2,2-tetrachloro-1,2-difluoroethane **I** 297–299
tetrachloroethene **III** 271
tetrachloroethylene **III** 271–297
2,4,5,6-tetrachloroisophthalonitrile **VI** 113
tetrachloromethane **XVIII** 81
tetrachloronaphthalenes **XIII** 101
*m*-tetrachlorophthalodinitrile **VI** 113
*m*-tetrachlorophthalonitrile **VI** 113
α,α,α,4-tetrachlorotoluene **X** 21
Tetradine **V** 165
tetraethoxysilane **III** 299
tetraethyl dithiopyrophosphate **XIII** 237
tetraethyllead **XV** 223–235
tetraethyl orthosilicate **III** 299
tetraethylplumbane **XV** 223
tetraethyl silicate **III** 299–305
tetraethylthioperoxydicarbonic diamide **V** 165
tetraethylthiram disulfide **V** 165
*N*,*N*,*N*′,*N*′-tetraethylthiuram disulfide **V** 165
Tetraetil **V** 165

1,1,1,2-tetrafluoroethane **XIII** 251–266
4,5,6,7-tetrahydro-1*H*-1,2,3-benzotriazole **II**
    317
4,5,6,7-tetrahydro-1*H*-benzotriazole **II** 317–
    319
[2*R*-(2α,6aα,12aα)]-1,2,12,12a-tetrahydro-
    8,9-dimethoxy-2-(1-methylethenyl)-[1]-
    benzopyrano[3,4-*b*]furo[2,3-*h*][1]benzo-
    pyran-6(6a*H*)-one **XIX** 261
tetrahydro-*p*-dioxin **XX** 105
3α,4,7,7α-tetrahydro-4,7-methano-1*H*-indene
    **V** 125
1,2,3,4-tetrahydrostyrene **XIV** 185
tetralit **XI** 179
tetralite **XI** 179
tetramethyldiaminodiphenylacetimine **IV** 11
*N,N,N',N'*-tetramethyl-*p,p'*-diaminodiphenyl-
    methane **I** 249
tetramethyl-*p*-diamino-imido-benzophenone
    **IV** 11
tetramethyllead **XV** 237–248
tetramethylplumbane **XV** 237
tetramethylthioperoxydicarbonic diamide **XV**
    163, **XIX** 141
tetramethylthiuram disulfide **XV** 163–164
tetranitromethane **IV** 323–330
Tetrosin OE® **II** 299
tetryl **XI** 179
teturamin **V** 165
thiazols **XV** 155–157
thimerosal **XV** 249–255
thioaniline **IV** 331
4,4'-thiobis(aniline) **IV** 331
4,4'-thiobisbenzenamine **IV** 331
1,1'-thiobis(2-chloroethane) **IV** 27
2,2'-thiobis(4,6-dichlorophenol) **IX** 245
thiobutyl alcohol **XXI** 161
thiocarbamide **I** 301, **XIV** 143
thiocarbonyl tetrachloride **I** 291
4,4'-thiodianiline **IV** 331–334
*p,p'*-thiodianiline **IV** 331
thiodi-*p*-phenylenediamine **IV** 331
thiodiphosphoric acid tetraethyl ester **XIII**
    237
thioethyl alcohol **XXI** 195
thioguanine **I** 4
Thiomersal **XV** 249
Thiomersalate **XV** 249
thiomethyl alcohol **XX** 217
thiopyrophosphoric acid tetraethyl ester **XIII**
    237
thiotepa **I** 3
thiotepp **XIII** 237
thiourea **I** 301–312, **XIV** 143–148

thiram **XV** 163
thiurams **XV** 159–164
Thiuranide **V** 165
THT **IX** 323
*Thuja* spp. **XIII** 288, **XVIII** 263–276
*Tieghemella africana* **XVIII** 277–281
*Tieghemella heckelii* **XIII** 288, **XVIII** 277–
    281
tin **XIV** 149–165
tin compounds, inorganic **XIV** 149–165
tin compounds, organic **I** 317
titanium dioxide **II** 199–204
titanium oxide **II** 199
TMTD **XV** 163
TNT **I** 359
tobacco smoke **I** 28, 39–62, **XIII** 3
2-tolidine **V** 153
3,3'-tolidine **V** 153
*o*-tolidine **V** 153
tolite **I** 359
toluene **VII** 257–318
2,4-toluenediamine **VI** 339
toluene-2,4-diamine **VI** 339–352, **XXI** 3
*m*-toluenediamine **VI** 339
toluene diisocyanate **XX** 291–338
toluene-2,4-diisocyanate **XX** 291
toluene-2,6-diisocyanate **XX** 291
toluene trichloride **VI** 80
2-toluidine **III** 307
4-toluidine **III** 323
*o*-toluidine **III** 307–322, 329, **VI** 144, **XXI** 3
*p*-toluidine **III** 323–330, **XXI** 3
toluol **VII** 257
2,4-toluylenediamine **VI** 339
*m*-toluylenediamine **VI** 339
*o*-tolylamine **III** 307
*p*-tolylamine **III** 323
tolyl chloride **VI** 79
tolyltriazole **II** 295
Topane® **II** 299
total dust **II** 52
toxaphene **XIX** 281–297
tremolite **II** 96, 182–187
treosulfan **I** 3
tretamine **I** 3
tri **X** 201
triatomic oxygen **X** 171
1,2,3-triazaindene **II** 231
1,3,5-triazin-1,3,5(2*H*,4*H*,6*H*)-triethanol **II**
    331
1*H*-1,2,4-triazol-3-amine **IV** 1, **XVIII** 19
tribromomethane **VII** 319–332
tributylchlorostannane **I** 315
tributylfluorostannane **I** 315

tributyl-(2-methyl-1-oxo-2-propenyl)oxy-
   stannane **I** 316
tributylmono(naphthenoyl-oxy)stannane
   derivatives **I** 316
tributyl-(1-oxo-9,12-octadecadienyl)oxy-
   (*Z,Z*)-stannane **I** 316
tributyl phosphate **XVII** 285–314
tri-*n*-butyl phosphate **XVII** 285
tri-*n*-butyltin benzoate **I** 315
tri-*n*-butyltin chloride **I** 315
tri-*n*-butyltin compounds **I** 315–331
tri-*n*-butyltin fluoride **I** 315
tri-*n*-butyltin linoleate **I** 316
tri-*n*-butyltin methacrylate **I** 316
tri-*n*-butyltin naphthenate **I** 316
tributyltin oxide **I** 315
1,2,3-trichlorobenzene **III** 331–363
1,2,4-trichlorobenzene **I** 72, **III** 331–363,
   **XIV** 167–172
1,3,5-trichlorobenzene **III** 331–363
1,1,2-trichloroethane **V** 66, 72
trichloroethene **X** 201
trichloroethylene **III** 273, 369, **X** 201–244
trichlorofluoromethane **I** 69, 335–344
trichlorohydrin **IX** 171
trichloromethane **XIV** 19
trichloromethanesulfenyl chloride **I** 291
(trichloromethyl)benzene **VI** 80
trichloromethylsulfenyl chloride **I** 291
trichloromonofluoromethane **I** 335
trichloronaphthalenes **XIII** 101
trichloronitromethane **VI** 105
trichlorooctylstannane **VII** 93
2,4,5-trichlorophenol **XII** 209–221
2,4,5-trichlorophenoxyacetic acid **XI** 243–
   278
2,4,5-trichlorophenoxyacetic acid salts and
   esters **XI** 243
trichlorophenylmethane **VI** 80
1,2,3-trichloropropane **IX** 171–192
*α,α,α*-trichlorotoluene **VI** 80
*ω,ω,ω*-trichlorotoluene **VI** 80
1,1,2-trichloro-1,2,2-trifluoroethane **III** 365–
   370
tridymite **XIV** 205
triethanolamine **I** 27
triethoxybutanol **IX** 315
triethylamine **I** 26, 33, **XIII** 267–280
triethylene glycol *n*-butyl ether **IX** 315–321
triethylene glycol monobutyl ether **IX** 315
1,3,5-triethylol-hexahydro-s-triazine **II** 331
trifluoroborane **XIII** 93
trifluorobromomethane **VI** 37
trifluorochloromethane **I** 69

1,1,1-trifluoro-2,2-dichloroethane **X** 53
trifluoromethyl chloride **I** 69
trifluoromonochlorocarbon **I** 69
trifluorotrichloroethane **III** 365
triglycol monobutyl ether **IX** 315
trihydroxymethylnitromethane **II** 287
trimethylamine **I** 26, 28
1,2,4-trimethyl-5-aminobenzene **IV** 335
2,4,5-trimethylaniline **IV** 335–340, **XXI** 3
2,4,5-trimethylbenzenamine **IV** 335
*sym*-trimethylbenzene **IV** 341
1,3,5-trimethylbenzene **IV** 341–348
2,4,5-trimethylbenzeneamine **IV** 335
trimethyl carbinol **XIX** 125
trimethylolmelamine **I** 7
trimethylolnitromethane **II** 287
trimethylolpropane triacrylate **XVI** 201–206
*N,N,N*-trimethyl-oxiranemethanaminium
   chloride **IV** 247
2,2,4-trimethylpentane **I** 196, 347–356
2,4,6-trinitrophenol **XVII** 273
2,4,6-trinitrophenylmethylnitramine **XI** 179
2,4,6-trinitrophenyl-*N*-methylnitramine **XI**
   179
trinitrotoluene (all isomers) **I** 148, 359–360
2,4,6-trinitrotoluene **XXI** 3
3,6,9-trioxa-1-tridecanol **IX** 315
triphenyl phosphate **II** 321–330
*Triplochiton scleroxylon* **XIII** 288, **XVIII**
   283–289
tris(dimethyldithiocarbamato)iron **XIX** 163
tris(2-hydroxyethyl)-hexahydro-1,3,5-triazine
   **II** 331
*N,N′,N″*-tris(*β*-hydroxyethyl)-hexahydro-
   1,3,5-triazine **II** 331–339, **IX** 323–324
tris(*ω*-hydroxyethyl)-hexahydro-1,3,5-triazine
   **II** 331
tris(hydroxymethyl)nitromethane **II** 287
Tris Nitro® **II** 287
tris(oxymethyl)nitromethane **II** 287
tritol **I** 359
triton **I** 359
trofosfamide **I** 3
trotyl **I** 359
Troysan® KK-108A **XVI** 247
Troysan® Polyphase® Anti-Mildew **XVI** 247
Troysan® Polyphase® P100 **XVI** 247
true teak **XIII** 288
TTD **V** 165
Tumescal OPE® **II** 299
turpentine **XIV** 173–179, **XVII** 315–332
turpentine oil **XIV** 173, **XVII** 315
*Turraeanthus africanus* **XIII** 288
ultrafine aerosol particles **XVI** 289–292

Uniclor® **III** 81
Uritone® **V** 355
Urotovet® **V** 355
Urotropin® **V** 355
Ursol D® **VI** 311
utile **XVIII** 229
δ-valerosultone **IV** 45
vanadic anhydride **IV** 349
vanadium compounds **IV** 349–361
vanadium(V) oxide **IV** 349
vanadium pentoxide **IV** 349–361
Vapona® **IV** 201
Varine® O **V** 373
VDC **VIII** 109
vinblastine sulfate **I** 6
vincristine sulfate **I** 6
vindesine **I** 6
vinegar naphtha **XII** 167
vinyl acetate **V** 229–239, **XXI** 271–294
vinyl alcohol 2,2-dichloro dimethyl phosphate **IV** 201
vinylbenzene **XX** 285
vinylbutyrolactam **V** 249
vinylcarbazole **XIV** 181–183
*N*-vinylcarbazole **XIV** 181
vinyl carbinol **XV** 31
vinyl chloride **I** 40, **II** 149–156, **V** 66, 71, 241–248
1-vinylcyclohex-3-ene **XIV** 185
4-vinylcyclohexene **XIV** 185–202
vinyl cyclohexene diepoxide **I** 391
4-vinyl-1,2-cyclohexene diepoxide **I** 391
4-vinyl-1-cyclohexene dioxide **I** 391–395
1-vinyl-3-cyclohexene dioxide **I** 391
vinyl ethanoate **V** 229
vinylethylene **XV** 51
vinylidene chloride **V** 66, 72, **VIII** 109–139
vinylidene dichloride **VIII** 109
vinylidene fluoride **V** 135
1-vinyl-2-pyrrolidinone **V** 249
*N*-vinylpyrrolidinone **V** 249
1-vinyl-2-pyrrolidone **V** 249

*N*-vinyl-2-pyrrolidone **V** 249–261
vitreous silica **II** 157
walnut **XIII** 286
wawa **XIII** 288, **XVIII** 283
western red cedar **XIII** 288, **XVIII** 263
wheat flour dust **XIII** 99–100
white acajou **XIII** 287
white afara **XIII** 288
white cedar **XIII** 288, **XVIII** 263
white peroba **XIII** 287, **XIV** 307
wishmore **XIII** 288
Witaclor® **III** 81
wollastonite **VIII** 141–338, **XVI** 297–315
wood dust **IV** 363–373, **XIII** 283–323, **XIV** 273–323, **XVIII** 221–227
wood ether **I** 125
woods **IV** 363–373, **XIII** 283–323, **XIV** 273–323, **XVIII** 229–289
wood turpentine **XIV** 173, **XVII** 315
xylene **V** 263–285, **XV** 257–288
xylidine isomers **XIX** 299–311, **XXI** 3
xylol **V** 263
yellow pine **XIII** 287
zebrano **XIII** 287
zebrawood **XIII** 287
zeolites **VIII** 141–338
zinc chloride **XVIII** 291–303
zinc chloride fume **XVIII** 291–303
zinc chromate **III** 101–122, **XV** 289–294
zinc chromate hydroxide **XV** 289
zinc dichloride **XVIII** 291
zincite **XVIII** 305
zinc molybdate **XVIII** 199
zinc oxide **XVIII** 305–323
zinc oxide fume **XVIII** 305–323
zinc potassium chromate **XV** 289
zinc white **XVIII** 305
zinc yellow **XV** 289
zirconium **XII** 223–236
zirconium alloys and compounds **XII** 223
Zoldine® ZE **IX**

# The MAK-Collection for Occupational Health and Safety

**MAK values (Maximum Concentrations at the Workplace)** and **BAT values (Biological Tolerance Values)** promote the protection of health at the workplace. They are an efficient indicator for the toxic potential of chemical compounds.

## List of MAK and BAT Values

Maximum Concentrations and Biological Tolerance Values at the Workplace
German DFG-Senate-Commission, Editor

**New feature: the complete list of MAK and BAT values now on a CD–ROM included in the book!**

Now you can reference the largest stock of carefully revised toxicological data currently available anywhere. This book contains a list of scientifically recommended threshold limit values for about 900 chemical compounds.

MAK Values set the standards for legal regulations in many countries of the world, e.g. they are the basis for at least 30% of the threshold limits valid in the European Union.

**Report 41 Paperback 260 pages August 2005**
**ISBN 3527-31357-5  approx € 75.00 /£ 55.00 /US$ 95.00**

## Part I: MAK Value Documentations

HELMUT GREIM, Editor

The volumes of this series present about 400 indispensable toxicological evaluation documents on important occupational toxicants and carcinogens. They describe the toxicological database which determines the level of a **MAK value.**

**Vol. 21 Hardcover 333 pages July 2005**
**ISBN 3527-31134-3**
**Series price € 99.00 /£ 70.00 /US$ 135.00**

■ Please contact the Customer Service for single volume prices

John Wiley & Sons, Ltd. • Customer Services Department
1 Oldlands Way • Bognor Regis • West Sussex
PO22 9SA England • Tel.: +44 (0) 1243-843-294
Fax: +44 (0) 1243-843-296 • www.wileyeurope.com

## Part II: BAT Value Documentations

HANS DREXLER, Editor

You will find detailed information on dozens of pharmacokinetics, critical toxicity, exposures and effects, selections of the indicators, methodologies, background exposures, interpretation of the data, and manifesto – the toxicological database that determins the **BAT value** of threshold limits for hazardous occupational toxicants in body fluids.

**Vol. 4 Hardcover approx 248 pages August 2005**
**ISBN 3527-27049-3**
**Series price approx € 109.00 /£ 80.00 /US$ 140.00**

## Part III: Air Monitoring Methods

Harun Palar, Editor

**Vol. 9  Hardcover 216 pages July 2005**
**ISBN 3527-31138-6  Series price € 109.00 /£ 55.00 /US$ 90.00**

## Part IV: Biomonitoring Methods

Jürgen Angerer et al., Editors

**Vol. 10 Hardcover approx. 354 pages January 2006**
**ISBN 3527-31137-8**

**Vol. 9 Hardcover 359 pages February 2004**
**ISBN 3527-27799-4  Series price € 109.00 /£ 55.00 /US$ 90.00**

Detailed procedures of the determination of occupational toxicants in body fluids and in air are provided in each volume of these two series.

Throughout, considerable emphasis is placed on sample collection methods and on analytical quality control: every method is checked by at least one examine.

MAK Online Database –
convenient, comprehensive, detailed

**Occupational Toxicants and MAK Values**

**www.mrw.interscience.wiley.com/makbat**

**DFG**  **www.dfg.wiley-vch.de**

Wiley-VCH • Customer Service Department
P.O. Box 101161 • D-69451 Weinheim • Germany
Tel.: (49) 6201 606-400 • Fax: (49) 6201 606-184
e-Mail: service@wiley-vch.de • www.wiley-vch.de

20870507_v0